Vascular Ion Channels in Physiology and Disease

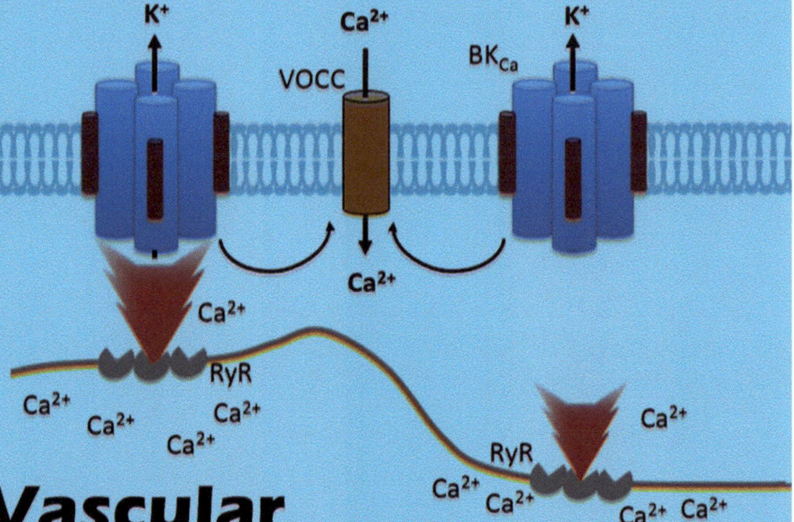

Vascular Ion Channels in Physiology and Disease

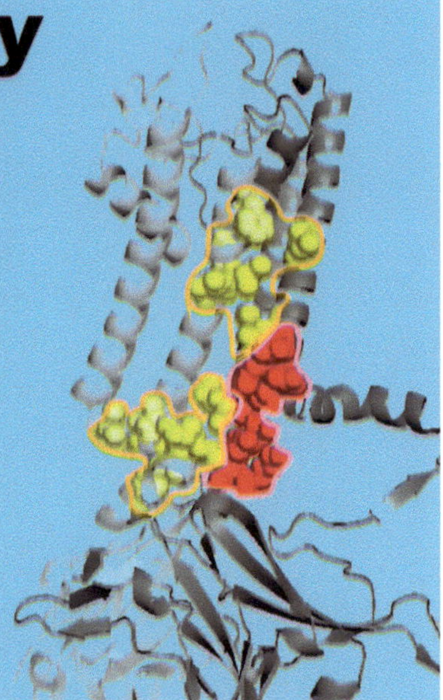

Editors:
Irena Levitan
Alex M. Dopico

Foreword by Mark T. Nelson

Springer

Irena Levitan • Alex M. Dopico
Editors

Vascular Ion Channels in Physiology and Disease

Foreword by Mark T. Nelson

Springer

Editors
Irena Levitan, PhD
University of Illinois at Chicago
Chicago, IL, USA

Alex M. Dopico, MD, PhD
The University of Tennessee
Memphis, TN, USA

ISBN 978-3-319-29633-3 ISBN 978-3-319-29635-7 (eBook)
DOI 10.1007/978-3-319-29635-7

Library of Congress Control Number: 2016940962

Printed on acid-free paper

This Springer imprint is published by Springer Nature
The registered company is Springer International Publishing AG Switzerland

Foreword

The central role of the vasculature in the function of the human body has been appreciated, at least in certain circles, since the time of ancient Greece, when the first known anatomical treatise that was not based on mysticism was produced. Early anatomists got much wrong, of course, crediting blood vessels with carrying tears and urine (but not sweat) in addition to blood, and suggesting the more fantastical role of conveying "breaths" of life and death (curiously these were said to be directed to right and left ears, respectively). Aristotle and his fourth century BC colleagues helped bring more empiricism to the subject, but their strictly anatomical approach based on animal dissections left much of vascular function to the imagination. Detailed studies of human cadavers, notably those performed by Leonardo da Vinci, the first to describe atherosclerosis, and later by Vesalius, considered by many to be the founder of modern human anatomy, dramatically advanced our understanding of the structure and basic responsibilities of the vascular system. However, medieval tools would prove to be no match for the twentieth century question of how various pieces and parts of the vascular system, though exquisitely detailed in exacting drawings, actually functioned.

If there is a consensus start to the modern era of functional vascular biology research, it is the publication in 1902 of a paper by William Bayliss, titled "On the local reactions of the arterial wall to changes of internal pressure," which reported the counterintuitive finding that intravascular pressure causes healthy arteries to constrict. More than 100 years later, this process, termed the myogenic response, remains an area of active investigation—a testament to the complexity of the underlying mechanism. The concept that ion channels are fundamentally involved in this and other vascular processes—the topic of this book—would be developed much later, after the seminal work of Hodgkin and Huxley on action potentials in the 1950s, made possible by the experimentally accessible neurons of their giant squid axon preparation, and after the confirmation of the existence of ion channels as specific molecular entities by Katz and Miledi in the 1970s. In fact, it wasn't until the development of the "patch-clamp" technique by Neher and Sakmann in the late 1970s and early 1980s that the study of ion channels in vascular cells would begin in earnest.

Since that time, the body of literature on the subject has grown at a rate that threatens our ability to keep pace. No single volume can possibly convey any more than a small subset of the available information, and this book is no exception. However, our hope is that those with different backgrounds will find something to like here, whether it be big-picture topics of potential interest to general readers, such as ion channel repertoires in endothelial cells (Chap. 1), regulation and function of calcium channels in smooth muscle cells (Chap. 2), regional variation in arterial myogenic responsiveness (Chap. 6), ion channel trafficking in the control of contractility (Chap. 7), and calcium mobilization from intracellular stores (Chap. 11), or more specific offerings, such cholesterol-potassium channel interactions (Chap. 15). As reflected in the title of this book, a number of chapters address the role of ion channels in the etiology of vascular diseases, including hypertension (Chaps. 8 and 14), cancer (Chap. 12), and metabolic diseases (Chaps. 17 and 18), or their potential as therapeutic targets in the treatment of disease (Chaps. 5 and 9). The book is organized into three main parts. Part I, "Ion channel regulation of vascular tone and blood flow," addresses aspects of endothelial cells (Part I) and smooth muscle cells (Part II) that regulate moment-to-moment changes in contractile state (tone) of arteries, and hence blood flow, as well as alterations associated with hypertension. Part III, "Ion channels in the regulation of cell proliferation, remodeling, hypertrophy and angiogenesis," focuses on longer-term dynamics, particularly vascular remodeling, under pathological (tumor angiogenesis, pulmonary hypertension) and physiological conditions. The final part (Part IV) presents selected topics in "Ion channel regulation by lipids and channel modifications in metabolic disease."

Each chapter is the product of scientists who are actively engaged in the cutting-edge work described and is a reflection of their individual research interests. The resulting book is thus an eclectic offering that allows each contributor to speak with his or her own voice, a feature that we like to think is part of its charm.

Burlington, VT, USA

Mark T. Nelson
University of Vermont

Preface

Ion channels are the major class of membrane proteins responsible for rapid and regulated transport of ions across biological membranes and for the generation and propagation of electrical signals in excitable tissues. Ion channels are also known to play critical roles in the regulation of cell proliferation, insulin secretion, and intracellular signaling in a variety of cell types. During the last decade, since the first atomic structure of a potassium-selective channel has been solved in 2000 by the group of Roderick MacKinnon, a discovery that led to a Nobel Prize in Chemistry shortly after, there has been an explosion of studies on the structure-function relationship of ion channels leading to an unprecedented level of mechanistic understanding of channel function. This new era of ion channel research provided the basis for detailed analysis of the roles of ion channels in cellular and tissue physiology. This book focuses on the roles of ion channels in vascular tissues under normal and pathological conditions.

Vascular abnormalities are known to underlie a plethora of severe pathological conditions, such as atherosclerosis, systemic and pulmonary hypertension, coronary or cerebral vasospasm, and diabetes. In addition, dysregulation of angiogenesis is one of the major contributors to tumor development. Therefore, it is clearly imperative to obtain a better understanding of the molecular mechanisms that contribute to vascular disorders. This book is the first comprehensive assembly of assays to present the studies that have been conducted during the last decade to elucidate the roles of ion channels in vascular physiology and in different vascular diseases.

Among the major concepts discussed in this volume are: the diversity of ion channels expressed in endothelial and vascular smooth muscle cells, and their roles in regulating arterial contractility and vascular tone in systemic, cerebral, and pulmonary circulation, as well as in the regulation of cell proliferation, barrier function, and vascular remodeling. It is also highlighted that both plasma membrane and intracellular ion channels, particularly in mitochondria, contribute significantly to the regulation of vascular functions. Emerging topics include the roles of channel trafficking in regulation of vascular tone, local control and co-ordination of ion channel function in different domains, and description of novel positive and negative gating modulators of the channels as innovative therapeutic tools to treat cardiac disorders and as anti-vasospastic agents. The last part of the volume describes the latest insights into the mechanisms underlying lipid regulation of ion channels and the implications of these mechanisms for cardiovascular dysfunction in metabolic disease.

Chicago, IL, USA Irena Levitan
Memphis, TN, USA Alex M. Dopico

Contents

Part I Ion Channel Regulation of Vascular Tone and Blood Flow. Changes with Hypertension: Endothelial Cells

1 **Endothelial Cell Ion Channel Expression and Function in Arterioles and Resistance Arteries** 3
William F. Jackson

2 **Contribution and Regulation of Calcium Channels in Endothelial Cells** 37
Kwong Tai Cheng, Avia Rosenhouse-Dantsker, and Asrar B. Malik

3 **Mitochondrial Depolarization in Endothelial and Other Vascular Cells and Their Role in the Regulation of Cerebral Vascular Tone** 63
David W. Busija, Ibolya Rutkai, and Prasad V. Katakam

4 **Ion Channels in Control of Blood Flow: Electrical Conduction Along Endothelium of Resistance Arteries** 79
Erik J. Behringer and Steven S. Segal

5 **Ca^{2+}/Calmodulin-Gated Small- and Intermediate-Conductance K_{Ca} Channels in Cardiovascular Regulation: Targets for Novel Pharmacological Treatments** 101
Ralf Köhler and Aida Olivan-Viguera

Part II Ion Channel Regulation of Vascular Tone and Blood Flow. Changes with Hypertension: Smooth Muscle Cell

6 **Regional Variation in Arterial Myogenic Responsiveness: Links to Potassium Channel Diversity/Function** 131
Michael A. Hill, Yan Yang, Zahra Nourian, Barry D. Kyle, Kwangseok Hong, and Andrew P. Braun

7 **Ion Channel Trafficking and Control of Arterial Contractility** 153
M. Dennis Leo and Jonathan H. Jaggar

8 **Abnormalities of Vascular Ion Channels During Hypertension** 169
Anup K. Srivastava, Lee Ann MacMillan-Crow, Sung W. Rhee, and Nancy J. Rusch

9 **Kv7 Potassium Channels as Therapeutic Targets in Cerebral Vasospasm** 191
Bharath K. Mani, Lyubov I. Brueggemann, Sarkis Morales-Vidal, Christopher M. Loftus, and Kenneth L. Byron

10 **Lysosomal Transient Receptor Potential Mucolipin (TRPML) Channels in Vascular Regulation and Diseases** 215
Fan Zhang and Pin-Lan Li

Part III Ion Channels in the Regulation of Cell Proliferation, Remodeling, Hypertrophy and Angiogenesis

11 **Calcium Mobilization via Intracellular Ion Channels, Store Organization and Mitochondria in Smooth Muscle** 233
John G. McCarron, Susan Chalmers, Calum Wilson, and Mairi E. Sandison

12 **Role of Mechanosensitive TRP Channels in Abnormal Vasculature of Tumors** 255
Holly C. Cappelli, Roslin J. Thoppil, Ravi K. Adapala, J. Gary Meszaros, Sailaja Paruchuri, and Charles K. Thodeti

13 **TRPC and Orai Channels in Store-Operated Calcium Entry and Vascular Remodelling** 275
David J. Beech, Jing Li, Lynn McKeown, and Hollie L. Appleby

14 **Smooth Muscle Cell Ion Channels in Pulmonary Arterial Hypertension: Pathogenic Role in Pulmonary Vasoconstriction and Vascular Remodeling** 295
Ramon J. Ayon, Haiyang Tang, Ruby A. Fernandez, Ayako Makino, and Jason X. -J. Yuan

Part IV Ion Channel Regulation by Lipids and Channel Modifications in Metabolic Disease

15 **Physiological Roles and Cholesterol Sensitivity of Endothelial Inwardly-Rectifying K^+ Channels: Specific Cholesterol-Protein Interactions Through Non Annular Binding Sites** 327
Irena Levitan, Sang Joon Ahn, Ibra Fancher, and Avia Rosenhouse-Dantsker

16 **Membrane Lipids and Modulation of Vascular Smooth Muscle Ion Channels** 349
Alex M. Dopico, Anna N. Bukiya, and Guruprasad Kuntamallappanavar

17 **Transient Receptor Potential Channels in Metabolic Syndrome-Induced Coronary Artery Disease** 381
Stacey L. Dineen, Zachary P. Neeb, Alexander G. Obukhov, and Michael Sturek

18 **Mitochondrial Ion Channels in Metabolic Disease** 397
Aaron H. Truong, Saravanakumar Murugesan, Katia D. Youssef, and Ayako Makino

Erratum to: Calcium Mobilization via Intracellular Ion Channels, Store Organization and Mitochondria in Smooth Muscle E1

Index 421

Contributors

Ravi K. Adapala Department of Integrative Medical Sciences, Northeast Ohio Medical University, Rootstown, OH, USA

School of Biomedical Sciences, Kent State University, Kent, OH, USA

Sang Joon Ahn Division of Pulmonary and Critical Care, Department of Medicine, University of Illinois at Chicago, Chicago, IL, USA

Hollie L. Appleby School of Medicine, University of Leeds, Leeds, UK

Ramon J. Ayon Division of Translational and Regenerative Medicine, The University of Arizona College of Medicine, The University of Arizona, Tucson, AZ, USA

Department of Medicine, The University of Arizona College of Medicine, Tucson, AZ, USA

David J. Beech School of Medicine, University of Leeds, Leeds, UK

Erik J. Behringer Medical Pharmacology and Physiology, University of Missouri, Columbia, MO, USA

Division of Pharmacology, Risley Hall, Loma Linda University, Loma Linda, CA, USA

Andrew P. Braun Department of Physiology and Pharmacology, Cumming School of Medicine, University of Calgary, Calgary, AB, Canada

Lyubov I. Brueggemann Department of Molecular Pharmacology & Therapeutics, Loyola University Chicago, Maywood, IL, USA

Anna N. Bukiya Department of Pharmacology, College of Medicine, The University of Tennessee Health Science Center, Memphis, TN, USA

David W. Busija Department of Pharmacology, Tulane University School of Medicine, New Orleans, LA, USA

Kenneth L. Byron Health Sciences Division, Loyola University Chicago, Maywood, IL, USA

Department of Molecular Pharmacology & Therapeutics, Loyola University Chicago, Maywood, IL, USA

Holly C. Cappelli Department of Integrative Medical Sciences, Northeast Ohio Medical University, Rootstown, OH, USA

School of Biomedical Sciences, Kent State University, Kent, OH, USA

Susan Chalmers Strathclyde Institute of Pharmacy and Biomedical Sciences, University of Strathclyde, Glasgow, UK

Kwong Tai Cheng Department of Pharmacology, College of Medicine, University of Illinois, Chicago, IL, USA

Stacey L. Dineen Department of Cellular & Integrative Physiology, Indiana University School of Medicine, Indianapolis, IN, USA

Alex M. Dowpico Department of Pharmacology, College of Medicine, The University of Tennessee Health Science Center, Memphis, TN, USA

Ibra Fancher Division of Pulmonary and Critical Care, Department of Medicine, University of Illinois at Chicago, Chicago, IL, USA

Ruby A. Fernandez Division of Translational and Regenerative Medicine, The University of Arizona College of Medicine, Tucson, AZ, USA

Department of Medicine, The University of Arizona College of Medicine, Tucson, AZ, USA

Michael A. Hill Dalton Cardiovascular Research Center and Department of Medical Pharmacology and Physiology, University of Missouri, Columbia, MO, USA

Kwangseok Hong Dalton Cardiovascular Research Center and Department of Medical Pharmacology and Physiology, University of Missouri, Columbia, MO, USA

William F. Jackson Department of Pharmacology and Toxicology, Michigan State University, East Lansing, MI, USA

Jonathan H. Jaggar Department of Physiology, University of Tennessee Health Science Center, Memphis, TN, USA

Prasad V. Katakam Department of Pharmacology, Tulane University School of Medicine, New Orleans, LA, USA

Ralf Köhler Fundación Agencia Aragonesa para la Investigación y Desarrollo (ARAID), Zaragoza, Spain

Unidad de Investigación Traslacional, Hospital Universitario Miguel Servet, Zaragoza, Spain

Aragon Institute of Health Sciences & IIS Aragon

Guruprasad Kuntamallappanavar Department of Pharmacology, College of Medicine, The University of Tennessee Health Science Center, Memphis, TN, USA

Barry D. Kyle Department of Physiology and Pharmacology, Cumming School of Medicine, University of Calgary, Calgary, AB, Canada

M. Dennis Leo Department of Physiology, University of Tennessee Health Science Center, Memphis, TN, USA

Irena Levitan Division of Pulmonary and Critical Care, Department of Medicine, University of Illinois at Chicago, Chicago, IL, USA

Department of Pharmacology, University of Illinois at Chicago, Chicago, IL, USA

Jing Li School of Medicine, University of Leeds, Leeds, UK

Pin-Lan Li Department of Pharmacology and Toxicology, Virginia Commonwealth University, School of Medicine, Richmond, VA, USA

Christopher M. Loftus Department of Neurosurgery, Loyola University Chicago, Maywood, IL, USA

Lee Ann MacMillan-Crow Department of Pharmacology and Toxicology, University of Arkansas for Medical Sciences, Little Rock, AR, USA

Ayako Makino Division of Translational and Regenerative Medicine, The University of Arizona, Tucson, AZ, USA

Department of Physiology, The University of Arizona College of Medicine, Tucson, AZ, USA

Asrar B. Malik Department of Pharmacology, College of Medicine, University of Illinois, Chicago, IL, USA

Bharath K. Mani Department of Molecular Pharmacology & Therapeutics, Loyola University Chicago, Maywood, IL, USA

John G. McCarron Strathclyde Institute of Pharmacy and Biomedical Sciences, University of Strathclyde, Glasgow, UK

Lynn McKeown School of Medicine, University of Leeds, Leeds, UK

J. Gary Meszaros Department of Integrative Medical Sciences, Northeast Ohio Medical University, Rootstown, OH, USA

School of Biomedical Sciences, Kent State University, Kent, OH, USA

Sarkis Morales-Vidal Department of Neurology, Loyola University Chicago, Maywood, IL, USA

Saravanakumar Murugesan Department of Physiology, University of Arizona, Tucson, AZ, USA

Zachary P. Neeb Department of Cellular & Integrative Physiology, Indiana University School of Medicine, Indianapolis, IN, USA

Eli Lilly and Company, Indianapolis, IN, USA

Zahra Nourian Dalton Cardiovascular Research Center and Department of Medical Pharmacology and Physiology, University of Missouri, Columbia, MO, USA

Alexander G. Obukhov Department of Cellular & Integrative Physiology, Indiana University School of Medicine, Indianapolis, IN, USA

Aida Olivan-Viguera Aragon Institute of Health Sciences & IIS Aragon, Zaragoza, Spain

Sailaja Paruchuri Department of Chemistry, University of Akron, Akron, OH, USA

Sung W. Rhee Department of Pharmacology and Toxicology, University of Arkansas for Medical Sciences, Little Rock, AR, USA

Avia Rosenhouse-Dantsker Division of Pulmonary and Critical Care, Department of Medicine, University of Illinois at Chicago, Chicago, IL, USA

Department of Pharmacology, University of Illinois at Chicago, Chicago, IL, USA

Nancy J. Rusch Department of Pharmacology and Toxicology, University of Arkansas for Medical Sciences, Little Rock, AR, USA

Ibolya Rutkai Department of Pharmacology, Tulane University School of Medicine, New Orleans, LA, USA

Mairi E. Sandison Strathclyde Institute of Pharmacy and Biomedical Sciences, University of Strathclyde, Glasgow, UK

Steven S. Segal Medical Pharmacology and Physiology, University of Missouri, Columbia, MO, USA

Dalton Cardiovascular Research Center, Columbia, MO, USA

Anup K. Srivastava Department of Pharmacology and Toxicology, University of Arkansas for Medical Sciences, Little Rock, AR, USA

Michael Sturek Department of Cellular & Integrative Physiology, Indiana University School of Medicine, Indianapolis, IN, USA

Haiyang Tang Division of Translational and Regenerative Medicine, The University of Arizona, Tucson, AZ, USA

Department of Medicine, The University of Arizona College of Medicine, Tucson, AZ, USA

Charles K. Thodeti Department of Integrative Medical Sciences, Northeast Ohio Medical University, Rootstown, OH, USA

School of Biomedical Sciences, Kent State University, Kent, OH, USA

Roslin J. Thoppil Department of Integrative Medical Sciences, Northeast Ohio Medical University, Rootstown, OH, USA

School of Biomedical Sciences, Kent State University, Kent, OH, USA

Aaron H. Truong Department of Physiology, University of Arizona, Tucson, AZ, USA

Calum Wilson Strathclyde Institute of Pharmacy and Biomedical Sciences, University of Strathclyde, Glasgow, UK

Yan Yang Dalton Cardiovascular Research Center and Department of Medical Pharmacology and Physiology, University of Missouri, Columbia, MO, USA

Katia D. Youssef Department of Physiology, University of Arizona, Tucson, AZ, USA

Jason X.-J. Yuan Division of Translational and Regenerative Medicine, The University of Arizona, Tucson, AZ, USA

Department of Physiology, The University of Arizona College of Medicine, Tucson, AZ, USA

Fan Zhang Department of Pharmacology and Toxicology, Virginia Commonwealth University, School of Medicine, Richmond, VA, USA

Part I
Ion Channel Regulation of Vascular Tone and Blood Flow. Changes with Hypertension: Endothelial Cells

Chapter 1
Endothelial Cell Ion Channel Expression and Function in Arterioles and Resistance Arteries

William F. Jackson

Abstract Ion channels importantly contribute to the function of endothelial cells. They serve as the major source of intracellular Ca^{2+}, which, in turn, controls the production of endothelium-derived vasodilators, the permeability of the endothelium, gene expression, and other properties of endothelial cells. In addition, the activity of ion channels determines the membrane potential of endothelial cells that serves as an important signal for cell-cell communication between endothelial cells and between endothelial cells and overlying smooth muscle cells, and may feedback to regulate the activity of the ion channels themselves. This review provides an overview of the expression and function of endothelial ion channels that contribute to Ca^{2+} and membrane potential signaling that is involved in the regulation and modulation of vasomotor tone of resistance arteries and arterioles. Channels discussed include inositol 1,4,5 trisphosphate receptors that mediate agonist-induced Ca^{2+} release from endoplasmic reticulum stores; members of the transient receptor potential family and other channels that mediate agonist-induced Ca^{2+} influx through the plasma membrane; Ca^{2+}-activated K^+ channels that mediate agonist-induced membrane hyperpolarization; and inward rectifier K^+ channels that serve as sensors for changes in extracellular K^+ and amplifiers of hyperpolarization induced by the activity of other ion channels. It is emphasized that all of these channels exist as members of macromolecular signaling complexes providing a rich environment for regulation of their activity and the function of endothelial cells in resistance arteries and arterioles.

Keywords Vascular endothelium • Ion channels • Endothelium-dependent vasodilatation • Conducted dilatation • Potassium channels • Transient receptor potential channels • Calcium ions • Inositol 1,4,5 trisphosphate receptors • Calcium waves • Calcium sparklets

---

W.F. Jackson (✉)
Department of Pharmacology and Toxicology, Michigan State University, 1355 Bogue Street, B429 Life Sciences, East Lansing, MI 48824, USA
e-mail: jacks783@msu.edu

I. Levitan, A.M. Dopico (eds.), *Vascular Ion Channels in Physiology and Disease*, DOI 10.1007/978-3-319-29635-7_1

Introduction

Endothelial cells express a diverse array of ion channels in their plasma membranes and in the membranes of intracellular organelles that contribute to the function of these cells. These channels provide the major source of intracellular Ca^{2+} that serves as an important second messenger controlling the activity of Ca^{2+}-dependent ion channels and cell membrane potential [40], endothelial cell production of NO, prostaglandins and epoxides of arachidonic acid (EETs) [40] and regulating barrier function of the endothelium [52, 90]. Intracellular Ca^{2+} is also an important signal controlling gene expression in [117, 130] and proliferation of [120, 123] endothelial cells. Ion channels also participate in cell volume regulation [70]. In addition, plasmalemmal ion channel activity importantly contributes to the membrane potential of endothelial cells that serves as a major signal for cell-cell communication between adjacent endothelial cells and as well as overlying smooth muscle cells due to the expression of homocellular and heterocellular gap junctions in the vascular wall [29]. Membrane potential may also feedback to affect Ca^{2+} influx through plasmalemmal Ca^{2+} permeable ion channels by influencing the electrochemical gradient for Ca^{2+} influx [65, 66, 69, 123], although this topic remains controversial [23, 34, 108, 113, 158]. Thus, ion channels importantly contribute to the function of endothelial cells in health and disease. This review will focus on the expression and function of endothelial ion channels involved in the regulation of vasomotor tone in resistance arteries and arterioles. Because there are considerable changes in ion channel expression and function during proliferation of cells in culture [11, 12, 20, 123, 135], emphasis will be placed on evidence from intact blood vessels and from freshly isolated endothelial cells from the peripheral circulation. The reader is also directed to a number of outstanding earlier reviews of ion channels in endothelial cells for access to earlier literature on this topic [123–125].

Setting the Stage

Most vasodilators that produce endothelium-dependent vasodilatation (see [79]) act on $G\alpha_q$-protein-coupled receptors which are linked to phospholipase C (PLC)-β producing inositol-1,4,5-trisphosphate (IP_3) and diacylglycerol (DAG) from membrane phospholipids [13] (Fig. 1.1). The released IP_3 activates IP_3R in the membranes of the smooth endoplasmic reticulum, releasing stored Ca^{2+} and increasing cytosolic Ca^{2+} [13]. The activation of IP_3R, loss of Ca^{2+} from intracellular Ca^{2+} stores and the DAG produced by PLC-β, activate plasma membrane Ca^{2+}-permeable ion channels [51], allowing Ca^{2+} to diffuse down its electrochemical gradient into the cells, producing a steady-state increase in intracellular Ca^{2+} (Figs. 1.1 and 1.2). The sum of these two major events (the release of ER-stored Ca^{2+} via IP_3R and the influx of Ca^{2+} via plasmalemmal ion channels) produces the well-described, agonist-induced cytosolic Ca^{2+} transient in endothelial cells (see Fig. 1.2). The increase in

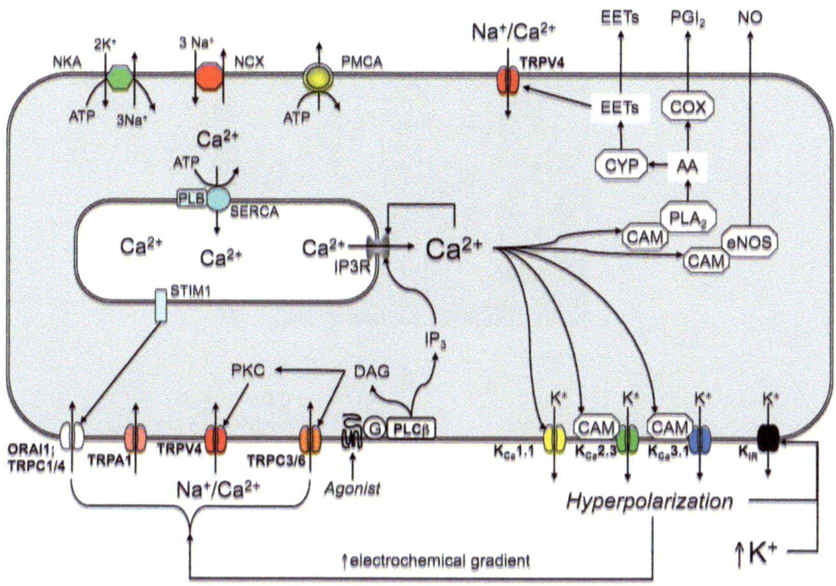

Fig. 1.1 Endothelial cell ion channel and Ca^{2+} signaling overview. Shown is a schematic representation of an endothelial cell and the ion channels and transporters relevant to agonist-induced Ca^{2+} signaling. Agonists of $G\alpha_q$-coupled receptors activate PLC-β producing $IP_3$and DAG. IP_3 activates IP_3R in the membrane of the endoplasmic reticulum (ER), releasing stored Ca^{2+} and raising cytosolic Ca^{2+} as shown. The released Ca^{2+} and Ca^{2+} entry through overlying plasma membrane Ca^{2+} permeable channels further stimulate Ca^{2+} release via Ca^{2+}-induced-Ca^{2+}-release. The elevated cytosolic Ca^{2+} will then activate plasma membrane K_{Ca} channels to produce membrane hyperpolarization, an important signal for cell-cell communication in resistance arteries and arterioles. This hyperpolarization also has the potential to increase the electrochemical gradient for diffusion of Ca^{2+} (and other cations) into the endothelial cell counter-acting the depolarizing effect of this cation influx. The DAG produced by the action of PLCs can activate TRPC3 and/or TRPC6 channels in the membrane, contributing to steady-state, agonist-induced Ca^{2+} influx into the cells. The DAG can also activate PKC, which phosphorylates TRPV4 channels increasing their activity, also contributing to Ca^{2+} influx. Loss of Ca^{2+} from the ER is sensed by STIM1, which clusters, interacts with and activates membrane ORAI1, TRPC1 and/or TRPC4 channels. The resultant Ca^{2+} influx contributes to steady-state, agonist-induced Ca^{2+} influx. The elevated cytosolic Ca^{2+} from these processes also activates nitric oxide synthase (eNOS) to stimulate NO production, and phospholipase A_2 (PLA_2) to produce arachidonic acid (AA) from membrane phospholipids. Arachidonic acid is then converted into vasodilator prostanoids, such as prostacyclin (PGI_2), by cyclooxygenase (COX), and epoxides (EETs) by cytochromes P450 (CYP). EETs may contribute to activation of TRPV4. Upon removal of agonist, Ca^{2+} is pumped back into the ER by the smooth endoplasmic reticulum Ca^{2+} ATPase (SERCA) and extruded from the cell by the Na^+/Ca^{2+} exchanger (NCX) and the plasma membrane Ca^{2+} ATPase (PMCA). Sodium that enters the cells via TRP channels is extruded by the Na^+/K^+ ATPase (NKA) and NCX. *CAM* calmodulin. *PLB* phospholamban

intracellular Ca^{2+} then activates plasmalemmal Ca^{2+}-activated K^+ channels to produce hyperpolarization of the endothelial cell membrane as shown in Fig. 1.2c, as well as activation of other Ca^{2+}-dependent processes such as NO production and release of arachidonic acid from membrane phospholipids resulting in increased production of prostacyclin and epoxides of arachidonic acid [40] (Fig. 1.1).

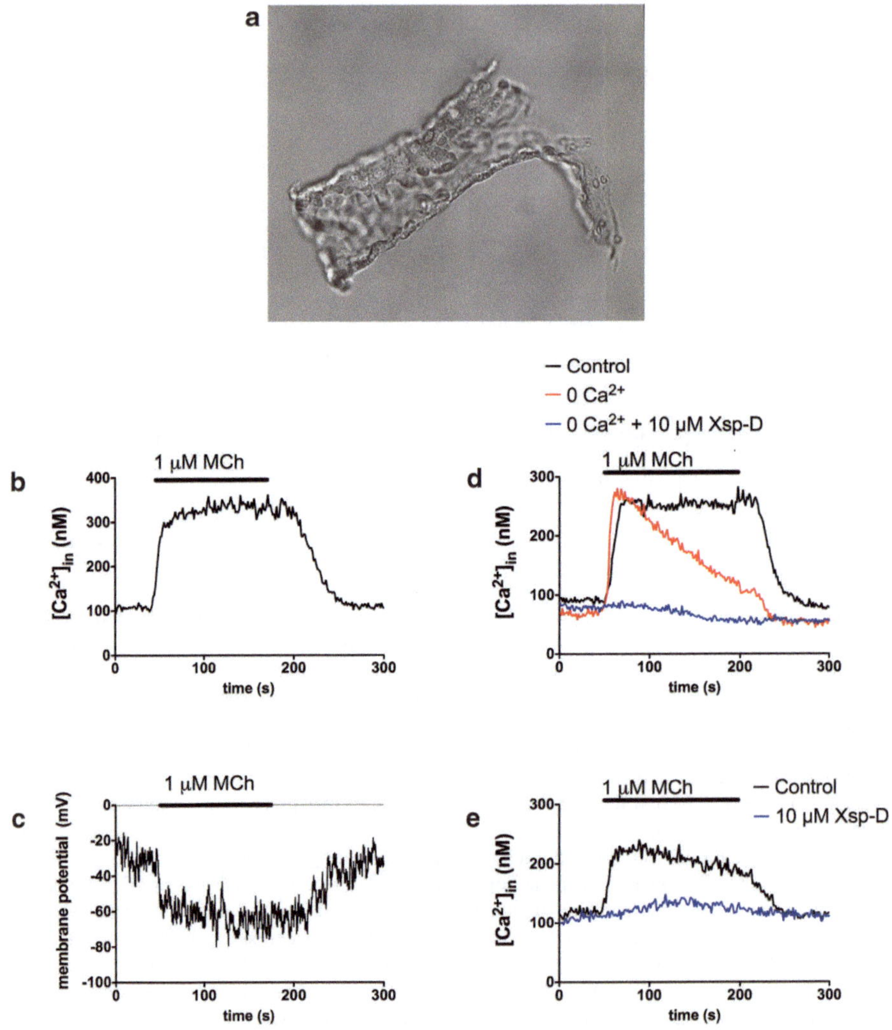

Fig. 1.2 Methacholine-induced global Ca^{2+} transients in arteriolar endothelial cells. Panel **a** shows an image of an endothelial cell tube enzymatically isolated from a second order hamster cremaster arteriole as described [23]. Panel **b** shows a representative Ca^{2+} transient elicited by the muscarinic receptor agonist, methacholine (MCh) from an endothelial cell tube loaded with the ratiometric calcium indicator, Fura-2AM. Panel **c** shows the MCh-induced membrane hyperpolarization of an endothelial cell tube loaded with the potentiometric indicator, di-8-ANEPPs as described [23]. Panel **d** shows MCh-induced Ca^{2+} transients, as in **a**, under Control conditions, after brief exposure to solutions containing 0 mM Ca^{2+} demonstrating the loss of the plateau phase of the Ca^{2+} transient. After return to Ca^{2+}-replete conditions (to obviate depletion of intracellular Ca^{2+}), subsequent exposure to 0 mM Ca^{2+} and the IP_3R-antagonist, xestospongin-D (Xsp-D; 10 μM) abolished the effects of MCh. Panel **e** shows inhibition of a MCh-induced Ca^{2+} transient by Xsp-D in the presence of extracellular Ca^{2+} (2 mM). Data shown in **a–e** are modified from [23]

The remainder of this review will focus on the expression and function of ion channels responsible for agonist-induced Ca^{2+} signals and membrane hyperpolarization in endothelial cells as shown in Fig. 1.2, as well as other endothelial cell ion channels that appear important to the regulation of myogenic tone in resistance arteries and arterioles.

What Endoplasmic Reticulum Ion Channels Mediate Agonist-Induced Ca^{2+} Signals?

Inositol-1,4,5-trisphosphate Receptors

Early studies demonstrated that agonists of $G\alpha_q$-coupled receptors increased intracellular Ca^{2+} in endothelial cells that was due to an initial release of Ca^{2+} from internal stores followed by Ca^{2+} influx [18, 25, 59, 139, 142, 143] (Figs. 1.1 and 1.2). Pharmacological studies subsequently identified IP_3Rs as the primary Ca^{2+} release channel responsible (see [23, 143] and Fig. 1.2 for examples). Inositol-1,4,5-trisphosphate receptors are large (350 kDa) tetrameric Ca^{2+} release channels found in the endoplasmic reticulum of all mammalian cells [47, 104]. Each monomer contains an IP_3-binding domain that is located in cytoplasmic N-terminus of the proteins [47, 104]. Calcium appears to be the trigger for gating IP_3 channels to open (see Table 1.1 for EC_{50} values for activation of the channels) [47, 104]. However, the response to increases in Ca^{2+} is biphasic, with higher concentrations of Ca^{2+} becoming inhibitory (Table 1.1) [47, 104]; Ca^{2+} activates the channels at low concentrations, but inhibits Ca^{2+} release at high concentrations (see Fig. 7 in [47] for examples). It has been proposed that the concentration of IP_3 determines the affinity of IP_3Rs for the inhibitory effects of elevated Ca^{2+}; as the concentration of IP_3 increases, higher levels of Ca^{2+} are required to inhibit the channels (Table 1.1) [47, 104]. Thus, the concentration of IP_3 effectively determines the range of cytosolic Ca^{2+} concentration over which the IP_3Rs will be active [47]. However, this has not been observed in all systems [171]. Nonetheless, in the presence of physiologically relevant concentrations of IP_3, IP_3R can undergo Ca^{2+}-induced-Ca^{2+} release (CICR) providing a positive feedback mechanism for release of Ca^{2+} from adjacent IP_3R (as in the case of Ca^{2+}waves), as well as amplification, with a limit, of Ca^{2+} signals originating from Ca^{2+} influx through overlying plasma membrane ion channels.

There are three isoforms of IP_3R, IP_3R1, 2 and 3 originating from three genes with modestly different characteristics (see Table 1.1); IP_3R2 and 3 appear to have the highest sensitivity for Ca^{2+}-induced activation with IP_3R3 having the lowest sensitivity for IP_3 [47, 104, 171]. In addition to IP_3 and Ca^{2+}, the activity of IP_3Rs is also sensitive to cytoplasmic ATP concentrations [47, 104, 171]. In IP_3R1 and 3, ATP produces a leftward shift in the Ca^{2+}-activity relationship increasing the affinity of the channels for Ca^{2+}, with little effect on the maximal open-state probability [47, 171]. In distinct contrast, ATP has no effect on Ca^{2+}-sensitivity of IP_3R2 receptors,

Table 1.1 Microvascular endothelial ion channels and their pharmacology

Channel	Gene	Alternative names	Accessory subunits	Antagonists (IC_{50})	Agonists (EC_{50})
CaV1.2	CACANA1C	L-type	β, $\alpha_2\delta$	Nifedipine (10–100 nM) [97] Nimodipine (139 nM) [183] Diltiazem (500 nM) [72] Verapamil (60 nM) [72] Mibefradil (1.4–13 μM) [67, 110] Cd^{2+} (7 μM) [121] Ni^{2+} (280 μM) [121]	BayK 8644 (6 nM) [188] FPL64176 (211 nM) [188]
CaV3.1	CACNA1G	T-type		Mibefradil (0.4–1.2 μM) [67] Cd^{2+} (160 μM) [94] Ni^{2+} (167–250 μM) [94]	
CaV3.2	CACNA1H	T-type		Mibefradil (1.1–1.2 μM) [67] Cd^{2+} (160 μM) [94] Ni^{2+} (5.7–12 μM) [94]	
$K_{Ca}1.1$	KCNMA1	BK_{Ca}, Slo1	β1-4 (KCNMB1-4)	Iberiotoxin (1.7 nM) [176] Charybdotoxin (2.9 nM) [176] Paxilline (1.9 nM) [176] TEA (0.14 mM) [176]	NS1619 BMS204352 DHS-1 Estradiol
$K_{Ca}2.3$	KCNN3	$SK_{Ca}3$, SK3	Calmodulin	Apamin (10 nM) [176] UCL1684 (9.5 nM) [176] TRAM-34 (20 μM) [181]	EBIO (87–600 μM) [181] NS309 (120–900 nM) [181] SKA-31 (3 μM) [181]
$K_{Ca}3.1$	KCNN4	$IK_{Ca}1$, IK1	Calmodulin	Charybdotoxin (5 nM) [181] Clotrimazole (70 nM) [176] TRAM-34 (10–25 nM) [181] NS6180 (11 nM) [181]	EBIO (24–80 μM) [181] NS309 (10–27 nM) [181] SKA-31 (260 nM) [181]
KIR2.1	KCNJ2			Ba^{2+} (2 μM at −100 mV; 19–30 μM at −40 mV) [6, 99] Intracellular Mg^{2+} and polyamines [68] ML133 (1.9 μM) [172]	Extracellular K^+ (3–20 mM) [101]

KIR2.2	KCNJ12			Ba^{2+} (0.5 μM at −100 mV; 9 μM at -40 mV) [99] ML133 (2.9 μM) [172] Intracellular Mg^{2+} and polyamines [68]	Extracellular K^+ (3–20 mM) [101]
KIR2.3	KCNJ4			Ba^{2+} (10.3 μM at −100 mV; 70 μM at −40 mV) [99] ML133 (4 μM) [172] Intracellular Mg^{2+} and polyamines [68]	Extracellular K^+ (3–20 mM) [101]
KIR6.1	KCNJ8		SUR2b	Glibenclamide (20–100 nM) [122, 129] Tolbutamide (350 μM) [129]	Diazoxide (32 μM) [106] Pinacidil (0.6 μM) [106] Levcromakalim (79 nM) [106]
KIR6.2	KCNJ11		SUR2b	Glibenclamide (20–100 nM) [122, 129] Tolbutamide (350 μM) [129] ML133 (7.7 μM) [172]	Diazoxide (32 μM) [106] Pinacidil (0.6 μM) [106] Levcromakalim (79 nM) [106]
IP_3R1	ITPR1		See [47] for list of interacting proteins	Ca^{2+} (1.3–52 μM) [47] Heparin (4.1 μg/ml) [133] Xestospongin C/D (358–844 nM) [49] 2-Aminoethoxydiphenyl borate (2-APB) (42 μM) [111]	Ca^{2+} (57–348 nM) [47, 171] IP_3 (34 nM) [133] Adenophostin A (4.5 nM) [133]
IP_3R2	ITPR2		See [47] for list of interacting proteins	Ca^{2+} (1.3–52 μM)[a] [47] Heparin (22 μg/ml) [133] 2-Aminoethoxydiphenyl borate (2-APB) (~100 μM) [133]	Ca^{2+} (58 nM) [171] IP_3 (151 nM) [133]
IP_3R3	ITPR3		See [47] for list of interacting proteins	Ca^{2+} (0.3–39 μM) [47] Heparin (2.8 μg/ml) [133] 2-Aminoethoxydiphenyl borate (2-APB) ($\gg$100 μM) [133]	Ca^{2+} (77 nM) [47] IP_3 (219 nM) [133] Adenophostin A (19.5 nM) [133]

(continued)

Table 1.1 (continued)

Channel	Gene	Alternative names	Accessory subunits	Antagonists (IC_{50})	Agonists (EC_{50})
RyR1	RYR1		See [46, 102] for list of interacting proteins	Ryanodine (100 nM to 1 μM)[b] [192] Tetracaine (100 μM) [192]	Ryanodine (>10 μM)[b] [192] Caffeine (0.2–0.5 mM) [192]
RyR2	RYR2		See above	See above	See above
RyR3	RYR3		See above	See above	See above
TRPA1	TRPA1	NAKTM1, TRPN1		Allyl isothiocyanate (AITC) (4.4–16.2 μM) [154]	HC-030031 (0.7–6.3 μM) [112]
TRPC1	TRPC1	TRP1	TRPC3, TRPC4 IP_3R1 STIM1, ORAI1	Gd^{3+} (1–10 μM) [10] La^{3+} (1–10 μM) [10] 2-Aminoethoxydiphenyl borate (2-APB) (80 μM) [10]	
TRPC3	TRPC3	TRP3	TRPC1 IP_3R1	Pyr-3 (0.7 μM) [84] Gd^{3+} (0.1 μM) [58] La^{3+} (4 μM) [58] SKF96365 (8 μM) [58]	OAG (100 μM) [71]
TRPC4	TRPC4	CCE1, TRP4	TRPC1, STIM1	ML-204 (0.96 μM) [116]	La^{3+} (100–300 μM peak effect) [138]
TRPC6	TRPC6	TRP6		La^{3+} (13–50 μM) [14, 42] SKF96365 (5–25 μM) [14, 42] ML-9 (36 μM) [64]	Hyperforin (1.5 μM) [96] OAG (117 μM) [71]
TRPV3	TRPV3		Calmodulin	Ruthenium red (10 μM) [38]	Carvacrol (4–34 μM) [38] 6-*Tert*-butyl-**m**-cresol (370 μM) [168]
TRPV4	TRPV4	TRP12, VRL-2	Calmodulin	HC-067047 (17–133 nM) [43]	GSK1016790A (18 nM) [161]

[a]The inhibitory effect of Ca^{2+} on Ca^{2+} release through IP_3R depends on the concentration of IP_3 to which the channel is exposed. The values shown are for $[IP_3]$ = 10–100 nM for IP_3R1 [105] and 20 nM to 10 μM for IP_3R3 [103]

[b]Concentrations required to block the channel

but has large effects on the maximum open-state probability of the channels; low ATP severely reduces the maximal activity of IP_3R2 [104, 171]. Given that IP_3Rs exist in signaling microdomains adjacent to ATPases in the ER and in the plasma membrane, local ATP concentrations could have a profound effect on IP_3R function in an isoform-specific fashion. Finally, IP_3R interact with a large number of proteins in the cytosol and in the lumen of the ER that can modulate the activity of these channels by protein-protein interactions, and phosphorylation/dephosphorylation in the case of protein kinases and phosphatases [47, 159].

Freshly isolated rat aortic endothelial cells express transcripts for all three isoforms of IP_3R with IP_3R1 being most highly expressed [118, 119]. Endothelial cells from mouse mesenteric arteries express transcripts for all three isoforms with IP_3R2 being most highly expressed [93]. Mouse cremaster arteriolar endothelial cells also express transcripts for all three isoforms, but IP_3R3 appears to be most highly expressed (Fig. 1.4a), and protein expression for all three isoforms has been reported [77]. All three IP_3R isoforms were detected in the endothelium of Wistar rat basilar and mesenteric arteries by immunocytochemistry [56]. Thus, it appears that there may be regional or species-dependent differences in the expression of IP_3R isoforms, and little is known about the localization or function of the individual IP_3R isoforms in native endothelial cells.

In co-cultures of smooth muscle and endothelial cells from mouse cremaster arterioles, and in intact vessels, it has been reported that IP_3R1 localizes at sites of myoendothelial gap junctions (MEJs) [77]. Similarly, in intact mouse mesenteric resistance vessels, there are clusters of IP_3Rs near holes in the internal elastic lamina [93], sites that have been correlated with projections of endothelial cells (myoendothelial projections, MEPs, Fig. 1.3) towards overlying smooth muscle cells and the localization of MEJs [136]. However, the IP_3R isoform expressed in these clusters was not identified. Importantly, these sites were shown to generate localized endothelial cell Ca^{2+} events that have been termed Ca^{2+} pulsars [93]. These events are too small and rapid to be detected by global Ca^{2+} measurements made with Fura 2 (Fig. 1.2, for example), but can be detected using Fluo-4 or genetic Ca^{2+}sensors such as GCaMP2 and high-speed confocal imaging [83, 93]. Previous studies have shown that $K_{Ca}3.1$ channels also are clustered in the same microdomain [137] providing a means for Ca^{2+} pulsars to be translated into, for example, changes in membrane potential (see below for more on this topic). In addition, a growing list of proteins congregate in the vicinity of MEJs including TRPA1 channels [37], TRPV4 channels [149, 150], anchoring proteins (e.g., AKAP150 [150]), protein kinases (e.g., PKC [150]), nitric oxide synthase [151], Na^+/K^+ ATPase [33] and other proteins [152].

Calcium pulsars [83, 93] and Ca^{2+}waves [36, 83] are present under resting conditions in endothelial cells of pressurized vessels [83, 93] and likely contribute to the resting activity of $K_{Ca}3.1$ channels (in the case of pulsars and waves) and $K_{Ca}2.3$ channels (Ca^{2+} waves) and endothelial cell membrane potential. Endothelium-dependent vasodilators, such as acetylcholine [93, 147] or adenosine [36], increase the number and frequency of Ca^{2+} pulsars [93], and also recruit IP_3R located throughout endothelial cells to produce asynchronous [93, 147] or synchronous [36,

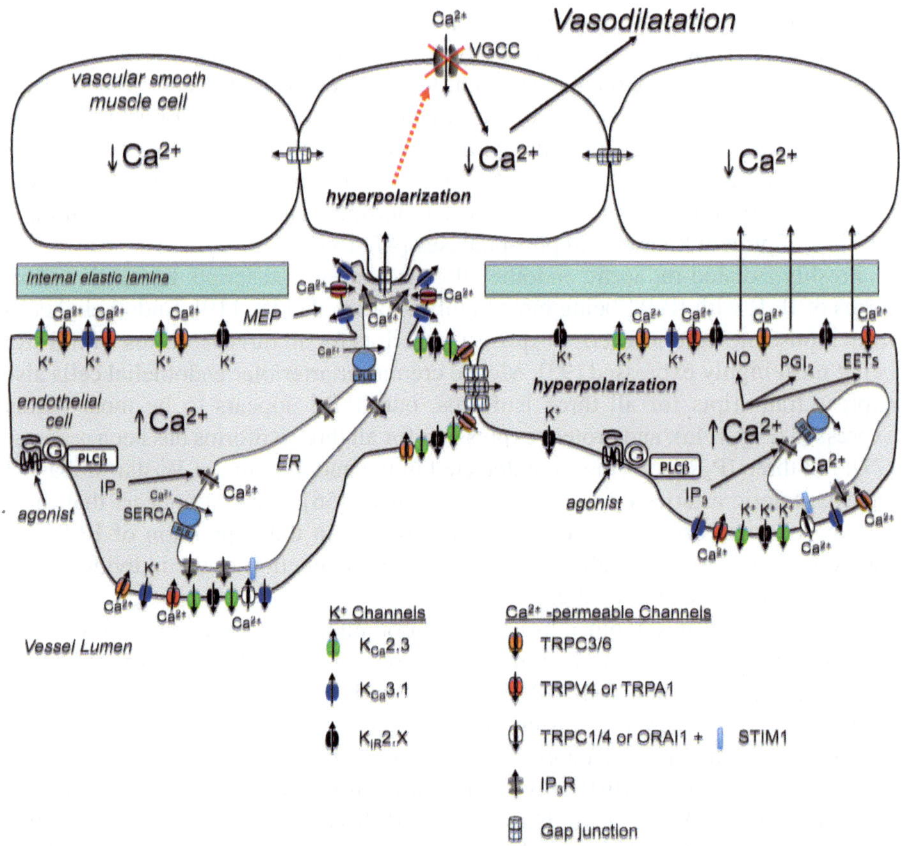

Fig. 1.3 Endothelial cell Ion channels and cell-cell communication in the vessel wall. Shown is a schematic representation of a longitudinal cross section through two endothelial cells and their relationship to overlying smooth muscle cells. Endothelial cells communicate with overlying smooth muscle cells via myoendothelial projections (MEPs), that pass through the internal elastic lamina to make contact with overlying smooth muscle cells, as shown. Gap junctions may form at MEPs to yield myoendothelial junctions (MEJs) allowing endothelial cell hyperpolarization to be conducted to the smooth muscle cells, closing smooth muscle voltage-gated Ca^{2+} channels (VGCCs) and leading to vasodilatation. Ion channels such as TRPV4, $K_{Ca}3.1$ and IP_3R (as shown) may cluster in MEPs to form signaling complexes to direct the endothelial cell responses to vasodilator agonists. Other ion channels such as TRPC3 and $K_{Ca}2.3$ may cluster elsewhere to form other signaling complexes. Abbreviations are as in Fig. 1.1

147] Ca^{2+}waves and increases in global Ca^{2+} (Fig. 1.2). Thus, the global Ca^{2+} signals that have been reported in native microvascular endothelial cells [23, 32, 107, 148] represent a complex mixture of local Ca^{2+} pulsars and Ca^{2+} waves in addition to more homogeneous increases in cytosolic Ca^{2+}. Additional studies are needed to define the precise localization of IP_3R isoforms and their function related to endothelium-dependent vasomotor activity.

Ryanodine Receptors

Ryanodine receptors (RyR) are composed of very large protein subunits (~500 kDa) that form Ca^{2+}-sensitive-Ca^{2+}-release channels in the endoplasmic reticulum [61]. Similar to IP_3R, they are tetramers with three isoforms from three distinct genes: RyR1, RyR2 and RyR3 [61, 62, 185]. Skeletal muscle expresses predominantly RyR1, the heart expresses predominantly RyR2 and RyR3 is expressed in the brain and other tissues [61, 62, 185]. Vascular smooth muscle cells may express all three isoforms, with RyR2 being predominant in resistance artery [166] and arteriolar smooth muscle cells [178]. Immunofluorescence demonstrated RyR expression in guinea pig endocardium and aortic endothelial cells [95], and ryanodine binding sites have been reported in porcine coronary artery endothelial cells [54]. Studies of cultured endothelial cells suggest expression of functional ryanodine receptors [191], and freshly isolated endothelial cells from rabbit aorta display caffeine-induced Ca^{2+} transients implying the presence of RyR in these cells [132]. In porcine coronary artery endothelial cells, caffeine elicits a Ca^{2+} transient in only 37 % cells studied suggesting heterogeneity of the distribution and function of endothelial RyR [55]. Transcripts for RyR3, but not RyR1 or RyR2 have been reported in endothelial cells freshly isolated from human mesenteric arteries [89]. Thus, RyR appear to be expressed and functional in macrovascular endothelial cells. However, there is little evidence for expression of ryanodine receptors in microvascular endothelial cells. In mouse mesenteric resistance arteries, where expression of IP3Rs are readily detected, no message for the three RyR isoforms were found, and ryanodine had no effect on basal or acetylcholine-stimulated Ca^{2+} events [93]. We have also found lack of expression of the three RyR isoforms in endothelial cells from mouse cremaster arterioles (Fig. 1.4b), and caffeine does not elicit a Ca^{2+} transient in freshly isolated hamster cremaster arteriolar endothelial cells [23] (Fig. 1.5a), although lack of effect of caffeine on global Ca^{2+} levels does not completely exclude a role for

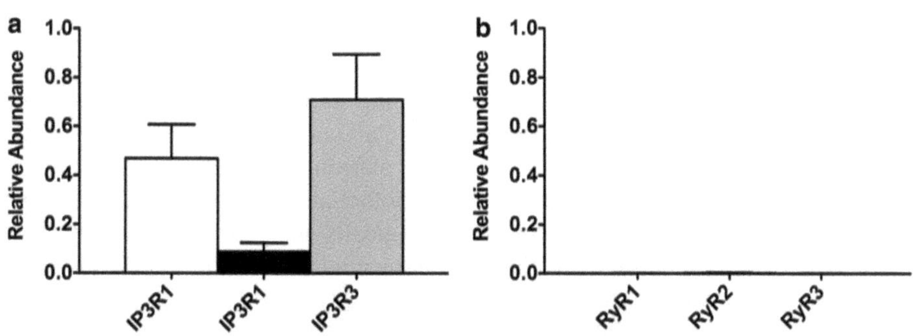

Fig. 1.4 Expression of transcripts for IP_3R and RyR in freshly isolated mouse cremaster arteriolar endothelial cells. Shown are means ± SE (n = 5 cell isolates for IP_3R and n = 7 for RyR) abundance of transcripts for IP_3R (panel **a**) and RyR (panel **b**) isoforms relative to eNOS in endothelial cell tubes isolated from second-order mouse cremaster arterioles. See [23, 178] for methodological details

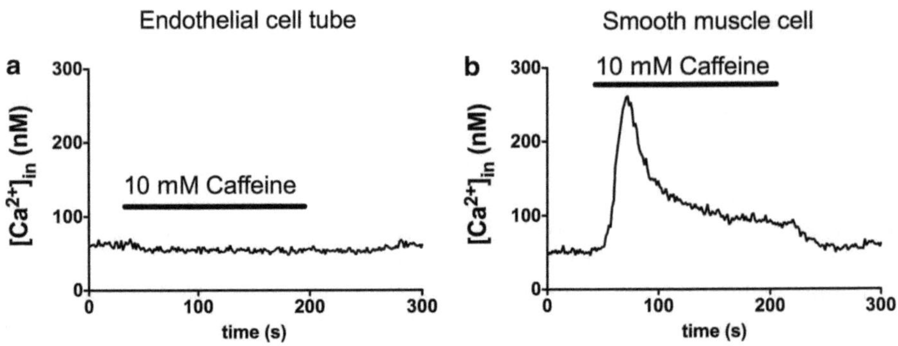

Fig. 1.5 Failure of caffeine to elicit Ca^{2+} transients in arteriolar endothelial cells but not smooth muscle cells. Shown are representative responses of an endothelial cell tube (panel **a**) and a smooth muscle cell (Panel **b**) isolated by the same method from hamster cremaster arterioles [23], to the RyR agonist, caffeine (10 mM). Consistent with the lack of expression of RyR in arteriolar endothelial cells (see Fig. 1.4b), caffeine failed to elicit a Ca^{2+} transient above baseline in endothelial cells in endothelial cells (n=5 isolates from five arterioles), but produced the expected response from smooth muscle cells isolated from the same vessels

RyR [126]. A lack of effect of the RyR antagonist, ryanodine, on Ca^{2+} signals in endothelial cells in rat mesenteric arteries also has been observed [83]. Although species and regional heterogeneity in the expression of RyR cannot be excluded, these data suggest that RyR do not play a major role in Ca^{2+} signaling in endothelial cells of resistance arteries and arterioles.

What Ion Channels Mediate Agonist-Induced Ca^{2+} Influx?

TRP Channels

Endothelium-dependent vasodilators not only increase the activity of IP_3R, by stimulating the production of IP_3, they also result in the activation of ion channels in the plasmalemma of endothelial cells that conduct Ca^{2+} and are responsible for steady-state increases in intracellular Ca^{2+} (i.e., the plateau phase of the Ca^{2+}transient shown in Fig. 1.2). Early studies in cultured endothelial cells provided evidence that agonist-induced Ca^{2+} entry was electrophysiologically and pharmacologically similar to the Ca^{2+} entry induced by depletion of intracellular Ca^{2+} stores [139, 142, 143, 164, 165]. In primary cultures of porcine coronary artery endothelial cells, substance P activates a non-selective, inward whole-cell cation current that can be completely inhibited by blocking IP_3-dependent activation of IP_3R with heparin [143]. Similarly, block of IP_3R with xestospongin-D abolishes methacholine-induced global Ca^{2+} transients in arteriolar endothelial cells [23] (Fig. 1.2e). These data suggest that, at least under the conditions of these experiments, agonist-induced activation of IP_3R, and likely release of Ca^{2+} from internal stores is required to activate the

Ca^{2+} influx pathway that is responsible for the plateau phase of the agonist-induced global Ca^{2+} transients, and hence the steady-state phase of agonist-induced endothelial cell hyperpolarization. This does not exclude the activation of receptor, or second-messenger activated Ca^{2+} influx as the currents activated may be too small to detect by conventional whole-cell methods, or may produce only local changes in Ca^{2+} that do not influence global Ca^{2+}, particularly as detected by Fura-2.

The ion channels that are responsible for agonist-induced Ca^{2+} influx in native endothelial cells remain in question. Several members of the transient receptor potential (TRP) family of ion channels, in particular TRPC1, TRPC3, TRPC4, TRPC6 and TRPV4, along with members of the stromal interaction molecule (STIM) and ORAI families appear to be likely candidates, and it is also likely that multiple channels are activated and contribute to the Ca^{2+} influx activated by endothelium-dependent vasodilators (Fig. 1.1).

The TRP channel family form, in general, cation channels that are weakly Ca^{2+} selective (permeability for Ca^{2+}/permeability for Na^{2+} <10) [179]. The channel monomers are assumed to have six membrane spanning domains with the pore between segments 5 and 6, and both the C- and N-termini of the channels located intracellularly, with four monomers forming a functional channel [179].

Endothelial cells express TRPC1, which may serve as store-operated Ca^{2+} channels in endothelial cells [124]. However, their function in agonist-induced endothelial hyperpolarization and vasodilatation remains unclear. This may partly be due to the observation that TRPC1 heteromultimerize with other members of the TRPC family (especially TRPC4) as well as STIM/ORAI containing channels [30]. Studies of cultured human pulmonary artery endothelial cells revealed expression of TRPC1, and that antisense oligonucleotide knockdown reduced Ca^{2+} influx induced by Ca^{2+} store depletion by about 50 % suggesting that TRPC1 contributes to the Ca^{2+} influx pathway in these cells [17]. A similar conclusion was drawn in cultured bovine aortic endothelial cells using a TRPC1 antibody to inhibit basic fibroblast growth factor-induced Ca^{2+} entry [7]. Studies of human and mouse pulmonary microvascular endothelial cells also support a role for TRPC1 in agonist- and Ca^{2+} store depletion-induced Ca^{2+} entry and microvascular permeability [5, 92, 114, 155, 156]. However, the roles played by TRPC1 in agonist-induced endothelial cell hyperpolarization and vasodilatation is not as clear. Study of carotid artery endothelial cells from TRPC1 knock-out mice reveal enhanced acetylcholine-induced hyperpolarization, rather than decreased responses predicted based on studies of cultured cells and lung models [140]. Although compensatory upregulation of expression of other channels might explain these results, prior experiments failed to detect upregulation of other TRPC channels [31]. At the least, these results indicate that TRPC1 is not essential for agonist-induce endothelial cell hyperpolarization and that additional channels contribute to the Ca^{2+} influx induced by agonists in native endothelial cells, although regional heterogeneity in expression and function cannot be excluded. In contrast, studies of native aortic endothelial cells from $TRPC1^{-/-}$ mice demonstrated a small reduction of endothelial cell Ca^{2+} transients induced by acetylcholine [87], suggesting that TRPC1 channels do participate, to a small extent, in agonist-induced Ca^{2+} signals. A small reduction in thrombin-induced

Ca^{2+} transients was also observed after siRNA knock down of TRPC1 in cultured pulmonary microvascular endothelial cells [155]. Taken together, these data do not support a major role for TRPC1 in agonist-induced Ca^{2+} entry into endothelial cells relevant to endothelial cell hyperpolarization and vasodilatation.

TRPC4 is another channel that has been implicated in agonist and store-depletion-induced Ca^{2+} entry into endothelial cells [48, 155, 162]. In cultured aortic endothelial cells isolated from TRPC4$^{-/-}$ mice, the plateau phase of agonist-induced Ca^{2+} transients and related endothelial cell hyperpolarization was substantially depressed (but not eliminated) [48], suggesting a major role for TRPC4 in agonist-induced Ca^{2+} transients in macrovascular endothelial cells. Similarly, use of cultured, pulmonary microvascular endothelial cells isolated from TRPC4$^{-/-}$ mice, as well as siRNA knock down of endogenous TRPC4 from cells isolated from wild-type mice demonstrated a major role for TRPC4 in thrombin- and Ca^{2+}-store-depletion-induced Ca^{2+} signals [155]. The authors also demonstrated that expression of STIM1 was necessary for normal store-operated Ca^{2+} entry and showed that STIM1 and TRPC4 interacted and that expression of both proteins was required for normal Ca^{2+} signaling [155]. Endothelium-dependent vasodilatation is also substantially reduced in vessels from TRPC4$^{-/-}$ mice [48]. These data support a role for TRPC4 as a Ca^{2+} influx pathway during agonist-induced Ca^{2+} signaling, with activity triggered by loss of Ca^{2+} from internal stores as sensed by STIM1 (Fig. 1.1).

In human umbilical vein endothelial cells, the Ca^{2+} influx induced by agonists or depletion of intracellular stores depends on expression of STIM1 and ORAI1 with STIM1 serving as the sensor of ER Ca^{2+} and ORAI1 forming the pore of the store-operated channels [1, 155, 190]. Furthermore, in contrast to the studies outlined above, it was shown that effective knock down of TRPC1 or TRPC4 had no effect on Ca^{2+} store depletion-induced Ca^{2+} signals in this model [1]. However, studies in murine pulmonary microvascular endothelial cells indicate that expression of ORAI1 is not required for normal endothelial cell Ca^{2+} signaling [155]. These data suggest that there may be regional or species-dependent differences in the ion channels responsible for agonist-induced Ca^{2+} signaling in endothelial cells. The role played by STIM1 and ORAI1 in agonist-induced endothelial cell hyperpolarization of endothelial cells in resistance arteries and arterioles has not been reported.

TRPV4 channels also participate in agonist- and shear-stress-induced Ca^{2+} signals in endothelial cells, and endothelium-dependent vasodilatation in intact vessels [8, 35, 39, 109, 149, 150, 187]. These channels are temperature sensitive [175], stretch-sensitive [170] and can be activated by EETs (Fig. 1.1) and phorbol ester derivatives [170]. Calmodulin interacts with the C-terminal domain and mediates Ca^{2+}-dependent activation of these channels [153]. Activation of TRPV4 channels in endothelial cells increases intracellular Ca^{2+} [109, 149, 150, 174, 187] and produces vasodilatation [39, 134, 149, 150, 187]. Importantly, the plateau-phase of vasodilator agonist-induced endothelial cell Ca^{2+} transients are reduced in the endothelium of vessels from TRPV4$^{-/-}$ mice [149, 150, 187]. In mouse mesenteric arteries, TRPV4 channels cluster in the same microdomain as IP_3R at MEPs (Fig. 1.3), and endothelium-dependent agonists activate these channels to produce TRPV4-Ca^{2+}sparklets at the sites of MEPs [149, 150]. Activation of Ca^{2+} influx through only a few TRPV4 channels per endothelial cell activates endothelial $K_{Ca}2.3$ and $K_{Ca}3.1$ channels and can produce maximal vasodilatation, with activation of $K_{Ca}3.1$ channels, which also cluster at MEJ's, occurring preferentially at low levels of TRPV4 activation [149]. Block of TRPV4 channels with HC-067047 in mouse mesenteric arteries substantially inhibits the component of acetylcholine-induced vasodilatation that is mediated by activation of endothelial $K_{Ca}2.3$ and $K_{Ca}3.1$ channels, suggesting that TRPV4 channels play a major role in agonist-induced Ca^{2+} influx that contributes to endothelial cell hyperpolarization and subsequent vasodilatation [149]. In mouse mesenteric arteries, muscarinic receptor agonists activate TRPV4 channels through a signaling pathway involving PLC-β and activation of PKC that is targeted to the channel by the scaffolding protein, AKAP150 [150] (Fig. 1.1). Thus, TRPV4 appears to play a major role in agonist-induced endothelial cell Ca^{2+} signaling that is related to hyperpolarization and vasodilatation.

Evidence also has been presented suggesting that TRPC3 is involved in agonist-induced Ca^{2+} signaling in endothelial cells [75, 86, 87, 98, 141, 184]. These TRP channels can be activated by DAG formed through the action of PLCs on membrane phospholipids [71] (Fig. 1.1). In cerebral vascular smooth muscle, activation of IP_3R1, independent from release of Ca^{2+}, results in activation of TRPC3 [4]. It is not known if a similar interaction occurs in endothelial cells. TRPC3 is involved in flow- and bradykinin-induced vasodilatation in rat small mesenteric arteries, but not dilation induced by histamine, ATP or cyclopiazonic acid [98]. As with TRPV4, TRPC3 appears to cluster near MEPs in rat mesenteric resistance artery endothelial cells [141]. In rat mesenteric arteries, the TRPC3 blocker, Pyr3, inhibited endothelial cell hyperpolarization and the portion of acetylcholine-induced endothelium-dependent vasodilatation mediated by activation of $K_{Ca}2.3$ and $K_{Ca}3.1$ channels and the consequent endothelial cell hyperpolarization [141]. These data suggest a close physical and functional relationship between TRPC3 and the K_{Ca} channels that mediate agonist-induced endothelial cell hyperpolarization. In contrast, in porcine coronary arteries, the TRPC3 antagonist Pyr3 inhibited the portion of bradykinin-induced vasodilatation that is mediated by NO, NO production in isolated endothelial cells and bradykinin-induced endothelial cell Ca^{2+} transients suggesting that TRPC3 channels contributed to Ca^{2+} signaling directed at NO production in this system [75]. In murine cerebral arteries, ATP-induced endothelial cell Ca^{2+} transients are reduced by Pyr3 or in cells isolated from $TRPC3^{-/-}$ mice, and Ca^{2+} entry through TRPC3 appears to selectively activate $K_{Ca}2.3$ channels during the plateau-phase of agonist-induced endothelial cell hyperpolarization [86]. Thus, there appear to be regional and likely species-dependent differences in the roles played by TRPC3 in endothelial cell Ca^{2+} signaling and hyperpolarization. This likely represents differences in the localization of the channels and the signaling microdomains in which they are expressed.

Endothelial cells in cerebral arteries also express TRPA1 channels that, when activated, produce endothelial cell hyperpolarization and endothelium-dependent vasodilatation [37, 154]; the endothelium of mouse, rat or human coronary, renal or mesenteric arteries do not express transcripts or protein for TRPA1 [154]. The TRPA1 subunits have 14–18 ankyrin repeats at their amino-terminal that give the channels their name, are activated by a diverse array of chemicals including pungent substances found in mustards and garlic and are heavily expressed in sensory nerves [9]. In endothelial cells of cerebral arteries, these channels cluster at MEPs and co-localize with $K_{Ca}3.1$ channels [37, 38, 154] (Fig. 1.3). Activation of TRPA1 leads to TRPA1-Ca^{2+}sparklets [154]. Vasodilatation induced by activation of TRPA1 channels with allyl isothiocyanate (AITC) in substantially inhibited by the $K_{Ca}3.1$ blocker, TRAM34 and abolished by the combination of TRAM34 and the $K_{Ca}2.3$ blocker, apamin, but is unaffected by blockade of nitric oxide synthase or cyclooxygenase [37, 154]. These data indicate that Ca^{2+} influx through TRPA1 primarily activates endothelial cell K_{Ca} channels to produce vasodilatation. Furthermore, as with TRPV4 channels [149], only a small number of active TRPA1 channel clusters per endothelial cell are required for maximal vasodilatation [154]. It was also shown that vasodilatation induced by AITC could be inhibited by Ba^{2+} suggesting that inward rectifier K^+ channels either amplify the hyperpolarization induced by activation of endothelial cell K_{Ca} channels, or transduce the endothelial cell K_{Ca} channel activation by detecting the K^+ released through the K_{Ca} channels [37] (Fig. 1.1). In the endothelium of cerebral arteries, TRPA1 co-localizes with NADPH oxidase (NOX) isoform 2, and lipid peroxides produced by NOX2 activate these channels to produce vasodilatation [154].

Expression and function of TRPV3 channels in rat cerebral artery endothelial cells also has been reported [38]. These channels appear to be more uniformly expressed than TRPV4 or TRPA1 [38]. Activation of TRPV3 with agents such as the oregano monoterpenoid phenol, carvacrol, increases endothelial cell Ca^{2+} and activates endothelial cell K_{Ca} channels to produce endothelial cell hyperpolarization and vasodilatation [38].

Finally, endothelial cells also express TRPC6 that appears to contribute to Ca^{2+} signaling [21, 60, 100, 127, 145]. As for TRPC3, TRPC6 is activated by DAG produced simultaneously with the formation of IP_3 by PLCs [71] (Fig. 1.1 and Table 1.1). Human pulmonary artery endothelial cells express TRPC6, and siRNA knockdown of this channel impairs Ca^{2+} signaling and increases monolayer permeability induced by thrombin [145]. Human dermal microvascular endothelial cells in culture express TRPC6, but not TRPC3, and the plateau phase of the histamine-induced increase in Ca^{2+} is reduced by SKF96365 in these cells. Furthermore, histamine-induced increases in microvascular permeability are abolished in TRPC6$^{-/-}$ mice supporting a role for this channel in regulation of microvascular Ca^{2+} signaling and permeability [21]. Increased microvascular permeability induced by vascular endothelial growth factor (VEGF) appears to be mediated by TRPC6 [127], and VEGF-induced Ca^{2+} transients are reduced in cultured human microvascular endothelial cells expressing a dominant-negative form of TRPC6 [60]. Mouse aortic endothelial cells express TRPC6 and TRPC3, but carbachol-induced Ca^{2+} transients in isolated endothelial cells as well as carbachol-induced endothelium-dependent relaxation of aortas are reduced only in vessels isolated from TRPC6$^{-/-}$ mice [100]. The role of TRPC6 in agonist-induced Ca^{2+} signaling related to vasodilatation of arterioles and resistance arteries remains to be established.

Taken together, these data suggest that it is likely that there are multiple ion channels in the plasma membrane of endothelial cells that conduct Ca^{2+} into the cells to activate endothelial cell K_{Ca} channels (and other processes) to produce endothelial cell hyperpolarization (Fig. 1.1). Agonists that activate $G\alpha_q$-coupled receptors appear to activate multiple channels including TRPC3, TRPC4, TRPV4 and possibly TRPC6 (Fig. 1.1). However, the roles played by each of these channels, together, in concert, remains to be established. Regional and species dependent heterogeneity in expression and function of these channels cloud our view of the larger picture. The use of isolated, intact blood vessels using multiple approaches (as for studies of TRPV4 [149, 150]) appears to be a step in the right direction in solving this puzzle.

Voltage-Gated Ca^{2+} Channels

The expression of voltage-gated Ca^{+} channels (VGCC) in endothelial cells is controversial. Capillary endothelial cells from bovine adrenal medullas were reported to express both L-type and T-type VGCC as assessed by patch clamp techniques [167]. However, the function of these channels was not defined. More recently, mouse and rat pulmonary microvascular endothelial cells have been shown to express CaV3.1-based T-type VGCC that appear to be involved in thrombin-induced Ca^{2+} signals, cell adhesion, exocytosis of von Willebrand's Factor and depolarization-induced Ca^{2+} signaling [177, 180, 189]. Endothelial cells of mesenteric arterioles label with antibodies for CaV3.2, but not CaV1.2 or CaV3.1 and these channels were proposed to participate in conducted vasomotor responses [16]. Immunofluorescence identified CaV3.1 and CaV3.2 in endothelial cells in rat middle cerebral artery and its branches, although their function was not studied [91]. In mouse mesenteric arteries, CaV3.1 has been detected in the endothelium by immunofluorescence, was colocalized with eNOS and appeared to play a role in depolarization-induced NO synthesis based on studies employing CaV3.1 knockout mice [157]. In mouse cremaster muscle arterioles, CaV3.2 has been implicated in electrically-induced, conducted vasodilatation based on the pharmacology of this response [45]. However, the location of the channel was not established. Thus, there appears to be some evidence for expression of T-type channels in endothelial cells of resistance arteries and arterioles. However, their electrophysiological function has not been adequately explored in other than cultured cells, particularly in the peripheral microcirculation. Whether hyperpolarization of endothelial cells induced by vasodilators can activate current through T-type channels by reducing voltage-dependent inactivation and recruiting a window-current has not been established. Depolarization of freshly isolated hamster arteriolar endothelial cells does not, in and of itself, elicit a global change in intracellular Ca^{2+} [23]. These data do not support a major role for VGCC in these endothelial cells.

Which Ion Channels Mediate Agonist-Induced Endothelial Cell Hyperpolarization?

Early studies in endothelial cells from conduit arteries and in cultured cells showed that agonists of $G\alpha_q$-protein-coupled receptors resulted in Ca^{2+} signals in endothelial cells [18, 25, 142], activation of endothelial cell Ca^{2+}-activated K^+ channels [18, 24, 25, 53, 142] and endothelial cell hyperpolarization [18, 142].

K_{Ca}2.3 and K_{Ca}3.1 Channels

Native endothelial cells in resistance arteries primarily express two types of K_{Ca} channels: K_{Ca}2.3, the small-conductance, K_{Ca} (SK_{Ca}) channel and K_{Ca}3.1, the intermediate-conductance K_{Ca} (IK_{Ca}) channel [41, 57, 88, 137, 144, 160]. The channels are the products of distinct genes (KCNN3 and KCNN4, respectively [176]) (Table 1.1). Both channels are voltage insensitive and use calmodulin as the Ca^{2+} sensor, which interacts with the intracellular C-terminus of both channels to gate channel opening [44, 182]. The concentration of free Ca^{2+} required for 50 % of maximal activation of both K_{Ca}2.3 [182] and K_{Ca}3.1 [78] is on the order of 300 nM, with the threshold for activity at approximately 100 nM and maximal activity at 1 μM [78, 182]. These channels display distinct pharmacology that has aided in elucidating their function in intact vessels (Table 1.1).

Both K_{Ca}2.3 and K_{Ca}3.1 can contribute to endothelial cell hyperpolarization induced by agonists of $G\alpha_q$-protein coupled receptors, such as acetylcholine acting at M_3-muscarinic receptors: for example, in guinea-pig carotid artery [26], rat mesenteric arteries preconstricted with phenylephrine [27] and porcine coronary arteries [19], inhibition of agonist-induced hyperpolarization requires block of both K_{Ca}2.3 and K_{Ca}3.1. However, in several systems, agonist-induced hyperpolarization of endothelial cells appears to be mediated by K_{Ca}3.1 alone. In rat middle cerebral arteries, block of K_{Ca}3.1 alone abolishes endothelial cell hyperpolarization and subsequent vasodilatation induced by UTP [108]. Similarly, in mouse carotid arteries and cremaster muscle microcirculation, K_{Ca}3.1 appear to dominate acetylcholine-induced vasodilatation based on studies using knockouts of K_{Ca}2.3 or K_{Ca}3.1 [15].

The distribution of K_{Ca}2.3 and K_{Ca}3.1 in the plasma membrane of endothelial cells is neither random nor uniform. In the endothelium of mouse [93] and rat [137] mesenteric resistance arteries and rat cerebral arteries [37], K_{Ca}3.1 appear to cluster at MEPs (Fig. 1.3), with some of these MEPs containing gap junction proteins (Connexin 40 and 37) and forming gap junctions with overlying vascular smooth muscle cells (MEJs) (see [136] and references therein). At MEPs and MEJ's, K_{Ca}3.1 likely exist in macromolecular signaling complexes with IP_3Rs [93], TRPA1 channels [37], TRPV4 channels [149, 150], anchoring proteins (e.g., AKAP150 [150]), protein kinases (e.g., PKC [150]) and likely G-protein coupled receptors [150]. This localization appears to facilitate activation of K_{Ca}3.1 channels by local Ca^{2+} signals [37, 93, 149, 150] and transmission of K_{Ca}3.1-mediated hyperpolarization from endothelial cells to overlying smooth muscle cells via MEJs (Fig. 1.3). In addition, it has been proposed that activation of $K_{Ca}3.1$ channels by Ca^{2+} influx through TRPV4 channels has a positive-feedback effect on Ca^{2+} influx through TRPV4 [128] (Fig. 1.1).

In contrast, $K_{Ca}2.3$ channels appear to localize on the periphery of endothelial cells [137]. These channels also likely exist in signaling microdomains. In rat mesenteric arteries $K_{Ca}2.3$ localizes in cholesterol-rich areas (caveolae or lipid rafts) and colocalizes with caveolin-1 [2]. In rat cerebral arteries, Ca^{2+} influx through TRPC3 channels selectively activates $K_{Ca}2.3$ [86], suggesting co-localization of the two channels. The localization of $K_{Ca}2.3$ in caveolae adjacent to gap junction plaques between endothelial cells also likely explains why shear-stress-induced dilation is severely attenuated in carotid arteries from mice with conditional knockout of $K_{Ca}2.3$ [15]. The differential cellular distribution of $K_{Ca}2.3$ and $K_{Ca}3.1$ and their respective signaling microdomains likely explain how these channels contribute to different aspects of agonist-induced endothelial cell hyperpolarization [27, 144].

In addition to agonist-mediated events, endothelial cell $K_{Ca}2.3$ and $K_{Ca}3.1$ appear to be active at rest contributing to endothelial cell membrane potential and the regulation of myogenic tone particularly in small resistance arteries and arterioles. For example, overexpression of $K_{Ca}2.3$ hyperpolarizes mouse mesenteric artery endothelial cells and reduces myogenic tone, whereas conditional knockout depolarizes the cells and increases tone [160]. Similarly, blockade of either $K_{Ca}2.3$ or $K_{Ca}3.1$, or both channels increases myogenic tone in rat cerebral parenchymal arterioles [22, 63]. It is noteworthy that in upstream middle cerebral arteries, under similar conditions, inhibition of $K_{Ca}2.3$ and $K_{Ca}3.1$ had a smaller effect on myogenic tone than was observed in downstream parenchymal arterioles, despite their importance in agonist-induced vasodilatation [22]. These data suggest that there are regional differences in the roles played by these channels with increased function in the microcirculation.

K_{Ca}1.1 Channels

The expression and function of Large-conductance Ca^{2+}-activated K^+ ($K_{Ca}1.1$) channels in endothelial cells remains controversial [135]. These voltage and Ca^{2+}-sensitive K^+ channels have a much larger conductance (~250 pS) than $K_{Ca}2.3$ or $K_{Ca}3.1$, are intrinsically sensitive to Ca^{2+} (i.e., they do not require association with calmodulin), and display distinct pharmacology [73] (Table 1.1). These channels are highly expressed in vascular smooth muscle cells [80, 122]. Studies of cultured endothelial cells have repeatedly demonstrated expression and function of $K_{Ca}1.1$ channels (see [135] for references). However, the function and expression of $K_{Ca}1.1$ in native and freshly isolated endothelial cells is not clear. For example, in patch-clamp studies of freshly isolated endothelial cells from bovine coronary arteries [50], mouse carotid arteries [15], and rat cerebral parenchymal arterioles [63] no currents characteristic of $K_{Ca}1.1$ are detected and essentially all Ca^{2+}-activated currents are inhibited by the combination of a $K_{Ca}2.3$ and a $K_{Ca}3.1$ blocker. However, currents through $K_{Ca}1.1$ have been recorded in endothelial cells isolated from rats exposed to chronic hypoxia [76, 131] or from endothelial cells acutely exposed to β-methylcyclodextrin to deplete membrane cholesterol [131], effects that could be reversed by exposure of cells to a membrane permeant version of caveolin-1 [131]. These data are consistent with an earlier study in cultured endothelial cells indicating that $K_{Ca}1.1$ are targeted to caveolae and that caveolin-1 inhibited the function of $K_{Ca}1.1$ in that microdomain [173]. These data suggest that, in some endothelial cells, $K_{Ca}1.1$ channels may be silent under normal conditions, but can be upregulated by conditions, such as chronic hypoxia, that disrupt membrane microdomains. This idea is consistent with the hypothesis that $K_{Ca}1.1$ expression and function in endothelial cells only occurs during stress or pathological conditions [135]. Immunohistochemical localization of $K_{Ca}1.1$ has been reported in rat cremaster arterioles [163]. There must be regional or species differences in expression of $K_{Ca}1.1$ because no transcripts or protein for $K_{Ca}1.1$ were detected in bovine coronary artery endothelial cells [50] excluding the possibility that there are silent $K_{Ca}1.1$ present in these cells. We also have not detected transcripts for $K_{Ca}1.1$ channels in endothelial cells isolated from mouse cremaster arterioles or upstream feed arteries (n = 5 endothelial cell isolates from five mice, Jackson, unpublished observations). Additional studies where message, protein expression, protein localization and electrophysiology are performed will be required to resolve this issue.

$K_{IR}2.X$ Channels

Endothelial cells also express inward rectifier K^+ (K_{IR}) channels [123]. In resistance arteries and arterioles, endothelial cells appear to express two to four isoforms of K_{IR} channels: strong inward rectifiers, $K_{IR}2.1$, $K_{IR}2.2$, and $K_{IR}2.3$ (Table 1.1) and weak inward rectifiers $K_{IR}6.1$ and $K_{IR}6.2$ that form ATP-sensitive K^+ (K_{ATP}) channels. The function of K_{ATP} channels will not be addressed. In general, K_{IR} channels are tetramers of K_{IR} subunits, with each subunit having two membrane-spanning domains with a pore-forming loop in between [68]. As shown in Fig. 1.6, at potentials more negative than the K^+ equilibrium potential, K_{IR} channels have a higher conductance and pass current in the inward direction, whereas at more positive potentials, channel conductance is reduced. This inward-rectification gives K_{IR} channels their name and results from voltage-dependent block of the channel pore by intracellular Mg^{2+} and polyamines [68]. The physiological function of K_{IR} channels is predominantly mediated by the outward "hump" in the current-voltage relationship (Fig. 1.6). Over membrane potentials that accompany this portion of the current-voltage relationship for K_{IR}, outward currents through these channels will tend to hyperpolarize cells, driving the membrane potential towards the K^+ equilibrium potential. It is this characteristic of the strong inward rectifiers like $K_{IR}2.1$ and 2.2 that sets the resting membrane potential of cardiac muscle, skeletal muscle and nerves to be quite negative and near the K^+ equilibrium potential [68]. Elevation of extracellular K^+ increases the conductance of K_{IR} channels (Fig. 1.6) and shifts the outward hump to more positive membrane potentials which allows these channels to transduce changes in

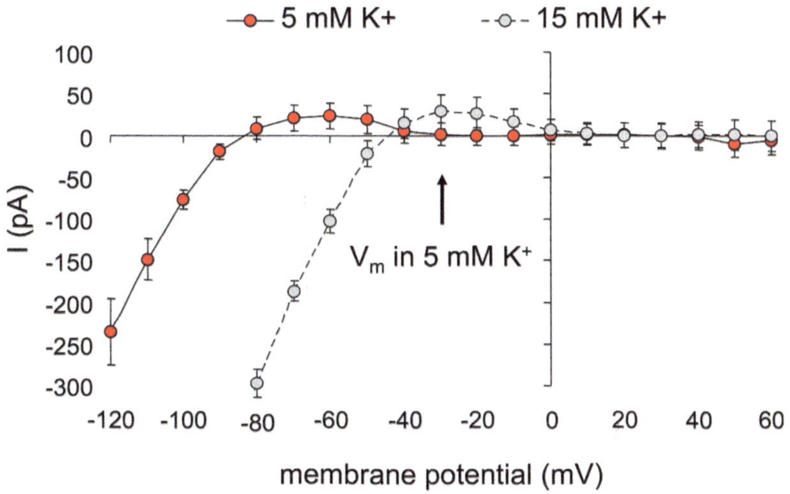

Fig. 1.6 K_{IR} channel currents in arteriolar endothelial cells. Shown are mean ± SE (n = 5) current-voltage relationships recorded from freshly isolated hamster cremaster arteriolar endothelial cells (see Fig. 1.2 and [23, 81] for method details) with 5 or 15 mM K^+ in the extracellular solution as indicated. At rest, these cells have a resting membrane potential of ~−30 mV [23], and K_{IR} channels contribute little to the resting membrane potential. However, note that membrane hyperpolarization from this potential, due to opening of K_{Ca} channels, for example, will recruit current through K_{IR} channels, amplifying the original hyperpolarization. K_{IR} channels are also sensitive to increases in extracellular [K^+] as shown, which increases the conductance of these channels and shifts the reversal potential and the outward "hump" to more positive potentials as shown. Thus, increases in extracellular K^+, due to release of K^+ from active skeletal muscle, or from release of K^+ from endothelial K_{Ca} channels, for example, will also recruit outward current through K_{IR} channels that will hyperpolarize endothelial cells. Figure modified from [81]

extracellular K^+ concentration into hyperpolarizing outward K^+ currents (more on this below). The strong inward rectifier K_{IR} channels can be blocked by micromolar concentration of Ba^{2+} (Table 1.1), which has been the primary pharmacological tool used to study the physiological function of these channels.

In cultured endothelial cells, currents through strong inward rectifier K_{IR} channels dominate the whole-cell, current-voltage relationship [3, 123], and from studies of cultured macrovascular and microvascular endothelial cells it was proposed that macrovascular, but not microvascular endothelial cells express K_{IR} channels [123]. However, studies of freshly isolated endothelial cells from porcine brain capillaries [74], guinea pig coronary capillaries [169], rat mesenteric resistance arteries [28], rat cerebral resistance arteries [86], mouse cerebral penetrating arterioles [101] and hamster cremaster arterioles [81] (Fig. 1.6) all display Ba^{2+}-sensitive K_{IR} currents. Expression of message for $K_{IR}2.1$, 2.2 and possibly 2.3 has been reported in guinea pig coronary capillary endothelial cells [99]. Similarly, in primary cultures of rat brain capillary endothelial cells $K_{IR}2.1$, 2.2 and a much lower level of $K_{IR}2.3$ have been detected [115]. In mouse and rat skeletal muscle arteriolar endothelial cells, we have detected the expression of $K_{IR}2.1$, but not $K_{IR}2.2$ at the protein level (Fig. 1.7).

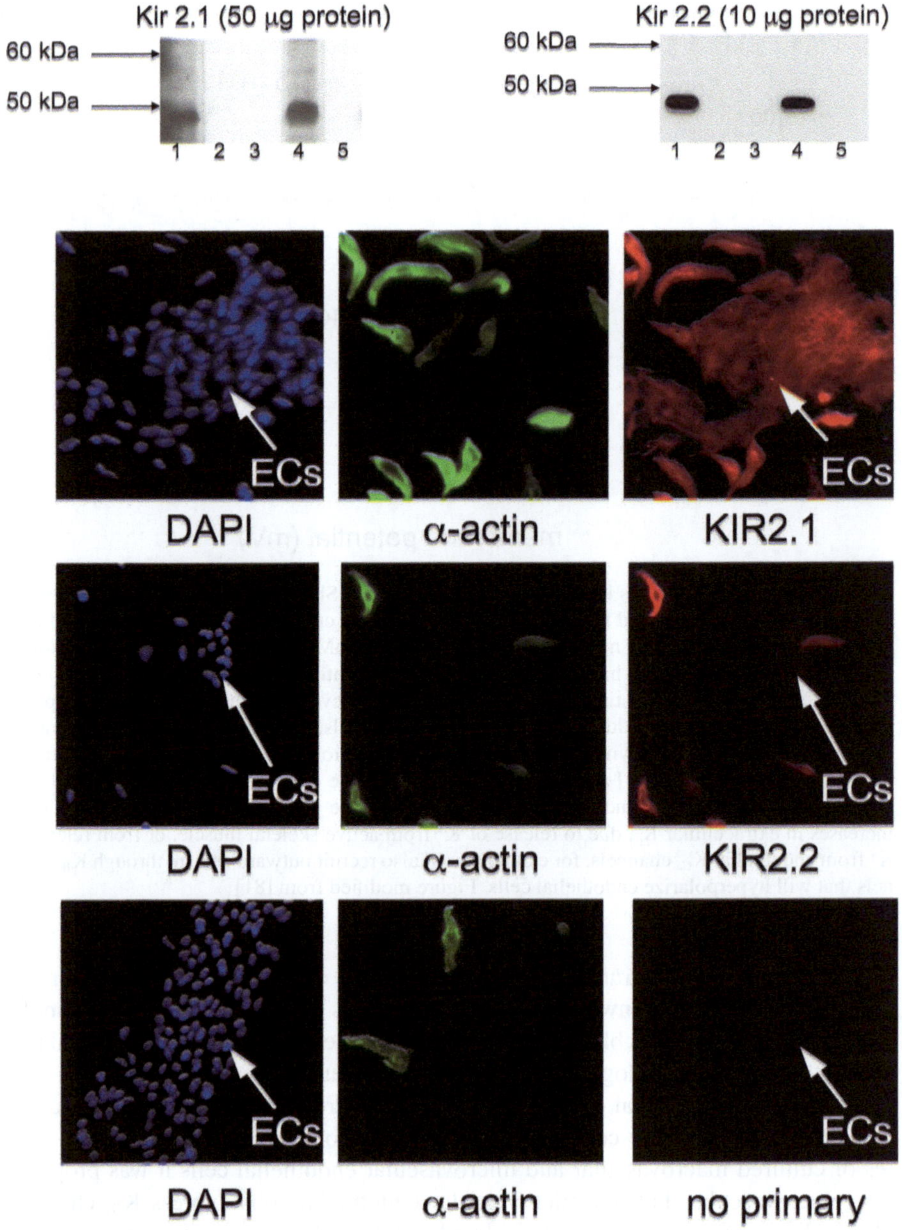

Fig. 1.7 Expression of K_{IR} protein in skeletal muscle arterioles. Top panels show western blots for $K_{IR}2.1$ and $K_{IR}2.2$ in mouse and rat skeletal muscle arteriolar whole homogenates demonstrating expression of both proteins. In both panels each lane is as follows: Lane 1 = 1° + 2° antibody for indicated protein, Lane 2 = 2° antibody only, Lane 3 = 1° antibody only, Lane 4 = 1° + 2° antibody for indicated protein and Lane 5 = 2° antibody only. Lanes 1–3 contain mouse abdominal arteriolar whole homogenate and Lanes 4 and 5 contain rat cremaster arteriolar whole homogenate. Representative of three experiments. Antibodies: 1° Kir2.1 (1:200) (Alomone) + 2° HRP-conjugated mouse anti-rabbit, light chain specific (1:25,000) (Jackson Immuno Research) 1° Kir2.2 (1:400) (Alomone) + 2° HRP-conjugated goat anti-rabbit (1:2000) (Cell Signaling).

The function of $K_{IR}2.X$ channels in endothelial cells of resistance arteries has been understudied. Theoretically, the expression of these channels in endothelial cells could serve to amplify endothelial cell hyperpolarization induced by the activation of other K^+ channels and also serve to transduce small increases in extracellular K^+ into cell hyperpolarization. In blood vessels where $K_{IR}2.X$ channels are expressed in the vascular smooth muscle, evidence for both of these functions has been demonstrated: smooth muscle K_{IR} channels serve to amplify smooth muscle hyperpolarization induced by other means [82, 146] and it is well established that elevated extracellular K^+ causes vasodilatation that is mediated, at least in part, by activation of smooth muscle $K_{IR}2.X$ channels [85, 101, 122, 186]. It has been argued that endothelial K_{IR} channels do not play similar roles, primarily based on studies of rat mesenteric arteries in which the endothelial cells, but not the smooth muscle cells express functional K_{IR} channels [146]. However, this hypothesis has not been adequately tested, particularly in arterioles with only a single layer of smooth muscle and strong expression of K_{IR} channels. Endothelial and smooth muscle-selective knockdown, as proposed by Longden and Nelson [101], would provide an appropriate system in which to better evaluate the roles played by endothelial K_{IR} channels.

Concluding Remarks

Endothelial cells importantly contribute to the regulation of the vasomotor function of resistance arteries and arterioles, and it is clear that a number of ion channels in these cells contribute to this function. As noted throughout this chapter, there appears to be substantial regional and species differences in the expression and function of endothelial cell ion channels that clouds a clear view of the function of individual channels and the interplay among and between specific channels. It appears that many, if not all of these channels exist in macromolecular signaling complexes. This further complicates analysis of the individual roles played by specific channels. While knockout approaches are certainly an important tool in studying the function of specific channels, given these macromolecular complexes, the likelihood of off target effects in disrupting other functions of such complexes is significant, but usually overlooked, and may explain why different answers are achieved using knockout approaches vs. pharmacology. The use of cell-specific

Fig. 1.7 (continued) Bottom panels show immunofluorescence localization of $K_{IR}2.1$ and $K_{IR}2.2$ in mouse abdominal arteriolar smooth muscle cells and endothelial cells. Vessels were enzymatically dissociated to yield single smooth muscle cells and endothelial cell tubes (labeled ECs) as described [23], the cells were fixed with 4 % paraformaldehyde and stained with the same antibodies as in top panels. Smooth muscle cells were identified by staining for α-smooth muscle actin (*Green* in *middle* panels). Note that endothelial cells expressed only $K_{IR}2.1$, whereas smooth muscle cells expressed both $K_{IR}2.1$ and 2.2 (*Red* in *right* panels). Data are representative of three isolates. Similar results were obtained in cells isolated from rat cremaster arterioles

conditional knockouts, as well as knockins of dominant negative channels coupled with careful pharmacological approaches may help to resolve the questions that still remain. Because of the inherent heterogeneity of expression and function of ion channels in different vascular beds, the detailed study of individual beds among species and particularly in humans is to be encouraged rather than stifled.

Acknowledgements Supported NIH grants RO1-HL32469, RO1-HL086483 and PO1-HL070687.

References

1. Abdullaev IF, Bisaillon JM, Potier M, Gonzalez JC, Motiani RK, Trebak M. Stim1 and Orai1 mediate CRAC currents and store-operated calcium entry important for endothelial cell proliferation. Circ Res. 2008;103:1289–99.
2. Absi M, Burnham MP, Weston AH, Harno E, Rogers M, Edwards G. Effects of methyl beta-cyclodextrin on EDHF responses in pig and rat arteries; association between SK(Ca) channels and caveolin-rich domains. Br J Pharmacol. 2007;151:332–40.
3. Adams DJ, Hill MA. Potassium channels and membrane potential in the modulation of intracellular calcium in vascular endothelial cells. J Cardiovasc Electrophysiol. 2004;15:598–610.
4. Adebiyi A, Zhao G, Narayanan D, Thomas-Gatewood CM, Bannister JP, Jaggar JH. Isoform-selective physical coupling of TRPC3 channels to IP3 receptors in smooth muscle cells regulates arterial contractility. Circ Res. 2010;106:1603–12.
5. Ahmmed GU, Mehta D, Vogel S, Holinstat M, Paria BC, Tiruppathi C, Malik AB. Protein kinase Calpha phosphorylates the TRPC1 channel and regulates store-operated Ca2+ entry in endothelial cells. J Biol Chem. 2004;279:20941–9.
6. Alagem N, Dvir M, Reuveny E. Mechanism of Ba2+ block of a mouse inwardly rectifying K+ channel: differential contribution by two discrete residues. J Physiol Lond. 2001;534:381–93.
7. Antoniotti S, Lovisolo D, Fiorio Pla A, Munaron L. Expression and functional role of bTRPC1 channels in native endothelial cells. FEBS Lett. 2002;510:189–95.
8. Bagher P, Beleznai T, Kansui Y, Mitchell R, Garland CJ, Dora KA. Low intravascular pressure activates endothelial cell TRPV4 channels, local Ca2+ events, and IKCa channels, reducing arteriolar tone. Proc Natl Acad Sci U S A. 2012;109:18174–9.
9. Baraldi PG, Preti D, Materazzi S, Geppetti P. Transient receptor potential ankyrin 1 (TRPA1) channel as emerging target for novel analgesics and anti-inflammatory agents. J Med Chem. 2010;53:5085–107.
10. Beech DJ, Xu SZ, McHugh D, Flemming R. TRPC1 store-operated cationic channel subunit. Cell Calcium. 2003;33:433–40.
11. Bergdahl A, Gomez MF, Wihlborg A-K, Erlinge D, Eyjolfson A, Xu S-Z, Beech DJ, Dreja K, Hellstrand P. Plasticity of TRPC expression in arterial smooth muscle: correlation with store-operated Ca2+ entry. Am J Physiol Cell Physiol. 2005;288:C872–80.
12. Berra-Romani R, Mazzocco-Spezzia A, Pulina MV, Golovina VA. Ca2+ handling is altered when arterial myocytes progress from a contractile to a proliferative phenotype in culture. Am J Physiol Cell Physiol. 2008;295:C779–90.
13. Berridge M. Inositol trisphosphate and calcium signalling. Nature. 1993;361:315–25.
14. Boulay G, Zhu X, Peyton M, Jiang M, Hurst R, Stefani E, Birnbaumer L. Cloning and expression of a novel mammalian homolog of Drosophila Transient Receptor Potential (Trp) involved in calcium entry secondary to activation of receptors coupled by the Gq class of G protein. J Biol Chem. 1997;272:29672–80.

15. Brahler S, Kaistha A, Schmidt VJ, Wolfle SE, Busch C, Kaistha BP, Kacik M, Hasenau AL, Grgic I, Si H, Bond CT, Adelman JP, Wulff H, de Wit C, Hoyer J, Kohler R. Genetic deficit of SK3 and IK1 channels disrupts the endothelium-derived hyperpolarizing factor vasodilator pathway and causes hypertension. Circulation. 2009;119:2323–32.
16. Braunstein TH, Inoue R, Cribbs L, Oike M, Ito Y, Holstein-Rathlou NH, Jensen LJ. The role of L- and T-type calcium channels in local and remote calcium responses in rat mesenteric terminal arterioles. J Vasc Res. 2009;46:138–51.
17. Brough GH, Wu S, Cioffi D, Moore TM, Li M, Dean N, Stevens T. Contribution of endogenously expressed Trp1 to a Ca2+-selective, store-operated Ca2+ entry pathway. FASEB J. 2001;15:1727–38.
18. Busse R, Fichtner H, Luckhoff A, Kohlhardt M. Hyperpolarization and increased free calcium in acetylcholine-stimulated endothelial cells. Am J Physiol. 1988;255:H965–9.
19. Bychkov R, Burnham MP, Richards GR, Edwards G, Weston AH, Feletou M, Vanhoutte PM. Characterization of a charybdotoxin-sensitive intermediate conductance Ca2+-activated K+ channel in porcine coronary endothelium: relevance to EDHF. Br J Pharmacol. 2002;137:1346–54.
20. Chatterjee S, Al-Mehdi AB, Levitan I, Stevens T, Fisher AB. Shear stress increases expression of a KATP channel in rat and bovine pulmonary vascular endothelial cells. Am J Physiol Cell Physiol. 2003;285:C959–67.
21. Chen W, Oberwinkler H, Werner F, Gassner B, Nakagawa H, Feil R, Hofmann F, Schlossmann J, Dietrich A, Gudermann T, Nishida M, Del Galdo S, Wieland T, Kuhn M. Atrial natriuretic peptide-mediated inhibition of microcirculatory endothelial Ca2+ and permeability response to histamine involves cGMP-dependent protein kinase I and TRPC6 channels. Arterioscler Thromb Vasc Biol. 2013;33:2121–9.
22. Cipolla MJ, Smith J, Kohlmeyer MM, Godfrey JA. SKCa and IKCa Channels, myogenic tone, and vasodilator responses in middle cerebral arteries and parenchymal arterioles: effect of ischemia and reperfusion. Stroke. 2009;40:1451–7.
23. Cohen KD, Jackson WF. Membrane hyperpolarization is not required for sustained muscarinic agonist-induced increases in intracellular Ca2+ in arteriolar endothelial cells. Microcirculation. 2005;12:169–82.
24. Colden-Stanfield M, Schilling WP, Possani LD, Kunze DL. Bradykinin-induced potassium current in cultured bovine aortic endothelial cells. J Membr Biol. 1990;116:227–38.
25. Colden-Stanfield M, Schilling WP, Ritchie AK, Eskin SG, Navarro LT, Kunze DL. Bradykinin-induced increases in cytosolic calcium and ionic currents in cultured bovine aortic endothelial cells. Circ Res. 1987;61:632–40.
26. Corriu C, Feletou M, Canet E, Vanhoutte PM. Endothelium-derived factors and hyperpolarization of the carotid artery of the guinea-pig. Br J Pharmacol. 1996;119:959–64.
27. Crane GJ, Gallagher N, Dora KA, Garland CJ. Small- and intermediate-conductance calcium-activated K+ channels provide different facets of endothelium-dependent hyperpolarization in rat mesenteric artery. J Physiol. 2003;553:183–9.
28. Crane GJ, Walker SD, Dora KA, Garland CJ. Evidence for a differential cellular distribution of inward rectifier K channels in the rat isolated mesenteric artery. J Vasc Res. 2003;40:159–68.
29. de Wit C, Griffith TM. Connexins and gap junctions in the EDHF phenomenon and conducted vasomotor responses. Pflugers Arch. 2010;459:897–914.
30. Dietrich A, Fahlbusch M, Gudermann T. Classical Transient Receptor Potential 1 (TRPC1): channel or channel regulator? Cells. 2014;3:939–62.
31. Dietrich A, Kalwa H, Storch U, Mederos Y, Schnitzler M, Salanova B, Pinkenburg O, Dubrovska G, Essin K, Gollasch M, Birnbaumer L, Gudermann T. Pressure-induced and store-operated cation influx in vascular smooth muscle cells is independent of TRPC1. Pflugers Arch. 2007;455:465–77.
32. Dora KA, Doyle MP, Duling BR. Elevation of intracellular calcium in smooth muscle causes endothelial cell generation of NO in arterioles. Proc Natl Acad Sci U S A. 1997;94: 6529–34.

33. Dora KA, Gallagher NT, McNeish A, Garland CJ. Modulation of endothelial cell KCa3.1 channels during endothelium-derived hyperpolarizing factor signaling in mesenteric resistance arteries. Circ Res. 2008;102:1247–55.
34. Dora KA, Garland CJ. Linking hyperpolarization to endothelial cell calcium events in arterioles. Microcirculation. 2013;20:248–56.
35. Du J, Ma X, Shen B, Huang Y, Birnbaumer L, Yao X. TRPV4, TRPC1, and TRPP2 assemble to form a flow-sensitive heteromeric channel. FASEB J. 2014;28:4677–85.
36. Duza T, Sarelius IH. Localized transient increases in endothelial cell Ca2+ in arterioles in situ: implications for the coordination of vascular function. Am J Physiol Heart Circ Physiol. 2004;286:H2322–31.
37. Earley S, Gonzales AL, Crnich R. Endothelium-dependent cerebral artery dilation mediated by TRPA1 and Ca2+-activated K+ channels. Circ Res. 2009;104:987–94.
38. Earley S, Gonzales AL, Garcia ZI. A dietary agonist of transient receptor potential cation channel V3 elicits endothelium-dependent vasodilation. Mol Pharmacol. 2010;77:612–20.
39. Earley S, Pauyo T, Drapp R, Tavares MJ, Liedtke W, Brayden JE. TRPV4-dependent dilation of peripheral resistance arteries influences arterial pressure. Am J Physiol Heart Circ Physiol. 2009;297:H1096–102.
40. Edwards G, Feletou M, Weston AH. Endothelium-derived hyperpolarising factors and associated pathways: a synopsis. Pflugers Arch. 2010;459:863–79.
41. Eichler I, Wibawa J, Grgic I, Knorr A, Brakemeier S, Pries AR, Hoyer J, Kohler R. Selective blockade of endothelial Ca2+-activated small- and intermediate-conductance K+-channels suppresses EDHF-mediated vasodilation. Br J Pharmacol. 2003;138:594–601.
42. Estacion M, Li S, Sinkins WG, Gosling M, Bahra P, Poll C, Westwick J, Schilling WP. Activation of human TRPC6 channels by receptor stimulation. J Biol Chem. 2004;279:22047–56.
43. Everaerts W, Zhen X, Ghosh D, Vriens J, Gevaert T, Gilbert JP, Hayward NJ, McNamara CR, Xue F, Moran MM, Strassmaier T, Uykal E, Owsianik G, Vennekens R, De Ridder D, Nilius B, Fanger CM, Voets T. Inhibition of the cation channel TRPV4 improves bladder function in mice and rats with cyclophosphamide-induced cystitis. Proc Natl Acad Sci U S A. 2010;107:19084–9.
44. Fanger CM, Ghanshani S, Logsdon NJ, Rauer H, Kalman K, Zhou J, Beckingham K, Chandy KG, Cahalan MD, Aiyar J. Calmodulin mediates calcium-dependent activation of the intermediate conductance KCa channel, IKCa1. J Biol Chem. 1999;274:5746–54.
45. Figueroa XF, Chen CC, Campbell KP, Damon DN, Day KH, Ramos S, Duling BR. Are voltage-dependent ion channels involved in the endothelial cell control of vasomotor tone? Am J Physiol Heart Circ Physiol. 2007;293:H1371–83.
46. Fill M, Copello JA. Ryanodine receptor calcium release channels. Physiol Rev. 2002;82:893–922.
47. Foskett J, White C, Cheung K, Mak D. Inositol trisphosphate receptor Ca2+ release channels. Physiol Rev. 2007;87:593–658.
48. Freichel M, Suh SH, Pfeifer A, Schweig U, Trost C, Weissgerber P, Biel M, Philipp S, Freise D, Droogmans G, Hofmann F, Flockerzi V, Nilius B. Lack of an endothelial store-operated Ca2+ current impairs agonist-dependent vasorelaxation in TRP4–/– mice. Nat Cell Biol. 2001;3:121–7.
49. Gafni J, Munsch JA, Lam TH, Catlin MC, Costa LG, Molinski TF, Pessah IN. Xestospongins: potent membrane permeable blockers of the inositol 1,4,5-trisphosphate receptor. Neuron. 1997;19:723.
50. Gauthier KM, Liu C, Popovic A, Albarwani S, Rusch NJ. Freshly isolated bovine coronary endothelial cells do not express the BK(Ca) channel gene. J Physiol. 2002;545:829–36.
51. Gees M, Colsoul B, Nilius B. The role of transient receptor potential cation channels in Ca2+ signaling. Cold Spring Harb Perspect Biol. 2010;2:a003962.
52. Goddard LM, Iruela-Arispe ML. Cellular and molecular regulation of vascular permeability. Thromb Haemost. 2013;109:407–15.
53. Gordon JL, Martin W. Endothelium-dependent relaxation of the pig aorta: relationship to stimulation of 86Rb efflux from isolated endothelial cells. Br J Pharmacol. 1983;79:531–41.

54. Graier WF, Paltauf-Doburzynska J, Hill BJ, Fleischhacker E, Hoebel BG, Kostner GM, Sturek M. Submaximal stimulation of porcine endothelial cells causes focal Ca2+ elevation beneath the cell membrane. J Physiol. 1998;506(Pt 1):109–25.
55. Graier WF, Simecek S, Bowles DK, Sturek M. Heterogeneity of caffeine- and bradykinin-sensitive Ca^{2+} stores in vascular endothelial cells. Biochem J. 1994;300:637–41.
56. Grayson TH, Haddock RE, Murray TP, Wojcikiewicz RJH, Hill CE. Inositol 1,4,5-trisphosphate receptor subtypes are differentially distributed between smooth muscle and endothelial layers of rat arteries. Cell Calcium. 2004;36:447.
57. Grgic I, Kaistha BP, Hoyer J, Kohler R. Endothelial Ca+-activated K+ channels in normal and impaired EDHF-dilator responses—relevance to cardiovascular pathologies and drug discovery. Br J Pharmacol. 2009;157:509–26.
58. Halaszovich CR, Zitt C, Jungling E, Luckhoff A. Inhibition of TRP3 channels by lanthanides. Block from the cytosolic side of the plasma membrane. J Biol Chem. 2000;275:37423–8.
59. Hallam TJ, Pearson JD. Exogenous ATP raises cytoplasmic free calcium in fura-2 loaded piglet aortic endothelial cells. FEBS Lett. 1986;207:95–9.
60. Hamdollah Zadeh MA, Glass CA, Magnussen A, Hancox JC, Bates DO. VEGF-mediated elevated intracellular calcium and angiogenesis in human microvascular endothelial cells in vitro are inhibited by dominant negative TRPC6. Microcirculation. 2008;15:605–14.
61. Hamilton S. Ryanodine receptors. Cell Calcium. 2005;38:253–60.
62. Hamilton S, Serysheva I. Ryanodine receptor structure: progress and challenges. J Biol Chem. 2009;284:4047–51.
63. Hannah RM, Dunn KM, Bonev AD, Nelson MT. Endothelial SK(Ca) and IK(Ca) channels regulate brain parenchymal arteriolar diameter and cortical cerebral blood flow. J Cereb Blood Flow Metab. 2011;31:1175–86.
64. Harteneck C, Gollasch M. Pharmacological modulation of diacylglycerol-sensitive TRPC3/6/7 channels. Curr Pharm Biotechnol. 2011;12:35–41.
65. He P, Curry FE. Depolarization modulates endothelial cell calcium influx and microvessel permeability. Am J Physiol Heart Circ Physiol. 1991;261:H1246–54.
66. He P, Curry FE. Endothelial cell hyperpolarization increases $[Ca^{2+}]_i$ and venular microvessel permeability. J Appl Physiol. 1994;76:2288–97.
67. Heady TN, Gomora JC, Macdonald TL, Perez-Reyes E. Molecular pharmacology of T-type Ca ²⁺ channels. Jpn J Pharmacol. 2001;85:339–50.
68. Hibino H, Inanobe A, Furutani K, Murakami S, Findlay I, Kurachi Y. Inwardly rectifying potassium channels: their structure, function, and physiological roles. Physiol Rev. 2010;90:291–366.
69. Himmel HM, Whorton AR, Strauss HC. Intracellular calcium, currents, and stimulus-response coupling in endothelial cells. Hypertension. 1993;21:112–27.
70. Hoffmann EK, Lambert IH, Pedersen SF. Physiology of cell volume regulation in vertebrates. Physiol Rev. 2009;89:193–277.
71. Hofmann T, Obukhov AG, Schaefer M, Harteneck C, Gudermann T, Schultz G. Direct activation of human TRPC6 and TRPC3 channels by diacylglycerol. Nature. 1999;397:259–63.
72. Hondeghem LM, Ayad MJ, Robertson RM. Verapamil, diltiazem and nifedipine block the depolarization-induced potentiation of norepinephrine contractions in rabbit aorta and porcine coronary arteries. J Pharmacol Exp Ther. 1986;239:808–13.
73. Hoshi T, Pantazis A, Olcese R. Transduction of voltage and Ca2+ signals by Slo1 BK channels. Physiology (Bethesda). 2013;28:172–89.
74. Hoyer J, Popp R, Meyer J, Galla HJ, Gogelein H. Angiotensin II, vasopressin and GTP[gamma-S] inhibit inward-rectifying K^+ channels in porcine cerebral capillary endothelial cells. J Membr Biol. 1991;123:55–62.
75. Huang JH, He GW, Xue HM, Yao XQ, Liu XC, Underwood MJ, Yang Q. TRPC3 channel contributes to nitric oxide release: significance during normoxia and hypoxia-reoxygenation. Cardiovasc Res. 2011;91:472–82.
76. Hughes JM, Riddle MA, Paffett ML, Gonzalez Bosc LV, Walker BR. Novel role of endothelial BKCa channels in altered vasoreactivity following hypoxia. Am J Physiol Heart Circ Physiol. 2010;299:H1439–50.

77. Isakson BE. Localized expression of an Ins(1,4,5)P3 receptor at the myoendothelial junction selectively regulates heterocellular Ca2+ communication. J Cell Sci. 2008;121:3664–73.
78. Ishii TM, Silvia C, Hirschberg B, Bond CT, Adelman JP, Maylie J. A human intermediate conductance calcium-activated potassium channel. Proc Natl Acad Sci U S A. 1997;94:11651–6.
79. Jackson WF. The endothelium-derived relaxing factor. J Reconstr Microsurg. 1989;5(3): 263–71.
80. Jackson WF. Ion channels and vascular tone. Hypertension. 2000;35(1 Pt 2):173–8.
81. Jackson WF. Potassium channels in the peripheral microcirculation. Microcirculation. 2005;12:113–27.
82. Jantzi MC, Brett SE, Jackson WF, Corteling RL, Vigmond EJ, Welsh DG. Inward rectifying potassium channels facilitate cell-to-cell communication in hamster retractor muscle feed arteries. Am J Physiol Heart Circ Physiol. 2006;291:H1319–28.
83. Kansui Y, Garland CJ, Dora KA. Enhanced spontaneous Ca2+ events in endothelial cells reflect signalling through myoendothelial gap junctions in pressurized mesenteric arteries. Cell Calcium. 2008;44:135–46.
84. Kiyonaka S, Kato K, Nishida M, Mio K, Numaga T, Sawaguchi Y, Yoshida T, Wakamori M, Mori E, Numata T, Ishii M, Takemoto H, Ojida A, Watanabe K, Uemura A, Kurose H, Morii T, Kobayashi T, Sato Y, Sato C, Hamachi I, Mori Y. Selective and direct inhibition of TRPC3 channels underlies biological activities of a pyrazole compound. Proc Natl Acad Sci U S A. 2009;106:5400–5.
85. Knot HJ, Zimmermann PA, Nelson MT. Extracellular K(+)-induced hyperpolarizations and dilatations of rat coronary and cerebral arteries involve inward rectifier K(+) channels. J Physiol. 1996;492(Pt 2):419–30.
86. Kochukov MY, Balasubramanian A, Abramowitz J, Birnbaumer L, Marrelli SP. Activation of endothelial transient receptor potential C3 channel is required for small conductance calcium-activated potassium channel activation and sustained endothelial hyperpolarization and vasodilation of cerebral artery. J Am Heart Assoc. 2014;3.
87. Kochukov MY, Balasubramanian A, Noel RC, Marrelli SP. Role of TRPC1 and TRPC3 channels in contraction and relaxation of mouse thoracic aorta. J Vasc Res. 2013;50:11–20.
88. Kohler R, Brakemeier S, Kuhn M, Behrens C, Real R, Degenhardt C, Orzechowski HD, Pries AR, Paul M, Hoyer J. Impaired hyperpolarization in regenerated endothelium after balloon catheter injury. Circ Res. 2001;89:174–9.
89. Kohler R, Brakemeier S, Kuhn M, Degenhardt C, Buhr H, Pries A, Hoyer J. Expression of ryanodine receptor type 3 and TRP channels in endothelial cells: comparison of in situ and cultured human endothelial cells. Cardiovasc Res. 2001;51:160.
90. Komarova Y, Malik AB. Regulation of endothelial permeability via paracellular and transcellular transport pathways. Annu Rev Physiol. 2010;72:463–93.
91. Kuo IY, Ellis A, Seymour VA, Sandow SL, Hill CE. Dihydropyridine-insensitive calcium currents contribute to function of small cerebral arteries. J Cereb Blood Flow Metab. 2010; 30:1226–39.
92. Kwiatek AM, Minshall RD, Cool DR, Skidgel RA, Malik AB, Tiruppathi C. Caveolin-1 regulates store-operated Ca2+ influx by binding of its scaffolding domain to transient receptor potential channel-1 in endothelial cells. Mol Pharmacol. 2006;70:1174–83.
93. Ledoux J, Taylor MS, Bonev AD, Hannah RM, Solodushko V, Shui B, Tallini Y, Kotlikoff MI, Nelson MT. Functional architecture of inositol 1,4,5-trisphosphate signaling in restricted spaces of myoendothelial projections. Proc Natl Acad Sci U S A. 2008;105:9627–32.
94. Lee JH, Gomora JC, Cribbs LL, Perez-Reyes E. Nickel block of three cloned T-type calcium channels: low concentrations selectively block alpha1H. Biophys J. 1999;77:3034–42.
95. Lesh RE, Marks AR, Somlyo AV, Fleischer S, Somlyo AP. Anti-ryanodine receptor antibody binding sites in vascular and endocardial endothelium. Circ Res. 1993;72:481–8.
96. Leuner K, Heiser JH, Derksen S, Mladenov MI, Fehske CJ, Schubert R, Gollasch M, Schneider G, Harteneck C, Chatterjee SS, Muller WE. Simple 2,4-diacylphloroglucinols as classic transient receptor potential-6 activators--identification of a novel pharmacophore. Mol Pharmacol. 2010;77:368–77.

97. Liao P, Yu D, Li G, Yong TF, Soon JL, Chua YL, Soong TW. A smooth muscle Cav1.2 calcium channel splice variant underlies hyperpolarized window current and enhanced state-dependent inhibition by nifedipine. J Biol Chem. 2007;282:35133–42.
98. Liu CL, Huang Y, Ngai CY, Leung YK, Yao XQ. TRPC3 is involved in flow- and bradykinin-induced vasodilation in rat small mesenteric arteries. Acta Pharmacol Sin. 2006;27:981–90.
99. Liu GX, Derst C, Schlichthorl G, Heinen S, Seebohm G, Bruggemann A, Kummer W, Veh RW, Daut J, Preisig-Muller R. Comparison of cloned Kir2 channels with native inward rectifier K+ channels from guinea-pig cardiomyocytes. J Physiol. 2001;532:115–26.
100. Loga F, Domes K, Freichel M, Flockerzi V, Dietrich A, Birnbaumer L, Hofmann F, Wegener JW. The role of cGMP/cGKI signalling and Trpc channels in regulation of vascular tone. Cardiovasc Res. 2013;100:280–7.
101. Longden TA, Nelson MT. Vascular inward rectifier k(+) channels as external k(+) sensors in the control of cerebral blood flow. Microcirculation. 2015;22:183–96.
102. Mackrill JJ. Ryanodine receptor calcium channels and their partners as drug targets. Biochem Pharmacol. 2010;79:1535–43.
103. Mak D, McBride S, Foskett J. Regulation by Ca2+ and inositol 1,4,5-trisphosphate (InsP3) of single recombinant type 3 InsP3 receptor channels. Ca2+ activation uniquely distinguishes types 1 and 3 insp3 receptors. J Gen Physiol. 2001;117:435–46.
104. Mak DD, Foskett JK. Inositol 1,4,5-trisphosphate receptors in the endoplasmic reticulum: a single-channel point of view. Cell Calcium. 2014;58:67–78.
105. Mak DO, McBride S, Foskett JK. Inositol 1,4,5-trisphosphate [correction of tris-phosphate] activation of inositol trisphosphate [correction of tris-phosphate] receptor Ca2+ channel by ligand tuning of Ca2+ inhibition. Proc Natl Acad Sci U S A. 1998;95:15821–5.
106. Mannhold R. KATP channel openers: structure-activity relationships and therapeutic potential. Med Res Rev. 2004;24:213–66.
107. Marrelli SP. Selective measurement of endothelial or smooth muscle [Ca(2+)](i) in pressurized/perfused cerebral arteries with fura-2. J Neurosci Methods. 2000;97:145–55.
108. Marrelli SP, Eckmann MS, Hunte MS. Role of endothelial intermediate conductance KCa channels in cerebral EDHF-mediated dilations. Am J Physiol Heart Circ Physiol. 2003;285:H1590–9.
109. Marrelli SP, O'Neil RG, Brown RC, Bryan Jr RM. PLA2 and TRPV4 channels regulate endothelial calcium in cerebral arteries. Am J Physiol Heart Circ Physiol. 2007;292: H1390–7.
110. Martin RL, Lee JH, Cribbs LL, Perez-Reyes E, Hanck DA. Mibefradil block of cloned T-type calcium channels. J Pharmacol Exp Ther. 2000;295:302–8.
111. Maruyama T, Kanaji T, Nakade S, Kanno T, Mikoshiba K. 2APB, 2-aminoethoxydiphenyl borate, a membrane-penetrable modulator of Ins(1,4,5)P3-induced Ca2+ release. J Biochem. 1997;122:498–505.
112. McNamara CR, Mandel-Brehm J, Bautista DM, Siemens J, Deranian KL, Zhao M, Hayward NJ, Chong JA, Julius D, Moran MM, Fanger CM. TRPA1 mediates formalin-induced pain. Proc Natl Acad Sci U S A. 2007;104:13525–30.
113. McSherry IN, Spitaler MM, Takano H, Dora KA. Endothelial cell Ca2+ increases are independent of membrane potential in pressurized rat mesenteric arteries. Cell Calcium. 2005;38:23–33.
114. Mehta D, Ahmmed GU, Paria BC, Holinstat M, Voyno-Yasenetskaya T, Tiruppathi C, Minshall RD, Malik AB. RhoA interaction with inositol 1,4,5-trisphosphate receptor and transient receptor potential channel-1 regulates Ca2+ entry. Role in signaling increased endothelial permeability. J Biol Chem. 2003;278:33492–500.
115. Millar ID, Wang S, Brown PD, Barrand MA, Hladky SB. Kv1 and Kir2 potassium channels are expressed in rat brain endothelial cells. Pflugers Arch. 2008;456:379–91.
116. Miller M, Shi J, Zhu Y, Kustov M, Tian JB, Stevens A, Wu M, Xu J, Long S, Yang P, Zholos AV, Salovich JM, Weaver CD, Hopkins CR, Lindsley CW, McManus O, Li M, Zhu MX. Identification of ML204, a novel potent antagonist that selectively modulates native TRPC4/C5 ion channels. J Biol Chem. 2011;286:33436–46.

117. Minami T. Calcineurin-NFAT activation and DSCR-1 auto-inhibitory loop: how is homoeostasis regulated? J Biochem. 2014;155:217–26.
118. Mountian I, Manolopoulos VG, De Smedt H, Parys JB, Missiaen L, Wuytack F. Expression patterns of sarco/endoplasmic reticulum Ca(2+)-ATPase and inositol 1,4,5-trisphosphate receptor isoforms in vascular endothelial cells. Cell Calcium. 1999;25:371–80.
119. Mountian II, Baba-Aissa F, Jonas JC, De Humbert S, Wuytack F, Parys JB. Expression of Ca(2+) transport genes in platelets and endothelial cells in hypertension. Hypertension. 2001;37:135–41.
120. Munaron L. Intracellular calcium, endothelial cells and angiogenesis. Recent Pat Anticancer Drug Discov. 2006;1:105–19.
121. Narahashi T, Tsunoo A, Yoshii M. Characterization of two types of calcium channels in mouse neuroblastoma cells. J Physiol. 1987;383:231–49.
122. Nelson M, Quayle J. Physiological roles and properties of potassium channels in arterial smooth muscle. Am J Physiol. 1995;268:C799–822.
123. Nilius B, Droogmans G. Ion channels and their functional role in vascular endothelium. Physiol Rev. 2001;81:1415–59.
124. Nilius B, Droogmans G, Wondergem R. Transient receptor potential channels in endothelium: solving the calcium entry puzzle? Endothelium. 2003;10:5–15.
125. Nilius B, Viana F, Droogmans G. Ion channels in vascular endothelium. Annu Rev Physiol. 1997;59:145–70.
126. Paltauf-Doburzynska J, Posch K, Paltauf G, Graier WF. Stealth ryanodine-sensitive Ca2+ release contributes to activity of capacitative Ca2+ entry and nitric oxide synthase in bovine endothelial cells. J Physiol. 1998;513(Pt 2):369–79.
127. Pocock TM, Foster RR, Bates DO. Evidence of a role for TRPC channels in VEGF-mediated increased vascular permeability in vivo. Am J Physiol Heart Circ Physiol. 2004;286: H1015–26.
128. Qian X, Francis M, Kohler R, Solodushko V, Lin M, Taylor MS. Positive feedback regulation of agonist-stimulated endothelial Ca2+ dynamics by KCa3.1 channels in mouse mesenteric arteries. Arterioscler Thromb Vasc Biol. 2014;34:127–35.
129. Quayle JM, Nelson MT, Standen NB. ATP-sensitive and inwardly rectifying potassium channels in smooth muscle. Physiol Rev. 1997;77:1165–232.
130. Quinlan KL, Naik SM, Cannon G, Armstrong CA, Bunnett NW, Ansel JC, Caughman SW. Substance P activates coincident NF-AT- and NF-kappa B-dependent adhesion molecule gene expression in microvascular endothelial cells through intracellular calcium mobilization. J Immunol. 1999;163:5656–65.
131. Riddle MA, Hughes JM, Walker BR. Role of caveolin-1 in endothelial BKCa channel regulation of vasoreactivity. Am J Physiol Cell Physiol. 2011;301:C1404–14.
132. Rusko J, Van Slooten G, Adams DJ. Caffeine-evoked, calcium-sensitive membrane currents in rabbit aortic endothelial cells. Br J Pharmacol. 1995;115:133–41.
133. Saleem H, Tovey SC, Molinski TF, Taylor CW. Interactions of antagonists with subtypes of inositol 1,4,5-trisphosphate (IP3) receptor. Br J Pharmacol. 2014;171:3298–312.
134. Saliez J, Bouzin C, Rath G, Ghisdal P, Desjardins F, Rezzani R, Rodella LF, Vriens J, Nilius B, Feron O, Balligand JL, Dessy C. Role of caveolar compartmentation in endothelium-derived hyperpolarizing factor-mediated relaxation: Ca2+ signals and gap junction function are regulated by caveolin in endothelial cells. Circulation. 2008;117:1065–74.
135. Sandow SL, Grayson TH. Limits of isolation and culture: intact vascular endothelium and BKCa. Am J Physiol Heart Circ Physiol. 2009;297:H1–7.
136. Sandow SL, Hill CE. Incidence of myoendothelial gap junctions in the proximal and distal mesenteric arteries of the rat is suggestive of a role in endothelium-derived hyperpolarizing factor-mediated responses. Circ Res. 2000;86:341–6.
137. Sandow SL, Neylon CB, Chen MX, Garland CJ. Spatial separation of endothelial small- and intermediate-conductance calcium-activated potassium channels (K(Ca)) and connexins: possible relationship to vasodilator function? J Anat. 2006;209:689–98.

138. Schaefer M, Plant TD, Obukhov AG, Hofmann T, Gudermann T, Schultz G. Receptor-mediated regulation of the nonselective cation channels TRPC4 and TRPC5. J Biol Chem. 2000;275:17517–26.
139. Schilling WP, Cabello OA, Rajan L. Depletion of the inositol 1,4,5-trisphosphate-sensitive intracellular Ca2+ store in vascular endothelial-cells activates the agonist-sensitive Ca2+-influx pathway. Biochem J. 1992;284:521–30.
140. Schmidt K, Dubrovska G, Nielsen G, Fesus G, Uhrenholt TR, Hansen PB, Gudermann T, Dietrich A, Gollasch M, de Wit C, Kohler R. Amplification of EDHF-type vasodilatations in TRPC1-deficient mice. Br J Pharmacol. 2010;161:1722–33.
141. Senadheera S, Kim Y, Grayson TH, Toemoe S, Kochukov MY, Abramowitz J, Housley GD, Bertrand RL, Chadha PS, Bertrand PP, Murphy TV, Tare M, Birnbaumer L, Marrelli SP, Sandow SL. Transient receptor potential canonical type 3 channels facilitate endothelium-derived hyperpolarization-mediated resistance artery vasodilator activity. Cardiovasc Res. 2012;95:439–47.
142. Sharma NR, Davis MJ. Mechanism of substance P-induced hyperpolarization of porcine coronary artery endothelial cells. Am J Physiol. 1994;266:H156–64.
143. Sharma NR, Davis MJ. Substance-P-induced calcium-entry in endothelial-cells is secondary to depletion of intracellular stores. Am J Physiol Heart Circ Physiol. 1995;268:H962–73.
144. Si H, Heyken WT, Wolfle SE, Tysiac M, Schubert R, Grgic I, Vilianovich L, Giebing G, Maier T, Gross V, Bader M, de Wit C, Hoyer J, Kohler R. Impaired endothelium-derived hyperpolarizing factor-mediated dilations and increased blood pressure in mice deficient of the intermediate-conductance Ca2+-activated K+ channel. Circ Res. 2006;99:537–44.
145. Singh I, Knezevic N, Ahmmed GU, Kini V, Malik AB, Mehta D. Galphaq-TRPC6-mediated Ca2+ entry induces RhoA activation and resultant endothelial cell shape change in response to thrombin. J Biol Chem. 2007;282:7833–43.
146. Smith PD, Brett SE, Luykenaar KD, Sandow SL, Marrelli SP, Vigmond EJ, Welsh DG. KIR channels function as electrical amplifiers in rat vascular smooth muscle. J Physiol. 2008;586:1147–60.
147. Socha MJ, Domeier TL, Behringer EJ, Segal SS. Coordination of intercellular Ca2+ signaling in endothelial cell tubes of mouse resistance arteries. Microcirculation. 2012;19:757–70.
148. Socha MJ, Hakim CH, Jackson WF, Segal SS. Temperature effects on morphological integrity and Ca2+ signaling in freshly isolated murine feed artery endothelial cell tubes. Am J Physiol Heart Circ Physiol. 2011;301:H773–83.
149. Sonkusare SK, Bonev AD, Ledoux J, Liedtke W, Kotlikoff MI, Heppner TJ, Hill-Eubanks DC, Nelson MT. Elementary Ca2+ signals through endothelial TRPV4 channels regulate vascular function. Science. 2012;336:597–601.
150. Sonkusare SK, Dalsgaard T, Bonev AD, Hill-Eubanks DC, Kotlikoff MI, Scott JD, Santana LF, Nelson MT. AKAP150-dependent cooperative TRPV4 channel gating is central to endothelium-dependent vasodilation and is disrupted in hypertension. Sci Signal. 2014;7:ra66.
151. Straub AC, Billaud M, Johnstone SR, Best AK, Yemen S, Dwyer ST, Looft-Wilson R, Lysiak JJ, Gaston B, Palmer L, Isakson BE. Compartmentalized connexin 43S-nitrosylation/denitrosylation regulates heterocellular communication in the vessel wall. Arterioscler Thromb Vasc Biol. 2011;31:399–407.
152. Straub AC, Zeigler AC, Isakson BE. The myoendothelial junction: connections that deliver the message. Physiology (Bethesda). 2014;29:242–9.
153. Strotmann R, Schultz G, Plant TD. Ca2+-dependent potentiation of the nonselective cation channel TRPV4 is mediated by a C-terminal calmodulin binding site. J Biol Chem. 2003;278:26541–9.
154. Sullivan MN, Gonzales AL, Pires PW, Bruhl A, Leo MD, Li W, Oulidi A, Boop FA, Feng Y, Jaggar JH, Welsh DG, Earley S. Localized TRPA1 channel Ca2+ signals stimulated by reactive oxygen species promote cerebral artery dilation. Sci Signal. 2015;8:ra2.
155. Sundivakkam PC, Freichel M, Singh V, Yuan JP, Vogel SM, Flockerzi V, Malik AB, Tiruppathi C. The Ca(2+) sensor stromal interaction molecule 1 (STIM1) is necessary and sufficient for

the store-operated Ca(2+) entry function of transient receptor potential canonical (TRPC) 1 and 4 channels in endothelial cells. Mol Pharmacol. 2012;81:510–26.
156. Sundivakkam PC, Kwiatek AM, Sharma TT, Minshall RD, Malik AB, Tiruppathi C. Caveolin-1 scaffold domain interacts with TRPC1 and IP3R3 to regulate Ca2+ store release-induced Ca2+ entry in endothelial cells. Am J Physiol Cell Physiol. 2009;296:C403–13.
157. Svenningsen P, Andersen K, Thuesen A, Shin H-S, Vanhoutte P, Skøtt O, Jensen B, Hill C, Hansen PL. T-type Ca2+ channels facilitate NO-formation, vasodilatation and NO-mediated modulation of blood pressure. Pflügers Arch. 2014;466:2205–14.
158. Takano H, Dora KA, Spitaler MM, Garland CJ. Spreading dilatation in rat mesenteric arteries associated with calcium-independent endothelial cell hyperpolarization. J Physiol. 2004;556(pt.3):887–903.
159. Taylor CW, Tovey SC, Rossi AM, Lopez Sanjurjo CI, Prole DL, Rahman T. Structural organization of signalling to and from IP3 receptors. Biochem Soc Trans. 2014;42:63–70.
160. Taylor MS, Bonev AD, Gross TP, Eckman DM, Brayden JE, Bond CT, Adelman JP, Nelson MT. Altered expression of small-conductance Ca2+-activated K+ (SK3) channels modulates arterial tone and blood pressure. Circ Res. 2003;93:124–31.
161. Thorneloe KS, Sulpizio AC, Lin Z, Figueroa DJ, Clouse AK, McCafferty GP, Chendrimada TP, Lashinger ES, Gordon E, Evans L, Misajet BA, Demarini DJ, Nation JH, Casillas LN, Marquis RW, Votta BJ, Sheardown SA, Xu X, Brooks DP, Laping NJ, Westfall TD. N-((1S)-1-{[4-((2S)-2-{[(2,4-dichlorophenyl)sulfonyl]amino}-3-hydroxypropanoyl)-1-piperazinyl] carbonyl}-3-methylbutyl)-1-benzothiophene-2-carboxamide (GSK1016790A), a novel and potent transient receptor potential vanilloid 4 channel agonist induces urinary bladder contraction and hyperactivity: part I. J Pharmacol Exp Ther. 2008;326:432–42.
162. Tiruppathi C, Freichel M, Vogel SM, Paria BC, Mehta D, Flockerzi V, Malik AB. Impairment of store-operated Ca2+ entry in TRPC4(–/–) mice interferes with increase in lung microvascular permeability. Circ Res. 2002;91:70–6.
163. Ungvari Z, Csiszar A, Koller A. Increases in endothelial Ca(2+) activate K(Ca) channels and elicit EDHF-type arteriolar dilation via gap junctions. Am J Physiol Heart Circ Physiol. 2002;282:H1760–7.
164. Vaca L, Kunze DL. Depletion and refilling of intracellular Ca^{2+} stores induce oscillations of Ca^{2+} current. Am J Physiol. 1993;264:H1319–22.
165. Vaca L, Kunze DL. Depletion of intracellular Ca^{2+} stores activates a Ca^{2+}-selective channel in vascular endothelium. Am J Physiol Cell Physiol. 1994;267:C920–5.
166. Vaithianathan T, Narayanan D, Asuncion-Chin MT, Jeyakumar LH, Liu J, Fleischer S, Jaggar JH, Dopico AM. Subtype identification and functional characterization of ryanodine receptors in rat cerebral artery myocytes. Am J Physiol Cell Physiol. 2010;299:C264–78.
167. Vinet R, Vargas FF. L- and T-type voltage-gated Ca2+ currents in adrenal medulla endothelial cells. Am J Physiol. 1999;276:H1313–22.
168. Vogt-Eisele AK, Weber K, Sherkheli MA, Vielhaber G, Panten J, Gisselmann G, Hatt H. Monoterpenoid agonists of TRPV3. Br J Pharmacol. 2007;151:530–40.
169. von Beckerath N, Dittrich M, Klieber HG, Daut J. Inwardly rectifying K+ channels in freshly dissociated coronary endothelial cells from guinea-pig heart. J Physiol. 1996;491(Pt 2):357–65.
170. Vriens J, Watanabe H, Janssens A, Droogmans G, Voets T, Nilius B. Cell swelling, heat, and chemical agonists use distinct pathways for the activation of the cation channel TRPV4. Proc Natl Acad Sci U S A. 2004;101:396–401.
171. Wagner II LE, Yule DI. Differential regulation of the InsP(3) receptor type-1 and -2 single channel properties by InsP(3), Ca(2)(+) and ATP. J Physiol. 2012;590:3245–59.
172. Wang H-R, Wu M, Yu H, Long S, Stevens A, Engers DW, Sackin H, Daniels JS, Dawson ES, Hopkins CR, Lindsley CW, Li M, McManus OB. Selective inhibition of the Kir2 family of inward rectifier potassium channels by a small molecule probe: the discovery, SAR, and pharmacological characterization of ML133. ACS Chem Biol. 2011;6:845–56.

173. Wang XL, Ye D, Peterson TE, Cao S, Shah VH, Katusic ZS, Sieck GC, Lee HC. Caveolae targeting and regulation of large conductance Ca(2+)-activated K+ channels in vascular endothelial cells. J Biol Chem. 2005;280:11656–64.
174. Watanabe H, Davis JB, Smart D, Jerman JC, Smith GD, Hayes P, Vriens J, Cairns W, Wissenbach U, Prenen J, Flockerzi V, Droogmans G, Benham CD, Nilius B. Activation of TRPV4 channels (hVRL-2/mTRP12) by phorbol derivatives. J Biol Chem. 2002;277:13569–77.
175. Watanabe H, Vriens J, Suh SH, Benham CD, Droogmans G, Nilius B. Heat-evoked activation of TRPV4 channels in a HEK293 cell expression system and in native mouse aorta endothelial cells. J Biol Chem. 2002;277:47044–51.
176. Wei AD, Gutman GA, Aldrich R, Chandy KG, Grissmer S, Wulff H. International Union of Pharmacology. LII. Nomenclature and molecular relationships of calcium-activated potassium channels. Pharmacol Rev. 2005;57:463–72.
177. Wei Z, Manevich Y, Al-Mehdi AB, Chatterjee S, Fisher AB. Ca2+ flux through voltage-gated channels with flow cessation in pulmonary microvascular endothelial cells. Microcirculation. 2004;11:517–26.
178. Westcott EB, Goodwin EL, Segal SS, Jackson WF. Function and expression of ryanodine receptors and inositol 1,4,5-trisphosphate receptors in smooth muscle cells of murine feed arteries and arterioles. J Physiol. 2012;590:1849–69.
179. Wu LJ, Sweet TB, Clapham DE. International Union of Basic and Clinical Pharmacology. LXXVI. Current progress in the mammalian TRP ion channel family. Pharmacol Rev. 2010;62:381–404.
180. Wu S, Haynes Jr J, Taylor JT, Obiako BO, Stubbs JR, Li M, Stevens T. Cav3.1 (alpha1G) T-type Ca2+ channels mediate vaso-occlusion of sickled erythrocytes in lung microcirculation. Circ Res. 2003;93:346–53.
181. Wulff H, Kohler R. Endothelial small-conductance and intermediate-conductance KCa channels: an update on their pharmacology and usefulness as cardiovascular targets. J Cardiovasc Pharmacol. 2013;61:102–12.
182. Xia XM, Fakler B, Rivard A, Wayman G, Johnson-Pais T, Keen JE, Ishii T, Hirschberg B, Bond CT, Lutsenko S, Maylie J, Adelman JP. Mechanism of calcium gating in small-conductance calcium-activated potassium channels. Nature. 1998;395:503–7.
183. Xu W, Lipscombe D. Neuronal Ca(V)1.3alpha(1) L-type channels activate at relatively hyperpolarized membrane potentials and are incompletely inhibited by dihydropyridines. J Neurosci. 2001;21:5944–51.
184. Yeon SI, Kim JY, Yeon DS, Abramowitz J, Birnbaumer L, Muallem S, Lee YH. Transient receptor potential canonical type 3 channels control the vascular contractility of mouse mesenteric arteries. PLoS One. 2014;9:e110413.
185. Zalk R, Lehnart S, Marks A. Modulation of the ryanodine receptor and intracellular calcium. Annu Rev Biochem. 2007;76:367–85.
186. Zaritsky JJ, Eckman DM, Wellman GC, Nelson MT, Schwarz TL. Targeted disruption of Kir2.1 and Kir2.2 genes reveals the essential role of the inwardly rectifying $K(^+)$ current in $K(^+)$-mediated vasodilation. Circ Res. 2000;87:160–6.
187. Zhang DX, Mendoza SA, Bubolz AH, Mizuno A, Ge ZD, Li R, Warltier DC, Suzuki M, Gutterman DD. Transient receptor potential vanilloid type 4-deficient mice exhibit impaired endothelium-dependent relaxation induced by acetylcholine in vitro and in vivo. Hypertension. 2009;53:532–8.
188. Zheng W, Rampe D, Triggle DJ. Pharmacological, radioligand binding, and electrophysiological characteristics of FPL 64176, a novel nondihydropyridine Ca2+ channel activator, in cardiac and vascular preparations. Mol Pharmacol. 1991;40:734–41.
189. Zhou C, Chen H, Lu F, Sellak H, Daigle JA, Alexeyev MF, Xi Y, Ju J, van Mourik JA, Wu S. Cav3.1 (alpha1G) controls von Willebrand factor secretion in rat pulmonary microvascular endothelial cells. Am J Physiol Lung Cell Mol Physiol. 2007;292:L833–44.

190. Zhou M-H, Zheng H, Si H, Jin Y, Peng JM, He L, Zhou Y, Muñoz-Garay C, Zawieja DC, Kuo L, Peng X, Zhang SL. Stromal Interaction Molecule 1 (STIM1) and Orai1 mediate histamine-evoked calcium entry and Nuclear Factor of Activated T-cells (NFAT) signaling in human umbilical vein endothelial cells. J Biol Chem. 2014;289:29446–56.
191. Ziegelstein RC, Spurgeon HA, Pili R, Passaniti A, Cheng L, Corda S, Lakatta EG, Capogrossi MC. A functional ryanodine-sensitive intracellular Ca^{2+} store is present in vascular endothelial cells. Circ Res. 1994;74:151–6.
192. Zucchi R, Ronca-Testoni S. The sarcoplasmic reticulum Ca2+ channel/ryanodine receptor: modulation by endogenous effectors, drugs and disease states. Pharmacol Rev. 1997;49: 1–51.

Chapter 2
Contribution and Regulation of Calcium Channels in Endothelial Cells

Kwong Tai Cheng, Avia Rosenhouse-Dantsker, and Asrar B. Malik

Abstract The endothelium is a highly metabolically active organ that plays a pivotal role in many physiological processes. Endothelial cells express a diversity of calcium-permeable ion channels that can be activated in response to a variety of stimuli including Ca^{2+} store depletion, oxidative stress, growth factors, and endotoxins. Emerging evidences have implicated the critical requirement of Ca^{2+} signaling in numerous vascular functions including vasomotor tone, barrier function, leukocyte homing and adhesion, inflammation, and hemostasis. The goal of this chapter is to present a comprehensive review of the expression and regulatory mechanisms of Ca^{2+} channels in endothelial cells, and discuss their contribution to vascular endothelial cell physiology and pathophysiology processes.

Keywords Endothelial cell • Endothelium • Permeability • SOCE • TRPC channels • TRPC1 • Orai1 • STIM1 • Ca^{2+} signaling • Cell function

Introduction

The endothelium is the largest organ in the body forming a highly specialized cellular network that is composed of $1–6 \times 10^{13}$ endothelial cells (ECs) [1, 2] with an estimated surface area that exceeds 1000 m^2 [3]. ECs line the inner wall of the entire vascular tree, forming a critical interface between the circulating blood and the surrounding tissues [4], thereby playing a key role in various physiological and pathological processes such as blood supply, nutrient delivery, metabolic homeostasis, immune cell trafficking and inflammation [5–7].

Under physiological conditions, ECs exert a number of functions that are important for normal homeostasis (see Fig. 2.1). The most important functions of the endothelium include regulating vasopermeability (i.e. barrier function) [8],

K.T. Cheng • A. Rosenhouse-Dantsker • A.B. Malik (✉)
Department of Pharmacology, College of Medicine, University of Illinois,
835 South Wolcott Ave, MSB, Chicago, IL 60610, USA
e-mail: abmalik@uic.edu

I. Levitan, A.M. Dopico (eds.), *Vascular Ion Channels in Physiology and Disease*, DOI 10.1007/978-3-319-29635-7_2

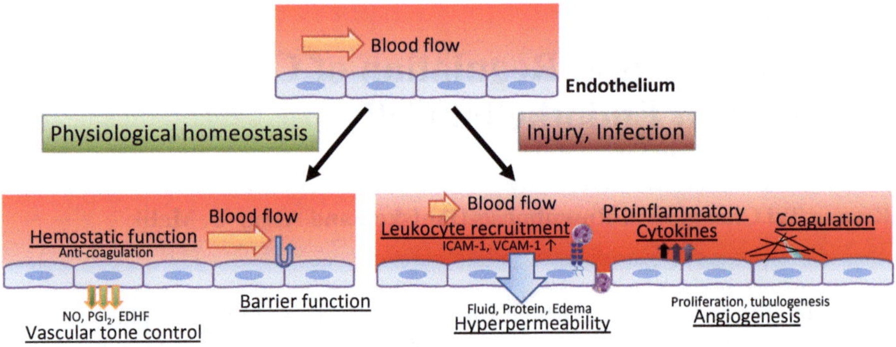

Fig. 2.1 Physiological and pathophysiology function of endothelial cells. Schematic overview of the various functions mediated by endothelial cells that impact physiological homeostasis, and their response under pathological conditions

prevention of inappropriate coagulation [9], regulation of microcirculation byproduction, and release of vasoactive mediators that can control the tones of underlying vascular smooth muscles and modulate EC functions themselves [10]. However, these functions can be significantly altered for adaptive responses under pathological conditions [7]. Specifically, ECs are the first cells exposed to invading pathogens circulating in the bloodstream, and have a prominent role in surveying the circulation of molecules of invading microbial pathogens. Furthermore, receptors of the innate immune system in EC scan be activated by microbial components and contribute to host-defense via the release of inflammatory mediators, recruit leukocytes and promote clotting as a measure to contain infections.

Ca^{2+} signaling in ECs has been shown to be critically required for the physiological and pathological processes described above [11]. In particular, it has been shown that release of vasoactive mediators is tightly governed by Ca^{2+} signaling, and that the increase in endothelial permeability depends on both Ca^{2+} release from the endoplasmic reticulum as well as Ca^{2+} entry through the plasma membrane (PM) [11]. The expression of adhesion molecules and the initiation of an inflammatory cascade have also been shown to be facilitated by Ca^{2+} signaling [12]. All these processes, which are vital for maintaining EC homeostasis, are interconnected via Ca^{2+} signaling, underpinning the importance of ion channels in the regulation of EC function. In addition, the gating and activation of ion channels are rapid processes that enable ECs to react to changes in the local environment by adjusting their function in a timely manner. Elucidating the mechanisms of Ca^{2+} signaling in ECs and the associated downstream signaling cascade promises to provide novel insights as to how these processes can be differentially regulated in a specific spatial and temporal manner as well as identify checkpoints that are critical for particular functions.

Thus, in this review, we will focus on Ca^{2+} entry pathways that have been reported in ECs. In particular, we will discuss store-operated calcium entry (SOCE) that has been shown to be critically required for multiple EC functions, and can therefore provide insights into the role of Ca^{2+} signaling in vasculature and shed light on the potential importance of ion channels as therapeutic targets for various cardiovascular diseases.

Ca^{2+} Entry Pathway in Endothelial Cells

Store-Operated Calcium Entry

Store-operated calcium entry (SOCE) is an ubiquitous Ca^{2+} entry pathway in all excitable and non-excitable cells that was first described by Putney almost three decades ago [13]. The primary trigger for SOCE activation is the depletion of the endoplasmic reticulum (ER) Ca^{2+} store. Conversely, refilling this store leads to SOCE inactivation. Under physiological conditions, SOCE is activated in response to stimulation of G-protein coupled receptors (GPCR) in the plasma membrane that results in an increase in phospholipase C (PLC) activity, and facilitates $PI(4,5)P_2$ hydrolysis, IP_3 generation, and IP_3-mediated Ca^{2+} release from the ER (see Fig. 2.2). This decrease in ER Ca^{2+} content subsequently leads to activation of SOCE through

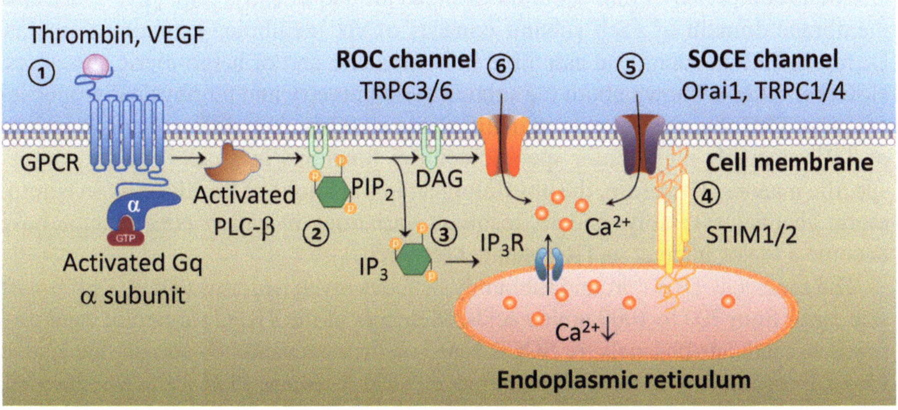

Fig. 2.2 Mechanism of Store-operated Ca^{2+} entry (SOCE) and Receptor-operated Ca^{2+} entry (ROCE) in endothelial cells. (1) An agonist (e.g. thrombin, VEGF) binds to G-protein coupled receptors or receptor tyrosine kinases. (2) Through the activation of phospholipase C (PLC), phosphatidylinositol 4,5-bisphosphate ($PI(4,5)P_2$) is hydrolyzed into the secondary messenger inositol 1,4,5-trisphosphate (InsP3) and diacylglycerol (DAG). (3) IP3 rapidly diffuse across the cytosol and interact with IP_3 receptors (IP_3R) located in the ER, which results in depletion of the calcium stored in the ER lumen through opening of the IP_3R. (4) Ca^{2+} dissociates from the EF-hand of STIM, facilitating STIM protein translocation and oligomerization in the plasma membrane ER junction. (5) The C-terminus of the STIM protein undergoes a conformational change to an extended structure, enabling it to directly gate plasma membrane SOCE channels (Orai1, TRPC1/4). (6) DAG directly activates receptor-operated calcium (ROC) channel(s) (TRPC3/TRPC6)

the opening of transmembrane store-operated calcium (SOC) channels. Ca^{2+} influx through these channels plays a primary role in refilling the ER Ca^{2+} store. Multiple studies have demonstrated that this Ca^{2+} influx mechanism can also regulate a broad spectrum of cellular responses such as EC migration, proliferation, T cell activation and mast cell degranulation [14]. Yet, only following the recent identification of the genes encoding the SOCE components, has it become possible to determine the structure of the fundamental units of SOCE. Furthermore, identification of the signaling mechanisms that underlie activation of SOC channels at the plasma membrane as a function of the filling status of the ER Ca^{2+} store, remain a major hurdle in SOCE research.

Molecular Components of the SOCE

Canonical Transient Receptor Potential (TRPC) Channels

The search for SOC channels led to the discovery of the mammalian transient receptor potential (TRP) channels. Among these, TRPC1 was the first mammalian TRP protein to be identified. Subsequent research revealed that mammalian TRP genes encode a superfamily of ion channels that can be grouped into seven subfamilies: TRPC, TRPV, TRPM, TRPA, TRPN, TRPP, and TRPML. All TRP channels are tetramers comprised of four subunits arranged around a central pore [15]. The transmembrane domain of each subunit consists of six membrane-spanning domains. Different TRP subunits can assemble in homomeric and/or heteromeric structures. However, little is known about the subunit stoichiometry and biophysical properties of native TRP channels expressed in different cell types. Yet, different combinations of TRP channels may confer specific cellular functions in a cell-type and tissue-specific manner. Therefore, the physiological implications of the diversity in heteromeric channel assembly, and its resultant variations in cation conductance have become a major interest in TRP channel research.

The canonical TRPC protein subfamily includes seven isoforms (TRPC1 to 7) with size range from 700 to 1000 amino acids. Members of the TRPC subfamily were proposed as candidate channels for SOCE based on their activation by stimuli that lead to $PI(4,5)P_2$ hydrolysis. Indeed, all mammalian TRPCs require PLC for activation [16]. However, there is variability in their selectivity for Ca^{2+} versus other cations, and in the mechanism coupling PLC activity to channel stimulation. Several TRPC proteins are stimulated through the SOCE mechanism including TRPC1, TRPC4, and TRPC5 in homomeric form, and TRPC3 and TRPC6 as a part of heteromultimers [17]. On the other hand, several mammalian TRPCs (TRPC3, TRPC6, and TRPC7) are coupled to PLC activation and activated in a store-independent manner by diacylglycerol (DAG). Whether these DAG-sensitive TRPC channels form homomultimers or heteromultimers remains unclear. It has also been reported that subunits from other TRP sub-families can form heteromeric channels with members of the TRPC family, and contribute to SOCE activity [18, 19]. However, this result is controversial, and additional studies are required to further investigate it.

Expression of TRPC Channels in Endothelial Cells

TRPC channels have been shown to be abundantly expressed in ECs obtained from a variety of sources. All TRPC isoforms were detected in primary ECs [10]. TRPC1 and TRPC3 to TRPC6 were identified in bovine aortic ECs [20]. In contrast, bovine pulmonary artery ECs did not express the TRPC3 isoform. Data from our group have demonstrated the expression of TRPC1, TRPC3, TRPC4 and TRPC6 in human umbilical vein ECs (HUVECs), human dermal microvascular ECs (HDMECs) and human pulmonary artery ECs (HPAECs) [21]. In addition to cultured ECs, several studies examined the expression profile of TRPC channels in ECs from intact vessels. TRPC1 and TRPC3 have been shown to be expressed in human mesenteric arteries by using single-cell RT-PCR [22] whereas in another study, in situ hybridization yielded strong labeling of TRPC1,3-6 in endothelial and smooth muscle cells of human coronary and cerebral arteries [23]. TRPC7 labeling was exclusively found in ECs but not in smooth muscle cells [23]. These studies have shown that the expression profile of TRPC channels may be different with respect to ECs derived from different vascular beds and/or from different animal species, suggesting that the function of ECs can be differentially regulated by expressing specific combinations of TRPC channels such that ECs from different vascular beds can response to environmental stimuli in a distinctive spatial and temporal manner.

STIM Proteins and Orai Channels

STIM was first identified in 2005 by an RNAi screen as a regulator of SOCE in Drosophila and HeLa cells [24, 25]. In human, there are two STIM homologs, STIM1 and STIM2. STIM1 encodes a single-pass transmembrane protein of 77 kDa, localized primarily in the ER membrane. Structural analysis of STIM1 revealed a Ca^{2+} binding EF-hand domain in the ER luminal residing portion implying that the protein functions as an ER Ca^{2+} sensor [24, 26]. The cytosolic C-terminus of STIM1 is organized in several distinct modules, the most critical of which is a region that includes roughly 100 amino acid, and is referred to as the STIM1-Orai Activation region (SOAR) [27], CRAC activation domain (CAD) [28] or CCb9 [29]. Overexpression of this region can facilitate spontaneous calcium induced activated calcium (CRAC) current (I_{CRAC}) in the absence of store depletion suggesting that this region is necessary for CRAC channel activation. The C terminal end of STIM1 contains a region with multiple lysine residues, which is referred to as the K-domain. The K-domain plays several roles. First, it interacts with $PI(4,5)P_2$ at the interface with the plasma membrane and stabilizes the STIM1 cluster at the ER-PM junction [30]. More importantly, the K-domain is required for gating TRPC channels via intermolecular electrostatic interactions [31]. Recently, the role of STIM1 in generating Ca^{2+} influx been further clarified and the underlying mechanism elucidated [32]. Specifically, when the ER Ca^{2+} store is filled, Ca^{2+} is bound to the EF-hand of STIM1. Under these conditions, STIM1 is uniformly distributed in the ER membrane. Upon agonist stimulation, generation of IP_3 will trigger IP_3R

opening and lead to depletion of the ER Ca^{2+} store. Ca^{2+} will then dissociate from the EF-hand of STIM1 and mediate a conformational change resulting in oligomerization of the STIM1 molecule. STIM1 oligomers then translocate to the ER-PM junction forming puncta. These localized STIM1 clusters interact and activate store-operated channels on the plasma membrane and generate Ca^{2+} influx.

STIM2, the second mammalian ER Ca^{2+} sensor, exhibits high structural similarity to STIM1. However, the Ca^{2+} sensitivity and activation kinetics of STIM2 have been shown to differ from those of STIM1 [33]. STIM2 also oligomerizes and translocates to the ER-PM junction in response to store depletion. However, STIM2 is a poor activator of Orai1 channel compared to STIM1. The role of STIM2 has not yet been conclusively established but recent data have shown that STIM2 can recruit STIM1 to the ER-PM junction under low stimulus intensities, thereby promoting assembly of the STIM1-Orai1 channel complex and fine tuning the agonist sensitivity of Ca^{2+} signaling. Further insights into the mechanisms that govern the function of the STIM protein will undoubtedly continue to emerge as research in this field continues.

The first store-operated Ca^{2+} current, I_{CRAC}, was measured in mast cells and T lymphocytes [34]. However, the molecular identity of this channel remains unknown after almost a decade of characterization. In 2006, Orai1 was identified as the core component of the CRAC channel through multiple studies using genome-wide RNAi screening in Drosophila [35, 36] and genetic linkage analysis in severe combined immune deficiency (SCID) patients [35]. It was demonstrated that a point mutation in the Orai1 gene (R91W) of SCID patients leads to the loss of I_{CRAC} and SOCE in SCID T-lymphocytes. Notably, expression of WT Orai1 in these cells rescued RAC channels function, thereby restoring store-operated Ca^{2+} influx [35]. Orai1 is a four-transmembrane domain protein primarily located on the plasma membrane. Early studies of the role of Orai1 in the SOCE mechanism suggested that Orai1 acts as a regulatory subunit of TRP channel [37] or that it forms a complex with STIM1 and TRPC1 [38]. However, extensive studies using site-directed mutagenesis led to the identification of a number of Orai1 mutants that were capable of altering the electrophysiological properties of I_{CRAC}, suggesting that Orai1 is the pore-forming subunit of the CRAC channel [39]. Following this observation, it is currently widely accepted that Orai1 alone constitute the I_{CRAC} channel pore and are responsible for SOCE.

Initial studies of Orai1 have primarily focused on lymphocyte and immune function. Surprisingly, Orai1 is not only expressed in haematopoietic cells, but is also widely expressed in other cell types and tissues. Earlier studies have shown that vascular endothelial growth factor (VEGF) receptor activation can facilitate Ca^{2+} release from the IP_3 receptor, and consequently induce a small Ca^{2+} current across the plasma membrane [40]. Moreover, using electrophysiology techniques, the Nilius group demonstrated that depletion of the ER calcium store can facilitate an inwardly rectifying CRAC-like current in bovine pulmonary artery ECs [41]. More recently, extensive research aimed to characterize the expression of Orai1 in the vasculature, was carried out. Our group and the groups of Trebak and Beech demonstrated the expression of Orai1 mRNA and protein in HUVEC [42–44]. Additionally, it has also been shown that Orai1 is expressed in human lung microvascular ECs [42], rat pulmonary microvessel ECs [45], and immortalized mouse lung ECs [42].

Channels Operated by Oxidative Stress

Oxidative stress can be generated in a variety of pathophysiological conditions including hypoxia, inflammatory response and injury. It has been shown that oxidative stress can modulate EC function through the activation of ion channels through the production of reactive oxygen species (ROS), subsequently leading to an increase in vascular endothelial permeability, thereby playing a crucial role in several lung diseases. The TRPM2 (melastatin) channel is among the best characterized oxidative stress sensitive channels in ECs. TRPM2 is a voltage-independent, oxidant-sensitive, nonselective cation channel that is uniquely gated by the binding of adenosine diphosphoribose (ADP-ribose) to its C-terminal domain. Recent studies demonstrated that TRPM2 may facilitate oxidative damage in the endothelium [46–50]. The proposed underlying mechanism suggested how oxidative stress could promote Ca^{2+} entry and specific Ca^{2+} dependent cellular processes. Accordingly, oxidative stress can lead to the production of hydrogen peroxide (H_2O_2) in the cytosol that would then result in nuclear and mitochondrial production of ADP-ribose. ADP-ribose would subsequently open the TRPM2 channel, resulting in increased cytosolic Ca^{2+} concentration and inducing endothelial dysfunction. In addition to the TRPM2 channel, there is emerging evidence suggesting that other TRPM channels can also be activated in response to oxidative stress and alter EC function. Knockdown of TRPM7 in HUVEC caused enhanced growth and proliferation, as well as increased expression of nitric oxide synthase and nitric oxide production [51]. In addition, knockdown of TRPM7 protects HUVEC from hyperglycemia-mediated injury [52]. Together these data suggest that TRPM7 has a prominent role in angiogenesis and smooth muscle tone control in the vasculature. In addition, a recent study has demonstrated that TRPM4 channel is critical to oxidative stress-enhanced EC migration [53]. It should be noted that the biophysical property of these various TRMP subtypes shows difference in ion selectivity, TRPM2 is a non-selective channel which show a linear current-voltage (I-V) relation that allow Na^+ and Ca^{2+} to pass through, while TRPM4 are only permeable to monovalent cations (Na^+, K^+, Cs^+) and TRPM7 carry divalent cations including Mg^{2+} as well as to a variable degree of Ca^{2+}, the difference in their ion selectivity thus may explain how these oxidative stress regulated channel can differentially affect multiple downstream EC function.

Cyclic Nucleotide-Gated Channels

The cyclic nucleotide-gated (CNG) channel is a family of Ca^{2+} permeable nonselective cation channels that open in response to elevated levels of cyclic nucleotides in the cytosol, particularly cAMP and cGMP. CNG channels were first identified in the visual and olfactory systems where they were shown to play pivotal roles in sensory detection.

Six CNG isoforms have been identified. In native cells, CNG channels usually form heterotetrameric complexes consisting of A and B subunits [54]. The A1–A3 subunits may also form functional channels on their own, whereas the B and A4 subunits serve modulatory functions. CNG channels are widely expressed in vascular tissues across species [55, 56]. CNGA1 channels have been shown to be abundantly expressed in the endothelium layer [56]. They also express in smooth muscle cells albeit at a much lower level [56]. In contrast, robust expression of CNGA2 channel was shown in both the endothelium and smooth muscle layers of human arteries [55]. Moreover, the expression of CNGA2 in endothelium can be upregulated by nitric oxide (NO), and it has been shown that this leads to elevated cytosolic Ca^{2+} concentration ($[Ca^{2+}]_i$) and SOCE in porcine PAEC [57].

Endothelial Cell Functions Regulated by Ca^{2+} Signaling

Permeability

The endothelium serves as a selective barrier for the exchange of fluid and macromolecules from the circulation to the tissue environment. Endothelial barrier function can be adjusted in response to tissue injury or infections. Inflammatory mediators (e.g. thrombin, TNF-α, histamine) bind to G-protein coupled receptors located on the EC plasma membrane, which then leads to an increase in $[Ca^{2+}]_i$ via the SOCE machinery. The increase in $[Ca^{2+}]_i$ facilitates cytoskeletal protein reorganization by myosin light chain-dependent EC contraction and disassembly of vascular endothelial cadherin (VE-cadherin) at adherens junctions, subsequently leading to gap formation and an increase in endothelial permeability [11]. Thrombin, the ligand of the proteinase-activated receptor (PAR), has been extensively used by us to study the mechanism of endothelial permeability *in vitro* and *in vivo* [8]. We have shown that thrombin receptor cleavage and the consequent combined elevation of intracellular Ca^{2+} and PKC activation is required for permeability increase in bovine pulmonary arterial ECs [58]. We also demonstrated that this process is mediated through activating the proteinase-activated receptor-1 (PAR-1) GPCR [59], and confirmed these observations by using a transgenic mouse model deficient in PAR-1 [60]. The question remaining to be answered is which Ca^{2+} entry pathway, and more specifically, which Ca^{2+} channel is responsible for this thrombin induced increase in endothelial permeability response.

Several reports from our group as well as others have demonstrated the involvement of Ca^{2+} entry through TRPC1, TRPC4 and TRPC6 channels in the disruption of the barrier function in pulmonary arteries [61, 62]. We have shown that TRPC1 channels make a major contribution to the increase in vascular permeability induced by thrombin and VEGF through the activation of store-operated Ca^{2+} influx [21, 63]. The expression of TRPC1 channel is upregulated by inflammatory mediators through the NF-κB pathway, and its activation can be controlled by Rho activation

[64] and PKCα phosphorylation [21, 65]. In addition, it has been reported that TRPC1 may have a role in cytoskeletal rearrangement [66], strongly supporting the role of TRPC1 in permeability regulation. We have also shown that TRPC4-deficient mice were protected from thrombin-induced permeability *in vivo* [11, 62]. ECs generated from TRPC4−/− mice did not respond to thrombin or PAR-1 agonist peptide (Figs. 2.3 and 2.4). Furthermore, *ex vivo* lung preparation from TRPC4−/− mice displayed a reduction in permeability increase in response to PAR-1 activation. In summary, TRPC4 channel deficiency led to reduced Ca^{2+} entry in ECs, suppressed thrombin-mediated actin-stress fiber formation and thus reduced permeability increase (Fig. 2.5). These observations were further confirmed by other research groups showing a direct interaction between TRPC4 and cytoskeletal proteins in ECs [67, 68].

TRPC6 is another non-selective Ca^{2+} permeable ion channel involved in vascular permeability control. In contrast to TRPC1 and TRPC4, which are regulated by the SOCE mechanism, activation of TRPC6 can also be triggered by Diacylglycerol (DAG), a second product of PLC activity, which is independent of store depletion [69]. It has been previously demonstrated that DAG or its analog 1-oleoyl-2-acetyl-sn-glycerol (OAG) can induce EC contraction and subsequently gap formation through protein kinase C-α (PKCα)-dependent activation of RhoA. TRPC6 siRNA attenuated Ca^{2+} entry induced by thrombin in concert with the decrease in permeability [61]. It was later shown that the phosphatase and tensin homolog (PTEN) exclusively interacts with TRPC6 and is necessary for OAG-induced calcium entry and the subsequent increases in monolayer permeability [70]. Consistent with this finding, ECs isolated from TRPC6−/− mice lungs displayed reduced Ca^{2+} signaling as well as reduced permeability increase in response to ischemia and OAG stimulation. Accordingly, TRPC6−/− mouse lungs are protected from ischemia/reperfusion-induced increases in lung permeability and edema [70].

Although it has been demonstrated in multiple studies that Orai1 and STIM1 form the core components of the CRAC channel, overexpression of these proteins are sufficient to generate I_{CRAC} and SOCE in T lymphocytes as well as other cell types. However, the involvement of Orai1 and STIM1 in generating SOCE in ECs as well as their role in permeability regulation yielded mixed results. The Trebak group first showed that STIM1 and Orai1 mediate CRAC currents and store-operated calcium entry in HUVECs. RNA silencing of either STIM1 or Orai1 essentially abolished SOCE and I_{CRAC} in ECs [44]. Surprisingly, in another report by the same group, they challenged the requirement for Ca^{2+} signaling in permeability control, and suggested that STIM1 alone controls endothelial barrier function independently of Orai1 and Ca^{2+} entry [71]. The authors' rationale was based on the observations that in HUVECs, STIM1 directly coupled the thrombin receptor to RhoA activation and myosin light chain phosphorylation, leading to formation of actin stress fibers, and loss of cell-cell adhesion. These data suggested that endothelial permeability regulated by STIM1 was independent of Ca^{2+} entry. In contrast to these studies, we observed that neither Orai1 knockdown nor expression of a dominant-negative Orai1 mutant could inhibit SOCE induced by either thrombin or Thapsigargin (Tg) in murine and human lung microvascular ECs. Whereas inhibition

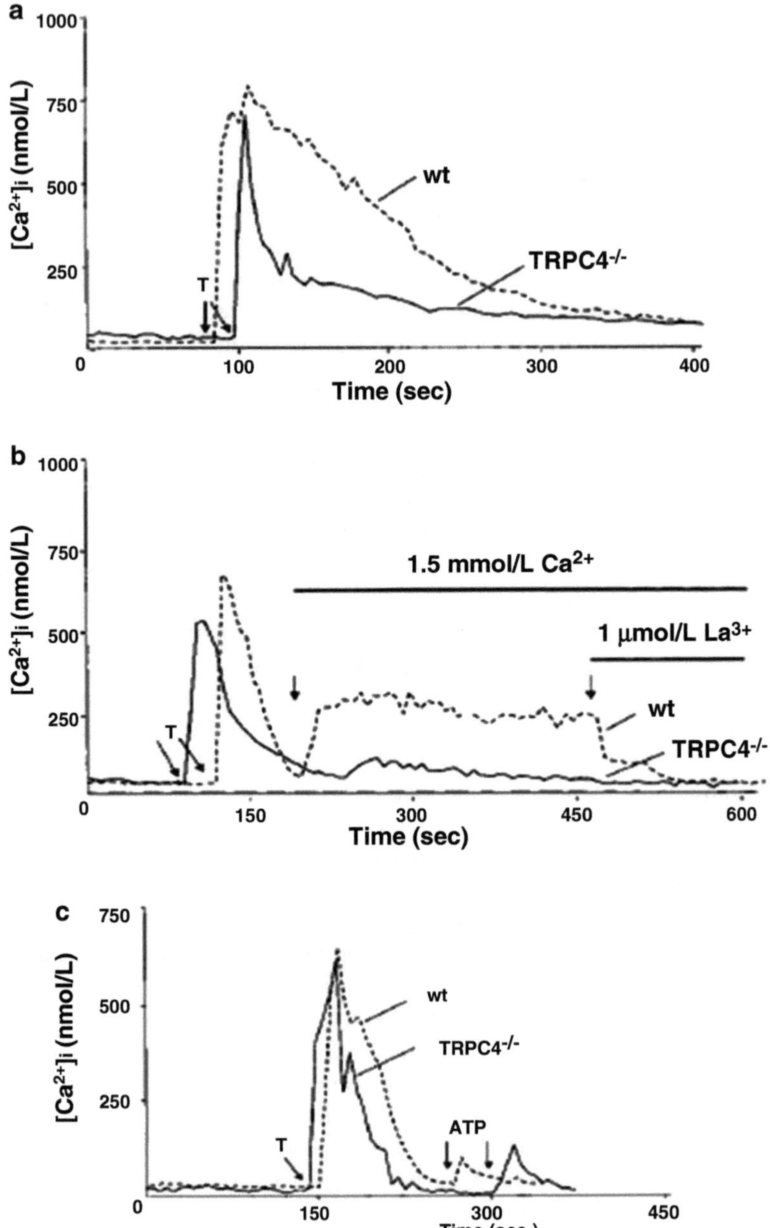

Fig. 2.3 (**a**) Effects of thrombin on $[Ca^{2+}]_i$ in LECs. Thrombin-induced increase in $[Ca^{2+}]_i$ was measured. In each experiment, 25–35 cells were selected to measure changes in $[Ca^{2+}]_i$. Extracellular Ca^{2+} concentration was 1.26 mmol/L. The *arrows* signify the time thrombin (5 U/mL) was added. Results are representative of four experiments. T indicates thrombin. (**b**) TRPC4−/− LECs fail to show thrombin-induced Ca^{2+} influx. Thrombin-induced Ca^{2+} influx was measured in wt and TRPC4−/− LECs. Fura 2-AM loaded cells were washed two times, placed in Ca^{2+}- and Mg^{2+}-free HBSS, and then stimulated with thrombin (5 U/mL). After return of $[Ca^{2+}]_i$ to baseline levels, cells were stimulated with $CaCl_2$ (1.5 mmol/L) and at the indicated time lanthanum chloride (1 μmol/L) was added to inhibit Ca^{2+} influx. The experiment was repeated three times with similar results. (**c**) Effects of thrombin on Ca^{2+} store depletion. Cells were first stimulated with thrombin (5 U/mL) in the absence of extracellular Ca^{2+}. After recovery to the base-line value, cells were challenged with ATP (5 μmol/L). Note that there was no significant increase in $[Ca^{2+}]_i$ in response to ATP. Experiment was repeated three times with similar results. [Adapted from Tiruppathi et al. [62]]

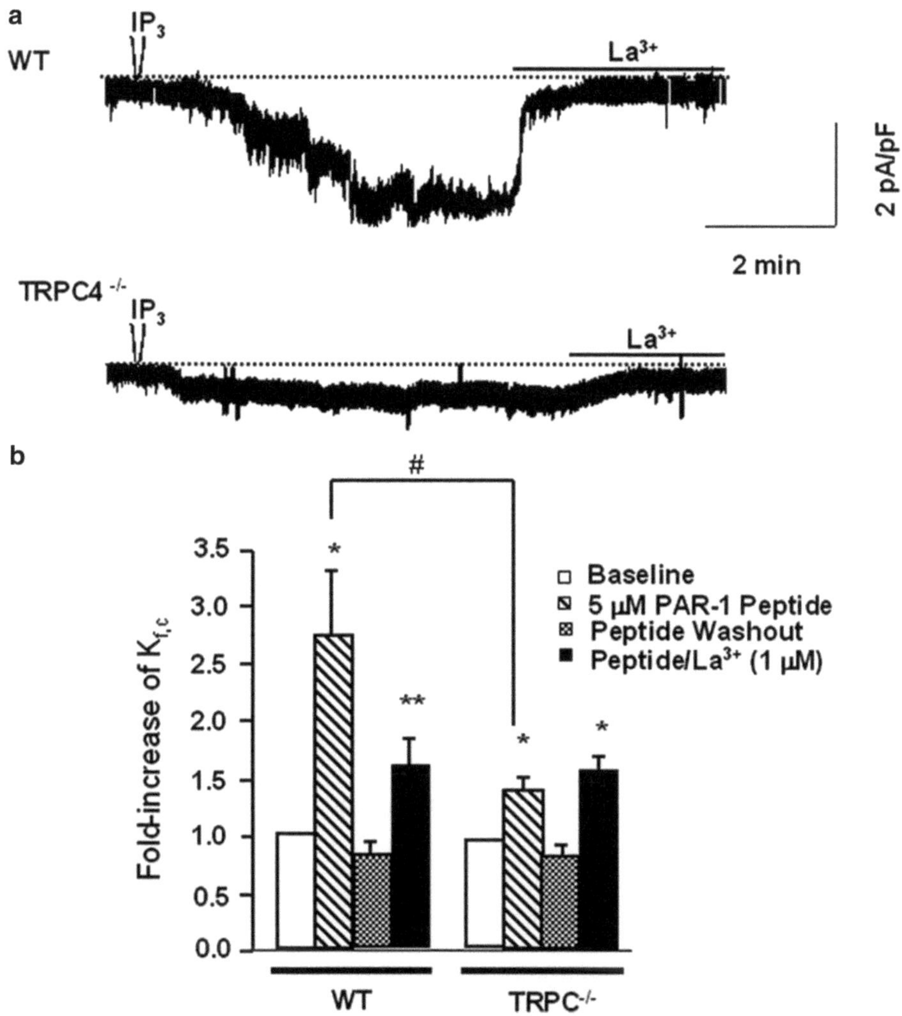

Fig. 2.4 TRPC4 activation of store-operated Ca^{2+} entry (SOCs) regulates microvessel permeability. (**a**) IP_3-induced activation of lanthanum (La^{3+})-sensitive SOCs is shown as whole cell continuous current recordings at a holding potential of −50 mV. The *traces* show that in endothelial cells isolated from wild-type (WT) mice, internal dialysis with IP_3 induced store-operated current, indicating Ca^{2+} entry. Deletion of TRPC4 in mice (TRPC4$^{-/-}$) prevented the Ca^{2+} entry in response to store depletion by IP_3. Scale represents picoampere (pA) and picofarad (pf). (**b**) In WT mice, PAR-1 peptide induced a 2.5- to 3-fold increase in lung microvessel permeability, which was reduced to 50 % by La^{3+}, the Ca^{2+} channel blocker, indicating Ca^{2+} entry contributes to the mechanism of increased lung vascular permeability. However, in TRPC4$^{-/-}$ mice, the PAR-1 peptide only induced a 1.5-fold increase in permeability, which was not affected by La^{3+}, indicating that Ca^{2+} entry via TRPC4 contributes to the regulation of microvessel permeability. *Different from baseline value. **Inhibitory effect of La^{3+}. #Significantly reduced response of TRPC4$^{-/-}$ for PAR-1 agonist peptide compared with WT. [Adapted from Tiruppathi et al. [62]]

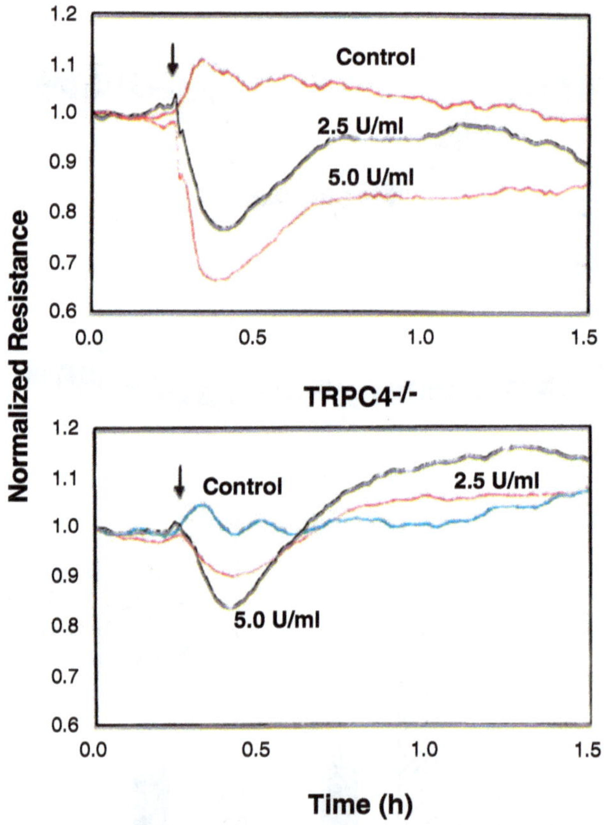

Fig. 2.5 Effects of thrombin on WT and TRPC4–/– LEC transendothelial monolayer electrical resistance values. LECs were grown to confluence on gold electrodes. Before the experiments, cells were washed twice and incubated with 1 % FBS containing growth medium for 2 h before the addition of thrombin. The *arrow* indicate the time thrombin was added. Experiment shown is representative of four trials. [Adapted from Tiruppathi et al. [62]]

of Orai1 only reduced the initial peak of SOCE, the effect was not statistically significant [42]. We further showed that knockdown of STIM1 or expression of the dominant-negative STIM1 mutant in ECs suppressed Ca^{2+} entry secondary to store depletion. Moreover, our data conclusively indicated that STIM1 could interact with TRPC1 and TRPC4 to form a functional SOC channel in ECs. Whereas both studies agreed regarding the requirement of STIM1 in permeability control, albeit through different regulatory pathways, our findings that Orai1 does not regulate SOCE in native ECs differs from the findings of Abdullaev et al. (2008) [44]. To address this difference, we also knocked-down Orai1 in HUVECs. Our data showed that in this cell type, and in agreement with the study of Abdullaev et al. [44], Orai1 plays a similar role in regulating SOCE, suggesting that the discrepancy was related to the origin of ECs being used in these studies.

Orai channels are highly Ca^{2+} selective, whereas TRPC proteins are nonselective cation channels permeable to Ca^{2+}, Na^{+}, and Cs^{+} [72]. We observed expression of Orai1, Orai2, and Orai3 proteins in HUVECs. Accordingly, Orai proteins may contribute to Ca^{2+}-selective SOC activity in HUVECs that may be required for specialized function of human umbilical veins. In contrast, exogenous expression of Orai isoforms or knockdown had no significant effect on thrombin-induced Ca^{2+} entry in both human and mouse primary ECs, indicating that TRPCs are the primary channels that conduct the nonselective cation current required for regulating native EC function.

As noted above, SOCE is activated in response to stimulation of G-protein coupled receptors (GPCRs). In addition, however, EC permeability can also increase in response to environmental stress due to, for example, ROS generation and shear stress. Specifically, ROS can directly activate plasma membrane ion channels and mobilize $[Ca^{2+}]_i$, and ECs exposed to shear stress show an increase in phospholipase activity that impacts membrane permeability [10]. Our group has also demonstrated that H_2O_2 elicited Ca^{2+} influx via TRPM2 channel activation leading to an increase in endothelial permeability [49] (see Fig. 2.6). This process could be attenuated by inhibiting TRPM2 using a pharmacological inhibitor or through the application of specific siRNA or a blocking antibody, thereby conclusively demonstrating a critical role for TRPM2 in the mechanism of endothelial barrier disruption following oxidative stress (Fig. 2.7). In addition to the full-length TRPM2 (TRPM2-L), we have also identified a short spice variant of TRPM2 (TRPM2-S) [47, 73] that could serve as a negative regulator of TRPM2 channel activity in ECs via PKCα phosphorylation upon H_2O_2 stimulation [48]. Further studies are needed to investigate the mechanism that underlies the regulation of TRPM2-mediated Ca^{2+} entry and endothelial injury by PKCα.

Permeability regulation through GPCR-dependent and GPCR-independent mechanisms merges at the calcium signaling level. An interesting question that remained to be elucidated is how ion channels can control multiple downstream processes through Ca^{2+} signaling. It has been demonstrated that the maximum increase in $[Ca^{2+}]_i$ occurs after the maximum increase in permeability [74], suggesting that regulation of EC permeability may not be determined by the magnitude of the increase in $[Ca^{2+}]_i$, but rather by the unique Ca^{2+} signature generated in distinct microdomains. For example, it has been shown that the expression of ion channels is highly compartmentalized in polarized exocrine acinar cells. Orai1 is localized in the lateral membrane towards the apical region whereas various TRPC subtypes spread from the apical region towards the lateral membrane, thereby overlapping with Orai1 in the luminal end of the lateral membrane [75, 76]. This implies that microdomains in different regions of the cell consist of specific compositions of ion channels. It is, therefore, not surprising that Ca^{2+} signals generated from these distinct microdomains display unequivocal biophysical properties (activation and regression kinetics, frequency, amplitude of the oscillations, spatial and temporal profile as well as the speed of the Ca^{2+} wave, etc.), and can differentially regulate downstream processes. In support of this notion, a recent study has shown that although Orai1, STIM1 and TRPC channels are all expressed in salivary gland

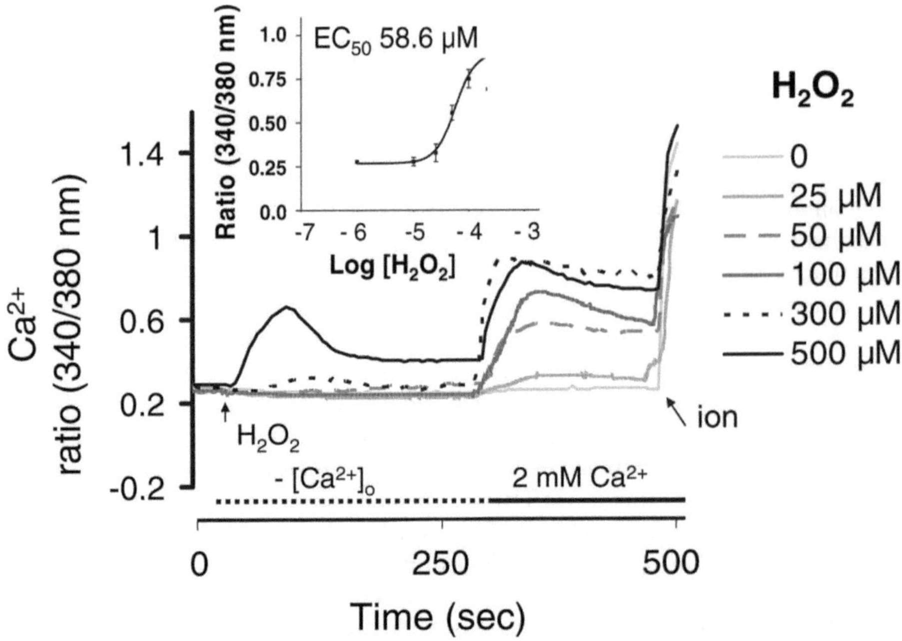

Fig. 2.6 Concentration-dependent effects of H_2O_2 on Ca^{2+} entry. HPAE cells in culture were loaded with fura-2, washed, and transferred to Ca^{2+}-free medium. H_2O_2 (0–500 μmol/L) was added at the *arrow* and $CaCl_2$ (2.0 mmol/L) was repleted at the fifth minute; the resulting Ca^{2+}-repletion transient reflects Ca^{2+} entry. The Ca^{2+} ionophore ionomycin (ion) was added at the end of the experimental recordings for calibration purposes. Each tracing is the average response of 60–99 cells in HPAE monolayers. The abscissa indicates time in seconds; the ordinate, relative $[Ca^{2+}]_i$ level. Experiments were repeated 3–5 times with similar results. The *inset* displays the dose–response curve of best fit for the calcium-repletion transient (EC_{50}, 58.6 μmol/L). The data points are mean values (n = 3 per point), and the bars indicate ± SEM. At >300 μmol/L, H_2O_2 mobilized stored intracellular Ca^{2+}. [Adapted from Hecquet et al. [48]]

acinar cells and associated with SOCE activity, the Ca^{2+} signal generated from these distinct channels can be differentially translated into diverse downstream processes [77]. Although the subcellular localization of these channels is not well characterized in ECs, the same rules are likely to apply. However, further studies are necessary to investigate the assembly and compartmentalization of these major Ca^{2+} signaling components in ECs, and determine how they control vascular endothelial permeability in a channel-specific manner.

Endothelium-Dependent Vascular Tone Control

Vascular tone, the contractile activity of vascular smooth muscle cells in the walls of small arteries and arterioles, is the major determinant of the resistance to blood flow through the circulation. Thus, vascular tone plays an important role in the

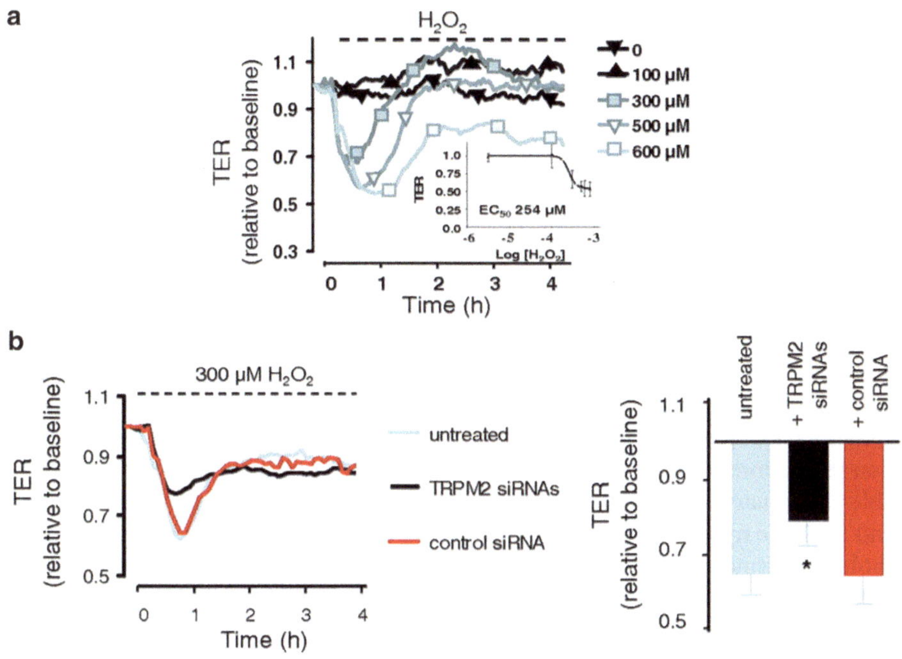

Fig. 2.7 H_2O_2-induced increase in endothelial barrier permeability depends on TRPM2 expression. (**a**) Concentration-dependent action of H_2O_2 on endothelial barrier function. HPAE cells were grown to confluence on gold microelectrodes, the cells were treated with H_2O_2 (concentration indicated), and TER was followed for 4 h. Each tracing is the average response of four wells. Experiments were repeated three times with similar results. The abscissa indicates time in hours; the ordinate, normalized resistance (relative to basal value). The *inset* shows the corresponding dose–response curve (n = 12; bars, ±SEM). H_2O_2 (0–600 μmol/L) caused a rapid, dose-dependent decrease in TER with an EC_{50} of 254 μmol/L. At <500 μmol/L, H_2O_2 effects were transitory. (**b**) TER decrease on H_2O_2 exposure (300 μmol/L). *Left*, Note that TRPM2 silencing inhibits H_2O_2 responses relative to the untreated group (no transfection) or the negative control group (scrambled siRNA; n = 4 per group). The abscissa indicates time in hours; the ordinate, normalized resistance. *Right*, Mean value (±SEM) of peak TER responses to H_2O_2 (n = 12). Experiments were repeated three times with similar results. H_2O_2-induced TER decrease was significantly attenuated in cells transfected by TRPM2-specific siRNAs compared with untreated control or negative control group transfected with a scrambled siRNA. [Adapted from Hecquet et al. [48]]

regulation of blood pressure and the distribution of blood flow between and within the tissues and organs of the body. Vascular endothelium plays a key role in controlling vascular tone by secreting a variety of endothelium-derived relaxing factors such as nitric oxide (NO), and prostacyclin (PGI2) [78–80]. Various circulating vasoactive agents (e.g. bradykinin, ATP, substance P and acetylcholine) can bind to EC surface GPCRs and lead to elevation of EC $[Ca^{2+}]_i$ and consequently activate NO synthase [81]. As noted above, TRPC channels play a critical role in agonist-induced Ca^{2+} entry in ECs. Several lines of evidence suggest that TRPC1, TRPC3 and TRPC4 channels are required for endothelium-dependent vascular tone modulation. Specifically, it has been shown that TRPC1 is responsible for bradykinin-mediated Ca^{2+} permeable currents in human mesenteric arteries ECs [22]. It was also demonstrated that in a bovine pulmonary artery EC line, overexpression of TRPC3 could enhance Ca^{2+} entry in response to ATP and bradykinin stimulation [82]. Furthermore, in primary aortic ECs isolated from TRPC4−/− mice, acetylcholine-induced Ca^{2+} entry was significantly reduced compare to WT, leading to suppression of endothelium-dependent NO-mediated vasorelaxation of blood vessels [83]. Together, these studies indicate that there is a direct functional link between endothelial TRPC channels and vascular tone control.

However, in addition to TRPC channels, other TRP isoforms may also contribute to endothelium dependent vasorelaxation. Endocannabinoids (AEA) and 2-AG as well as the 5′,6′-epoxyeicosatrienoic acid (EET) are potent vasorelaxants [84]. The Nilius group first showed that TRPV4 channels are a molecular target for these vasorelaxants, and could be directly activated and generate Ca^{2+}-permeable channel activity in ECs, thus suggesting a potential role for TRPV4 in vascular tone modulation [85]. More recently, Sonkusare et al. demonstrated that TRPV4 channels were able to mediate local calcium sparks that activate Ca^{2+} sensitive K^+ channels (BK_{Ca}) and facilitate hyperpolarization, and induced maximal dilation of resistance arteries [86]. Furthermore, it has also been shown that TRPV4 forms heteromeric channels with TRPC1 in vascular ECs and contributes to SOCE and NO synthesis [19]. These data further support the notion that other TRP isoforms may also be involved in the regulation of vascular function. In addition to TRP channel, other Ca^{2+} entry mechanisms in ECs may also contribute to vasorelaxation. For example, endothelial cell CNG channels have been shown to play an important role in endothelium-dependent vascular dilation in response to a number of cAMP-elevating agents including adenosine, adrenaline and ATP [87–89].

Vascular Remodeling, Migration and Angiogenesis

Vascular endothelial growth factor (VEGF), basic fibroblast growth factor (bFGF) and platelet-derived growth factor (PDGF) promote angiogenesis by enhancing EC proliferation and migration. These angiogenic factors can enhance the increase in intracellular Ca^{2+} concentration in ECs through the activation of plasma membrane receptors and the release of Ca^{2+} from the endoplasmic reticulum, which leads to activation of plasma membrane Ca^{2+} channels via the SOCE mechanism. In addition, receptor-operated Ca^{2+} entry (ROCE) that does not involve depletion of the ER Ca^{2+} store has also been shown to play an important role. Several studies have highlighted the importance of TRP channel-mediated Ca^{2+} signaling in the angiogenesis process [90]. TRP channels may play a major role in the initiation of vascular remodeling through two different mechanisms. First, proliferative factors may activate TRP channels and mediated TRP-dependent Ca^{2+} entry, thereby modulating downstream signal transduction pathways and leading to vascular remodeling of ECs. Second, Ca^{2+} influx through TRP channels may stimulate ECs to produce and release proliferative factors such as VEGF and PDGF, which subsequently impact the growth, migration and differentiation of the underlying smooth muscle cells, which is critical for arterial remodeling. The interplay of these two distinct signaling pathways allows control of angiogenesis in a vascular bed specific manner. Data generated by our group as well as by others have provided a detailed description of the mechanism of VEGF-stimulated Ca^{2+} entry in human microvascular ECs (HMVECs) [63, 91, 92]. VEGF binding to the VEGF Receptor 2 (VEGFR2) facilitates tyrosine phosphorylation of the receptor, leading to the generation of IP_3 and diacylglycerol (DAG), which in turn results in the activation of TRPC3 and TRPC6 via a store-independent mechanism [93, 94]. It has been reported that TRPC6 is required for the VEGF-mediated increase in cytosolic Ca^{2+} and the subsequent downstream signaling that leads to processes associated with angiogenesis [95, 96]. Overexpression of a dominant negative TRPC6 construct in HMVECs inhibited the VEGF-mediated increases EC $[Ca^{2+}]_i$, migration, sprouting, and its proliferation in matrigel. Conversely, overexpression of a wild-type TRPC6 construct increased the proliferation and migration of HMVECs [96]. The requirement for TRPC6 in angiogenesis was further explored in another study that demonstrated that inhibition of TRPC6 in HUVECs resulted in cell cycle arrestment at the G2/M phase and suppressed VEGF-induced HUVEC proliferation and tube formation [95]. Furthermore, inhibition of TRPCs abolished VEGF-, but not FGF-induced angiogenesis in the chick embryo chorioallantoic membrane. Together, these results suggest that TRPC6 plays a critical role in VEGF-mediated angiogenesis.

Hypoxia is another major factor that promotes angiogenesis. Hypoxic conditions induce pulmonary artery smooth muscle cell (PASMC) proliferation by promoting transcriptional activation of genes encoding vasoactive agonists such as endothelin-1 (ET-1), and mitogens such as the VEGF and PDGF growth factors. Hypoxia upregulates the mRNA and protein expression of TRPC4 in ECs resulting in elevated SOCE together with increased synthesis and secretion of mitogenic and vasoactive factors. This suggests that TRPC4 may play an important role in the development of hypoxia-mediated pulmonary vascular remodeling [97].

Following the identification of Orai1 as the pore forming subunit of the I_{CRAC} channel, its role in angiogenesis has been extensively studied. As noted above, it has been shown that VEGF, the primary growth factor that drives EC migration, proliferation and angiogenesis, can evoke Ca^{2+} signaling through the SOCE mechanism. Notably, it has been shown that VEGF-mediated Ca^{2+} entry in HUVECs was attenuated by knockdown of Orai1 using specific siRNA, by its inhibition using a dominant negative Orai1 mutant, and by application of a pharmacological inhibitor. It has also been shown that Orai1 is necessary for HUVEC proliferation [44] as well as *in vitro* tube formation [43]. These results were challenged by Antigny et al. [98] who showed that in a HUVEC cell line (EA.hy926), TRPC3, TRPC4 and TRPC5 generated spontaneous Ca^{2+} oscillation upon plating on matrigel and that Ca^{2+} entry through these channels was necessary for tubulogenesis. In contrast, knockdown of Orai1 and STIM1 had no effect. Interestingly, the only phenotype associated with Orai1 knockdown in this cell type was reduced proliferation, which is similar to the effect of knockdown of other TRPC channels. Antigny et al also studied primary HUVECs using the same approach and demonstrated the involvement of STIM1, TRPC1, and TRPC4, but not Orai1 in endothelial tube formation [98]. The requirement of STIM1 was further corroborated in another study that showed that suppression of STIM1 inhibited angiogenesis *in vivo* [99]. In a different study using endothelial progenitor cells (EPCs), it was shown that SOCE was associated with Orai1, STIM1 and TRPC1 activity. Suppression of SOCE using the SOCE inhibitor BTP-2 or by chelating intracellular Ca^{2+} using BAPTA resulted in the inhibition of EPC proliferation and tubulogenesis [100]. These data collectively demonstrated the requirement for SOCE in angiogenetic processes. The discrepancy in the involvement of Orai1 in vascular remodeling may be due to the use of HUVECs from different sources or due to differences in experimental conditions (e.g. VEGF concentration). In addition, ECs generated from various vascular beds may display different functional phenotypes as well as channel expression profiles due to heterogeneity [101]. Together, these factors may potentially contribute to the inconsistency in the findings. Other than the contributions of SOCE and TRPC channels to the angiogenic process, it has also been suggested that other TRP channels may play an important role in this process. In particular, the potential importance of TRPM6 and TRPM7 channels, and especially of TRPM7 in regulating angiogenesis and vascular remodeling has been proposed [102]. The underlying rationale in these cases is that both TRPM6 and TRPM7 are highly permeable to Mg^{2+} [103], and Mg^{2+} homeostasis has been directly linked to EC proliferation [104, 105].

ECs derived from human breast carcinomas (BTEC) exhibit enhanced angiogenic properties and are thus a useful tool for studying the mechanisms involved in vascular remodeling. Arachidonic acid (AA), a lipid second messenger released by different pro-angiogenic factors (e.g. bFGF and VEGF), is actively involved in the early stages of the angiogenic process *in vitro*, and mediates its effects by activating Ca^{2+} entry via Ca^{2+}-permeable plasma membrane channels [106]. It has been demonstrated that the expression of endogenous TRPV4 is significantly higher in BTECs compared with 'normal' ECs (HMVEC) [107]. Moreover, knockdown of TRPV4 using siRNA resulted in decreased $[Ca^{2+}]_i$ responses and a complete inhibition of AA-induced migration of BTECs. These data suggest that AA regulates angiogenesis by inducing remodeling of the cytoskeleton and stimulating TRPV4 channel activity, both of which are critical for tumor derived EC migration.

Immune Function and Inflammation

The endothelium, positioned at the interface between blood and tissue, is equipped to react quickly to local changes in biological needs caused by trauma or inflammation. Under injury or infection, circulating inflammatory mediators (e.g. endotoxins, cellular content released by damaged cells) can activate ECs and significantly modify their normal physiological functions. This may result in increased leakiness, modification of the vascular tone, enhanced procoagulant activity, and formation of new vessels [9]. In addition, ECs can also directly contribute to the immune response by expressing cell surface-molecules that direct the trafficking of circulating blood cells. These cell-adhesion molecules determine the homing and migration of leucocytes into specific organs that are under stress or inflammatory conditions. ECs express a variety of pattern recognition receptors [108] that recognize pathogen-associated molecular patterns (PAMPs). Among these, Toll-like receptors (TLRs) are surface molecules that trigger signals resulting in proinflammatory gene expression, leukocyte chemotaxis, phagocytosis, cytotoxicity and activation of adaptive immune responses.

The immune response is tightly regulated by Ca^{2+} signaling. Indeed, the I_{CRAC} channel and Orai1 were discovered during the course of studying T lymphocyte function [35, 109]. SOCE, TRP, Orai1 and STIM1-mediated Ca^{2+} entry have been closely associated with adaptive immunity, and the molecular basis has been described in detail in numerous studies [32]. Although there are considerably less data describing the role of Ca^{2+} signaling in innate immunity, particularly the involvement of Ca^{2+} signaling in EC-mediated immune response, there is a strong indication that SOCE are involved in the pathogenesis of many inflammatory diseases [110]. In addition, a variety of inflammatory substances cause an increase in intracellular Ca^{2+}, which then leads to permeability changes in the microvascular endothelium, and hence, to tissue edema [11]. Data generated by our group as well as others support the notion that Ca^{2+} entry in ECs can modulate immunological function and the postulated involvement of ECs in contributing to the host-defense mechanism during pathological conditions.

TLR4 expression has been demonstrated in various ECs, and was found to be significantly upregulated under inflammatory conditions [111]. An increasing number of studies have indicated that TLR-mediated immune response is associated with Ca^{2+} signaling. Tauseef et al. [112] have shown that endotoxin (lipopolysaccharide; LPS) induces DAG generation in a Toll-like receptor 4 (TLR4)-dependent manner, and triggers Ca^{2+} increase through the TRPC6 channel. Importantly, the authors showed that Ca^{2+} entry through TRPC6 participates in mediating both lung vascular barrier disruption and inflammation induced by endotoxin [112].

The role of STIM1 in TLR signaling was reported in another study [113]. By using EC-specific STIM1-knockout mice, the authors showed that LPS-mediated Ca^{2+} oscillations are ablated in STIM1-deficient ECs. ECs lacking STIM1 failed to trigger SOCE and NFAT nuclear translocation as well as receptor-interacting protein 3 (RIP3)-dependent cell death. Interestingly, LPS-induced vascular permeability changes were reduced in EC-specific STIM1−/− mice. Furthermore, application of a CRAC channel inhibitor in WT animals also halted LPS-induced vascular leakage and pulmonary edema. Taken together, these results demonstrate that Ca^{2+} signaling plays a critical role in promoting vascular barrier dysfunction during infection, and indicate that the SOCE machinery may provide crucial therapeutic targets to limit edema and acute lung injury (ALI) [113]. The role of STIM1-dependent signaling in EC-dependent immune response was further explored recently by using a mouse sepsis model [114]. It was shown that systemic administration of bacterial endotoxin (LPS) enhances STIM1 expression in ECs. LPS also induced STIM1 mRNA and protein expression in human and mouse lung ECs. The induced STIM1 expression was associated with augmented SOCE as well as a permeability increase in both *in vitro* and *in vivo* models. Remarkably, inhibition of either NF-κB or p38 MAPK activation by pharmacological agents or siRNA prevented LPS-induced STIM1 expression and increased SOCE in ECs. These data suggest that TLR4-signaling mediated through cooperative signaling of NF-κB and AP1 (via p38α) amplifies STIM1 expression in ECs, and thereby contributes to the lung vascular hyperpermeability response during sepsis [114].

In addition to endotoxin, histamine and chemokines are important immunomodulators involved in allergic reactions, leukocyte homing and inflammatory responses [111]. Notably, both histamine and chemokines can bind to specific GPCRs and induce Ca^{2+} signaling through SOCE mechanism [115, 116]. It has been shown that the CXC chemokine stromal cell-derived factor 1 (SDF-1), also known as CXCL12, can bind to CXCR4 in ECs, leading to an increase in $[Ca^{2+}]_i$, thus initiating signals related to chemotaxis, cell survival, proliferation and gene transcription [115]. Histamine has also been shown to be able to facilitate SOCE in HUVEC, and is sensitive to SOCE inhibitor blockage, to knockdown of Orai1 or STIM1, and to dominant negative expression of Orai1 mutant. It was therefore suggested that histamine-mediated Ca^{2+} entry through SOCE in HUVECs can facilitate NFAT activation as well as interleukin 8 production [116]. These results highlight the central role of Ca^{2+} signaling in general, and of Ca^{2+} entry through TRPC, STIM1 and Orai1 in particular, in mediating Ca^{2+} mobilization linked to inflammatory signaling of ECs upon inflammatory mediator stimulation.

Conclusion and Future Perspective

ECs express a variety of Ca^{2+} channels, which are involved in a variety of physiological and pathophysiological processes (see Fig. 2.8). Despite remarkable progress in the understanding of the role of Ca^{2+} signaling in regulating EC function, numerous aspects of the activation of Ca^{2+} channels remain unclear. In particular, the involvement of the newly identified SOCE core components, the Orai channels and the STIM proteins, in EC function needs to be further elucidated. SOCE has been linked to multiple physiological processes in ECs. Yet, the mechanisms that underlie the differential contribution of SOCE generated through specific microdomains to distinct cellular function have only recently been identified [77]. Elucidating the roles of downstream molecules associated with specific ion channels is essential for refining our understanding of vasculature response to environmental stimuli. It is important to note that the endothelial lining of blood vessels shows remarkable heterogeneity in structure and function in time and space, and in health and disease. An interesting question is whether the

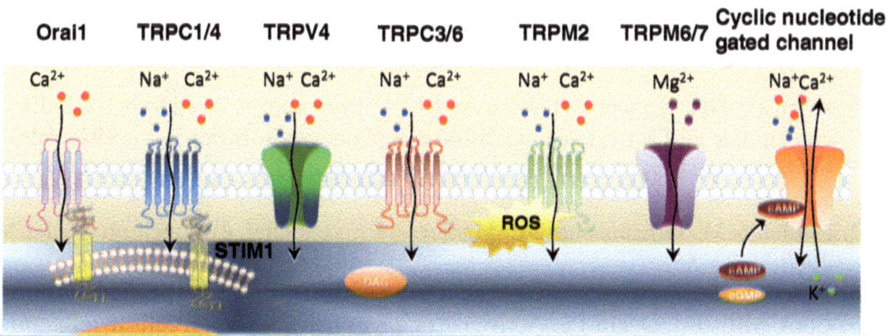

	Orai1, STIM1	TRPC1/4	TRPV4	TRPC3/6	TRPM2	TRPM6/7	CNG
Operating mechanism	SOCE	SOCE	SOCE/AA	ROCE, DAG	ROS, ADP-ribose	ROCE	cGMP, cAMP
Permeability	VEGF, Tg (34. 36, 60) LPS (104, 105)	Thrombin, VEGF, TNFα (15, 34, 45, 51-57)		Bradykinin Thrombin (50, 59) LPS (103)	H_2O_2 (65,67)		
Vascular tone		Bradykinin (16), acetylcholine (73), Tg (77)	AEA, EET , GSK, Tg (75, 76, 77)	ATP, Bradykinin (72)			ATP, Adenosine, Epinephrine, Thromboxane A2 (78-80)
Angiogenesis	VEGF (35, 36, 88, 89, 90)	Hypoxia, ECM (87,88, 90)	bFGF, VEGF, AA (96, 97)	VEGF, ECM (86, 88)		Mg^{2+} homeostasis (93-95)	
Immune response	LPS (104, 105), Histamine (108)			LPS			

Fig. 2.8 Schematic representation of Ca^{2+} channels and their functional role in endothelial cells. A variety of Ca^{2+} permeable ion channels express in endothelial cells where they regulate different physiological processes such as increase in permeability, vascular tone control, vessel growth and angiogenesis as summarized in the Table. The *numbers* in *parenthesis* indicate the respective reference. Abbreviations: *SOCE* Store-operated Ca^{2+} entry, *ROCE* receptor-operated calcium entry, *AA* Arachidonic acid, *DAG* Diacylglycerol, *ADP-ribose* adenosine diphosphoribose, *VEGF* vascular endothelial growth factor, *Tg* Thapsigargin, *LPS* lipopolysaccharide, *bFGF* basic fibroblast growth factor, *ECM* extracellular matrix, *AEA* Endocannabinoids, *EET* 5′,6′-epoxyeicosatrienoic acid

phenotypic and functional variation of ECs from different vascular beds is due to their distinct ion channel expression profile. Emerging data support the notion that Ca^{2+} signaling can be generated in a cell type and tissue dependent manner through distinctive expression patterns of Ca^{2+} channels. The majority of the studies that addressed this aspect have been carried out in cultured EC lines or in primary cultured cells *in vitro* only. Furthermore, *in situ* detection of ion channels in vessels across tissues has only been convincingly demonstrated in a few studies. It is thus of crucial importance to elucidate both the candidate channels architecture across different vascular tissues and their physiological function *in vivo*. Indeed, understanding the molecular basis for phenotypic heterogeneity is a prerequisite for developing vascular bed-specific therapies for modulating specific EC responses in a clinical setting.

Acknowledgement We thank Kit Man Tsang for manuscript editing and Katrian Cheng for support. This work was supported by NIH grant HL077806 to A.B.M.

References

1. Augustin HG, Kozian DH, Johnson RC. Differentiation of endothelial cells: analysis of the constitutive and activated endothelial cell phenotypes. Bioessays. 1994;16:901–6.
2. Fishman AP. Endothelium: a distributed organ of diverse capabilities. Ann N Y Acad Sci. 1982;401:1–8.
3. Muller MM, Griesmacher A. Markers of endothelial dysfunction. Clin Chem Lab Med. 2000;38:77–85.
4. Jaffe EA. Cell biology of endothelial cells. Hum Pathol. 1987;18:234–9.
5. Cook-Mills JM, Deem TL. Active participation of endothelial cells in inflammation. J Leukoc Biol. 2005;77:487–95.
6. Santiago-Delpin EA. The endothelium and early immune activation: new perspective and interactions. Transplant Proc. 2004;36:1709–13.
7. Biedermann BC. Vascular endothelium: checkpoint for inflammation and immunity. News Physiol Sci. 2001;16:84–8.
8. Lum H, Malik AB. Regulation of vascular endothelial barrier function. Am J Physiol. 1994;267:L223–41.
9. Cines DB, et al. Endothelial cells in physiology and in the pathophysiology of vascular disorders. Blood. 1998;91:3527–61.
10. Yao X, Garland CJ. Recent developments in vascular endothelial cell transient receptor potential channels. Circ Res. 2005;97:853–63.
11. Tiruppathi C, Ahmmed GU, Vogel SM, Malik AB. Ca2+ signaling, TRP channels, and endothelial permeability. Microcirculation. 2006;13:693–708.
12. Ahmmed GU, Malik AB. Functional role of TRPC channels in the regulation of endothelial permeability. Pflugers Arch. 2005;451:131–42.
13. Putney Jr JW. A model for receptor-regulated calcium entry. Cell Calcium. 1986;7:1–12.
14. Parekh AB, Putney Jr JW. Store-operated calcium channels. Physiol Rev. 2005;85:757–810.
15. Amiri H, Schultz G, Schaefer M. FRET-based analysis of TRPC subunit stoichiometry. Cell Calcium. 2003;33:463–70.
16. Montell C. The TRP superfamily of cation channels. Sci STKE. 2005;2005:re3.
17. Yuan JP, Zeng W, Huang GN, Worley PF, Muallem S. STIM1 heteromultimerizes TRPC channels to determine their function as store-operated channels. Nat Cell Biol. 2007;9:636–45.
18. Du J, et al. TRPV4, TRPC1, and TRPP2 assemble to form a flow-sensitive heteromeric channel. FASEB J. 2014;28:4677–85.
19. Ma X, et al. Heteromeric TRPV4-C1 channels contribute to store-operated Ca(2+) entry in vascular endothelial cells. Cell Calcium. 2011;50:502–9.
20. Garcia RL, Schilling WP. Differential expression of mammalian TRP homologues across tissues and cell lines. Biochem Biophys Res Commun. 1997;239:279–83.
21. Paria BC, et al. Tumor necrosis factor-alpha-induced TRPC1 expression amplifies store-operated Ca2+ influx and endothelial permeability. Am J Physiol Lung Cell Mol Physiol. 2004;287:L1303–13.
22. Kohler R, et al. Expression of ryanodine receptor type 3 and TRP channels in endothelial cells: comparison of in situ and cultured human endothelial cells. Cardiovasc Res. 2001;51:160–8.
23. Yip H, et al. Expression of TRPC homologs in endothelial cells and smooth muscle layers of human arteries. Histochem Cell Biol. 2004;122:553–61.
24. Liou J, et al. STIM is a Ca2+ sensor essential for Ca2+-store-depletion-triggered Ca2+ influx. Curr Biol. 2005;15:1235–41.
25. Roos J, et al. STIM1, an essential and conserved component of store-operated Ca2+ channel function. J Cell Biol. 2005;169:435–45.

26. Zhang SL, et al. STIM1 is a Ca2+ sensor that activates CRAC channels and migrates from the Ca2+ store to the plasma membrane. Nature. 2005;437:902–5.
27. Yuan JP, et al. SOAR and the polybasic STIM1 domains gate and regulate Orai channels. Nat Cell Biol. 2009;11:337–43.
28. Park CY, et al. STIM1 clusters and activates CRAC channels via direct binding of a cytosolic domain to Orai1. Cell. 2009;136:876–90.
29. Kawasaki T, Lange I, Feske S. A minimal regulatory domain in the C terminus of STIM1 binds to and activates ORAI1 CRAC channels. Biochem Biophys Res Commun. 2009;385:49–54.
30. Liou J, Fivaz M, Inoue T, Meyer T. Live-cell imaging reveals sequential oligomerization and local plasma membrane targeting of stromal interaction molecule 1 after Ca2+ store depletion. Proc Natl Acad Sci U S A. 2007;104:9301–6.
31. Zeng W, et al. STIM1 gates TRPC channels, but not Orai1, by electrostatic interaction. Mol Cell. 2008;32:439–48.
32. Hogan PG, Lewis RS, Rao A. Molecular basis of calcium signaling in lymphocytes: STIM and ORAI. Annu Rev Immunol. 2010;28:491–533.
33. Stathopulos PB, Ikura M. Structural aspects of calcium-release activated calcium channel function. Channels (Austin). 2013;7:344–53.
34. Parekh AB, Penner R. Store depletion and calcium influx. Physiol Rev. 1997;77:901–30.
35. Feske S, et al. A mutation in Orai1 causes immune deficiency by abrogating CRAC channel function. Nature. 2006;441:179–85.
36. Vig M, et al. CRACM1 is a plasma membrane protein essential for store-operated Ca2+ entry. Science. 2006;312:1220–3.
37. Liao Y, et al. Orai proteins interact with TRPC channels and confer responsiveness to store depletion. Proc Natl Acad Sci U S A. 2007;104:4682–7.
38. Ong HL, et al. Dynamic assembly of TRPC1-STIM1-Orai1 ternary complex is involved in store-operated calcium influx. Evidence for similarities in store-operated and calcium release-activated calcium channel components. J Biol Chem. 2007;282:9105–16.
39. Prakriya M, et al. Orai1 is an essential pore subunit of the CRAC channel. Nature. 2006;443:230–3.
40. Faehling M, et al. Essential role of calcium in vascular endothelial growth factor A-induced signaling: mechanism of the antiangiogenic effect of carboxyamidotriazole. FASEB J. 2002;16:1805–7.
41. Fasolato C, Nilius B. Store depletion triggers the calcium release-activated calcium current (ICRAC) in macrovascular endothelial cells: a comparison with Jurkat and embryonic kidney cell lines. Pflugers Arch. 1998;436:69–74.
42. Sundivakkam PC, et al. The Ca(2+) sensor stromal interaction molecule 1 (STIM1) is necessary and sufficient for the store-operated Ca(2+) entry function of transient receptor potential canonical (TRPC) 1 and 4 channels in endothelial cells. Mol Pharmacol. 2012;81:510–26.
43. Li J, et al. Orai1 and CRAC channel dependence of VEGF-activated Ca2+ entry and endothelial tube formation. Circ Res. 2011;108:1190–8.
44. Abdullaev IF, et al. Stim1 and Orai1 mediate CRAC currents and store-operated calcium entry important for endothelial cell proliferation. Circ Res. 2008;103:1289–99.
45. Sayner SL, Balczon R, Frank DW, Cooper DM, Stevens T. Filamin A is a phosphorylation target of membrane but not cytosolic adenylyl cyclase activity. Am J Physiol Lung Cell Mol Physiol. 2011;301:L117–24.
46. Dietrich A, Gudermann T. Another TRP to endothelial dysfunction: TRPM2 and endothelial permeability. Circ Res. 2008;102:275–7.
47. Groschner K, Rosker C, Lukas M. Role of TRP channels in oxidative stress. Novartis Found Symp. 2004;258:222–30. discussion 231–225, 263–226.
48. Hecquet CM, Ahmmed GU, Vogel SM, Malik AB. Role of TRPM2 channel in mediating H_2O_2-induced Ca2+ entry and endothelial hyperpermeability. Circ Res. 2008;102:347–55.

49. Hecquet CM, Malik AB. Role of H(2)O(2)-activated TRPM2 calcium channel in oxidant-induced endothelial injury. Thromb Haemost. 2009;101:619–25.
50. Kwan HY, Huang Y, Yao X. TRP channels in endothelial function and dysfunction. Biochim Biophys Acta. 2007;1772:907–14.
51. Inoue K, Xiong ZG. Silencing TRPM7 promotes growth/proliferation and nitric oxide production of vascular endothelial cells via the ERK pathway. Cardiovasc Res. 2009;83:547–57.
52. Sun H, et al. Role of TRPM7 channels in hyperglycemia-mediated injury of vascular endothelial cells. PLoS One. 2013;8:e79540.
53. Sarmiento D, et al. Increases in reactive oxygen species enhance vascular endothelial cell migration through a mechanism dependent on the transient receptor potential melastatin 4 ion channel. Microvasc Res. 2014;98:187–96.
54. Kaupp UB, Seifert R. Cyclic nucleotide-gated ion channels. Physiol Rev. 2002;82:769–824.
55. Cheng KT, Chan FL, Huang Y, Chan WY, Yao X. Expression of olfactory-type cyclic nucleotide-gated channel (CNGA2) in vascular tissues. Histochem Cell Biol. 2003;120:475–81.
56. Yao X, Leung PS, Kwan HY, Wong TP, Fong MW. Rod-type cyclic nucleotide-gated cation channel is expressed in vascular endothelium and vascular smooth muscle cells. Cardiovasc Res. 1999;41:282–90.
57. Zhang J, Xia SL, Block ER, Patel JM. NO upregulation of a cyclic nucleotide-gated channel contributes to calcium elevation in endothelial cells. Am J Physiol Cell Physiol. 2002;283:C1080–9.
58. Lum H, Del Vecchio PJ, Schneider AS, Goligorsky MS, Malik AB. Calcium dependence of the thrombin-induced increase in endothelial albumin permeability. J Appl Physiol (1985). 1989;66:1471–6.
59. Ellis CA, et al. Thrombin induces proteinase-activated receptor-1 gene expression in endothelial cells via activation of Gi-linked Ras/mitogen-activated protein kinase pathway. J Biol Chem. 1999;274:13718–27.
60. Vogel SM, et al. Abrogation of thrombin-induced increase in pulmonary microvascular permeability in PAR-1 knockout mice. Physiol Genomics. 2000;4:137–45.
61. Singh I, et al. Galphaq-TRPC6-mediated Ca2+ entry induces RhoA activation and resultant endothelial cell shape change in response to thrombin. J Biol Chem. 2007;282:7833–43.
62. Tiruppathi C, et al. Impairment of store-operated Ca2+ entry in TRPC4(−/−) mice interferes with increase in lung microvascular permeability. Circ Res. 2002;91:70–6.
63. Jho D, et al. Angiopoietin-1 opposes VEGF-induced increase in endothelial permeability by inhibiting TRPC1-dependent Ca2 influx. Circ Res. 2005;96:1282–90.
64. Mehta D, et al. RhoA interaction with inositol 1,4,5-trisphosphate receptor and transient receptor potential channel-1 regulates Ca2+ entry. Role in signaling increased endothelial permeability. J Biol Chem. 2003;278:33492–500.
65. Paria BC, et al. Ca2+ influx induced by protease-activated receptor-1 activates a feed-forward mechanism of TRPC1 expression via nuclear factor-kappaB activation in endothelial cells. J Biol Chem. 2006;281:20715–27.
66. Moore TM, et al. Store-operated calcium entry promotes shape change in pulmonary endothelial cells expressing Trp1. Am J Physiol. 1998;275:L574–82.
67. Cioffi DL, et al. Activation of the endothelial store-operated ISOC Ca2+ channel requires interaction of protein 4.1 with TRPC4. Circ Res. 2005;97:1164–72.
68. Odell AF, Van Helden DF, Scott JL. The spectrin cytoskeleton influences the surface expression and activation of human transient receptor potential channel 4 channels. J Biol Chem. 2008;283:4395–407.
69. Hofmann T, et al. Direct activation of human TRPC6 and TRPC3 channels by diacylglycerol. Nature. 1999;397:259–63.
70. Kini V, Chavez A, Mehta D. A new role for PTEN in regulating transient receptor potential canonical channel 6-mediated Ca2+ entry, endothelial permeability, and angiogenesis. J Biol Chem. 2010;285:33082–91.

71. Shinde AV, et al. STIM1 controls endothelial barrier function independently of Orai1 and Ca2+ entry. Sci Signal. 2012;6:ra18.
72. Salido GM, Sage SO, Rosado JA. TRPC channels and store-operated Ca(2+) entry. Biochim Biophys Acta. 2009;1793:223–30.
73. Zhang W, et al. A novel TRPM2 isoform inhibits calcium influx and susceptibility to cell death. J Biol Chem. 2003;278:16222–9.
74. Glass CA, Pocock TM, Curry FE, Bates DO. Cytosolic Ca2+ concentration and rate of increase of the cytosolic Ca2+ concentration in the regulation of vascular permeability in Rana in vivo. J Physiol. 2005;564:817–27.
75. Hong JH, et al. Polarized but differential localization and recruitment of STIM1, Orai1 and TRPC channels in secretory cells. Traffic. 2011;12:232–45.
76. Petersen OH, Tepikin AV. Polarized calcium signaling in exocrine gland cells. Annu Rev Physiol. 2008;70:273–99.
77. Cheng KT, Liu X, Ong HL, Swaim W, Ambudkar IS. Local Ca(2)+entry via Orai1 regulates plasma membrane recruitment of TRPC1 and controls cytosolic Ca(2)+signals required for specific cell functions. PLoS Biol. 2011;9:e1001025.
78. Fleming I, Busse R. NO: the primary EDRF. J Mol Cell Cardiol. 1999;31:5–14.
79. Feletou M, Vanhoutte PM. The alternative: EDHF. J Mol Cell Cardiol. 1999;31:15–22.
80. Drexler H, Hornig B. Endothelial dysfunction in human disease. J Mol Cell Cardiol. 1999;31:51–60.
81. Boulanger CM, Vanhoutte PM. G proteins and endothelium-dependent relaxations. J Vasc Res. 1997;34:175–85.
82. Kamouchi M, et al. Properties of heterologously expressed hTRP3 channels in bovine pulmonary artery endothelial cells. J Physiol. 1999;518(Pt 2):345–58.
83. Freichel M, et al. Lack of an endothelial store-operated Ca2+ current impairs agonist-dependent vasorelaxation in TRP4−/− mice. Nat Cell Biol. 2001;3:121–7.
84. Randall MD, Kendall DA. Anandamide and endothelium-derived hyperpolarizing factor act via a common vasorelaxant mechanism in rat mesentery. Eur J Pharmacol. 1998;346:51–3.
85. Watanabe H, et al. Anandamide and arachidonic acid use epoxyeicosatrienoic acids to activate TRPV4 channels. Nature. 2003;424:434–8.
86. Sonkusare SK, et al. Elementary Ca2+ signals through endothelial TRPV4 channels regulate vascular function. Science. 2012;336:597–601.
87. Cheng KT, et al. CNGA2 channels mediate adenosine-induced Ca2+ influx in vascular endothelial cells. Arterioscler Thromb Vasc Biol. 2008;28:913–8.
88. Shen B, et al. Epinephrine-induced Ca2+ influx in vascular endothelial cells is mediated by CNGA2 channels. J Mol Cell Cardiol. 2008;45:437–45.
89. Kwan HY, et al. CNGA2 contributes to ATP-induced noncapacitative Ca2+ influx in vascular endothelial cells. J Vasc Res. 2009;47:148–56.
90. Nilius B, Droogmans G. Ion channels and their functional role in vascular endothelium. Physiol Rev. 2001;81:1415–59.
91. Brock TA, Dvorak HF, Senger DR. Tumor-secreted vascular permeability factor increases cytosolic Ca2+ and von Willebrand factor release in human endothelial cells. Am J Pathol. 1991;138:213–21.
92. Faehling M, Koch ED, Raithel J, Trischler G, Waltenberger J. Vascular endothelial growth factor-A activates Ca2+ -activated K+ channels in human endothelial cells in culture. Int J Biochem Cell Biol. 2001;33:337–46.
93. Cheng HW, James AF, Foster RR, Hancox JC, Bates DO. VEGF activates receptor-operated cation channels in human microvascular endothelial cells. Arterioscler Thromb Vasc Biol. 2006;26:1768–76.
94. Pocock TM, Foster RR, Bates DO. Evidence of a role for TRPC channels in VEGF-mediated increased vascular permeability in vivo. Am J Physiol Heart Circ Physiol. 2004;286:H1015–26.

95. Ge R, et al. Critical role of TRPC6 channels in VEGF-mediated angiogenesis. Cancer Lett. 2009;283:43–51.
96. Hamdollah Zadeh MA, Glass CA, Magnussen A, Hancox JC, Bates DO. VEGF-mediated elevated intracellular calcium and angiogenesis in human microvascular endothelial cells in vitro are inhibited by dominant negative TRPC6. Microcirculation. 2008;15:605–14.
97. Fantozzi I, et al. Hypoxia increases AP-1 binding activity by enhancing capacitative Ca2+ entry in human pulmonary artery endothelial cells. Am J Physiol Lung Cell Mol Physiol. 2003;285:L1233–45.
98. Antigny F, Girardin N, Frieden M. Transient receptor potential canonical channels are required for in vitro endothelial tube formation. J Biol Chem. 2012;287:5917–27.
99. Chen YF, et al. Calcium store sensor stromal-interaction molecule 1-dependent signaling plays an important role in cervical cancer growth, migration, and angiogenesis. Proc Natl Acad Sci U S A. 2011;108:15225–30.
100. Lodola F, et al. Store-operated Ca2+ entry is remodelled and controls in vitro angiogenesis in endothelial progenitor cells isolated from tumoral patients. PLoS One. 2012;7:e42541.
101. Aird WC. Endothelial cell heterogeneity. Cold Spring Harb Perspect Med. 2012;2:a006429.
102. Touyz RM. Transient receptor potential melastatin 6 and 7 channels, magnesium transport, and vascular biology: implications in hypertension. Am J Physiol Heart Circ Physiol. 2008;294:H1103–18.
103. Owsianik G, Talavera K, Voets T, Nilius B. Permeation and selectivity of TRP channels. Annu Rev Physiol. 2006;68:685–717.
104. Wolf FI, Cittadini A. Magnesium in cell proliferation and differentiation. Front Biosci. 1999;4:D607–17.
105. Sgambato A, Wolf FI, Faraglia B, Cittadini A. Magnesium depletion causes growth inhibition, reduced expression of cyclin D1, and increased expression of P27Kip1 in normal but not in transformed mammary epithelial cells. J Cell Physiol. 1999;180:245–54.
106. Fiorio Pla A, et al. Arachidonic acid-induced Ca2+ entry is involved in early steps of tumor angiogenesis. Mol Cancer Res. 2008;6:535–45.
107. Fiorio Pla A, et al. TRPV4 mediates tumor-derived endothelial cell migration via arachidonic acid-activated actin remodeling. Oncogene. 2012;31:200–12.
108. Iwasaki A, Medzhitov R. Toll-like receptor control of the adaptive immune responses. Nat Immunol. 2004;5:987–95.
109. Hoth M, Penner R. Depletion of intracellular calcium stores activates a calcium current in mast cells. Nature. 1992;355:353–6.
110. Nilius B, Owsianik G. Transient receptor potential channelopathies. Pflugers Arch. 2010;460:437–50.
111. Danese S, Dejana E, Fiocchi C. Immune regulation by microvascular endothelial cells: directing innate and adaptive immunity, coagulation, and inflammation. J Immunol. 2007;178:6017–22.
112. Tauseef M, et al. TLR4 activation of TRPC6-dependent calcium signaling mediates endotoxin-induced lung vascular permeability and inflammation. J Exp Med. 2012;209:1953–68.
113. Gandhirajan RK, et al. Blockade of NOX2 and STIM1 signaling limits lipopolysaccharide-induced vascular inflammation. J Clin Invest. 2013;123:887–902.
114. DebRoy A, et al. Cooperative signaling via transcription factors NF-kappaB and AP1/c-Fos mediates endothelial cell STIM1 expression and hyperpermeability in response to endotoxin. J Biol Chem. 2014;289:24188–201.
115. Lombardi L, et al. Chemokine receptor CXCR4: role in gastrointestinal cancer. Crit Rev Oncol Hematol. 2013;88:696–705.
116. Zhou MH, et al. Stromal interaction molecule 1 (STIM1) and Orai1 mediate histamine-evoked calcium entry and nuclear factor of activated T-cells (NFAT) signaling in human umbilical vein endothelial cells. J Biol Chem. 2014;289:29446–56.

Chapter 3
Mitochondrial Depolarization in Endothelial and Other Vascular Cells and Their Role in the Regulation of Cerebral Vascular Tone

David W. Busija, Ibolya Rutkai, and Prasad V. Katakam

Abstract Mitochondrial initiated events of the diverse cells types comprising the neurovascular unit promote changes in cerebrovascular tone through multiple signaling pathways. Activation of the adenosine triphosphate (ATP)-dependent potassium channels on the inner mitochondrial membrane (mitoK_{ATP} channels) leads to mitochondrial depolarization as well as activation of cell-specific signaling mechanisms in endothelium, vascular smooth muscle (VSM), and perivascular and parenchymal nerves resulting in an integrated dilator response of cerebral arteries. Activation of mitoK_{ATP} channels relaxes VSM via generation of calcium sparks and subsequent downstream signaling mechanisms, and this relaxation can be augmented by nitric oxide (NO) produced by mitoK_{ATP} channel activation in endothelium and adjacent neurons. Some research suggests that calcium activated potassium channels may also be present in mitochondria (mitoK_{Ca} channels) and may affect cerebral vascular tone, but more research is needed to support this view. Pre-existing chronic conditions such as insulin resistance (IR) and/or diabetes impair mitoK_{ATP} channel-relaxation of cerebral arteries. Surprisingly, mitoK_{ATP} channel function after intense stress such as ischemia appears to be retained in large cerebral arteries despite generalized cerebral vascular dysfunction. Production of vasoactive factors following activation of mitochondria in response to physiological stimuli in one or more of the cells comprising the neurovascular unit may represent the elusive signaling link between metabolic rate and blood flow. In addition, our data indicate that mitoK_{ATP} channels represent an important, but underutilized target toward improving vascular dysfunction and decreasing brain injury in stroke patients.

Keywords ATP sensitive potassium channels • Calcium sparks • Sarcoplasmic reticulum • Calcium activated potassium channels • Insulin resistance • Strokes • Ischemia • Nitric oxide • Neurons • Vascular smooth muscle

---

D.W. Busija, Ph.D., M.D. [Hon] (✉) • I. Rutkai • P.V. Katakam
Department of Pharmacology, Tulane University School of Medicine,
1430 Tulane Avenue, #8683, New Orleans, LA, USA
e-mail: dbusija@tulane.edu

I. Levitan, A.M. Dopico (eds.), *Vascular Ion Channels in Physiology and Disease*, DOI 10.1007/978-3-319-29635-7_3

Introduction

The neurovascular axis/unit is an integrated system of vascular and parenchymal cells, including endothelium, vascular smooth muscle (VSM), astroglia, perivascular neurons, parenchymal neurons, pericytes, and microglia which contributes to the coupling of blood flow to both metabolic demand and to cerebral vascular responses to physiological and pathological stimuli [9, 13]. Recent, but limited, evidence indicates that mitochondria in each of these cell types are important elements in cerebral vascular control. The purpose of this review is to explore the evidence for an important role of mitochondrial-based mechanisms independent of ATP production on the tone of cerebral resistance vessels. Of particular interest are the potassium channels which regulate mitochondrial membrane potential. Specifically, we will examine how the selective targeting of mitochondrial ATP sensitive potassium (mitoK_{ATP}) channels in endothelium, VSM, and neurons leads to changes in cerebral vascular tone. This discussion is limited to the systemic circulation; mitochondrial mechanisms in the pulmonary circulation are very different [28]. We also will discuss the possible presence of other potassium channels such as calcium activated potassium (mitoK_{Ca}) channels in regulating mitochondrial membrane potential and affecting cerebral vascular tone. Additionally, we will show that even mild metabolic stress, such as occurs with insulin resistance (IR), a component of the metabolic syndrome, can dramatically affect mitochondrial related events in the cerebral vasculature. Lastly, we will present our recent unexpected findings which indicate that mitochondrial based dilator pathways are intact following ischemic stress, and targeting these mechanisms may benefit patients following strokes. Several excellent reviews on mitochondria in the peripheral endothelium [57, 89] or VSM [15, 73] have been published recently and emphasize the importance of this emerging research area. However, this is the first review to explore the integration of mitochondrial based signals ranging from multiple cell types to the final vascular diameter response in any regional circulation in normal and disease states. Previously, we have shown in neurons, astroglia, and cerebral vascular endothelium that activation of mitoK_{ATP} channels protects cells of the neurovascular unit against potentially lethal stresses such as anoxia or experimental stroke [10, 23–26, 69, 70, 83].

Mitochondrial Essentials

Mitochondria are double membrane organelles which generate chemical energy in the form of ATP which promotes the varied activities of the diverse cell types of the neurovascular unit [27, 47]. Typical morphological features of mitochondria in cells of the neurovascular unit are presented in Figs. 3.1, 3.2, and 3.3. We have shown morphological features of mitochondria in cultured cortical neurons previously [88]. The mitochondrial membrane potential ($\Delta\psi m$) across the inner membrane, which is normally maintained at −120 to −180 mV, provides the proton difference used to drive the synthesis of ATP by the electron transport chain. Rather than being

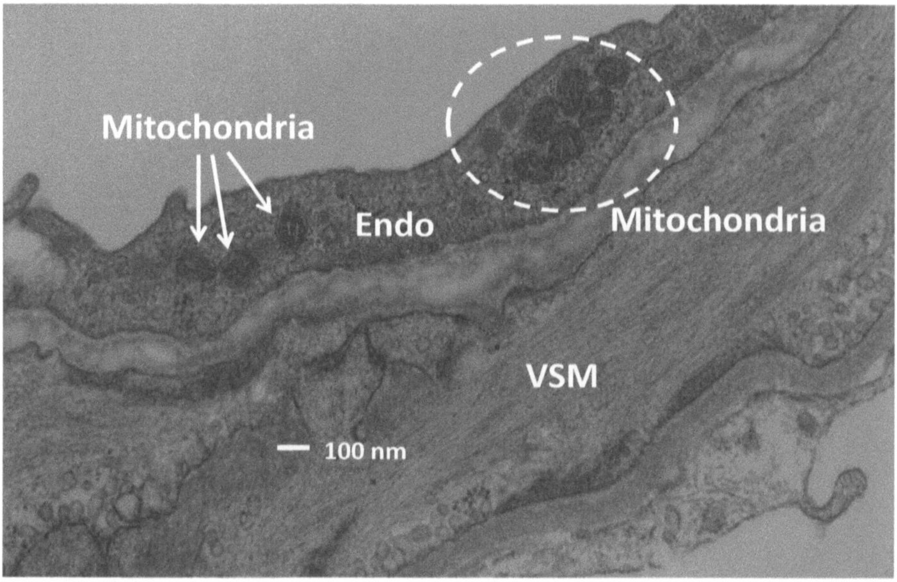

Fig. 3.1 Electron microscopy showing mitochondria in cerebral vascular endothelium (Endo). Mitochondria are located singly or in groups across the endothelial cells. Arteries were fixed in 2.5 % glutaraldehyde and post-fixed in 1 % osmium tetroxide and embedded in Epon812. Ultrathin sections (80–90 nm) were mounted on formvar-coated nickel grids (200 mesh), air dried, and stained with 4.7 % uranyl acetate and lead citrate (at 10 min and 2 min, respectively). The sections were put on grids and viewed using a FEI Tecnai BioTwin 120 keV TEM with a digital imaging setup (Wake Forest University Health Sciences, Winston-Salem, NC). *VSM* vascular smooth muscle

static entities, mitochondria are dynamic structures. Mitochondria can move from one location to another within cells, form networks with other mitochondria or cellular structures and undergo replication and fission/fusion to increase efficiency of ATP production in response to physiological and pathological stimuli [21, 27, 29, 30, 88, 93]. In some cell types, such as VSM, mitochondria do not appear to be mobile in the normal state, but this situation changes during VSM propagation and migration following vascular damage [16]. In addition to individual mitochondria forming networks within cells, the mitochondrial outer membranes are often connected or closely juxtapositioned with the membranes of the endoplasmic/sarcoplasmic reticulum (ER/SR) (Fig. 3.2); cellular structures which are involved in calcium storage and release [84]. The complex interplay among mitochondrial, ER/SR, and cytosolic calcium has been described in several recent reviews [15, 19]. In normal VSM, mitochondria and SR often appear to be "bundled" together (Fig. 3.2) and this structural feature may account for the lack of significant movement of mitochondria under normal conditions in this vascular cell type as well as the functional interactions between mitochondria and SR. A systematic study is needed to compare the mitochondrial morphology and function in the various cell types within the neurovascular unit.

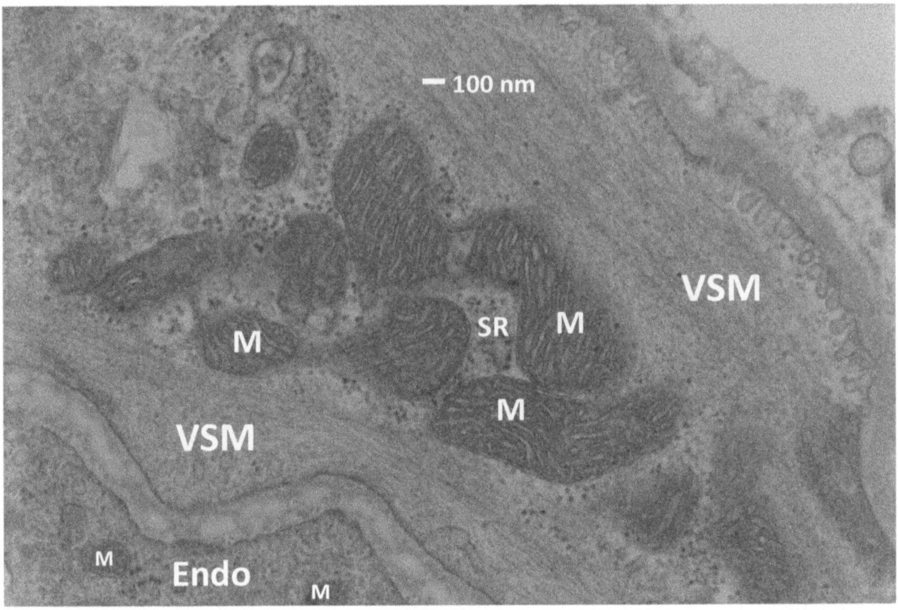

Fig. 3.2 Electron microscopy showing mitochondria vascular smooth muscle (VSM) from cerebral arteries from normal rats showing clustering of mitochondria (M) and sarcoplasmic reticulum (SR) in distinct locations. Mitochondria in VSM are much larger and of different configurations than in endothelial cells. Arteries were prepared as indicated in Fig. 3.1. *Endo* endothelium

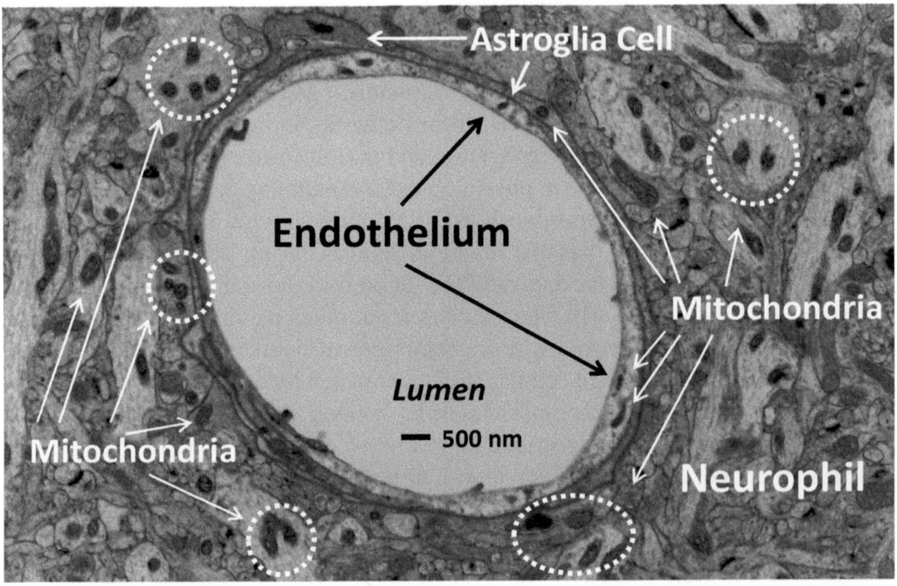

Fig. 3.3 Electron microscopy showing a capillary in cross section surrounded by an astroglial cell. In close proximity to the capillary are neurons as well as glial cells which are rich in mitochondria. All parenchymal cerebral blood vessels are in close proximity to astroglia and neurons. Arteries were prepared as indicated in Fig. 3.1

Mitochondria in quiescent neurons, VSM, and cerebral vascular endothelium are constant producers of ROS by the electron transport chain in the form of superoxide anion [36, 51, 78] and mitochondrial ROS appear to be important, if not essential, signaling agents involved in the maintenance of basal cell functions. Superoxide anion produced within the matrix from the respiratory complexes can leave the matrix [42, 45] following enzymatic conversion to hydrogen peroxide through aquaporin-like channels in the inner mitochondrial membrane [7, 14, 68]. Hydrogen peroxide in the intermembrane space, or superoxide anion released into the intermembrane space under certain conditions by Complex III, can transverse the mitochondrial outer membrane into the cytosol via voltage dependent anion channels (VDACs) [41, 67, 75]. Under certain conditions such as shear stress, a physical distortion of the relationship between the two mitochondrial membranes in the endothelium leads to even greater ROS release by mitochondria resulting in coronary artery dilation [93].

We were the first to show that mitochondrial depolarization and enhanced ROS production are not linked events in cells. In neurons, cerebral vascular endothelium and cerebral VSM, BMS-191095 causes mitochondrial depolarization without ROS production; thus, mitochondrial depolarization alone is sufficient to activate distinct cellular signaling pathways [11, 12, 26, 55]. Subsequent studies have supported our findings [87]. The previous concept linking mitochondrial membrane potential and ROS release arose from experiments using inhibitors of the electron transport chain which reduced the ATP levels necessary for maintaining the very negative transmembrane potential. The independent signaling pathways arising from mitochondrial depolarization as well as ROS release underscore the importance of this organelle in mediating diverse cellular activities including regulation of vascular tone.

Mitochondrial Depolarization and Related Signaling Events

The most efficient and reproducible approach to depolarize mitochondria is by activating mitoK_{ATP} channels on the inner mitochondrial membrane with drugs such as BMS-191095 or diazoxide. The physical structure of the mitoK_{ATP} channel is not known with certainty but appears to differ substantially from the previously described plasmaK_{ATP} channels [1, 35, 80, 92]. Foster et al. [4, 79] provided evidence that the renal outer medullary potassium channel (ROMK) is a component of the K^+ channel of the cardiac mitoK_{ATP}. Succinate dehydrogenase might also be involved in either the assembly or function of the mitoK_{ATP} channel [79]. Additional studies are needed to elucidate the complete structure of the mitoK_{ATP} channels.

The consequences to the cell of activating mitoK_{ATP} or plasmaK_{ATP} channels are quite different. The matrix of the mitochondria has a very negative potential and opening of the mitoK_{ATP} channels allows potassium to enter the matrix making the potential less negative and thereby "depolarizing" the mitochondrion. In contrast, opening the plasmaK_{ATP} channels allows potassium to leave the cytosol into the extracellular space and makes the cytosol more negative and thereby "hyperpolarizes" the cell.

Isolated mitochondria and mitochondria *in situ* in cultured endothelial cells and neurons as well as in endothelial and VSM cells in isolated cerebral arteries depolarize in a dose-dependent manner to selective mitoK_{ATP} channel openers such as diazoxide and BMS-191095 [10, 11, 13]. The classical K_{ATP} channel antagonist glibenclamide as well as 5-hydroxydecanoic acid (5-HD), which needs to be metabolized before becoming active [43], block the actions of diazoxide or BMS-191095 [11, 36]. The mitoK_{ATP} channel activity also is affected by endogenous factors such as the ADP/ATP ratio [1], peroxynitrite [61, 62], superoxide anion [60, 62], and cytosolic protein kinase C epsilon (PKCε) [77]. Other, yet unidentified, physiological and pathological factors likely will be found to directly or indirectly activate or block the mitoK_{ATP} channel. Diazoxide, a drug previously used in the treatment of acute hypertension or hypoglycemia in people, is the most commonly used mitoK_{ATP} channel opener [34], but it has the additional effect of inhibiting succinate dehydrogenase (SDH; complex II), especially at high doses [18, 54]. Diazoxide readily crosses the blood-brain barrier (BBB) and thus is effective in the brain when given intravenously [66].

Although applications of diazoxide or BMS-191095 depolarize mitochondria, diazoxide, but not BMS-191095, also causes the liberation of ROS [14], which may be secondary to SDH inhibition. This view is supported by examination of the effects of the specific inhibitor of SDH, 3-nitropropionic acid (3-NPA), which increases ROS production by mitochondria [11] and changes cerebral vascular tone [49]. Nonetheless, the primary actions of diazoxide on the cells of the neurovascular unit *ex vivo* are still specific to mitochondria [12, 54], and the associated ROS increase appears to enhance the degree of depolarization [61, 62]. In contrast, BMS-191095 is very selective for mitoK_{ATP} channels and has no known non-specific effects to complicate the interpretation of the results [12, 39, 40].

A potential role for mitochondrial calcium activated potassium (mitoK_{Ca}) channels in depolarizing mitochondria has been suggested based primarily on the use of the multiple target drug NS1619 [6, 59]. NS1619 is vasoactive in cerebral and peripheral [5, 44] arteries. Although NS1619 results in mitochondrial depolarization, it seems likely that, at least in neurons, mitochondrial effects are due to initiating factors such as inhibition of Complex I and subsequent increased release of ROS [37, 38, 53]. Given the multiple potential sites of action of NS1619 within various cell types, including mitoK_{Ca} channels, it is also possible that mitochondrial effects to this drug are secondary to non-mitochondrial events. Nonetheless, more research in this area is warranted and the development of a more specific agonist would aid these efforts.

mitoK_{ATP} Channels and Cerebral Vascular Tone

Although it is known that extracellular ATP, from either glycolysis or oxidative phosphorylation, as well as the metabolites adenosine, AMP, and ADP [2, 74] can alter cerebrovascular tone via plasma purinergic receptors [22], only a few studies have examined direct mitochondrial influences via opening of mitoK_{ATP} channels on the diameter of cerebral resistance vessels [16, 49–51, 91]. Thus, the study of direct mitochondrial effects on the cerebral vasculature is a relatively new field. Even fewer studies have examined mitochondrial influences on the cerebral vasculature during disease states, but increasing amounts of information are rapidly becoming available in other systemic circulations [39, 58, 89] and VSM [73].

Vascular Smooth Muscle

Jaggar and colleagues [17, 91] were the first to show that activation of mitochondria by diazoxide promoted relaxation of VSM cells in endothelium-denuded cerebral arteries or freshly dissociated VSM via a mechanism primarily involving production and localized effects of ROS. Thus, diazoxide application enhanced the generation of ROS from mitochondria, which sequentially caused the activation of ryanodine-sensitive Ca^{2+} channels on the SR, the localized generation of Ca^{2+} transients called "Ca^{2+} sparks", and the opening of adjacent large-conductance Ca^{2+}-activated K^+ (BK_{Ca}) channels on the plasma membrane. Calcium sparks by themselves are insufficient in the absence of BK_{Ca} channels to affect global intracellular Ca^{2+}. The resulting K^+ efflux due to opening of BK_{Ca} channels by calcium sparks leads to VSM hyperpolarization, decreased global intracellular Ca^{2+}, and VSM relaxation. We have reported similar findings in endothelium denuded arteries with diazoxide and BMS-191095 [49, 51]. The role of mitochondrial ROS in promoting VSM relaxation of cerebral arteries is also shown from experiments using 3-nitropropionic acid (3-NPA), an inhibitor of SDH [49].

We recently found that BMS-191095 has a similar effect to diazoxide in VSM without the involvement of ROS-mediated influences [50]. Thus, BMS-191095 does not increase vascular cytosolic or mitochondrial levels of ROS in VSM and dilation of endothelium-denuded cerebral arteries to BMS-191095 is not affected by ROS scavengers. However, BMS-191095 via mitochondrial depolarization increases calcium spark activity in VSM, the opening of adjacent BK_{Ca} channels on the plasma membrane application, and relaxation of endothelium-denuded cerebral arteries. The reasons for these differences in experimental findings are unclear, but cumulatively underscore the importance and diversity of mitochondrial mechanisms in promoting relaxation of VSM. We speculate that there is direct electrophysiological coupling between mitochondria and SR in VSM. The close physical association with the possibility of direct coupling of mitochondria and SR are clearly seen with electron microscopy in cerebral VSM (Fig. 3.2).

Endothelium

Mitochondrial content in the cerebral vascular endothelium is higher than in endothelium in other peripheral circulations (Fig. 3.1), probably due to the augmented transport requirements of the BBB [72, 76]. We investigated the contribution of

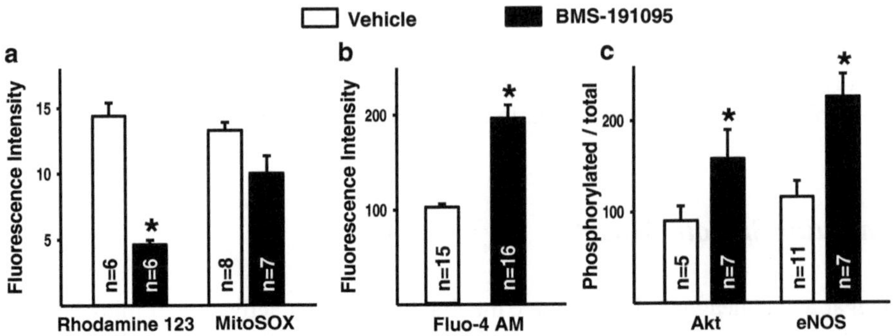

Fig. 3.4 Experiments conducted in cultured cerebral vascular endothelium or endothelium-intact cerebral arteries show that BMS-191095 depolarizes mitochondria without ROS production (**a**), and subsequently elicits an increase in global calcium (**b**) as well as the phosphorylation of Akt and eNOS (**c**). From [51]. *$p<0.05$, compared to vehicle treated

mitochondrial factors arising within the endothelium on the integrated response of intact cerebral arteries using several approaches. First, removal of the endothelium or inhibition of eNOS with L-NAME reduced the vasodilation to diazoxide and BMS-191095, implying that traditional endothelium-derived factors such as NO contribute to changes in vascular tone [51]. Second, fluorescence and electron spin resonance measurements of NO in intact arteries or cultured cerebral microvascular endothelial cells confirmed the production of NO in response to diazoxide and BMS-191095. Third, fluorescence measurements showed that a global increase in free cytosolic calcium was temporally associated with increased NO production with diazoxide or BMS-191095 application. Fourth, BMS-191095 or diazoxide increase phosphorylation of eNOS, an event which has been shown to activate eNOS and enhance NO production. Thus, BMS-191095 and diazoxide application led to mitochondrial depolarization, activation of eNOS by increased cytosolic calcium as well as the phosphorylation of eNOS, enhanced production of NO and its diffusion to VSM, and relaxation of VSM [51; Fig. 3.4]. The increased production and diffusion of NO from endothelium augments the relaxation due to mitochondrial activation in the VSM (Fig. 3.5). Mitochondrial depolarization also can lead to the production of constrictor prostanoids [49], but the NO effects appear to be more important. Although it has been suggested that mitochondria contain a variant of NOS, there is significant evidence against this view [63–65].

Gutterman and colleagues [92] showed that a physiological stimulus such as increased shear stress is able to directly activate mitochondria in coronary endothelial cells via filamentous connections and dilate coronary arteries via effects of hydrogen peroxide. Whether a similar endothelium-specific mechanism operates in response to increases in flow or shear stress in cerebral arteries is unclear, since alternative mechanisms promoting either dilation or constriction in cerebral resistance vessels of different sizes and locations have been proposed by Koller and Toth [58]. Nonetheless, the common feature is that stimulation of mitochondria is able to liberate substances by endothelium which can secondarily affect VSM tone.

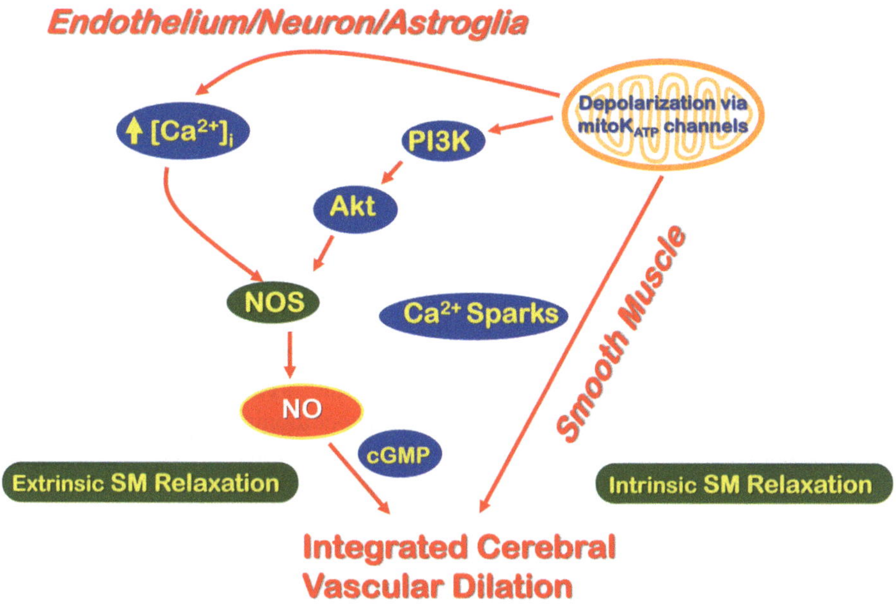

Fig. 3.5 Schematic illustration showing signaling events and ultimate mechanisms resulting in cerebral vasodilation in response to depolarization of one or more of the cells types comprising the neurovascular unit. Factors produced following activation of mitochondria by physiological, pharmacological, and pathological stimuli in any of these cell types can affect VSM and the interaction of these factors will determine the final, integrated arterial tone. Abbreviations: $[Ca^{2+}]_i$ cytosolic calcium, *PI3K* phosphoinositide 3-kinase, *Akt* protein kinase B, *NO* nitric oxide, *NOS* nitric oxide synthase

Perivascular and Parenchymal Nerves

New findings indicate that mitochondrial influences from perivascular nerves or adjacent parenchymal neurons could have major effects on cerebral vascular responses. Cerebral arteries receive dense innervation from sympathetic, parasympathetic, and nitroxidergic nerves [3, 9, 20, 86]. Additionally, the entire cerebral vasculature is in close contact with adjacent parenchymal neurons or astroglia (Fig. 3.3). BMS-191095 and diazoxide depolarize mitochondria in cultured cortical neurons and enhance NO production by these cells. As in endothelial cells, opening of mito-K_{ATP} channels increases intracellular calcium levels and phosphorylates neuronal NOS. Furthermore, inhibition of nNOS with 7-NI in perivascular nerves inhibits dilation in both intact and denuded cerebral arteries to BMS-191095 [52]. Therefore, perivascular nerves, parenchymal neurons, and possibly astroglia as well [46, 78] can provide mitochondrial-initiated vasoactive signals to VSM [82–84] for the final determination of integrated changes in cerebral vascular diameter in response to physiological and pathological challenges (Fig. 3.5).

Vascular Mitochondria in Insulin Resistance

Insulin resistance often precedes the development of type II diabetes by years or decades, and is considered a relatively quiet phase of the metabolic syndrome despite vascular dysfunction and moderate arterial hypertension. We and others have characterized many of the general cardiovascular effects of this disease [31–33, 49, 56, 90] and have shown that tissue and vascular inflammation and increased basal levels of ROS from both mitochondrial and non-mitochondrial sources are responsible for vascular dysfunction [33, 34].

We were among the first investigators to demonstrate that insulin resistant is able to disrupt functioning of plasma K_{ATP} channels in cerebral and peripheral arteries [4, 31, 32] and therefore it is not surprising that mitoK_{ATP} channels could be affected in a similar manner. Thus, diazoxide-induced preconditioning fails to protect hearts from ischemia/reperfusion in IR, Zucker obese rats, whereas diazoxide limited infarct size in hearts from non-IR rats [48]. An additional finding from this study was that infarct size was enhanced in the hearts from Zucker obese compared with lean rats, which is similar to that which we [71] and others [8] have found in brains of IR, obese mice following transient occlusion of the middle cerebral artery (MCAO). These findings may indicate that normal protective mechanisms initiated at the level of the mitoK_{ATP} channels are impaired in many common disease states and thus the brain and other organs are more at risk during ischemic episodes.

Similar to ineffective preconditioning, mitochondrial-dependent vasoactive responses in cerebral arteries are impaired in IR rats [49]. This attenuation of dilation appears to be due to reduced mitochondrial depolarization of VSM as well as reduced ROS generation and calcium sparks generation in response to diazoxide. In addition, decreased NO production and/or bioavailability appears to contribute to diminished relaxation of VSM. We have found similar impairment of cerebral vascular dilation to BMS-191095 in IR animals [51]. The potential for the uncoupling of the tight relationship between metabolic need and blood flow in the brain due to impairment of mitoK_{ATP} channel function associated with IR may account for the increased risk and severity of neurological diseases and strokes in patients suffering from the metabolic syndrome [85]. A possible new therapeutic approach would involve the early detection and treatment of IR individuals with the aim toward reducing vascular inflammation and mitochondrial dysfunction prior to the further progression of this chronic disease. We have shown that treating IR rats with statins was effective in reducing cerebral vascular ROS levels and restoring normal plasma K_{ATP} and possibly mitoK_{ATP} channel function as well as vascular responsiveness [33].

Mitochondria and Cerebral Ischemia

Although mitochondrial failure is a major cause of cell death due to ischemia, mitochondria in surviving endothelium may represent a useful target to prevent further cell death and restore cerebral blood flow even after the onset of stroke. In a recent study, we examined the effects of experimental stroke on middle cerebral artery

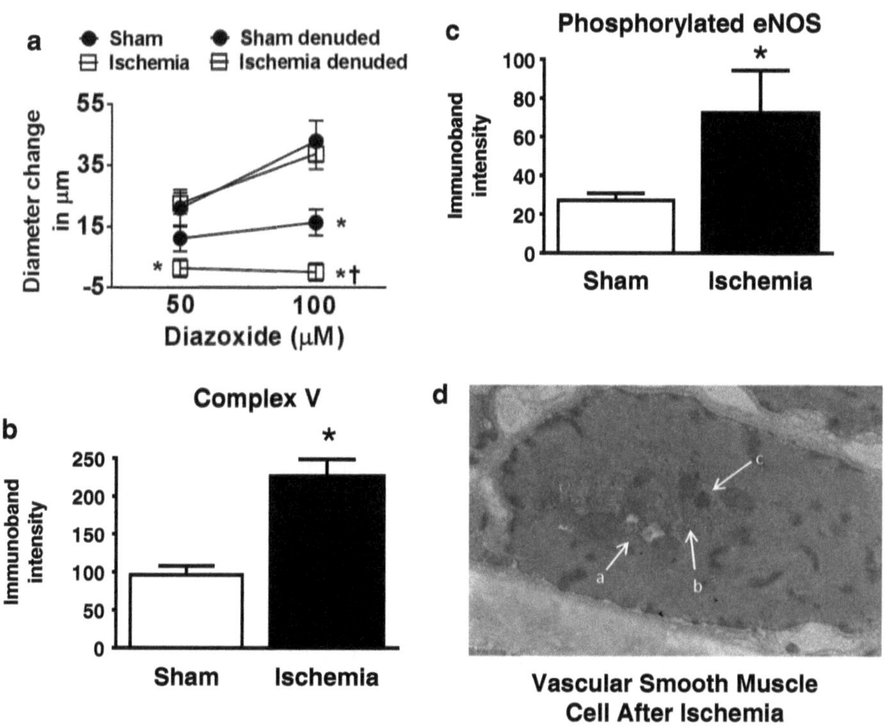

Fig. 3.6 Ischemia alters the relative contributions of endothelium and VSM to the final integrated vasodilation following mitochondrial activation. Dilation of cerebral arteries to diazoxide is similar in arteries previously occluded (Ipsi; arteries on the side ipsilateral to ischemia) for 90 min to control arteries from sham (Sham) operated animals (**a**). However, the entire dilation following ischemia in Ipsi arteries is due to endothelial factors compared to roughly equal endothelial and VSM contributions in Sham arteries. The enhanced contribution of endothelial factors is confirmed by a doubling of eNOS phosphorylation after ischemia (**b**). While total mitochondrial mass is increased following ischemia (**c**), the normal profile of mitochondria in VSM including positioning and physical relationship to SR, is disrupted by ischemia (**d**). From [81]. *$p<0.05$, compared to endothelium intact arteries; †$p<0.05$, compared to Sham denuded arteries

responses 2 days following reperfusion, and found that dilation to mitoK_{ATP} channel opening with diazoxide was retained on the side ipsilateral to the (MCAO) [81]. Mitochondrial mass and DNA were maintained or enhanced in middle cerebral arteries from the stroke side. At this time, dilator and constrictor responses to other, non-mitochondrial stimuli were reduced in middle cerebral arteries on the stroke side. Retention of dilator responses to mitoK_{ATP} channel opening appeared to be due primarily to enhanced contributions of endothelium (Fig. 3.6). Additionally, in separate studies, we have found that post-conditioning of cultured cerebral vascular endothelium increases viability following anoxic stress [13]. Thus, not only are endothelial cells responsive to mitochondrial-induced post-conditioning after anoxic stress [13], but cerebral vascular endothelium responsiveness to mitochondrial activators is also retained or enhanced [81]. Therefore, because strokes occur unpredictably, therapeutic targeting of mitochondria after the onset of ischemic stroke, especially after delayed clot resolution, might be a clinically useful approach to prevent further injury to the endothelium and VSM, correct cerebral hypoperfusion, restore normal cerebral vascular responsiveness and the BBB, and restrict further neuronal and glia cell death.

Conclusions, Significance, and Perspective

A review of the available literature indicates that mitochondrial mechanisms, including the opening of mitoK_{ATP} channels, are important modulators of cerebral vascular function. *First*, individual mitochondrial influences from endothelium, VSM, perivascular nerves, and parenchymal neurons as well as from other cell types such as astroglia contribute to mitochondrial-initiated relaxation effects of VSM into a final, integrated change in cerebral vascular tone. Thus, production of vasoactive factors following activation of mitochondria in response to physiological stimuli in one or several of the cells comprising the neurovascular unit may represent the elusive signaling link between cerebral metabolic rate and blood flow. *Second*, mitochondrial-derived mechanisms such as those that induce changes in cerebral vascular tone are disrupted by chronic disease processes such as IR or diabetes. Thus, direct or indirect targeting of mitochondria in the early stages of the metabolic syndrome might prove to be an effective treatment to reduce morbidity and mortality in these patients to reduce subsequent risk of cardiovascular disease, strokes, and dementias such as Alzheimer's disease. *Third*, mitochondrial mechanisms promoting vasodilation and endothelial preconditioning are present following experimental strokes. Thus, targeting mitochondria following an ischemic injury, especially after delayed resolution of an occlusion, may restrict further neuronal, vascular, and glial cell death.

Acknowledgments The authors thank Nancy Busija, M.A. for the help with editing the manuscript. We also thank Ken Grant of the Cellular Imaging Shared Resource at Wake Forest University Health Sciences for assistance with electron microscopy. This work was supported by National Institutes of Health grants (D.W.B.: HL-077731, HL-030260, HL093554 and HL-065380), Louisiana Board of Regents Support Fund-Research Competitiveness Subprogram (P.V.K.: LEQSF(2014-17)-RD-A-11), American Heart Association National Center NRCP Scientist Development Grant (P.V.K.: 14SDG20490359), and American Heart Association Post-Doctoral Fellowship Grant (I.R.: 15POST23040005). This research was supported in whole or in part by the Louisiana Board of Regents Endowed Chairs for Eminent Scholars program to D.W.B.

References

1. Ardehali H, O'Rourke B. Mitochondrial K(ATP) channels in cell survival and death. J Mol Cell Cardiol. 2005;39:7–163.
2. Auchampach JA, Gross GJ. Adenosine A1 receptors, K_{ATP} channels, and ischemic preconditioning in dogs. Am J Physiol. 1993;264:H1327–36.

3. Ayajiki K, et al. Evidence for nitroxidergic innervation in monkey ophthalmic arteries in vivo and in vitro. Am J Physiol. 2000;279:H2006–12.
4. Bari F, et al. Global ischemia impairs ATP-sensitive K+ channel function in cerebral arterioles in piglets. Stroke. 1996;27:1874–81.
5. Bari F, Louis TM, Busija DW. Calcium-activated K+ channels in cerebral arterioles in piglets are resistant to ischemia. J Cereb Blood Flow Metab. 1997;17:1152–6.
6. Bednarczyk P, et al. Large-conductance Ca^{2+}-activated potassium channel in mitochondria of endothelial EA.hy926 cells. Am J Physiol. 2013;304:H1415–27.
7. Bienert GP, et al. Specific aquaporins facilitate the diffusion of hydrogen peroxide across membranes. Biochim Biophys Acta. 2007;1717:1–10.
8. Brown KA, et al. Effects of aging, MnSOD deficiency, and genetic background on endothelial function: evidence for MnSOD haploinsufficiency. Arterioscler Thromb Vasc Biol. 2007;27:1941–6.
9. Busija DW, Heistad DD. Factors involved in the physiological regulation of the cerebral circulation. Rev Physiol Biochem Pharmacol. 1984;101:161–211.
10. Busija DW, et al. Targeting mitochondrial ATP-sensitive potassium channels—a novel approach to neuroprotection. Brain Res Rev. 2004;46:282–94.
11. Busija DW, et al. Effects of ATP-sensitive potassium channel activators diazoxide and BMS-191095 on membrane potential and reactive oxygen species production in isolated piglet mitochondria. Brain Res Bull. 2005;66:85–90.
12. Busija DW, et al. Mitochondrial-mediated suppression of ROS production upon exposure of neurons to lethal stress: mitochondrial targeted preconditioning. Adv Drug Deliv Rev. 2008;16:1471–7.
13. Busija DW, Katakam PV. Mitochondrial mechanisms in cerebral vascular control: shared signaling pathways with preconditioning. J Vasc Res. 2014;51:175–89.
14. Calamita G, et al. The inner mitochondrial membrane has aquaporin-8 water channels and is highly permeable to water. J Biol Chem. 2005;280:17149–53.
15. Chalmers S, et al. Ion channels in smooth muscle: regulation by the sarcoplasmic reticulum and mitochondria. Cell Calcium. 2007;42:447–66.
16. Chalmers S, et al. Mitochondrial motility and vascular smooth muscle proliferation. Arterioscler Thromb Vasc Biol. 2012;32:3000–11.
17. Cheranov SY, Jaggar JH. Mitochondrial modulation of Ca^{2+} sparks and transient K_{Ca} currents in smooth muscle cells of rat cerebral arteries. J Physiol. 2004;556:755–71.
18. Coetzee WA. Multiplicity of effectors of the cardioprotective agent, diazoxide. Pharmacol Ther. 2013;140:167–75.
19. Dai J, et al. Rearrangement of the close contact between the mitochondria and the sarcoplasmic reticulum in airway smooth muscle. Cell Calcium. 2005;37:333–40.
20. Dauphin F, MacKenzie ET. Cholinergic and vasoactive intestinal polypeptidergic innervation of the cerebral arteries. Pharmacol Ther. 1995;67:385–417.
21. Davidson SM. Endothelial mitochondria and heart disease. Cardiovasc Res. 2010;88:58–66.
22. Dietrich HH, et al. Mechanism of ATP-induced local and conducted vasomotor responses in isolated rat cerebral penetrating arterioles. J Vasc Res. 2009;46:253–64.
23. Domoki F, et al. Mitochondrial potassium channel opener diazoxide preserves neuronal-vascular function after cerebral ischemia in newborn pigs. Stroke. 1999;30:2713–8.
24. Domoki F, et al. Diazoxide prevents mitochondrial swelling and Ca (2+) accumulation in CAI pyramidal cells after cerebral ischemia in newborn pigs. Brain Res. 2004;1019:97–104.
25. Domoki F, et al. Diazoxide preserves hypercapnia-induced arteriolar vasodilation after global cerebral ischemia in piglets. Am J Physiol. 2005;289:H368–73.
26. Domoki F, et al. Rosuvastatin induces delayed preconditioning against oxygen-glucose deprivation in cultured cortical neurons. Am J Physiol. 2009;296:C97–105.
27. Dorn GW, Scorrano L. Two close, too close: sarcoplasmic reticulum-mitochondrial crosstalk and cardiomyocyte fate. Circ Res. 2010;107:689–99.
28. Dromparis P, Sutendra G, Michelakis ED. The role of mitochondria in pulmonary vascular remodeling. J Mol Med (Berl). 2010;88:1003–10.

29. Duckles SP, Krause DN. Cerebrovascular effects of oestrogen: multiplicity of action. Clin Exp Pharmacol Physiol. 2007;34:801–8.
30. Duckles SP, Krause DN. Mechanisms of cerebrovascular protection: oestrogen, inflammation and mitochondria. Acta Physiol. 2011;203:149–54.
31. Erdős B, et al. Cerebrovascular dysfunction in Zucker obese rats is mediated by oxidative stress and protein kinase C. Diabetes. 2004;53:1352–9.
32. Erdős B, et al. Subtype specific potassium channel dysfunction in cerebral arteries of insulin-resistant rats is mediated by reactive oxygen species. Stroke. 2004;35:964–9.
33. Erdős B, et al. Rosuvastatin improves cerebrovascular function in Zucker obese rats by inhibiting NAD(P)H-oxidase-dependent superoxide anion production. Am J Physiol. 2006;290: H1264–70.
34. Farkas E, et al. Neuroprotection by diazoxide in animal models for cerebrovascular disorders. Vasc Dis Prev. 2006;3:253–63.
35. Foster DB, et al. Mitochondrial ROMK channel is a molecular component of mitoK(ATP). Circ Res. 2012;111:446–54.
36. Gaspar T, et al. ROS-independent preconditioning in neurons via activation of mitoK_{ATP} channels by BMS-191095. J Cereb Blood Flow Metab. 2008;28:1090–103.
37. Gaspar T, et al. Delayed neuronal preconditioning by NS1619 is independent of calcium activated potassium channels. J Neurochem. 2008;105:1115–28.
38. Gaspar T, et al. Immediate neuronal preconditioning with NS1619. Brain Res. 2009;1285:196–207.
39. Grover GJ, et al. Pharmacologic characterization of BMS-191095, a mitochondrial K(ATP) opener with no peripheral vasodilator or cardiac action potential shortening activity. J Pharmacol Exp Ther. 2001;297:1184–92.
40. Grover GJ, et al. Protective effect of mitochondrial K_{ATP} activation in an isolated gracilis model of ischemia and reperfusion in dogs. J Cardiovasc Pharmacol. 2003;42:790–2.
41. Han D, et al. Voltage-dependent anion channels control the release of the superoxide anion from mitochondria to cytosol. Arterioscler Thromb Vasc Biol. 2012;32:2531–9.
42. Han D, William E, Cadenas E. Mitochondrial respiratory chain-dependent generation of superoxide anion and its release into the intermembrane space. Biochem J. 2001;353:411–6.
43. Hanley PJ, et al. Beta-oxidation of 5-hydroxydecanoate, a putative blocker of mitochondrial ATP-sensitive potassium channels. J Physiol. 2003;547:387–93.
44. Holland M, et al. Effects of the BKCa channel activator, NS1619, on rat cerebral artery smooth muscle. Br J Pharmacol. 1996;117:119–29.
45. Jastroch M, et al. Mitochondrial proton and electron leaks. Essays Biochem. 2010;47: 53–67.
46. Jiang K, et al. Regulation of gap junctional communication by astrocytic mitochondrial K(ATP) channels following neurotoxin administration in in vitro and in vivo models. Neurosignals. 2011;19:63–74.
47. Jonckheere AI, Smeitink JA, Rodenburg RJ. Mitochondrial ATP synthase: architecture, function and pathology. J Inherit Metab Dis. 2012;35:211–25.
48. Katakam PV, et al. Myocardial preconditioning against ischemia-reperfusion is abolished in Zucker obese rats with insulin. Am J Physiol. 2007;292:R920–6.
49. Katakam PV, et al. Impaired mitochondria-dependent vasodilation in cerebral arteries of Zucker obese rats with insulin resistance. Am J Physiol. 2009;296:R289–98.
50. Katakam PV, et al. Diversity of mitochondria-dependent dilator mechanisms in vascular smooth muscle of cerebral arteries from normal and insulin-resistant rats. Am J Physiol. 2014;307:H493–503.
51. Katakam PV, et al. Depolarization of mitochondria in endothelial cells promotes cerebral vascular vasodilation by activation of nitric oxide synthase. Arterioscler Thromb Vasc Biol. 2013;33:752–9.
52. Katakam PVG, et al. Mitochondria-dependent cerebral artery vasodilation is mediated by the activation of neuronal nitric oxide synthase following mitochondrial depolarization of perivascular nerves. FASEB J. 2012;26:1058.1059. Abstract.

53. Kicinska A, Szewczyk A. A Large-conductance potassium cation channel opener NS1619 inhibits cardiac mitochondria respiratory chain. Toxicol Mech Methods. 2004;14:59–61.
54. Kis B, et al. Diazoxide induces delayed preconditioning in cultured rat cortical neurons. J Neurochem. 2003;87:969–80.
55. Kis B, et al. The mitochondrial K_{ATP} channel opener BMS191095 induces neuronal preconditioning. Neuroreport. 2004;15:345–9.
56. Kizhakekuttu TJ, et al. Adverse alterations in mitochondrial function contribute to type 2 diabetes mellitus-related endothelial dysfunction in humans. Biochem Biophys Res Commun. 2012;422(3):515–21.
57. Kluge MA, Fetterman JL, Vita JA. Mitochondria and endothelial function. Circ Res. 2013;12:1171–88.
58. Koller A, Toth P. Contribution of flow-dependent vasomotor mechanisms to the autoregulation of cerebral blood flow. J Vasc Res. 2012;49:375–89.
59. Korper S, et al. The K+ channel openers diazoxide and NS1619 induce depolarization of mitochondria and have differential effects on cell Ca^{2+} in CD34+ cell line KG-1a. Exp Hematol. 2003;31:815–23.
60. Kubli DA, Gustafsson ÅB. Mitochondria and mitophagy: the yin and yang of cell death control. Circ Res. 2012;12(111):1208–21.
61. Lacza Z, et al. Investigation of the subunit composition and the pharmacology of the mitochondrial ATP-dependent K^+ channel in the brain. Brain Res. 2003;19:27–36.
62. Lacza Z, et al. Heart mitochondria contain functional ATP-dependent K^+ channels. J Mol Cell Cardiol. 2003;35:1339–47.
63. Lacza Z, et al. Lack of mitochondrial nitric oxide production in the brain. J Neurochem. 2004;90:942–51.
64. Lacza Z, et al. Mitochondrial NO and reactive nitrogen species production: does mtNOS exist? Nitric Oxide. 2006;14:162–8.
65. Lacza Z, et al. Mitochondria produce reactive nitrogen species via an arginine-independent pathway. Free Radic Res. 2006;40:369–78.
66. Lenzser G, et al. Diazoxide preconditioning attenuates global cerebral ischemia-induced blood-brain barrier permeability. Brain Res. 2005;105:72–80.
67. Lustgarten MS, et al. Complex I generated, mitochondrial matrix-directed superoxide is released from the mitochondria through voltage dependent anion channels. J Biol Chem. 2003;278:5557–63.
68. Marchissio MJ, et al. Mitochondrial aquaporin-8 knockdown in human hepatoma HepG2 cells causes ROS-induced mitochondrial depolarization and loss of viability. Toxicol Appl Pharmacol. 2012;64:246–54.
69. Mayanagi K, et al. The mitochondrial K(ATP) channel opener BMS-191095 reduces neuronal damage after transient focal cerebral ischemia in rats. J Cereb Blood Flow Metab. 2007;27:348–55.
70. Mayanagi K, et al. Systemic administration of diazoxide induces delayed preconditioning against transient focal cerebral ischemia in rats. Brain Res. 2007;1168:106–11.
71. Mayanagi K, et al. Acute treatment with rosuvastatin protects insulin resistant (C57BL/6J ob/ob) mice against transient cerebral ischemia. J Cereb Blood Flow Metab. 2008;28:1927–35.
72. McBride HM, Neuspiel M, Wasiak S. Mitochondria: more than just a powerhouse. Curr Biol. 2006;16:R551–60.
73. McCarron JG, et al. From structure to function: mitochondrial morphology, motion and shaping in vascular smooth muscle. J Vasc Res. 2013;50(5):357–71.
74. McIntosh VJ, Lasley RD. Adenosine receptor-mediated cardioprotection: are all 4 subtypes required or redundant? J Cardiovasc Pharmacol Ther. 2012;17:21–33.
75. Muller FL, Liu Y, Van Remmem H. Complex III releases superoxide to both sides of the inner mitochondrial membrane. J Biol Chem. 2004;279:49064–73.
76. Oldendorf WH, Cornford ME, Brown WJ. The large apparent work capacity of the blood-brain barrier: a study of the mitochondrial content of capillary endothelial cells in brain and other tissues of the rat. Ann Neurol. 1977;1:409–17.

77. Perez-Pinzon MA, Dave KR, Raval AP. Role of reactive oxygen species and protein kinase C in ischemic tolerance in the brain. Antioxid Redox Signal. 2005;7:1150–7.
78. Rajapakse N, et al. Diazoxide pretreatment induces delayed preconditioning in astrocytes against oxygen glucose deprivation and hydrogen peroxide-induced toxicity. J Neurosci Res. 2003;73:206–14.
79. Rines AK, Bayeva M, Ardehali H. A new pROM king for the MitoK_{ATP} dance ROMK takes the lead. Circ Res. 2012;111:392–3.
80. Robin E, et al. Postconditioning in focal cerebral ischemia: role of the mitochondrial ATP-dependent potassium channel. Brain Res. 2011;1375:137–46.
81. Rutkai I, et al. Sustained mitochondrial functioning in cerebral arteries after transient ischemic stress in the rat: a potential target for therapies. Am J Physiol. 2014;307:H958–66.
82. Seharaseyon J, et al. Molecular composition of mitochondrial ATP-sensitive potassium channels probed by viral Kir gene transfer. J Mol Cell Cardiol. 2000;32:1923–30.
83. Shimizu K, Lacza Z, Rajapakse N, Horiguchi T, Snipes J, Busija DW. MitoK_{ATP} opener, diazoxide, reduces neuronal damage after middle cerebral artery occlusion in the rat. Am J Physiol. 2002;283:H1005–11.
84. Somlyo AV, Somlyo AP. Strontium accumulation by sarcoplasmic reticulum and mitochondria in vascular smooth muscle. Science. 1971;174:955–8.
85. Terao S, et al. Inflammatory and injury responses to ischemic stroke in obese mice. Stroke. 2008;39:943–50.
86. Toda N, Okamura T. Nitroxidergic nerve: regulation of vascular tone and blood flow in the brain. J Hypertens. 1996;14:423–34.
87. Votyakova TV, Reynolds IJ. DeltaPsi(m)-dependent and -independent production of reactive oxygen species by rat brain mitochondria. J Neurochem. 2001;79:266–77.
88. Wappler EA, et al. Mitochondrial dynamics associated with oxygen-glucose deprivation in rat primary neuronal cultures. PLoS One. 2013;8(5):e63206.
89. Widlansky ME, Gutterman DD. Regulation of endothelial function by mitochondrial reactive species. Antiox Redox Signal. 2011;15:1517–30.
90. Wojtovich AP, Smith CO, Haynes CM, Nehrke KW, Brookes PS. Physiological consequences of complex II inhibition for aging, disease, and the mK_{ATP} channel. Biochim Biophys Acta. 2013;1827(5):598–611.
91. Xi Q, Cheranov SY, Jaggar JH. Mitochondria-derived reactive oxygen species dilate cerebral arteries by activating Ca^{2+} sparks. Circ Res. 2005;97:354–62.
92. Zhang DX, Gutterman DD. Mitochondrial reactive oxygen species-mediated signaling in endothelial cells. Am J Physiol. 2007;292:H2023–31.
93. Zick M, Rabl R, Reichert AS. Cristae formation-linking ultrastructure and function of mitochondria. Biochim Biophys Acta. 2008;1793:5–19.

Chapter 4
Ion Channels in Control of Blood Flow: Electrical Conduction Along Endothelium of Resistance Arteries

Erik J. Behringer and Steven S. Segal

Abstract The regulation of tissue blood flow in response to the metabolic demand of parenchymal cells is effected through changes in vascular resistance as governed by arteriolar networks and their proximal feed arteries. Vasodilation and vasoconstriction must be coordinated among downstream and upstream segments to optimize blood flow distribution within the tissue and to attain maximal perfusion of the vascular supply. Such coordinated vasomotor activity is promoted by the transmission of electrical signals (e.g., hyperpolarization and depolarization) through gap junctions from cell-to-cell along the vessel wall. Based upon underlying structural and functional relationships, we explore the biophysical basis of intercellular electrical signaling along the endothelium of resistance arteries. The endothelium is presented as a cable, whereby electrical signals decay passively with distance from the site of initiation. Key to our findings is how K^+ channels expressed constitutively in endothelial cell membranes [e.g., $K_{Ca}2.3$ (SK_{Ca}) and $K_{Ca}3.1$ (IK_{Ca})] regulate the spatial domain of electrical signal transmission and how this role is effected during advanced age through the actions of hydrogen peroxide. New insights into the regulation of electrical conduction along microvascular endothelium advance our understanding of how blood flow is governed by ion channels while providing mechanistic insight into how such processes can be affected during vascular disease.

Keywords Electrical conduction • Endothelium • Gap junctions • Oxidative stress • Potassium channels • Vascular aging

---

E.J. Behringer
Medical Pharmacology and Physiology, University of Missouri, MA415 Medical Science Building, Columbia, MO 65212, USA

Division of Pharmacology, Risley Hall, Loma Linda University, Loma Linda, CA 92354, USA

S.S. Segal, Ph.D. (✉)
Medical Pharmacology and Physiology, University of Missouri, MA415 Medical Science Building, Columbia, MO 65212, USA

Dalton Cardiovascular Research Center, Columbia, MO 65211, USA
e-mail: segalss@health.missouri.edu

I. Levitan, A.M. Dopico (eds.), *Vascular Ion Channels in Physiology and Disease*, DOI 10.1007/978-3-319-29635-7_4

Introduction

Blood flow control requires dynamic responses of the resistance vasculature, perhaps best exemplified by skeletal muscle, which can increase blood flow in excess of 100-fold as it transitions from inactivity to maximal aerobic exercise [1, 48, 60] and vasodilation ascends from arterioles embedded within the muscle into their proximal feed arteries [69]. This chapter is concerned with the intrinsic properties of the endothelium as the principal cellular pathway for coordinating vasomotor responses along and among the branches of vascular resistance networks. Following a brief review of the ultrastructural properties and spatial relationships that underlie intercellular conduction, we consider basic electrophysiological properties and cable theory in light of electrical signal transmission along the vessel wall. The use of endothelial "tubes" freshly isolated from resistance arteries of mouse skeletal muscle is presented as a model with which the effect of ion channel activation on electrical signaling has been resolved. Our discussion concludes with consideration for how advanced age can affect these relationships through the actions of hydrogen peroxide. These recent insights provide novel perspective for therapeutic interventions aimed at restoring and sustaining tissue blood flow.

Functional Morphology

The intima of arterioles and resistance arteries is comprised of an endothelial cell (EC) monolayer that is well suited for the conduction of electrical signals in a manner analogous to an electrical cable [5, 21]. Thus, ECs are aligned with the direction of blood flow, parallel to the vessel axis. Though only ~0.5 μm thick and ~5 μm wide, individual ECs can exceed 100 μm in length [29]. Each EC can thereby span ~20 SMCs, an ideal morphological arrangement for signal transmission along and among vessel branches [16] while minimizing the number of cells required to span each vessel segment. A complementary feature that facilitates direct intercellular signal transmission along the endothelium is the robust expression of gap junctions at EC borders [51, 85, 30].

The media of resistance microvessels consists of a single layer of adjacent smooth muscle cells (SMCs) that wrap around the intima while oriented near-perpendicular to the vessel axis. When relaxed, individual SMCs are ~5 μm in diameter and up to ~100 μm long. Under resting conditions in vivo, SMCs exist in a partially activated state (i.e., myogenic tone) and must relax (vasodilation) or contract (vasoconstriction) in unison along the vessel axis to increase or decrease vessel diameter, respectively, to effect functional changes in vascular resistance. While physically separated from SMCs by the internal elastic lamina (IEL), cellular projections through holes (fenestrae) in the IEL enable ECs to directly contact SMCs at discrete, specialized signaling domains recognized as myoendothelial junctions. In addition to the bidirectional exchange of ions and small molecules (<1 kDa), the expression of gap junctions between ECs and SMCs enables electrical charge originating in the endothelium to spread radially into SMCs [20, 67, 78]. For example, by transmitting a hyperpolarizing signal to successive SMCs through gap myoendothelial gap junctions, intercellular conduction along the endothelium coordinates SMC relaxation over distances spanning several millimeters [19]. The high input resistance of SMCs promotes effective charge transfer [i.e., changes in membrane potential (V_m)] from ECs while minimizing charge loss from the endothelium, thereby preserving the axial direction of current flow [16]. This key biophysical property of direct coupling through homocellular and heterocellular gap junctions enables electrical signals to travel rapidly along arterioles and their feed arteries, thereby coordinating vasomotor responses. Thus, conduction of hyperpolarization along the endothelium and into consecutive SMCs leads to vasodilation and an increase in tissue blood flow [11, 21, 27, 50]. Exemplary modeling studies [16, 31, 43] and in-depth reviews [9, 45, 73] have addressed the role of myoendothelial communication in greater detail.

In the context of exercise, vasodilation originating within arteriolar networks of active skeletal muscle can ascend into their proximal feed arteries, thereby increasing the volume of blood flowing through arteriolar branches and into capillary networks supplying oxygen to active skeletal muscle fibers. Complementary mechanisms of "local" and "remote" coordination of vasoreactivity have also been developed for other tissue and organ systems including the brain [13, 36], mesentery [8, 33, 75], retina [87], kidney [53, 72], lung [82] and heart [23]. While recognizing that there are multiple mechanisms for relaxing vascular smooth muscle to produce vasodilation, and that the endothelium does more than conduct electrical signals, the present discussion is based on the nature of our electrophysiological studies of native intact endothelium freshly isolated from resistance (superior epigastric) arteries of mouse skeletal muscle. With the functional role of dynamically coordinating changes in vascular resistance in mind, the present discussion is focused on answering the question: What is the biophysical basis for the initiation and transmission of electrical signals along the endothelium of resistance microvessels? The modulation of gap junctions to regulate cell-to-cell coupling has been the focus of previous studies [26, 85]. In contrast, the present discussion centers on regulating the activity of plasma membrane K^+ channels in light of their ability to modulate the electrical resistance of EC membranes and the effects of oxidative stress during advanced age.

Basis for Electrical Communication in the Endothelium of Resistance Arteries

The flux of ions through plasma membranes underlies electrical signaling in cells. Despite the endothelium typically being considered electrically unexcitable, signaling through ion flux—particularly hyperpolarization—is integral to endothelial function and its ability to both govern and coordinate vasodilation. The net flux of an ion across the plasma membrane is determined by its electrical driving force (i.e., V_m relative to the Nernst potential of the ion), its chemical driving force (i.e., transmembrane concentration gradient), the number and conductance of permeant ion channels as well as their probability of being open (i.e., nP_o). At any instant, the integration of these variables for all permeant ions determines the plasma membrane potential (V_m), which defines the electrical driving force of the cell interior (V_{in}) with reference to outside (V_{out}) the cell in the extracellular fluid [see Fig. 4.1]. Resting V_m (i.e., $V_{in} - V_{out}$) of ECs ranges from −25 mV to −50 mV as recorded in native endothelium freshly isolated from resistance arteries [5, 6], intact resistance arteries [21, 20], or arterioles controlling blood flow *in vivo* [83, 84].

The plasma membrane of ECs behaves as a capacitor for storing electrical charge in parallel with a resistor for dissipating electrical charge (i.e., current flow) through ion-permeant channels. In practice, membrane capacitance adds a temporal delay to the change in V_m in accord with the membrane time constant, tau (~240 ms; [19]). This effect is apparent during the transition from resting V_m to a more negative or more positive V_m in response to injecting square-wave pulses of current into the cell interior (Fig. 4.1). It will be recognized that the injection of hyperpolarizing current (i.e., of electrons into the cell interior) is analogous to the activation of K^+ channels with K^+ efflux resulting in a more negative V_m. Because of the constitutive expression of K^+ channels in EC membranes coupled with high mobility of K^+ in solution and the electrochemical potential for K^+, these monovalent cations are the predominant charge carrier for hyperpolarization (and thereby electrical signaling) in ECs. The monovalent anion Cl^- may also contribute to charge transfer along the endothelium, though it remains to be defined in such context as its effect is likely to be passive in response to local shifts in the electrochemical potential of K^+. During a square-wave hyperpolarizing stimulus of sufficient duration (e.g., several times longer than tau), the absolute change in V_m is achieved when membrane capacitance has fully charged and all current flow through the membrane (I_m) occurs through the resistance of the membrane (R_m), with V_m stabilized at the new steady-state (i.e., $V_m = I_m R_m$). In turn, when the hyperpolarizing current is removed (or K^+ channels close), the charge stored in the membrane capacitance is dissipated through the membrane until resting V_m is restored (Fig. 4.1). In a passive, non-rectifying system (as applies here and discussed below), equal and opposite responses occur with injection of positive current.

Simple Cable Theory: Electrical Conduction and the Role of Ion Channels

Cell-to-cell conduction of electrical charge (e.g., ion flux) through gap junctions is highly effective as a mechanism for coordinating vasomotor responses along arterioles and their feed (resistance) arteries. Such dynamic signaling in blood flow regulation effectively "tunes" local and regional perfusion according to the prevailing metabolic demands. Linear cable theory (for complete texts dedicated to principles of electrical signaling in biology consult [34] and [42]) encompasses the passive flow

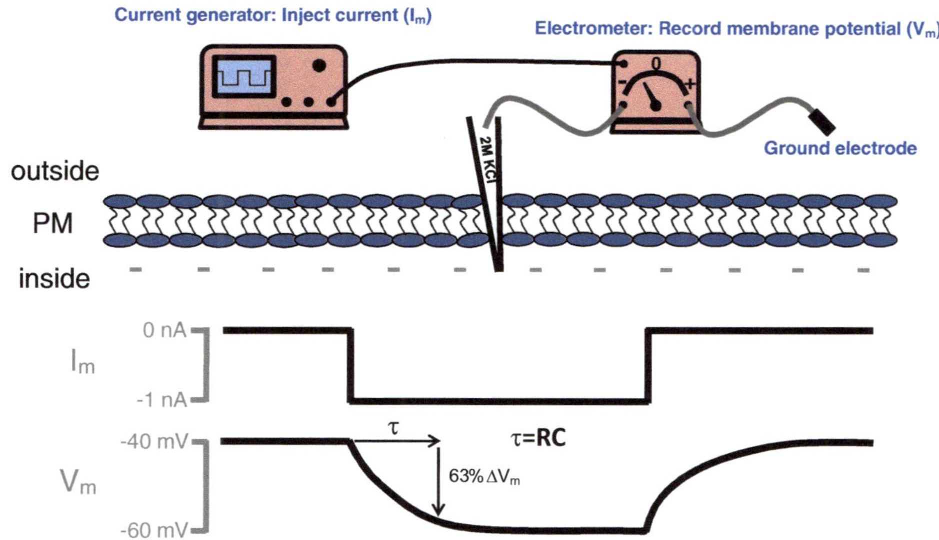

Fig. 4.1 The endothelial plasma membrane as a capacitor and resistor. The illustration shows details of the sharp electrode technique and the behavior of an endothelial plasma membrane. The sharp electrode approach entails a current generator and an electrometer to simultaneously inject current and record V_m. The onset of a square wave current injection pulse (e.g., −1 nA) elicits a change in V_m. Rather than an abrupt transition, charge separation (i.e., hyperpolarization) entails charging a capacitance (proportional to membrane surface area) as well as current flow through the membrane (proportional to electrochemical driving force and ion flux). Once capacitance is fully charged, V_m stabilizes at a new steady state as determined by the resistive properties of the membrane. Upon termination of the current pulse, charge stored in membrane capacitance (c_m) discharges through the membrane resistance (r_m) as V_m returns to the initial value. The time constant, tau ($\tau = r_m c_m$) reflects the time required for the flow of current (I_m) to alter V_m 63 % of the total change from the initial steady state to the new steady state value; τ also applies to discharging the capacitance upon removal of the current pulse

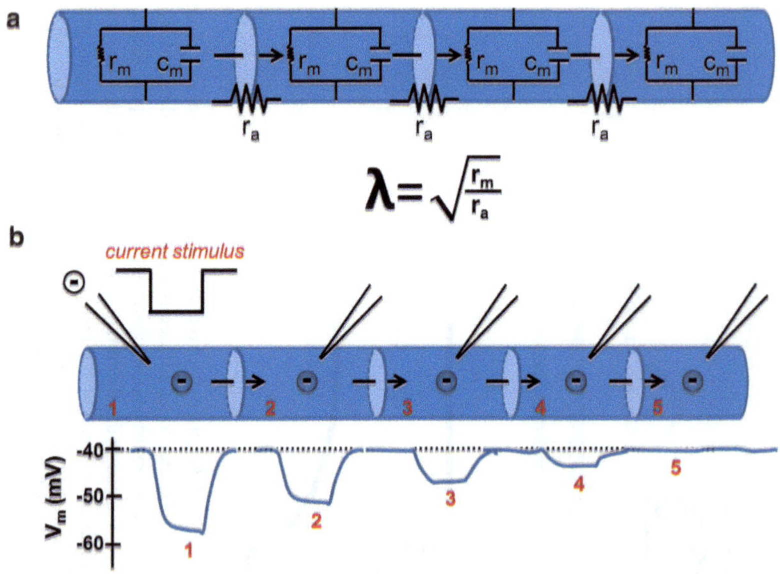

Fig. 4.2 Electrical length constant and "passive" conduction. (**a**) The illustration is a simplified schematic of a cable as reflected by an electrically coupled cellular network. The length constant [$\lambda = (r_m/r_a)^{1/2}$] is biophysical property analogous to the conduction of electricity along a cable. It reflects the square root of the ratio of membrane resistance for current flow (r_m, determined by the nP_o of ion channels) to the internal resistance for current flow (r_a) along the cable. For the endothelium, r_a is determined primarily by gap junctions between neighboring cells. A relatively high r_m and low r_a result in a longer λ. Activation of ion channels will alter r_m, whereas modulation of gap junctions will alter r_a (See Fig. 4.8). (**b**) As a model of electrical signaling along the endothelium of resistance arteries, the illustration is a schematic of a cable indicating "passive" electrical conduction. Site "1" indicates cell receiving negative (hyperpolarizing, e.g., −1 nA) current injection and eliciting a local V_m response. Note that with increasing distance (approaching further towards the right of the figure, sites 2–5), the magnitude of the change in V_m decreases progressively. Embedded "disks" (individual cells) comprising the cable with associated arrows signifying signal transmission through gap junctions. Note that the plasma membrane of cells acts as a capacitor (c_m) and resistor (r_m) in parallel. Thus, the V_m response to a current stimulus becomes stable after the plasma membrane as a capacitor has been charged (see actual V_m2 recording of Fig. 4.3b)

of electrical current along microvascular endothelium. The effective distance for non-regenerative electrical conduction is defined by the length constant, lambda (λ; $=\sqrt{r_m/r_a}$), whereby r_m is membrane resistivity (i.e., "leakiness" to ion flux, determined primarily by nP_o of ion channels spanning the lipid bilayer) and r_a is the axial resistivity to current flow along the endothelium. Assuming negligible resistance imposed by the cytoplasm, r_a is determined primarily by the number and patency of gap junctions (see Fig. 4.2a, b). Thus, in light of the functional morphology discussed above, long ECs with fewer gap junctions to traverse per unit distance effectively promotes electrical conduction along the vessel wall. Nevertheless, independent of changes in (or regulation of) cell-to-cell coupling through gap junctions, the distance that electrical current effectively travels along the vessel wall can be modulated by

nP_o [5] as discussed in the following section. An array of ion channels may be expressed in the plasma membrane of ECs [38, 62]. Here we focus on those concerned with the initiation of hyperpolarization and its conduction along the vessel wall leading to the coordinated relaxation of SMCs and vasodilation.

In resistance vessels, SK_{Ca} ($K_{Ca}2.3$) and IK_{Ca} ($K_{Ca}3.1$) channels are expressed abundantly in the plasma membrane of ECs [50, 78]. The SK_{Ca}/IK_{Ca} proteins contains six transmembrane domains (pore between S5 and S6) and intracellular amino and carboxy termini [50]. In response to increases in intracellular calcium concentration ($[Ca^{2+}]_i$), the Ca^{2+} sensor protein calmodulin interacts with the carboxy terminus of SK_{Ca}/IK_{Ca} to open the channel pore and thereby allow K^+ efflux according to its electrochemical potential [50]. In contrast to the large conductance Ca^{2+}-activated K^+ channels (BK_{Ca}) typically located in SMC membranes, SK_{Ca}/IK_{Ca} do not contain a voltage sensor to transduce changes in V_m towards affecting nP_o [49]. For ECs that contain the inward-rectifying K^+ channel (K_{IR}; dominant isoforms are of the $K_{IR}2.x$ family [18, 22, 62]), K_{IR} can serve as an extracellular K^+ sensor to modulate K^+ efflux [61]. It should also be recognized that, with electrical coupling between ECs and SMCs via myoendothelial gap junctions, activation of K_{IR} in SMCs can contribute additional outward current flow and thereby enhance the conduction of hyperpolarization and vasodilation [40, 70].

To enable charge transfer (i.e., current flow) between neighboring ECs, functional gap junction channels bridge respective plasma membranes. For a comprehensive review regarding role of gap junctions in regulating cellular communication along blood vessels, the confounding actions imposed by gap junction pharmacological manipulation of gap junctions and the genetic deletion of endothelial connexins consult [9]. Each EC contributes a hemichannel (i.e., a connexon) which dock to form a pore that is physically shielded from the extracellular fluid. Each connexon contains six connexin (Cx) subunits with Cx37, Cx40 and Cx43 the primary isoforms present in resistance microvessels [30, 51, 66]. The pore diameter of gap junctions is ~9 nm, which allows ions and second messengers having a mass <1 kDa to pass directly from within one cell to the next [79]. Thus, ions with high mobility (e.g., Na^+, K^+, Cl^-), second messengers (e.g., Ca^{2+}, 1,4,5-inositol trisphosphate, cyclic adenosine monophosphate), or fluorescent dye (e.g., propidium iodide; ~668 Da) can pass between ECs that are effectively coupled through gap junctions.

Given the integral role of the endothelium in conducting hyperpolarization, vasodilation and increased tissue perfusion [20, 21, 69], we developed freshly isolated endothelial "tubes" as an experimental preparation that is free from potentially confounding influences that include transmural pressure and blood flow, SMCs and perivascular nerves (via myoendothelial coupling), as well as signaling molecules contained within the blood or surrounding tissue. Using this preparation, the spread of electrical charge along native intact endothelium of resistance arteries was evaluated for the first time [5, 7]. Given the integral role of gap junctions in electrical signaling between coupled cells, previous studies of conducted vasodilation and signal transmission have focused on manipulating r_a through altering the expression profile of connexin subunits [15, 26, 47, 85]. However, as alluded to above, r_m is a critical determinant of the length constant for electrical conduction and is governed by ion channels in the EC plasma membrane. We have used the endothelial tube preparation to investigate whether—and if so, how—manipulation of SK_{Ca}/IK_{Ca} (and thereby r_m) can affect the nature of electrical conduction along the endothelium.

Electrical Conduction Along Endothelial Tubes

Given its role in vasodilation, the conduction of hyperpolarization is typically emphasized when discussing endothelial signal transmission. However, experiments have shown that the endothelium is equally proficient in the conduction of depolarization. Thus, intracellular injection of positive current evokes a change in V_m equal and opposite to that obtained with negative current of the same magnitude and the amplitude of the V_m response changes directly with the magnitude and polarity of injected current through the range of ±3 nA (Figs. 4.3b and 4.4a). For example, injection of ±1 nA into a single EC changes V_m by ±8 mV when recorded at a distance of 500 μm, which corresponds to the distance of ~15 ECs (each ~35 μm long) positioned end-to-end along the vessel axis. At a separation distance of 500 μm,

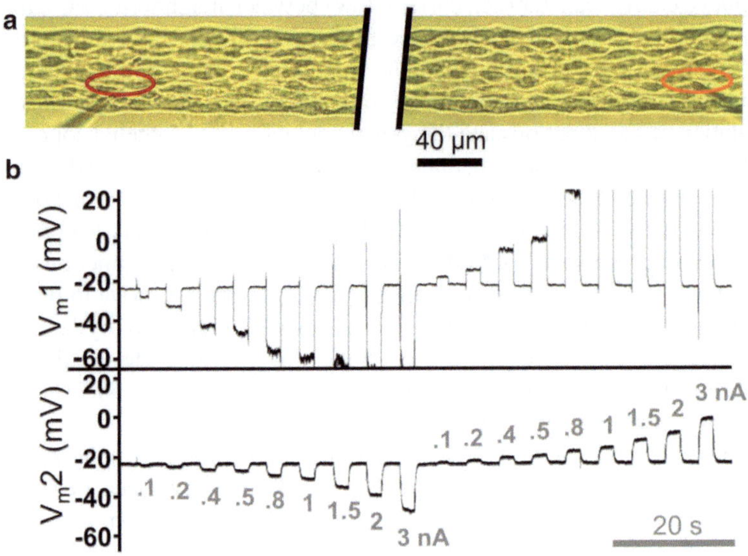

Fig. 4.3 Sharp electrode recordings of electrical conduction along endothelium of resistance arteries. (**a**) Brightfield image of endothelial tube showing dual intracellular electrodes ("Site 1", *left* (*red oval*) and "Site 2", *right* (*orange oval*)) positioned in an endothelial tube isolated from a mouse superior epigastric artery. Illustration is from [5]. (**b**) Representative dual simultaneous intracellular recording illustrating V_m traces at Site 1 (V_m1; where current is injected) and Site 2 (V_m2) separated by 500 μm to evaluate electrical conduction along the endothelium. Current pulses (±0.1–3 nA, ~2 s each) were injected at Site 1 while V_m was recorded at Sites 1 and 2. Note that duration of current pulse was sufficient for obtaining stable ΔV_m2 responses following charging of the plasma membrane capacitance (see Figs. 4.1 and 4.2). These data are from [7]; with permission

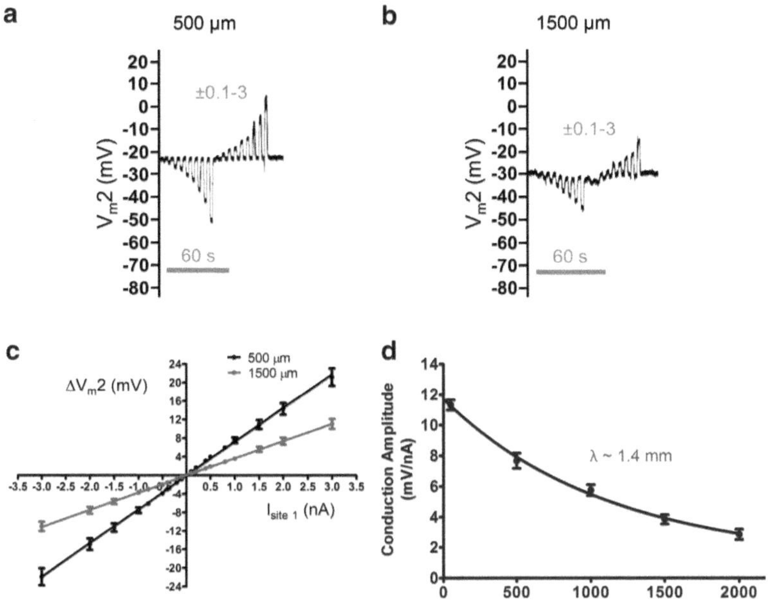

Fig. 4.4 Electrical conduction along the endothelium of resistance arteries. (**a**) Representative V_m recording at site 2 during ±0.1–3 nA injected at site 1 (distance = 500 μm). (**b**) As in Panel **a**, with a separation distance of 1500 μm. Note reduction in V_m2 responses compared with 500 μm distance illustrated in Panel **a**. (**c**) Plot of summary data for current injected into Site 1 versus ΔV_m at Site 2. *Black line*: recordings at 500 μm separation represents "Conduction Amplitude" ~8 mV/nA. *Grey line*: recordings at 1500 μm separation. Slopes (linear regression, $R^2 > 0.99$) of respective lines are the "Conduction Amplitude." (**d**) Conduction amplitude with increasing separation distance between Sites 1 and 2 (50–2000 μm). Length constant (λ) ~ 1.4 mm calculated from mono-exponential fit ($R^2 = 0.99$). Summary data in **c** and **d** reflect means ± standard error (n = 11 at 50–1500 μm; n = 7 at 2000 μm). These data are taken from [5, 7] with permission

the linearity in V_m responses (i.e., ΔV_m from the initial V_m at rest) encompasses a dynamic range of ~50 mV. As the distance between the local site of current injection and the remote site of recording V_m increases, the amplitude of bidirectional V_m responses decays but remains proportional to the level of current injected (compare Figs. 4.4a and b). It should be recognized that such linear behavior in the V_m response to current injection at each distance away from the site of current injection (Fig. 4.4c) indicates a lack of voltage-gated ion channels [49]. In contrast, if voltage-activated ion channels were present, rectification (i.e., abrupt non-linearity in the V_m response) would occur for either inward (e.g., per K_{IR}) or outward [e.g., per voltage-gated K^+ channel (K_v)] ionic currents. Thus, at each distance from the site of current injection, the linear slope of the current-voltage relationship is evaluated (e.g., in response to injecting −3 to +3 nA and recording V_m at distances 50–2000 μm from the injection site) and referred to as "Conduction Amplitude" [5, 7].

Consistent with the passive spread of electrical signals along a cable, Conduction Amplitude decays mono-exponentially with distance along the endothelium (Fig. 4.4d) with a λ of ~1.4 mm. Empirically, λ corresponds to the distance over which the electrical signal decayed to 37 % [(1/e) of the "local" ΔV_m]. Remarkably, this value for λ is similar to what has been determined for intact arterioles isolated from the guinea pig submucosa (1.1–1.6 mm [35]) and intact feed arteries isolated from the hamster cheek pouch retractor muscle (λ: 1.2 mm [19]). Our finding that the amplitude of the electrical signal decays by 4.4 % per 100 μm [5, 7] is also consistent with the decay of ~5 % per 100 μm observed in retinal arterioles of the rat [87]. Thus, whether isolated or coupled with SMCs through myoendothelial gap junctions, the endothelium of resistance microvessels effectively serves as the predominant cellular pathway for the transmission of electrical signals that underlie conducted vasodilation. This consistency across preparations supports a property stated earlier in this discussion; i.e., that the electrical signal is transmitted to SMCs with negligible charge loss due to the conductance of myoendothelial gap junctions in conjunction with the high input resistance of individual SMCs [16].

Difficulties in Resolving the Role of Gap Junctions in Modulating Electrical Conduction

The use of pharmacologic agents to manipulate gap junctions and genetic deletion of connexin subunits may have confounding effects (see [9, 41]). For example, several laboratories have identified non-selective actions of the putative gap junction blockers carbenoxolone and β-glycyrrhetinic acid [7, 54, 65, 76, 80]. While these agents do block gap junctions, their unintended effects include the inhibition of K_{Ca} channels that initiate hyperpolarization in response to acetylcholine, as shown for guinea pig and rat arteries [76] as well as endothelial tubes freshly isolated from mouse resistance arteries [7]. The implications of such findings are considerable because these pharmacological agents have been used widely to assess the presence of endothelium-derived hyperpolarization and relaxation of vascular smooth muscle in response to acetylcholine [17, 54, 55, 76, 86]. Such non-junctional effects [e.g., transient hyperpolarization and/or depolarization; suppressed activation of SK_{Ca}/IK_{Ca}] at concentrations less than or equal to what is required for blocking gap junctions [7, 54, 76] indicates that glycyrrhetinic acid derivatives may not effectively resolve the role of gap junctions from other ion-permeant channels in shaping electrical conduction along the endothelium. Nevertheless, intercellular conduction of electrical signals is blocked by carbenoxolone and β-glycyrrhetinic acid as indicated by the inhibition of electrical conduction and loss of dye transfer (e.g., propidium iodide) from one cell to several neighboring cells (see Fig. 4.5c) [7].

It is reassuring that the inhibition of gap junctions with glycyrrhetinic acid derivatives in our experiments is consistent with the findings of others [17, 54, 76, 86]. Nevertheless, resolving the role of r_a in regulating electrical signaling along vascular endothelium remains a difficult challenge. Our experiments on endothelial tubes have shown that these agents can inhibit endothelium-derived hyperpolarization whether it is evoked indirectly with acetylcholine (via G_q protein-coupled muscarinic receptors) or via the direct action of the SK_{Ca}/IK_{Ca} opener NS309 [7].

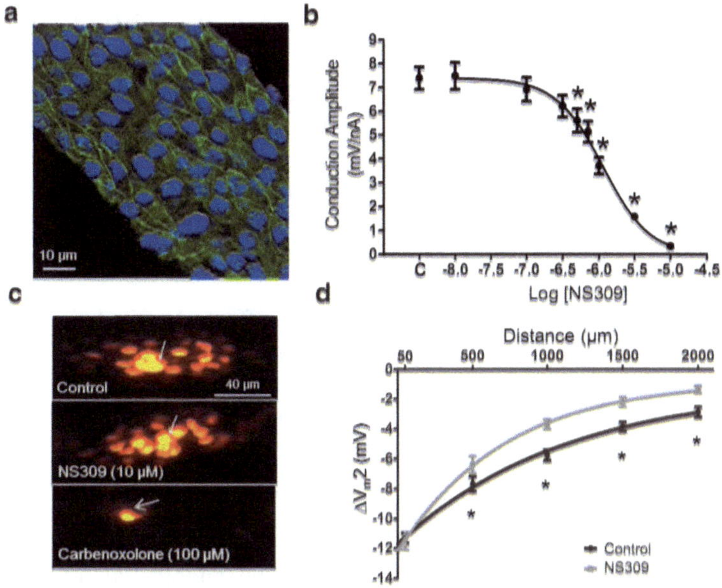

Fig. 4.5 The role of K^+ channel activation and r_m in governing electrical conduction. (**a**) *Robust expression of IK_{Ca} in endothelial tube.* Image depicts confocal slice image of an endothelial tube isolated from a male mouse superior epigastric artery indicating presence of IK_{Ca} (*green*) and nuclei of individual cells (*blue*). Expression of SK_{Ca} has been confirmed in the same manner (not shown). (**b**) *Effect of progressive activation of SK_{Ca}/IK_{Ca} on electrical conduction.* Plot of summary data (means ± standard error) for Conduction Amplitude (Site 2 located 500 μm from current injection at Site 1; refer to Fig. 4.3) during increasing [NS309]. Note that electrical conduction decreases with increasing [NS309] up to complete inhibition with 10 μM. *P<0.05, significantly different from control indicated as "C" (n=6–8). (**c**) *Intercellular dye transfer through gap junctions is maintained during maximal SK_{Ca}/IK_{Ca} activation with NS309.* Propidium iodide dye (0.1 % in 2 mol/L KCl) was included in microelectrode during intracellular recording to evaluate intercellular coupling through gap junction channels. *Arrows* indicate impaled cell in each panel and recordings lasted 30 min. Dye transfer occurred during control conditions (*top*) or following 5 min of 10 μM NS309 treatment (*middle*) where V_m was ~−80 mV and conduction was abolished. In contrast, treatment with the gap junction blocker carbenoxolone (100 μM, 5 min) prevented dye transfer between cells. (**d**) *Effect of SK_{Ca}/IK_{Ca} activation on spatial decay of electrical conduction.* Electrical conduction versus separation distance between intracellular microelectrodes before and during treatment with NS309. At each distance (50–2000 μm), continuous (paired) recordings were obtained under control conditions and during NS309 (1 μM, ~EC_{50}). With current injection adjusted to produce the same local V_m (control: −1 nA; NS309: −2 nA), the ΔV_m2 (resting V_m minus peak response V_m) with distance indicates greater decay of hyperpolarization with NS309 (Control λ: ~1.4 mm, NS309 λ: ~0.9 mm). *P<0.05, control significantly different from NS309. Summary data illustrate mean ± standard error (n=11 at 50–1500 μm; n=7 at 2000 μm). These data are taken from [5, 7] with permission

Thus the endothelial tube preparation may serve as a useful experimental tool for characterizing the actions of agents purported to act selectively on gap junctions. It may also prove useful in assessing how the genetic manipulation of connexin expression affects the intercellular transmission of electrical signals, as such studies to date have focused on changes in the nature of conducted vasomotor responses along intact arteriolar networks [15, 25, 85].

Effects of SK_{Ca}/IK_{Ca} Activation on Electrical Conduction

The activation of SK_{Ca}/IK_{Ca} is integral to endothelial-derived hyperpolarization [3, 27]. Once hyperpolarization is initiated, the endothelium serves as an effective cellular pathway for electrical conduction and coordinating vasodilatation as discussed above. It should also be recognized that the activity of SK_{Ca}/IK_{Ca} has a direct effect on r_m and can thereby regulate the transmission of electrical signals along the endothelium. This role is now explored, with particular attention given to the effects of aging and oxidative stress.

The possibility of investigating how r_m could affect electrical signal transmission led us to the question: Can the efficacy of electrical conduction along the endothelium be tuned according to the level of activation of ion channels in the plasma membrane? To provide an answer we focused on manipulating SK_{Ca}/IK_{Ca} activity as these channels are highly expressed in the endothelial tubes studied thus far (Fig. 4.5a). Importantly, the open probability of SK_{Ca}/IK_{Ca} can be controlled acutely during experiments via the addition of pharmacological agents that selectively activate (e.g., NS309) or inhibit (apamin + charybdotoxin) respective channels [5]. The experimental approach entailed impaling endothelial tubes with dual simultaneous intracellular microelectrodes to inject current into one EC while recording V_m in a remote EC at a defined separation distance, typically 500 μm (see Fig. 4.3). The progressive activation of SK_{Ca}/IK_{Ca} (e.g., with increasing concentrations of NS309 added to the tissue bath) resulted in progressive hyperpolarization. Concomitantly recording V_m from the remote EC demonstrated a progressive reduction in the efficacy of electrical signal transmission (i.e., conduction amplitude; Fig. 4.5b). Remarkably, when the remote response was completely abolished (using 10 μM NS309), gap junction patency remained fully intact as confirmed with dye (propidium iodide) transfer from the injected EC to its surrounding neighbors as seen under control conditions (Fig. 4.5c, compare top and middle panels). Because cells hyperpolarized as electrical conduction was impaired during SK_{Ca}/IK_{Ca} activation with NS309, experiments were repeated under conditions that prevented hyperpolarization. This was achieved through changing the Nernst potential for K^+ by raising its bath concentration from 5 mM (control) to 60 mM, effectively clamping V_m at its resting level under control conditions. Remarkably, despite negligible change in V_m, the activation SK_{Ca}/IK_{Ca} consistently impaired electrical conduction along the endothelium, attributable to the increase in nP_o augmenting the ability of K^+ diffuse through plasma membranes and dissipate longitudinal signal transmission. Thus, hyperpolarization *per se* is not required for attenuating cell-to-cell electrical signaling through changes in r_m [5]. It should be recognized that (1) under physiological conditions, activating SK_{Ca}/IK_{Ca} within a discrete location (analogous to injecting current into a single EC) will initiate hyperpolarization without altering signal transmission along the endothelium; and that (2) the effect of ion channel activation along the endothelium on the efficacy of electrical signal transmission would apply to any permeant ion channel expressed by ECs.

The application of NS309 at a concentration that was half-maximally effective (~1 μM) decreased λ from the control value of 1.4 mm to 0.9 mm (Fig. 4.5d). While this effect was reproducible with another synthetic opener of SK_{Ca}/IK_{Ca} (e.g., SKA 31), a critical question was whether the regulation of electrical conduction through changing r_m could be achieved using a physiological agent. Thus, acetylcholine, activates SK_{Ca}/IK_{Ca} secondarily to an increase in $[Ca^{2+}]_i$, also inhibited electrical conduction (e.g., by ~70 % in response to 3 μM, a concentration of acetylcholine that evokes maximal dilation) and did so when hyperpolarization was prevented by 60 mM K^+ as we had confirmed for NS309. Further, when constitutive SK_{Ca}/IK_{Ca} activity was blocked during resting conditions with apamin+charybdotoxin the efficacy of electrical conduction increased by ~30 % [5], an effect attributable to a reduction in nP_o and enhanced r_m. Collectively these findings were the first to illustrate that, aside from initiating hyperpolarization, opening of SK_{Ca}/IK_{Ca} also regulates the efficacy of electrical conduction along the endothelium. Thus, the activity of ion channels in the plasma membrane can govern both the initiation and the conduction of electrical signals in the endothelium of resistance vessels. In turn, these properties are likely to prove integral to the spatial and temporal regulation of tissue blood flow in accord with local and regional metabolic demand.

The Effect of Advanced Age on Electrical Conduction: Regulation by Hydrogen Peroxide

Older humans (e.g., >60 years) exhibit endothelial dysfunction [24, 68, 77, 81], a disorder characterized by impaired vasodilation in response to acetylcholine [39, 46, 74], to physical activity [46, 63] or to heating the skin [44]. Complementary findings in comparatively older rats and mice (e.g., >20 months) support the use of animal models when investigating mechanisms underlying vascular dysfunction with advancing age [4, 14, 37, 58]. Nitric oxide bioavailability decreases in advanced age, with greater reliance on reactive oxygen species in mediating endothelium-dependent vasodilation [57, 58, 77]. Moreover, the function of SK_{Ca}/IK_{Ca} in ECs may be altered in pathological states [10, 28, 52]. However, the impact of aging on SK_{Ca}/IK_{Ca} function, particularly in light of impaired blood flow, is poorly understood. To address a translational application of our observations regarding SK_{Ca}/IK_{Ca} activity and the regulation of conduction through changes in r_m, we tested the hypothesis that electrical conduction along the endothelium would be impaired in advanced age [6].

As shown when comparing endothelial tubes from Young (4–6 months) versus Old (24–26 months) mice, the ability to evoke hyperpolarization via SK_{Ca}/IK_{Ca} activation with either acetylcholine or NS309 is sustained in advanced age [Fig. 4.6a, b]. However, more SK_{Ca}/IK_{Ca} (primarily IK_{Ca}) were open at rest in the endothelium of Old compared to that of Young mice. This conclusion was based upon V_m being 5–10 mV more negative in the endothelium of Old mice at rest and depolarizing by an additional 5–10 mV during application of apamin+charybdotoxin when compared to endothelium of Young mice. Further, the efficacy of electrical conduction was reduced significantly in endothelium of Old compared to Young mice [Figs. 4.6c and 4.6d, control]. We hypothesized that these subtle alterations in SK_{Ca}/IK_{Ca} activity during advanced age may be due, at least in part, to elevated oxidative stress.

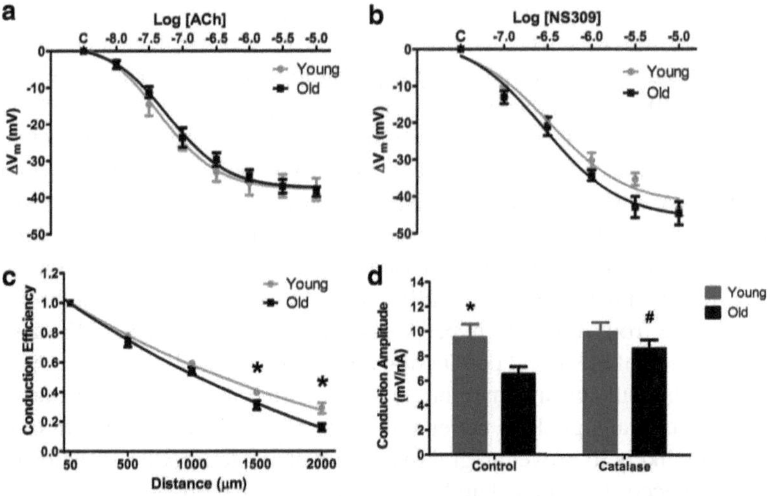

Fig. 4.6 The role of age-related oxidative stress in governing electrical conduction. (**a**, **b**) *The ability to hyperpolarize the endothelium either indirectly (acetylcholine, ACh) or directly (NS309) through SK_{Ca}/IK_{Ca}activation is sustained with advanced age.* Plots of summary data indicating effects of ACh (**a**) and NS309 (**b**) on the change in membrane potential [ΔV_m from control V_m] in endothelial tubes isolated from Young (4–6 months) and Old (24–26 months) mice (n=6 per group). (C) *Electrical conduction along the endothelium of resistance arteries is impaired in Old vs. Young mice:* Plot of summary data indicating "conduction efficiency" [conduction amplitude normalized to local (50 μm) site for each respective age group] in Young vs. Old with increasing distance. Note relatively greater decay with distance in Old (at 1500–2000 μm; Young λ: ~1.6 mm, Old λ: ~1.3 mm). *P<0.05, Young (n=12) vs Old (n=9). (**d**) *Catalase improves electrical conduction along endothelium of Old vs. Young mice.* Plot of summary data for conduction amplitude at 500 μm distance under respective conditions for Young (n=6) and Old (n=7). *P<0.05 vs. Old under Control condition; #P<0.05 vs. Old Control. All data indicate means ± standard error. These data are taken from [6]

Aging and endothelial dysfunction have been attributed to heightened oxidative stress [14, 32, 58] including excess production of hydrogen peroxide [2, 58, 59, 71]. The effect of aging and oxidative stress on impaired NO signaling is well documented [57, 58, 68, 77]. However, it was unclear how age-related oxidative stress may impact electrical signaling along the endothelium. Consistent with earlier reports [12, 56, 64], we determined that acute exposure to hydrogen peroxide (200 μM for 20 min) hyperpolarized ECs of Young mice to levels approaching the Nernst potential for K^+ (−90 mV) and did so through the activation of SK_{Ca}/IK_{Ca} [6] (Fig. 4.7). Ensuing experiments revealed that inhibiting SK_{Ca}/IK_{Ca} (or even IK_{Ca} alone) as well as decomposing hydrogen peroxide with catalase restored the efficacy of electrical conduction in the endothelium of Old to that recorded in Young (Fig. 4.6d). Thus, constitutive production of hydrogen peroxide in microvascular endothelium during advancing age can modulate SK_{Ca}/IK_{Ca} activity in a manner similar to NS309 treatment [5], whereby resting V_m is greater (i.e., more negative) and the ability to spread vasodilatory electrical signals is depressed (Fig. 4.8). A key question for future studies is to

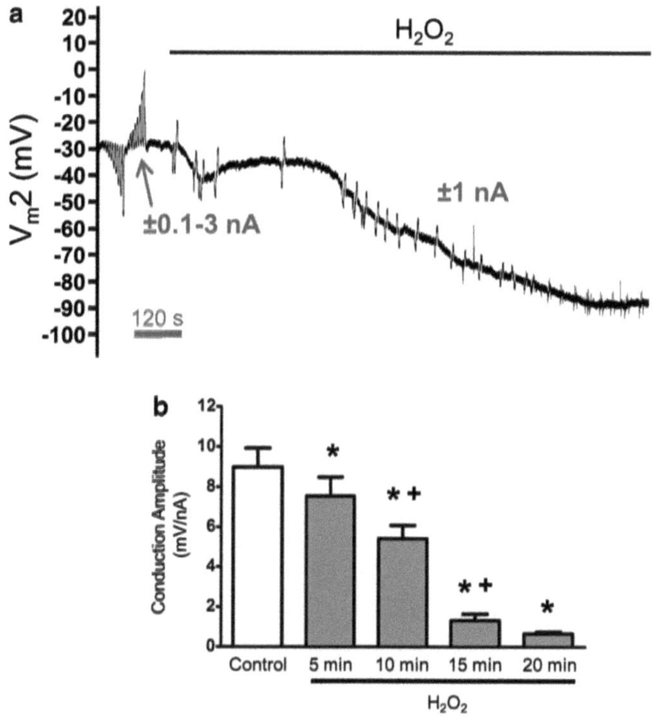

Fig. 4.7 The role of hydrogen peroxide and oxidative stress in governing electrical conduction. (**a**) Representative recording of V_m responses at 500 μm (V_m2) from current injected at Site 1 before and during hydrogen peroxide (H_2O_2, 200 μM) exposure. Note progressive hyperpolarization and loss of V_m2 responses with residual capacitance spikes. (**b**) Plot of summary data (means ± standard error) before (Control) and during exposure to 200 μM hydrogen peroxide on Conduction Amplitude (Site 2 located 500 μm away from current injection at Site 1). Note that electrical conduction diminishes progressively over time during the 20 min exposure. $^*P<0.05$ vs. control; $^+P<0.05$ vs. preceding time point (n = 6–8). These data are taken from [6]

determine whether the actions of hydrogen peroxide on SK_{Ca}/IK_{Ca} activity are direct (e.g., via oxidation of cysteine residues within channel subunits) and/or indirect due to local increases in $[Ca^{2+}]_i$. Consistent with the latter effect, our recent findings illustrate that acute exposure to hydrogen peroxide increases $[Ca^{2+}]_i$ [71] with a time course that mimics its effects on V_m via SK_{Ca}/IK_{Ca} activation [6].

Summary and Conclusions

Electrical signaling is integral to the control of blood flow to tissues and organs throughout the body. A key determinant of electrical signaling is the V_m of ECs, which is regulated by the number and type of open channels and the electrochemical potential driving ion flux across the plasma membrane. Local changes in V_m give

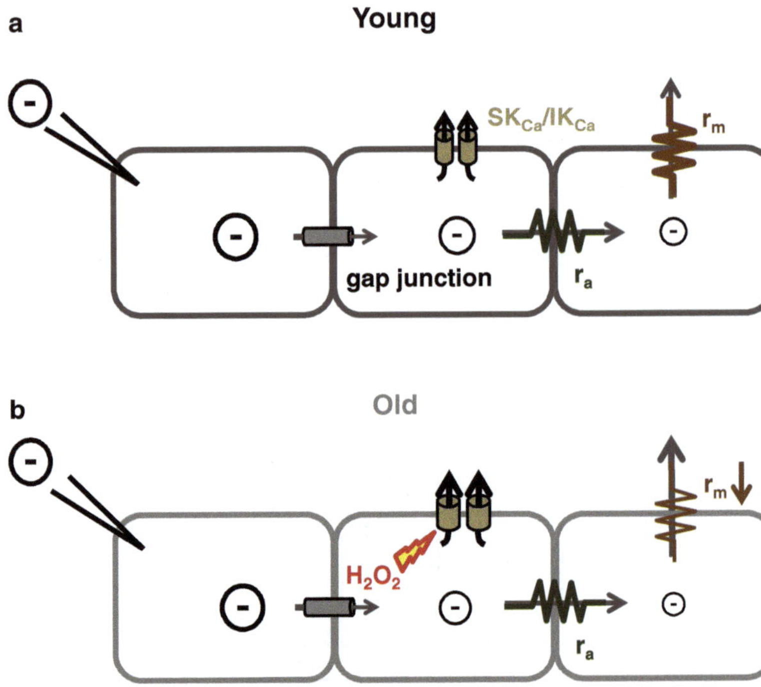

Fig. 4.8 Biophysical determinants of electrical transmission along the endothelium of resistance arteries: role of oxidative signaling and enhanced K^+channel activity during aging. Hyperpolarization [*circled negative symbols*; initiated by activation of negative current injection or activation of SK_{Ca}/IK_{Ca} spreads to neighboring cells through gap junctions. The efficacy of conducted hyperpolarization between cells depends upon both gap junction patency (i.e., r_a) and "leakiness" of the plasma membrane (i.e., r_m) defined by SK_{Ca}/IK_{Ca} activity. (**a**) In endothelium from Young, only a few SK_{Ca}/IK_{Ca} are open and gap junction patency is high, enabling effective conduction of hyperpolarization from cell to cell along the endothelium to govern conducted vasodilation. (**b**) During advanced age and constitutively elevated hydrogen peroxide (H_2O_2), SK_{Ca}/IK_{Ca} activity is enhanced; sustaining local signaling but reducing the strength of electrical conduction through gap junctions due to charge loss (leakage) through the plasma membrane. This effect may contribute to the diminished spread of vasodilation and blood flow limitations with advanced age

rise to the conduction of electrical signals along the endothelium, analogous to the spread of electricity along a cable. Charge spreads to other cells via ion flux through gap junctions or leaks through the plasma membrane when its properties as an insulator are compromised; e.g., through the opening of SK_{Ca}/IK_{Ca}. Herein, we have highlighted the unique morphological and electrical properties of microvascular endothelium that underscore its role as the principal pathway for initiating and transmitting changes in V_m along the intima of resistance vessels. We suggest that the ability of the endothelium to transmit the electrical signal for vasodilation can be regulated temporally and spatially in accord with the metabolic demand of tissue parenchymal cells. Underlying these physiological properties is the molecular organization of endothelial plasma membranes as manifested through the expression of SK_{Ca}/IK_{Ca} and connexin subunits of gap junctions. Further, the longitudinal orientation of ECs in contact with multiple SMCs encircling the intima favors intercellular coordination of the electrical signals that are transduced (via myoendothelial coupling to SMCs) into effective changes in vessel diameter. While the endothelium behaves electrically as a passive cable, the efficacy of charge transfer can be finely tuned via graded activation of ion channels in the plasma membrane. Thus, even in the presence of fully patent gap junctions between neighboring cells, electrical signaling along and between vessel branches can be abolished when SK_{Ca}/IK_{Ca} are maximally activated. Conversely, blocking constitutively active SK_{Ca}/IK_{Ca} channels will enhance electrical conduction and promote the spread of vasodilation. Although the lack of selective gap junction inhibitors has limited our ability to acutely manipulate cell-to-cell coupling (i.e., r_a) in the manner we have done for r_m via K^+ channel interventions, genetic deletion of connexin 40 (i.e., a primary protein subunit that forms EC gap junctions) has been shown to impair conducted vasodilation in arterioles [15, 25, 26, 85]. The latter findings illustrate that longitudinal conduction can also be governed through changes in r_a. In turn, acute (i.e., physiological) regulation of r_a via changes in gap junction conductance may occur through phosphorylation and/or nitrosylation of connexin subunits [9].

Our recent findings confirm the role of SK_{Ca}/IK_{Ca} for initiating of EC hyperpolarization [11, 50] while revealing their role in governing the longitudinal transmission of electrical signals [5]. Thus, SK_{Ca}/IK_{Ca} channels serve to initiate smooth muscle relaxation (vasodilation) through myoendothelial coupling and to modulate the spread of vasodilation along resistance networks. With advancing age, attenuating the spatial domain of electrical conduction can restrict the spread of vasodilation and contribute to blood flow limitations [37]. This effect is attributable to oxidative stress associated with greater availability of hydrogen peroxide, which activates SK_{Ca}/IK_{Ca}. In turn, enhanced charge dissipation restricts the spatial domain of electrical signaling. Thus, we suggest that impaired tissue blood flow during endothelial dysfunction with advanced age reflects enhanced ion channel activation (i.e., "leaky" plasma membranes) that impairs effective control of vascular resistance. Impaired longitudinal conduction associated with endothelial dysfunction may also reflect alterations in gap junction expression and/or regulation in accord with the nature of disease (e.g., hypertension, diabetes) and the underlying signaling pathways involved. Therapeutic interventions designed to restore tissue blood flow through modulating the activity of ion channels may prove most effective when striking a balance between generating local signals governing vasodilation and regulating the transmission of electrical signals along the intima.

Acknowledgements This research was supported by National Institutes of Health grants 1K99-AG047198 to E.J.B. and R37-HL041026 to S.S.S. The content of this article is solely the responsibility of the authors and does not necessarily represent the official views of the National Institutes of Health.

References

1. Andersen P, Saltin B. Maximal perfusion of skeletal muscle in man. J Physiol. 1985; 366:233–49.
2. Bachschmid MM, Schildknecht S, Matsui R, Zee R, Haeussler D, Cohen RA, Pimental D, Loo B. Vascular aging: chronic oxidative stress and impairment of redox signaling-consequences for vascular homeostasis and disease. Ann Med. 2013;45:17–36.
3. Bagher P, Segal SS. Regulation of blood flow in the microcirculation: role of conducted vasodilation. Acta Physiol (Oxf). 2011;202:271–84.
4. Bearden SE, Payne GW, Chisty A, Segal SS. Arteriolar network architecture and vasomotor function with ageing in mouse gluteus maximus muscle. J Physiol. 2004;561:535–45.
5. Behringer EJ, Segal SS. Tuning electrical conduction along endothelial tubes of resistance arteries through Ca^{2+}-activated K^+ channels. Circ Res. 2012;110:1311–21.
6. Behringer EJ, Shaw RL, Westcott EB, Socha MJ, Segal SS. Aging impairs electrical conduction along endothelium of resistance arteries through enhanced Ca^{2+}-activated K^+ channel activation. Arterioscler Thromb Vasc Biol. 2013;33:1892–901.
7. Behringer EJ, Socha MJ, Polo-Parada L, Segal SS. Electrical conduction along endothelial cell tubes from mouse feed arteries: confounding actions of glycyrrhetinic acid derivatives. Br J Pharmacol. 2012;166:774–87.
8. Beleznai TZ, Yarova PL, Yuill KH, Dora KA. Smooth muscle Ca^{2+}-activated and voltage-gated K^+ channels modulate conducted dilation in rat isolated small mesenteric arteries. Microcirculation. 2011;18:487–500.
9. Billaud M, Lohman AW, Johnstone SR, Biwer LA, Mutchler S, Isakson BE. Regulation of cellular communication by signaling microdomains in the blood vessel wall. Pharmacol Rev. 2014;66:513–69.
10. Brahler S, Kaistha A, Schmidt VJ, Wolfle SE, Busch C, Kaistha BP, Kacik M, Hasenau AL, Grgic I, Si H, Bond CT, Adelman JP, Wulff H, de Wit C, Hoyer J, Kohler R. Genetic deficit of SK3 and IK1 channels disrupts the endothelium-derived hyperpolarizing factor vasodilator pathway and causes hypertension. Circulation. 2009;119:2323–32.
11. Busse R, Edwards G, Feletou M, Fleming I, Vanhoutte PM, Weston AH. EDHF: bringing the concepts together. Trends Pharmacol Sci. 2002;23:374–80.
12. Bychkov R, Pieper K, Ried C, Milosheva M, Bychkov E, Luft FC, Haller H. Hydrogen peroxide, potassium currents, and membrane potential in human endothelial cells. Circulation. 1999;99:1719–25.
13. Chen BR, Kozberg MG, Bouchard MB, Shaik MA, Hillman EM. A critical role for the vascular endothelium in functional neurovascular coupling in the brain. J Am Heart Assoc. 2014;3:e000787.
14. Csiszar A, Labinskyy N, Zhao X, Hu F, Serpillon S, Huang Z, Ballabh P, Levy RJ, Hintze TH, Wolin MS, Austad SN, Podlutsky A, Ungvari Z. Vascular superoxide and hydrogen peroxide production and oxidative stress resistance in two closely related rodent species with disparate longevity. Aging Cell. 2007;6:783–97.
15. de Wit C, Roos F, Bolz SS, Kirchhoff S, Kruger O, Willecke K, Pohl U. Impaired conduction of vasodilation along arterioles in Connexin40-deficient mice. Circ Res. 2000;86:649–55.
16. Diep HK, Vigmond EJ, Segal SS, Welsh DG. Defining electrical communication in skeletal muscle resistance arteries: a computational approach. J Physiol. 2005;568:267–81.
17. Dora KA, Gallagher NT, McNeish A, Garland CJ. Modulation of endothelial cell $K_{Ca}3.1$ channels during endothelium-derived hyperpolarizing factor signaling in mesenteric resistance arteries. Circ Res. 2008;102:1247–55.
18. Edwards G, Weston AH. Potassium and potassium clouds in endothelium-dependent hyperpolarizations. Pharmacol Res. 2004;49:535–41.
19. Emerson GG, Neild TO, Segal SS. Conduction of hyperpolarization along hamster feed arteries: augmentation by acetylcholine. Am J Physiol Heart Circ Physiol. 2002;283:H102–9.
20. Emerson GG, Segal SS. Electrical coupling between endothelial cells and smooth muscle cells in hamster feed arteries: role in vasomotor control. Circ Res. 2000;87:474–9.

21. Emerson GG, Segal SS. Endothelial cell pathway for conduction of hyperpolarization and vasodilation along hamster feed artery. Circ Res. 2000;86:94–100.
22. Fang Y, Schram G, Romanenko VG, Shi C, Conti L, Vandenberg CA, Davies PF, Nattel S, Levitan I. Functional expression of Kir2.x in human aortic endothelial cells: the dominant role of Kir2.2. Am J Physiol Cell Physiol. 2005;289:C1134–44.
23. Feher A, Broskova Z, Bagi Z. Age-related impairment of conducted dilation in human coronary arterioles. Am J Physiol Heart Circ Physiol. 2014;306:H1595–601.
24. Feletou M, Vanhoutte PM. Endothelial dysfunction: a multifaceted disorder (The Wiggers Award Lecture). Am J Physiol Heart Circ Physiol. 2006;291:H985–1002.
25. Figueroa XF, Duling BR. Dissection of two Cx37-independent conducted vasodilator mechanisms by deletion of Cx40: electrotonic versus regenerative conduction. Am J Physiol Heart Circ Physiol. 2008;295:H2001–7.
26. Figueroa XF, Paul DL, Simon AM, Goodenough DA, Day KH, Damon DN, Duling BR. Central role of connexin40 in the propagation of electrically activated vasodilation in mouse cremasteric arterioles in vivo. Circ Res. 2003;92:793–800.
27. Garland CJ, Hiley CR, Dora KA. EDHF: spreading the influence of the endothelium. Br J Pharmacol. 2011;164:839–52.
28. Grgic I, Kaistha BP, Hoyer J, Kohler R. Endothelial Ca^{2+}-activated K^{+} channels in normal and impaired EDHF-dilator responses--relevance to cardiovascular pathologies and drug discovery. Br J Pharmacol. 2009;157:509–26.
29. Haas TL, Duling BR. Morphology favors an endothelial cell pathway for longitudinal conduction within arterioles. Microvasc Res. 1997;53:113–20.
30. Hakim CH, Jackson WF, Segal SS. Connexin isoform expression in smooth muscle cells and endothelial cells of hamster cheek pouch arterioles and retractor feed arteries. Microcirculation. 2008;15:503–14.
31. Hald BO, Jensen LJ, Sorensen PG, Holstein-Rathlou NH, Jacobsen JC. Applicability of cable theory to vascular conducted responses. Biophys J. 2012;102:1352–62.
32. Heitzer T, Schlinzig T, Krohn K, Meinertz T, Munzel T. Endothelial dysfunction, oxidative stress, and risk of cardiovascular events in patients with coronary artery disease. Circulation. 2001;104:2673–8.
33. Hill CE, Hickey H, Sandow SL. Role of gap junctions in acetylcholine-induced vasodilation of proximal and distal arteries of the rat mesentery. J Auton Nerv Syst. 2000;81:122–7.
34. Hille B. Ion channels of excitable membranes. 3rd ed. Sunderland, MA: Sinauer Associates; 2001. ISBN 978-0878933211.
35. Hirst GD, Neild TO. An analysis of excitatory junctional potentials recorded from arterioles. J Physiol. 1978;280:87–104.
36. Iadecola C, Yang G, Ebner TJ, Chen G. Local and propagated vascular responses evoked by focal synaptic activity in cerebellar cortex. J Neurophysiol. 1997;78:651–9.
37. Jackson DN, Moore AW, Segal SS. Blunting of rapid onset vasodilatation and blood flow restriction in arterioles of exercising skeletal muscle with ageing in male mice. J Physiol. 2010;588:2269–82.
38. Jackson WF. Potassium channels in the peripheral microcirculation. Microcirculation. 2005;12:113–27.
39. James MA, Tullett J, Hemsley AG, Shore AC. Effects of aging and hypertension on the microcirculation. Hypertension. 2006;47:968–74.
40. Jantzi MC, Brett SE, Jackson WF, Corteling R, Vigmond EJ, Welsh DG. Inward rectifying potassium channels facilitate cell-to-cell communication in hamster retractor muscle feed arteries. Am J Physiol Heart Circ Physiol. 2006;291:H1319–28.
41. Juszczak GR, Swiergiel AH. Properties of gap junction blockers and their behavioural, cognitive and electrophysiological effects: animal and human studies. Prog Neuropsychopharmacol Biol Psychiatry. 2009;33:181–98.
42. Kandel ER. Principles of Neural Science. 5th ed. New York: McGraw-Hill; 2013. ISBN 978-0071390118.
43. Kapela A, Bezerianos A, Tsoukias NM. A mathematical model of vasoreactivity in rat mesenteric arterioles: I. Myoendothelial communication. Microcirculation. 2009;16:694–713.

44. Kenney WL, Morgan AL, Farquhar WB, Brooks EM, Pierzga JM, Derr JA. Decreased active vasodilator sensitivity in aged skin. Am J Physiol Heart Circ Physiol. 1997;272:H1609–14.
45. Kerr PM, Tam R, Ondrusova K, Mittal R, Narang D, Tran CH, Welsh DG, Plane F. Endothelial feedback and the myoendothelial projection. Microcirculation. 2012;19:416–22.
46. Kirby BS, Voyles WF, Simpson CB, Carlson RE, Schrage WG, Dinenno FA. Endothelium-dependent vasodilatation and exercise hyperaemia in ageing humans: impact of acute ascorbic acid administration. J Physiol. 2009;587:1989–2003.
47. Kurjiaka DT, Bender SB, Nye DD, Wiehler WB, Welsh DG. Hypertension attenuates cell-to-cell communication in hamster retractor muscle feed arteries. Am J Physiol Heart Circ Physiol. 2005;288:H861–70.
48. Laughlin MH, Armstrong RB. Muscular blood flow distribution patterns as a function of running speed in rats. Am J Physiol Heart Circ Physiol. 1982;243:H296–306.
49. Ledoux J, Bonev AD, Nelson MT. Ca^{2+}-activated K^+ channels in murine endothelial cells: block by intracellular calcium and magnesium. J Gen Physiol. 2008;131:125–35.
50. Ledoux J, Werner ME, Brayden JE, Nelson MT. Calcium-activated potassium channels and the regulation of vascular tone. Physiology (Bethesda). 2006;21:69–78.
51. Looft-Wilson RC, Payne GW, Segal SS. Connexin expression and conducted vasodilation along arteriolar endothelium in mouse skeletal muscle. J Appl Physiol. 2004;97:1152–8.
52. Luksha L, Agewall S, Kublickiene K. Endothelium-derived hyperpolarizing factor in vascular physiology and cardiovascular disease. Atherosclerosis. 2009;202:330–44.
53. Marsh DJ, Toma I, Sosnovtseva OV, Peti-Peterdi J, Holstein-Rathlou NH. Electrotonic vascular signal conduction and nephron synchronization. Am J Physiol Renal Physiol. 2009;296:F751–61.
54. Matchkov VV, Rahman A, Peng H, Nilsson H, Aalkjaer C. Junctional and nonjunctional effects of heptanol and glycyrrhetinic acid derivates in rat mesenteric small arteries. Br J Pharmacol. 2004;142:961–72.
55. Mather S, Dora KA, Sandow SL, Winter P, Garland CJ. Rapid endothelial cell-selective loading of connexin 40 antibody blocks endothelium-derived hyperpolarizing factor dilation in rat small mesenteric arteries. Circ Res. 2005;97:399–407.
56. Miura H, Bosnjak JJ, Ning G, Saito T, Miura M, Gutterman DD. Role for hydrogen peroxide in flow-induced dilation of human coronary arterioles. Circ Res. 2003;92:e31–40.
57. Muller-Delp JM. Aging-induced adaptations of microvascular reactivity. Microcirculation. 2006;13:301–14.
58. Muller-Delp JM, Gurovich AN, Christou DD, Leeuwenburgh C. Redox balance in the aging microcirculation: new friends, new foes, and new clinical directions. Microcirculation. 2012;19:19–28.
59. Murphy MP. How mitochondria produce reactive oxygen species. Biochem J. 2009;417:1–13.
60. Musch TI, Friedman DB, Pitetti KH, Haidet GC, Stray-Gundersen J, Mitchell JH, Ordway GA. Regional distribution of blood flow of dogs during graded dynamic exercise. J Appl Physiol. 1987;63:2269–77.
61. Nelson MT, Quayle JM. Physiological roles and properties of potassium channels in arterial smooth muscle. Am J Physiol Cell Physiol. 1995;268:C799–822.
62. Nilius B, Droogmans G. Ion channels and their functional role in vascular endothelium. Physiol Rev. 2001;81:1415–59.
63. Proctor DN, Parker BA. Vasodilation and vascular control in contracting muscle of the aging human. Microcirculation. 2006;13:315–27.
64. Reeve HL, Weir EK, Nelson DP, Peterson DA, Archer SL. Opposing effects of oxidants and antioxidants on K^+ channel activity and tone in rat vascular tissue. Exp Physiol. 1995;80:825–34.
65. Rouach N, Segal M, Koulakoff A, Giaume C, Avignone E. Carbenoxolone blockade of neuronal network activity in culture is not mediated by an action on gap junctions. J Physiol. 2003;553:729–45.
66. Sandow SL, Hill CE. Incidence of myoendothelial gap junctions in the proximal and distal mesenteric arteries of the rat is suggestive of a role in endothelium-derived hyperpolarizing factor-mediated responses. Circ Res. 2000;86:341–6.

67. Sandow SL, Neylon CB, Chen MX, Garland CJ. Spatial separation of endothelial small- and intermediate-conductance calcium-activated potassium channels (K_{Ca}) and connexins: possible relationship to vasodilator function? J Anat. 2006;209:689–98.
68. Seals DR, Jablonski KL, Donato AJ. Aging and vascular endothelial function in humans. Clin Sci (Lond). 2011;120:357–75.
69. Segal SS, Jacobs TL. Role for endothelial cell conduction in ascending vasodilatation and exercise hyperaemia in hamster skeletal muscle. J Physiol. 2001;536:937–46.
70. Smith PD, Brett SE, Luykenaar KD, Sandow SL, Marrelli SP, Vigmond EJ, Welsh DG. KIR channels function as electrical amplifiers in rat vascular smooth muscle. J Physiol. 2008;586:1147–60.
71. Socha MJ, Boerman EM, Behringer EJ, Shaw RL, Domeier TL, Segal SS. Advanced age protects microvascular endothelium from aberrant Ca^{2+} influx and cell death induced by hydrogen peroxide. J Physiol. 2015;593:2155–69.
72. Steinhausen M, Endlich K, Nobiling R, Parekh N, Schutt F. Electrically induced vasomotor responses and their propagation in rat renal vessels in vivo. J Physiol. 1997;505:493–501.
73. Straub AC, Zeigler AC, Isakson BE. The myoendothelial junction: connections that deliver the message. Physiology (Bethesda). 2014;29:242–9.
74. Taddei S, Virdis A, Mattei P, Ghiadoni L, Gennari A, Fasolo CB, Sudano I, Salvetti A. Aging and endothelial function in normotensive subjects and patients with essential hypertension. Circulation. 1995;91:1981–7.
75. Takano H, Dora KA, Spitaler MM, Garland CJ. Spreading dilatation in rat mesenteric arteries associated with calcium-independent endothelial cell hyperpolarization. J Physiol. 2004;556:887–903.
76. Tare M, Coleman HA, Parkington HC. Glycyrrhetinic derivatives inhibit hyperpolarization in endothelial cells of guinea pig and rat arteries. Am J Physiol Heart Circ Physiol. 2002;282:H335–41.
77. Toda N. Age-related changes in endothelial function and blood flow regulation. Pharmacol Ther. 2012;133:159–76.
78. Tran CH, Taylor MS, Plane F, Nagaraja S, Tsoukias NM, Solodushko V, Vigmond EJ, Furstenhaupt T, Brigdan M, Welsh DG. Endothelial Ca^{2+} wavelets and the induction of myoendothelial feedback. Am J Physiol Cell Physiol. 2012;302:C1226–42.
79. Unger VM, Kumar NM, Gilula NB, Yeager M. Three-dimensional structure of a recombinant gap junction membrane channel. Science. 1999;283:1176–80.
80. Vessey JP, Lalonde MR, Mizan HA, Welch NC, Kelly ME, Barnes S. Carbenoxolone inhibition of voltage-gated Ca channels and synaptic transmission in the retina. J Neurophysiol. 2004;92:1252–6.
81. Vita JA. Endothelial function. Circulation. 2011;124:e906–12.
82. Wang L, Yin J, Nickles HT, Ranke H, Tabuchi A, Hoffmann J, Tabeling C, Barbosa-Sicard E, Chanson M, Kwak BR, Shin HS, Wu S, Isakson BE, Witzenrath M, de Wit C, Fleming I, Kuppe H, Kuebler WM. Hypoxic pulmonary vasoconstriction requires connexin 40-mediated endothelial signal conduction. J Clin Invest. 2012;122:4218–30.
83. Welsh DG, Segal SS. Endothelial and smooth muscle cell conduction in arterioles controlling blood flow. Am J Physiol Heart Circ Physiol. 1998;274:H178–86.
84. Wolfle SE, Chaston DJ, Goto K, Sandow SL, Edwards FR, Hill CE. Non-linear relationship between hyperpolarisation and relaxation enables long distance propagation of vasodilatation. J Physiol. 2011;589:2607–23.
85. Wolfle SE, Schmidt VJ, Hoepfl B, Gebert A, Alcolea S, Gros D, de Wit C. Connexin45 cannot replace the function of connexin40 in conducting endothelium-dependent dilations along arterioles. Circ Res. 2007;101:1292–9.
86. Yamamoto Y, Fukuta H, Nakahira Y, Suzuki H. Blockade by 18beta-glycyrrhetinic acid of intercellular electrical coupling in guinea-pig arterioles. J Physiol. 1998;511:501–8.
87. Zhang T, Wu DM, Xu GZ, Puro DG. The electrotonic architecture of the retinal microvasculature: modulation by angiotensin II. J Physiol. 2011;589:2383–99.

Chapter 5
Ca^{2+}/Calmodulin-Gated Small- and Intermediate-Conductance K_{Ca} Channels in Cardiovascular Regulation: Targets for Novel Pharmacological Treatments

Ralf Köhler and Aida Olivan-Viguera

Abstract In vascular biology, the Ca^{2+}/calmodulin-gated K^+ channels, $K_{Ca}3.1$ and $K_{Ca}2.3$, produce membrane hyperpolarization in response to Ca^{2+} mobilization events and thereby initiate endothelium-derived hyperpolarization (EDH)-type of arterial dilation. The physiological relevance of this system in-vivo is evidenced by the observation that genetically encoded loss of $K_{Ca}3.1$ and $K_{Ca}2.3$ caused channel-subtype specific cardiovascular phenotypes characterized by endothelial dysfunction to receptor stimulation or mechanical stress and blood pressure alterations. From the translational perspective, $K_{Ca}3.1$ and $K_{Ca}2.3$ dysfunctions are a feature of idiopathic cardiovascular disease, chronic inflammation, atherosclerosis and organ fibrosis and $K_{Ca}2.3$ has been implicated in atrial fibrillation. Accordingly, $K_{Ca}3.1$ and $K_{Ca}2.3$ emerge as possible drug targets. In this chapter, we would like to highlight our recent advances in $K_{Ca}3.1$ and $K_{Ca}2$ biology, pharmacology, as well as consequences of pharmacological manipulating $K_{Ca}3.1$ and $K_{Ca}2.3$ for systemic cardiovascular regulation and cardiovascular health. Moreover, we explore impacts of innovative channel modulators on cardiac function, physical activity and behavior in keeping with the expression of $K_{Ca}2$-subtypes in the heart and neurons.

Keywords Endothelium • EDH • $K_{Ca}2$ • $K_{Ca}3.1$ • KCNN3 • KCNN4 • Vasodilation • Hypertension • Hypotension • Heart • Behavior

R. Köhler (✉)
Fundación Agencia Aragonesa para la Investigación y Desarrollo (ARAID), Zaragoza, Spain

Unidad de Investigación Traslacional, Hospital Universitario Miguel Servet, Paseo Isabel la Católica, 1-3, 50009 Zaragoza, Spain

Aragon Institute of Health Sciences & IIS Aragon, 50009 Zaragoza, Spain
e-mail: rkohler.iacs@aragon.es

A. Olivan-Viguera
Aragon Institute of Health Sciences & IIS Aragon, 50009 Zaragoza, Spain

I. Levitan, A.M. Dopico (eds.), *Vascular Ion Channels in Physiology and Disease*, DOI 10.1007/978-3-319-29635-7_5

Introduction

Ca^{2+}-activated K^+ channels (K_{Ca}) are familiar to most electrophysiologists because K_{Ca} channels produce afterhyperpolarizations following an action potential (AP) in neurons and thereby influence firing pattern and neurotransmission [1]. Cardiovascular researchers recognize K_{Ca} as important effector proteins in endothelial control of arterial tone [2–5]. This is based on observations that subtype-selective inhibition of the major vascular K_{Ca}, **K_{Ca}3.1**, **K_{Ca}2.3**, and **K_{Ca}1.1** impair the ***endothelium-derived hyperpolarization factor*** (**EDHF**)-type vasodilation (involving smooth muscle K_{Ca}1.1 as target of a diffusible factor) and the ***endothelium-derived hyperpolarization*** (**EDH**)-type (involving endothelial K_{Ca}3.1/K_{Ca}2.3 as initiators and gap-junctional coupling of endothelium and smooth muscle) (Fig. 5.1). Both mechanisms have in common that they converge on hyperpolarization of smooth muscle leading to closure of voltage-gated Ca^{2+} channels and the resulting drop in intracellular Ca^{2+} causes finally myocyte relaxation and vasodilation. Hence, EDH or EDHF pathways are mechanistically different from pathways involving other

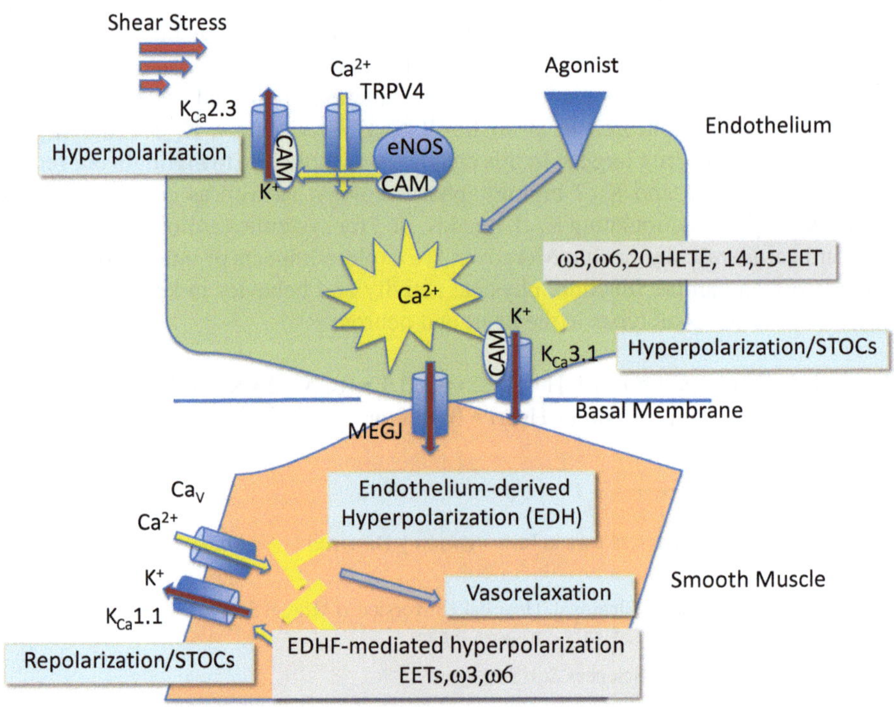

Fig. 5.1 Schematic overview of putative mechanisms of endothelium-derived-hyperpolarization mediated dilation(EDH) and endothelium-derived hyperpolarizing Factor (EDHF)-mediated dilation. *CAM* calmodulin, *eNOS* endothelial nitric oxide synthase, *MEGJ* myo-endothelial gap junction, *STOCs* spontaneous transient outward currents

diffusible molecules like NO or prostaglandins, although the latter paracrine factors are also capable in some cases to induce smooth muscle hyperpolarization and thus to act as *EDHFs* and endothelial hyperpolarization stimulates Ca^{2+}-dependent NO-synthesis by enhancing Ca^{2+} dynamics after receptor stimulations.

At present, the relative physiological significance of either EDHF or EDH systems is still debated, when discussing the input of the respective systems to human vasculature and in cardiovascular disease [4]. Yet, it is clear that the EDHF- or EDH-triggered hyperpolarization of smooth muscle is a potent mechanism that can stand on its own. However, it is not fully understood when and under what conditions EDHF and EDH mechanisms contribute alone, in parallel, or co-operate (or not) with NO-signaling or other diffusible dilating or contracting factors to vasoregulation in the different vascular beds. But it is likely that they integrate in a very complex fashion with NO being without any doubt a major player (for recent extensive review see [6]).

Here, we review mainly the contributions of the authors to their specific field according to the Editors' wish and start with a short introduction into K_{Ca} genes and their expression in the vascular wall, defects of EDH signaling and blood pressure alterations caused by $K_{Ca}3.1$/$K_{Ca}2.3$ deficiency, followed by K_{Ca}-pharmacology. In the following, the main focus will be on the cardiovascular actions of novel small molecule modulators (activators and inhibitors) targeting endothelial $K_{Ca}3.1$ and $K_{Ca}2.3$ channels in arteries and systemically. Finally, we explore potential therapeutic possibilities and briefly elaborate on possible impacts on cardiac and neurological functions. For additional in-depth and complete reviews and to not diminish previous work by others and us, we wish to re-direct the reader to [4, 7, 8].

K_{Ca} Genes, General Biophysical Properties of K_{Ca}, and Genetic Variations Relevant for Cardiovascular Disease

The gene family of K_{Ca} channels is divided into two subgroups according to phylogenetic relationships, small or large unitary conductance, and the mechanisms of Ca^{2+} activation [9] (Fig. 5.2a). The first group comprises four subtypes, of which three have a small unitary conductance of 5–10 pS (gene names: KCNN1, KCNN2, and KCNN3, protein names SK1-SK3, or KCa2.1-KCa2.3) and one member with an intermediate conductance of 20–40 pS (gene name: KCNN4; protein name: IK, IK_{Ca}, or KCa3.1). The functional channels are tetramers formed by four subunits and the four pore loops located between S5 and S6 of each subunit form the ion conduction pathway. Ca^{2+} sensitivity is conferred by four calmodulin molecules constitutively bound to the four CAM-bindings domains located in the cytosolic c-terminus.

An important feature to understand the roles of these channels in cardiovascular biology is that the channels lack a voltage sensor and are thus voltage-independent, which means that the channels do not inactivate at negative membrane potential like e.g. voltage-gated K_V channels. This feature enables $K_{Ca}2$ and $K_{Ca}3.1$ channels to produce long-standing K^+-outward currents at potentials more negative than the

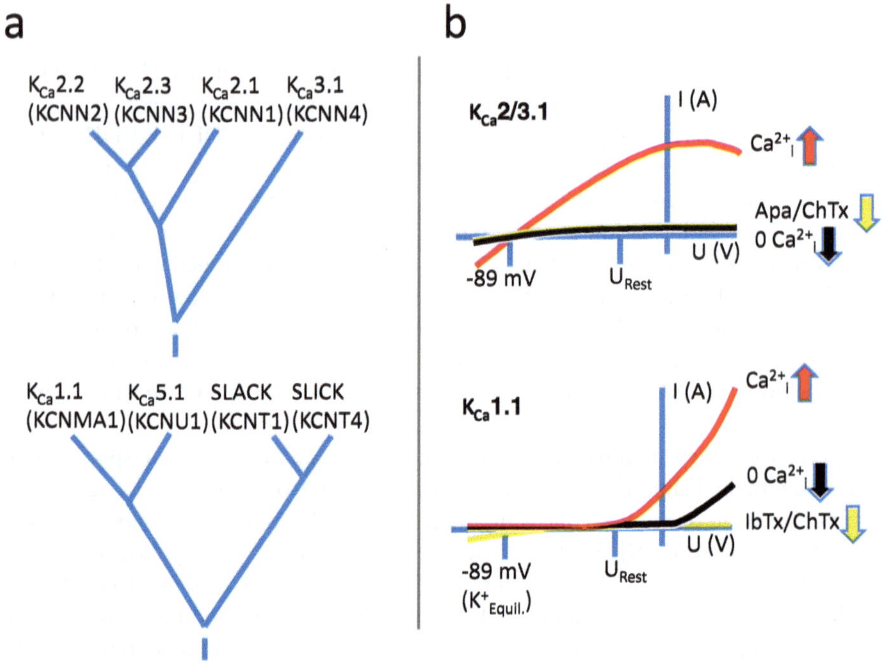

Fig. 5.2 (**a**) Phylogenetic relationships of Ca^{2+}-activated K^+ channel genes. (**b**) Schematic representation of K_{Ca} currents and their sensitivity to peptide toxins. *Upper* panel: Illustration of voltage-independent Ca^{2+}-activated $K_{Ca}2.3/K_{Ca}3.1$ currents in endothelial cells and inhibition by the peptide toxins, apamin (Apa) and charybdotoxin (ChTx). U_{Rest} resting membrane potential; –89 mV is the K equilibrium potential. Note that the $K_{Ca}2.3/K_{Ca}3.1$ channel activation produces voltage-independent currents shifting the reversal potential towards the K equilibrium potential. *Lower* panel: Ca^{2+} activation of voltage-dependent $K_{Ca}1.1$ currents and inhibition by iberiotoxin (IbTx) or ChTx. Note that dual Ca^{2+} and voltage activation occurs at near the U_{Rest}, stabilizing the membrane potential

resting potential[1] and by this they produce persisting membrane hyperpolarization (e.g. to –50 up to max. –89 mV (K^+ equilibrium potential) (Fig. 5.2b). In vascular biology, such a durable hyperpolarization is required for complete closure of smooth muscle voltage-gated Ca^{2+} channels and thereby for the endothelium-dependent hyperpolarization (EDH)-type of arterial dilation [8].

In other tissues, $K_{Ca}2$ (and possibly $K_{Ca}3.1$) produce afterhyperpolarization in neurons and thereby regulate firing frequency [1] and—in the heart—co-determine the length of the atrial action potential and thus atrial repolarization and the contraction cycle [10]. In particular, the link between $K_{Ca}2.3$ and the heart is of interest

[1] Values for endothelial resting potentials vary considerably (from 0 to –89 mV), which very much depends on the preparation (intact vessel vs. cultured cells; observation by the authors' group). In current-clamp experiments on morphologically intact endothelium of murine and human arteries we found values ranging from –25 to –45 mV that mirror the potential in smooth muscle of the same preparation (measured by sharp electrode techniques).

since a genome-wide association study (GWAS) identified an association of the $K_{Ca}2.3$-encoding KCNN3 gene with lone atrial fibrillation in humans [11]. In addition, pharmacological blocking of atrial $K_{Ca}2$ is currently studied as a way to prolong selectively atrial action potential length and thereby to stop atrial fibrillation [12].

In respect to $K_{Ca}3.1$, single-nucleotide polymorphisms (SNP) in the KCNN4 gene have been associated with cardiac infarction in a Japanese cohort [13] and in Crohn's disease in the Australian and New Zealand population [14, 15]. Moreover, there are several reports, which have been reviewed in depth elsewhere, on reduced, increased, or conserved functions of $K_{Ca}2.3$ and $K_{Ca}3.1$ in cardiovascular disease and other diseases such as chronic inflammation, allergy, organ fibrosis, and neurodegeneration [2–4, 16–18]. This assigns $K_{Ca}2.3$ and $K_{Ca}3.1$ as disease-relevant channels in idiopathic disease and targets in translational medicine.

The second subgroup of K_{Ca} channels, which is phylogenetically distantly related to the first subgroup [9, 19], consists of the well known K_{Ca} with large unitary conductance (200–300 pS; gene name KCNMA1, protein name: BK_{Ca}, BigK, Maxi K, or $K_{Ca}1.1$), and three other members (KCNU1, KCNT1, and KCNT2), of which KCNT1 and KCNT2 (protein names: SLACK and SLICK) are in fact more distantly related Na-activated K channels (K_{Na}) with large conductance (150–300 pS) [20].

In contrast to $K_{Ca}2$ and $K_{Ca}3.1$ channels, $K_{Ca}1.1$ contains a voltage-sensor in S4 producing gating by depolarization, particularly in a membrane potential range, in which voltage-dependent L-type Ca^{2+} channels operate (Fig. 5.2b). The $K_{Ca}1.1$ channel complex recruits four beta-1 subunits that sensitize channel gating to Ca^{2+} and voltage by favoring channel opening at physiological internal Ca^{2+} levels reached in the vicinity of the $K_{Ca}1.1$ channel Ca^{2+} sensors.

In respect to gene expression of channels of this subfamily in the vasculature, $K_{Ca}1.1$ is predominant in smooth muscle and the CNS while other members (SLICK and SLACK) are predominantly expressed in neuronal tissues. $K_{Ca}1.1$ forms complex functional units ("signalosomes") with physical interactions with voltage-regulated L-type Ca^{2+} channels [21, 22]. The electrical membrane events of such co-operations are spontaneous-transient-outward-currents (STOCs, mediated by $K_{Ca}1.1$) that are triggered by spatially co-localized Ca^{2+} influx through L-type Ca^{2+} channels and small Ca^{2+} release events from the sarcoplasmic reticulum, termed "sparks" [22, 23]. Physiologically, these $K_{Ca}1.1$-mediated STOCs are considered to provide a negative feedback on Ca^{2+} influx through L-type Ca^{2+} channels and thereby to limit depolarization and contractions or vasospasm [21, 24]. Presumably, diffusible EDHFs, like epoxyeicosanoids augment $K_{Ca}1.1$-mediated STOCs and by this produce their vasorelaxing actions.

In contrast to $K_{Ca}3.1$ and $K_{Ca}2.3$ channels, $K_{Ca}1.1$ does not seem to be expressed under physiological conditions in human, rat, and murine arterial endothelium, although porcine endothelium could be an exception [25, 26].

From the epidemiological perspective, SNPs in the $K_{Ca}1.1$-encoding KCNMA1 gene causing "loss-of-function" or in the associated beta-subunit gene (KCNMB1) causing "gain-of-function" are associated with systolic hypertension (KCNMA1) or resistance to hypertension, respectively [14, 27, 28].

K_{Ca} and Their Functions in Arterial Endothelium

In early nineties, the genes encoding for channels were still awaiting cloning ($K_{Ca}3.1$ in 1997 [29] and $K_{Ca}2$ in 1996, [30]) and K_{Ca} channels were pharmacologically characterized in endothelium and differentiated according to their sensitivity to the peptide toxins, apamin ($K_{Ca}2$), charybdotoxin ($K_{Ca}3.1$ and $K_{Ca}1.1$), and iberiotoxin ($K_{Ca}1.1$), and the small molecules, clotrimazole ($K_{Ca}3.1$) [31], and paxilline ($K_{Ca}1.1$) [32]. $K_{Ca}2$-blocking compounds like UCL-1684 [33] have not been invented yet. Notably, in 1996, Marchenko and Sage were first on providing comprehensive data on functional $K_{Ca}3.1$ and $K_{Ca}2$ by elegant in-situ patch-clamp electrophysiology on intact endothelium of isolated rat aorta [34]. Almost contemporaneously, $K_{Ca}3.1$ currents were identified in endothelia from different species and different vascular beds [2, 3, 35], for instance, in native bovine and rat aortic endothelial cells [36, 37], and rat brain capillaries [35]. In general, expression of $K_{Ca}3.1$ and $K_{Ca}2.3$ is a common finding for endothelia of several species and is documented throughout the entire vascular tree, with no considerable differences (at least qualitatively) between arterioles, arterioles of the microcirculation, and veins.

Moreover, alterations of $K_{Ca}3.1$ currents have been documented in a model of experimental hypertension, the spontaneously hypertensive rat [37]. This provided early electrophysiological evidence that endothelial $K_{Ca}3.1$ channels were altered in a hypertension-associated fashion. Moreover, these findings point to a capacity of endothelial channels to respond in a probably compensatory way to hypertension in this model. A few years later, the genes encoding for $K_{Ca}3.1$ and $K_{Ca}2.3$ currents were identified by others [26, 38] and our group as KCNN4 and KCNN3 using a single-cell RT-PCR approach combined with patch-clamp in endothelium from rats [39] and—importantly from the translational perspective—in human mesenteric endothelium [40, 41].

K_{Ca} and Endothelium-Dependent Vasodilation

In the eighties, combinations of peptide toxins in cardiovascular pharmacology were widely used to study endothelium-dependent vasodilator mechanisms by classical in-vitro myography on large rodent arteries. A contribution of $K_{Ca}3.1$ and $K_{Ca}2$ to the "EDHF" has been deduced from the blocking actions of the combination of charydotoxin ($K_{Ca}3.1$ and $K_{Ca}1.1$) and apamin ($K_{Ca}2$). A contribution of $K_{Ca}1.1$ has been discarded as iberiotoxin ($K_{Ca}1.1$) cannot replace charybdotoxin [4]. Yet, these peptide blockers are not fully selective and e.g. charybdotoxin—besides $K_{Ca}3.1$ and $K_{Ca}1.1$—blocks some Kv channels that are present in arterial tissue. Interestingly, in coronary arteries of pigs, "EDHF" required reportedly smooth muscle $K_{Ca}1.1$ as the response was abrogated by iberiotoxin, which suggested $K_{Ca}1.1$ as target of a diffusible EDHF (Fig. 5.1) that has been later identified as a P450-derived eicosanoids [42]. Still, vasorelaxation that involves endothelial $K_{Ca}3.1$ and $K_{Ca}2.3$ is also present in porcine coronary artery as concluded from more recent studies by others and us [43, 44].

It is noteworthy that P450-products and ω3 and ω6 fatty acids that activate $K_{Ca}1.1$ and have been used as antagonist of EDHF-signaling—like 20-HETE [45]—inhibit

$K_{Ca}3.1$ [46], suggesting interactions between putative EDHF and EDH mechanisms (Fig. 5.1). Despite this evidence from in-vitro experimentation, the true contribution of EDH or possibly EDHF to circulatory regulation in humans is not established at present and a matter of ongoing debate. A main reason for the lack of knowledge is that we miss sufficiently selective, potent, safe, and ethically approved pharmacological tools for in-vivo investigation. Yet, in the following paragraphs we explore potential utilities of novel small molecule activators and inhibitors.

Endothelial Dysfunction Caused by Genetic $K_{Ca}3.1/K_{Ca}2.3$-Deficiency

Since 2003, the availability of gene-targeted mice as tools provided more definite experimental evidence for the significance of $K_{Ca}3.1$ and $K_{Ca}2.3$ channels for EDH-type dilation in-vitro and in-vivo [47–49]. Indeed, myography on conduit arteries and intra-vital microscopy on resistance-size skeletal muscle arterioles of $K_{Ca}3.1$−/−/$K_{Ca}2.3^{T/T}$+Dox mice with genetically encoded deficiency of both endothelial channels revealed that loss of $K_{Ca}3.1$ and $K_{Ca}2.3$ impairs strongly endothelial and smooth muscle membrane hyperpolarization and EDH-type dilations in both vessel types [50–52] (Fig. 5.3a). Interestingly, also NO-type dilations were disturbed in the mice, suggesting that $K_{Ca}3.1$ and $K_{Ca}2.3$-mediated hyperpolarization promotes Ca^{2+} influx and thereby stimulates also Ca^{2+}-dependent NO production [50, 51, 53]. A candidate channel mediating the Ca^{2+} influx during hyperpolarization is the TRPV4 channel [54–56], which conducts considerably Ca^{2+} at clearly negative potentials while it shows negative-feedback regulation by Ca^{2+} at potentials near zero. In support of the above-mentioned, loss of $K_{Ca}3.1$ in $K_{Ca}3.1$−/− and pharmacological inhibition of TRPV4 in mesenteric artery impaired Ca^{2+} dynamics to acetylcholine [55].

Moreover, our studies on single and double $K_{Ca}3.1/K_{Ca}2.3$-deficient mice [50] revealed subtype specific roles of the channels for EDH-type dilation since $K_{Ca}3.1$ is of greater importance than $K_{Ca}2$ for acetylcholine-induced dilations while $K_{Ca}2.3$ is more important in dilations caused by mechanical stress acting on the endothelium (shear stress), for hyperemia during skeletal muscle twitching [52] and for a basal tonic dilator input from the endothelium [49]. Thus, a $K_{Ca}2.3$-EDH system is likely needed to regulate basal blood flow and the higher blood flow demand during exercise (metabolic demand) while $K_{Ca}3.1$-EDH is linked to muscarinic receptor stimulation and ER Ca^{2+} release and to possibly sympathetic drive on resistance arteries.

Some of these subtype specific roles can be explained by a differential intracellular localization of $K_{Ca}3.1$ and $K_{Ca}2.3$ in the endothelial cell (Fig. 5.1). However, there are some caveats because of the uncertain specificity of antibodies used to localize $K_{Ca}3.1$ and $K_{Ca}2.3$ and of lack of approach validation in KO-controls. Regardless of these uncertainties, $K_{Ca}3.1$ has been localized in proximity to protrusions of the endoplasmic reticulum (ER) [57] where it senses Ca^{2+} release events after e.g. muscarinic receptor (M3) stimulation by acetylcholine. Moreover, $K_{Ca}3.1$ has been reported to co-localize with myo-endothelial gap-junction proteins thus forming a functional unit for electro-tonic spread of EDH to smooth muscle (for review [2, 58–60]).

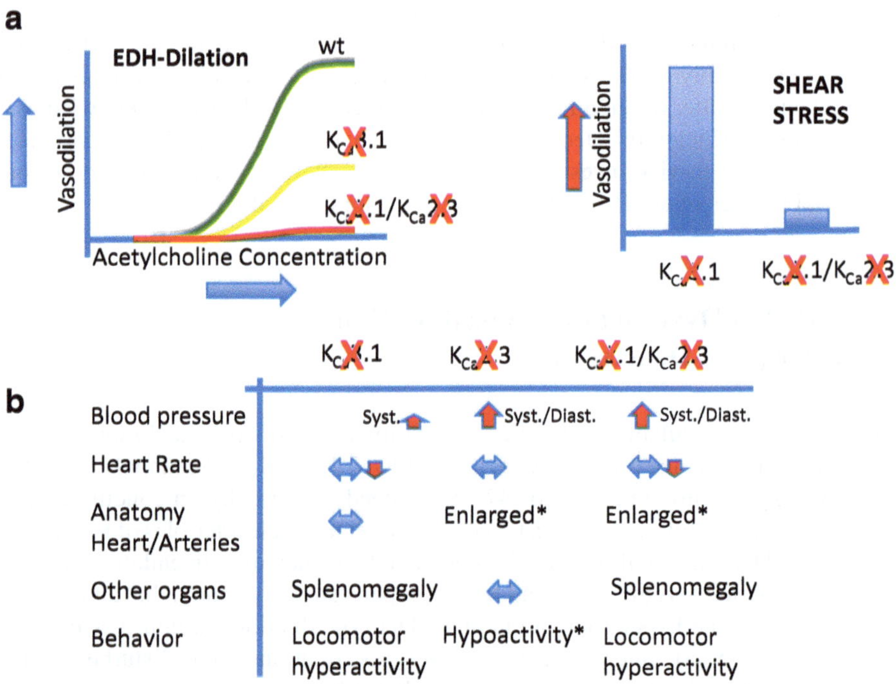

Fig. 5.3 (**a**) On *left*: Schematic illustration of the impaired EDH-type arterial dilation to acetylcholine in $K_{Ca}3.1$ and/or $K_{Ca}2.3$-deficient mice. On *right*: Impaired shear stress-induced EDH-type dilation in $K_{Ca}2.3$-deficient mice. (**b**) Major cardiovascular, anatomical, and behavioral phenotypes. The *asterisks* (*) indicate phenotypes that are unaltered by subchronic treatment with doxycycline to suppress $K_{Ca}2.3$-gene expression. Note that $K_{Ca}3.1-/-$ exhibit mild systolic hypertension only during locomotor activity

$K_{Ca}2.3$ has been found at intercellular junctions [61] where it is suggested to form a functional unit with TRPV4 [62] or possibly with another Ca^{2+}-permeable channels. Interestingly, such a functional unit has recently been suggested to occur in the tubular system of the kidney [63] as well as in pulmonary and mesenteric arteries [62, 64], where TRPV4 uses $K_{Ca}2.3$ to produce vasodilation. Nevertheless, we need more definite visualization of $K_{Ca}3.1$ and $K_{Ca}2.3$ using gold-labeled AB, KO-controls, and transmission electron microscopy to ascertain the present view.

Interplay of $K_{Ca}3.1/K_{Ca}2.3$ Channels with Ca^{2+}-Permeable Channels

Evidence from patch-clamp studies point at close functional interactions of endothelial K_{Ca} with Ca^{2+}-permeable channels (Fig. 5.1): While patch-clamping the endothelium of intact vessel preparations or bovine endothelial cells in a flow chamber, we frequently observed that $K_{Ca}3.1$ gating and membrane hyperpolarization occurs following stimulation of shear stress-activated or hyposmotic stress-activated Ca^{2+} permeable channels [65]. Moreover, these studies showed temporally fluctuating Ca^{2+} release events that amplified $K_{Ca}3.1$ activation. These fluctuating Ca^{2+} release events and concomitant $K_{Ca}3.1$-mediated currents and hyperpolarization were modulated in frequency and amplitude by the degree of shear force acting on the cells. Interestingly, these fluctuating currents shared some similarities with spontaneous transient outward currents (STOC) seen in smooth muscle, although the endothelial STOCS were of longer duration and largely voltage-independent and did not involve $K_{Ca}1.1$ and $K_{Ca}1.1$-interacting L-type Ca^{2+} channels as in smooth muscle. Today, one of the candidate MSCs triggering Ca^{2+} signaling events and co-activation of $K_{Ca}3.1$ (and $K_{Ca}2.3$) to flow or shear stress stimulation is (again) TRPV4 as concluded from elegant in-situ imaging in cerebral and mesenteric arteries [66] and by the loss of shear stress-induced dilation in carotid arteries from TRPV4−/− mice [67].

Thus, from the perspective of integrative cardiovascular physiology, the interplay of TRPV4 and $K_{Ca}2.3$ ($K_{Ca}3.1$) can be mechanistically important to fine-tune EDH and arterial tone in response to alterations of blood flow in e.g. the working skeletal muscle or the brain with their metabolic demands [52]. In terms of cross-reactivity with other vasodilator systems, TRPV4 has been proposed to physically interact with the scaffold protein caveolin-1 and thus the endothelial nitric oxide synthase (eNOS) [68] and the Ca^{2+} influx through TRPV4 further stimulates Ca^{2+}-dependent NO synthesis [54] in addition to EDH, producing some redundancy in endothelial vasoregulator capacity and during exercise.

It should be noted that $K_{Ca}3.1$ and $K_{Ca}2$ channels are not considered “smooth muscle” channels. In fact, according to our electrophysiological studies, $K_{Ca}3.1$ channels are absent in contractile i.e., differentiated, vascular smooth muscle of rodents and this makes some physiological sense because the channels are highly Ca^{2+}-sensitive because of the Ca^{2+} sensor, calmodulin, and, moreover, are voltage-independent. Functioning of $K_{Ca}3.1$ in smooth muscle would almost immediately abrogate any contraction because the relatively high Ca^{2+} signaling in contracting smooth muscle would cause strong $K_{Ca}3.1$ activation and lasting hyperpolarization, which in turn causes cessation of Ca^{2+} influx. In other words, $K_{Ca}3.1$ would provide a too strong negative feedback on Ca^{2+}-dependent contraction. In contrast, the voltage-dependent and less Ca^{2+}-sensitive $K_{Ca}1.1$—known as smooth muscle channel—is the better channel to mediate negative feedback during depolarization. With respect to $K_{Ca}2$ it is—because of the same theoretical considerations—rather unlikely that these closely related and similarly voltage-independent $K_{Ca}2$ channels are functionally expressed in contractile smooth muscle. We have not been successful in detecting $K_{Ca}2$ channel functions in vascular smooth muscle of the rat and mouse.

Systemic Cardiovascular Alterations in $K_{Ca}3.1$/$K_{Ca}2.3$-Deficient Mice

That $K_{Ca}3.1$ and $K_{Ca}2.3$ channels add significantly to systemic cardiovascular regulation has been suggested by the higher blood pressure in $K_{Ca}3.1$−/− mice [50] and $K_{Ca}2.3$-deficient mice ($K_{Ca}2.3^{T/T}$+Dox) [49, 50] as found by blood pressure telemetry in the freely moving animals (Fig. 5.3b). The hypertension has been reported to be, however, mild and overt only during physical activity (for review see [7]). Still, the latter observation was particularly interesting as it could mean that $K_{Ca}3.1$ and $K_{Ca}2.3$ and the EDH-system are needed to adjust tone during exercise and sympathetic input while they are of less or no importance during rest. Recently, $K_{Ca}3.1$−/− and connexin(Cx)37−/− mice—Cx37 that has been considered a major constituent of myo-endothelial gap junctions—have been shown to exhibit lower ADP/ATP-receptor ($P2Y_{2/4}$)-mediated fast depressor responses and secondary increases in pressure [69] in anesthetized mice, suggesting decreased responsiveness to evoked endothelial function. Responses to acetylcholine-infusion has been reported to also be reduced at low but not high concentrations, suggesting endothelial dysfunction in vivo as it has been observed so far only in isolated $K_{Ca}3.1$−/− arteries.

Regarding the contribution of nitric oxide in these mice, inhibition by L-NAME has been shown to be still capable to increase blood pressure in $K_{Ca}3.1$−/−/ $K_{Ca}2.3^{T/T}$+Dox mice [7], indicating that NO signaling is largely intact in these mice and it is therefore likely that NO- and EDH mechanisms act independently from each other to a large extent but are likely additive.

Other Phenotypes

It is worth to mention that $K_{Ca}3.1$−/− mice have been shown to exhibit spontaneous physical hyperactivity (but without signs of general distress) and exhibited complex alteration in monoamine levels, with increased noradrenalin turnover in frontal cortex and lower serotonin-turnover in frontal cortex, striatum and brain stem, and no change in dopamine turnover [70]. Plasma corticosterone has been found to be normal but increased to higher levels in $K_{Ca}3.1$−/− under stress as reported by others [71]. This suggests substantial alteration of neurotransmission and behavior in $K_{Ca}3.1$, which can add to the cardiovascular phenotypes in these mice. Another solid phenotype in $K_{Ca}3.1$−/− is progressive splenomegaly (Fig. 5.3b) caused presumably by defects in erythrocyte volume regulation and a resulting increased erythrocyte degradation and turnover [72], producing a higher workload for the spleen and secondary adaptation of size to meet the demand.

Also $K_{Ca}2.3^{T/T}$ (+Dox/−Dox) have complex behavioral and morphological phenotypes (Fig. 5.3b) [1, 70], which can be linked in $K_{Ca}2.3^{T/T}$−Dox to larger afterhyperpolarization, refractory periods, and thus firing frequencies. Locomotor hypoactivity is evident in $K_{Ca}2.3^{T/T}$+Dox and −Dox (Fig. 5.3b). Moreover, there are significant anatomical (developmental) changes in the vasculature and the heart, i.e. visible enlargement of diameters of mesenteric and uterine arteries and an enlarged heart with increased right ventricular wall thickness and altered atrioventricular morphology [49, 73, 74]. Postnatal development of retinal vessels has been found to be normal, although deficiency of $K_{Ca}2.3$ ($K_{Ca}2.3^{T/T}$+Dox) increased branching (D. Rappert, 2011, German MD thesis; archiv.ub.uni-marburg.de/diss/z2011/0306/

pdf/ddr.pdf). In sum, these anatomical changes suggest that alterations of $K_{Ca}2.3$ expression levels can have a substantial impact on the development of the vasculature. As possible pathophysiological consequences, the mice experience parturition defects [48] and sudden cardiac death [75] when over-expressing the channels ($K_{Ca}2.3^{T/T}$–Dox).

$K_{Ca}3.1$ and $K_{Ca}2.3$ in Cardiovascular and Other Disease States

There is growing evidence that $K_{Ca}3.1$ and $K_{Ca}2.3$ channels are differentially regulated in disease. In fact, up-regulated $K_{Ca}3.1$ functions have been found to promote many disease states characterized by abnormal proliferative organ remodeling and inflammation such as neointima formation [76, 77], atherosclerosis[78], fibrosis of lungs [79], fibrosis in damaged and diabetic kidneys [80, 81], possibly liver cirrhosis and portal hypertension [82], allograft vasculopathy [83], kidney allograft rejection [84], as well as in obliterated trachea transplants [85]. Such up-regulation of $K_{Ca}3.1$ in diseased tissues reflected infiltration of $K_{Ca}3.1$-expressing T cells and macrophages and on de-novo mRNA-expression of $K_{Ca}3.1$ in the tissue itself. The de-novo mRNA-expression was induced by actions of classical growth factors on smooth muscle, fibroblast, and endothelial cells *in vitro* [80, 86, 87]. Moreover, shear stress has been shown to induce $K_{Ca}3.1$ in human endothelial cells [88]. In general, the induction of $K_{Ca}3.1$-mRNA synthesis requires activation of the MEK/ERK MAP kinase cascade and transcriptional mechanisms involving activator protein-1 (c-jun) and repressor protein-1 as negative regulator of gene expression [87]. Also the Ca^{2+}/calmodulin-dependent protein kinase kinase/Akt/p300 cascade has recently been reported to mediate the up-regulation of $K_{Ca}3.1$ in endothelial cells during shear stress stimulation [89].

With respect to endothelial functions and cardiovascular disease, endothelial $K_{Ca}2.3$ function and EDH were reduced in regenerated endothelium [90], renal insufficiency [91], ageing [92], and ovariectomy [93], which could explain some aspects of the endothelial dysfunction reported in these conditions. However, up-regulation of gene expression of the channel has been documented in obese rats [94, 95], in pulmonary hypertensive mice [74], and by shear stress [89] in vitro, suggesting here compensatory roles of the channel to maintain/improve endothelial vasodilator function under these conditions. The precise mechanisms of the altered gene expression in endothelium are not understood. However, estrogen is considered a major regulator of $K_{Ca}2.3$ expression and this positive regulation can be an additional explanation for pre-menopausal cardiovascular protection in women.

Interestingly, low $K_{Ca}2.3$ mRNA expression in mammary arteries correlated negatively ($r > -0.5$) with blood pressure in a small Danish cohort of patients with or without renal failure (uremia) or of diabetic and non-diabetic patients (all, $n = 55$) (unpublished data provided by Dr. Lars Melholt Rasmussen, Director of the Odense Artery Biobank SDU, Denmark).

In contrast, $K_{Ca}3.1$ expression did not correlate with blood pressure.

Regarding other tissue and disease states, it is worth to mention that, both, $K_{Ca}2.3$ and $K_{Ca}3.1$ are highly expressed in some cancers [96–99] and have been implicated in cancer progression and metastasis (for review of the field see also [100, 101]). In addition, $K_{Ca}3.1$ has been found to be up-regulated during immune responses and in chronic inflammation such as in T cells within the chronically inflamed mucosa in ulcerative colitis [102] and in microglia after ischemic stroke [103], which is in line with the notion that $K_{Ca}3.1$ is an important regulator of immune cell function [18].

Endothelial $K_{Ca}3.1$ and $K_{Ca}2.3$, Their Pharmacological Manipulation, and Systemic Cardiovascular Consequences

From the pharmacological viewpoint, the introduction of TRAM-34 as a potent and selective $K_{Ca}3.1$ blocker in early 2000 by the laboratory of Heike Wulff provided a "modernized" pharmacological tool to define mechanisms of EDH-type or EDHF-type dilation, in addition to TRAM-34's value for testing roles of $K_{Ca}3.1$ in inflammatory and autoimmune disease [104]. TRAM-34 derived from the fungicide and P450-blocker, BayerAG compound clotrimazole that was the first well studied small molecule blocker of $K_{Ca}3.1$ (Fig. 5.4a). TRAM-34 is more selective for $K_{Ca}3.1$ since the imidazole ring in clotrimazole—being required for P450-blockade—was replaced by a pyrazole ring. TRAM-34 was initially developed as possible immunosuppressant as it blocks $K_{Ca}3.1$ in T cells and macrophages and thereby proliferation, migration, and cytokine production of lymphocytes [104].

The introduction of TRAM-34 into the cardiovascular field greatly helped to define with more precision the requirement of $K_{Ca}3.1$ for EDH-type dilation in several vascular beds and from different species, including human arteries. While it is clear that TRAM-34 reduces acetylcholine-induced vasodilations, this does not have any systemic consequences in vivo since TRAM-34 injections into mice or rats did not cause a change in blood pressure. Interestingly, blocking $K_{Ca}3.1$ in the blood-brain-barrier by TRAM-34 has been reported to reduce water movements into the brain and early edema formation after ischemic stroke [105]. Another structurally different $K_{Ca}3.1$-blocker developed by the BayerAG was effective in reducing brain edema caused by trauma [106].

Senicapoc (ICA-17043; Fig. 5.4a), another $K_{Ca}3.1$-blocker similar to TRAM-34 advanced into clinical trials as a treatment for sickle cell disease [107] and has been found to be cardiovascular-safe since it did not increase blood pressure in humans. Senicapoc has been proposed to be efficient in sickle cell disease because blocking of the over-active $K_{Ca}3.1$ may hinder irreversible erythrocyte sickling (shrinkage) by blocking volume reduction by $K_{Ca}3.1$-mediated K efflux. In Phase IIb and III clinical trials, Senicapoc was found to improve erythrocyte parameters and hemoglobin content, but, unfortunately, did not prevent but rather increased painful vaso-occlusive crisis as primary end point for unknown reasons. The disappointing outcome terminated the development of Senicapoc at least for this indication. Whether endothelial

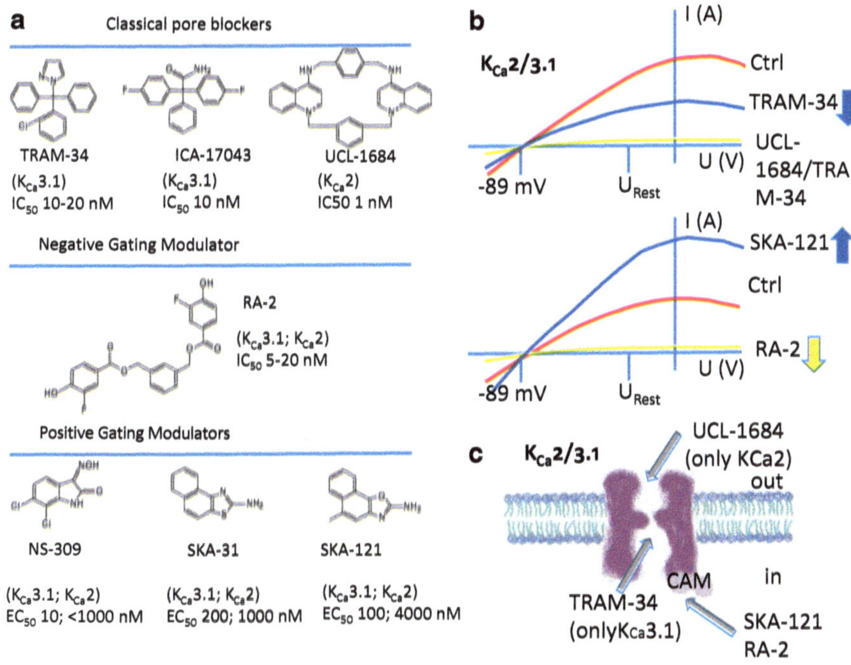

Fig. 5.4 (**a**) Chemical structures of selected $K_{Ca}2/3.1$ modulators with nanomolar potencies. (**b**) *Upper* panel: Schematic illustration of inhibition of mixed voltage-independent Ca^{2+}-activated $K_{Ca}2.3/K_{Ca}3.1$ currents by TRAM-34 ($K_{Ca}3.1$) and UCL-1684 ($K_{Ca}2.3$). *Lower* panel: SKA-121-evoked potentiation of mixed $K_{Ca}2.3/K_{Ca}3.1$ currents and antagonism by the pan-negative gating modulator, RA-2, at equimolar concentration. (**c**) Hypothetical interaction sides of blockers and gating modulators

dysfunction was one reason why Senicapoc failed is not known and has not been studied further.

From a more general perspective, we may consider that—during $K_{Ca}3.1$-blockade—the other potent vasodilator system, the NO system and $K_{Ca}2.3$ channels are still functional and have the capability to compensate the inhibition of $K_{Ca}3.1$. Still, in the case of reduced NO-synthesis and/or NO-availability, inhibition of endothelial $K_{Ca}3.1$ (and $K_{Ca}2.3$) may aggravate the impaired endothelial vasodilator influence on tone and could represent a cardiovascular safety issue.

Senicapoc has been also tested as treatment for exercise and allergic asthma and failed. Still, this $K_{Ca}3.1$ blocker—albeit safe—may have other utilities as mild in the case of immunosuppressant in chronic inflammatory disease, e.g. chronic inflammatory bowel disease, in reducing astrocyte-mediated secondary damage after spinal cord injury [108] and microglia-mediated neurotoxicity [109] and neurodegenerative disease such as Alzheimer's disease [110].

Recent Developments: Positive-Gating Modulators

From the perspective of cardiovascular translational pharmacology, activators of $K_{Ca}2.3$ and $K_{Ca}3.1$ may have more therapeutic utilities in cardiovascular disease characterized by endothelial dysfunction than obviously blockers. However, the existing activators with reasonable potency have not been tested in humans. At present all experimental evidence for their potential therapeutic utilities in cardiovascular disease relies on in-vitro experimentation and few in-vivo cardiovascular monitoring in rats, mice, and dogs [111–114].

The Neurosearch compound NS309 (Fig. 5.4a) has widespread use in in-vitro pharmacological experimentation. It has a high potency but a disadvantageous selectivity profile as the compound, besides activating $K_{Ca}3.1$ and $K_{Ca}2$ channels, has additional blocking effects on voltage-gated Ca^{2+} channels. Moreover, NS309 blocks cardiac hERG channels that may impair ventricular repolarization, which can be regarded a severe safety issue.

The development of SKA-31 (Fig. 5.4a), using the ALS-drug Riluzole (a mixed glutamate antagonist, Na^{+} channel blocker and unselective K^{+} channel activator with low potency) as template, provides first evidence that systemic activation of $K_{Ca}3.1$/$K_{Ca}2.3$ improves endothelial vasodilator function, particularly in vivo. In fact, SKA-31 displays fivefold selectivity for $K_{Ca}3.1$ over $K_{Ca}2$, channels and has been shown to potentiate endothelial $K_{Ca}3.1$/$K_{Ca}2.3$ currents as well as EDH-vasodilation to acetylcholine in carotid arteries of mice [53, 111]. In the microcirculation of skeletal muscle (cremaster muscle) SKA-31 has be shown to induce vasodilation in its own right [115], suggesting that there is some basal Ca^{2+} signaling or "Ca^{2+} pulsar" activity that gives rise to some $K_{Ca}3.1$ activation, which can be potentiated by SKA-31. Moreover, SKA-31 has been found to improve coronary blood flow in rats [112], to potentiate bradykinin-induced relaxation and to reduce serotonin-induced contraction in large porcine coronary artery [44]. In a model of severe fatal hypertension, SKA-31 has been found to increase renal blood flow and to increase survival in mouse model of fatal hypertension [16, 116]. In other tissues, SKA-31 modulated bladder contractility and SKA-31 decreased human detrusor muscle excitability and contractility, suggesting utility of the activator in overactive bladder [117].

Mechanistically, it should be noted, that SKA-31 and recently developed other SKAs [114] are not simple channel openers like other channel openers that activate the channel in the closed configuration (no gating). Rather, SKAs act as positive-gating modulators that keep the Ca^{2+}-gated channel in the open-configuration and thereby shift Ca^{2+} dependence of the channel to the left [114] (Fig. 5.5a). In other words, the channel has a higher activity (open probability) at Ca^{2+} concentrations that are normally not sufficient to produce large channel activation on its own.

The very elegant work by Michael Zhang's group on crystals of the $K_{Ca}2.2$ c-terminus has provided first insight into molecular details and potential bindings domains. They showed that 1-EBIO, an early less potent precursor of DC-EBIO, fits into a pocket between calmodulin and the c-terminus [118] (Fig. 5.4c). Moreover, they showed that NS309, the other activator of this compound class, stabilizes such interactions by stabilizing an intrinsically unordered linker between the CAM-binding

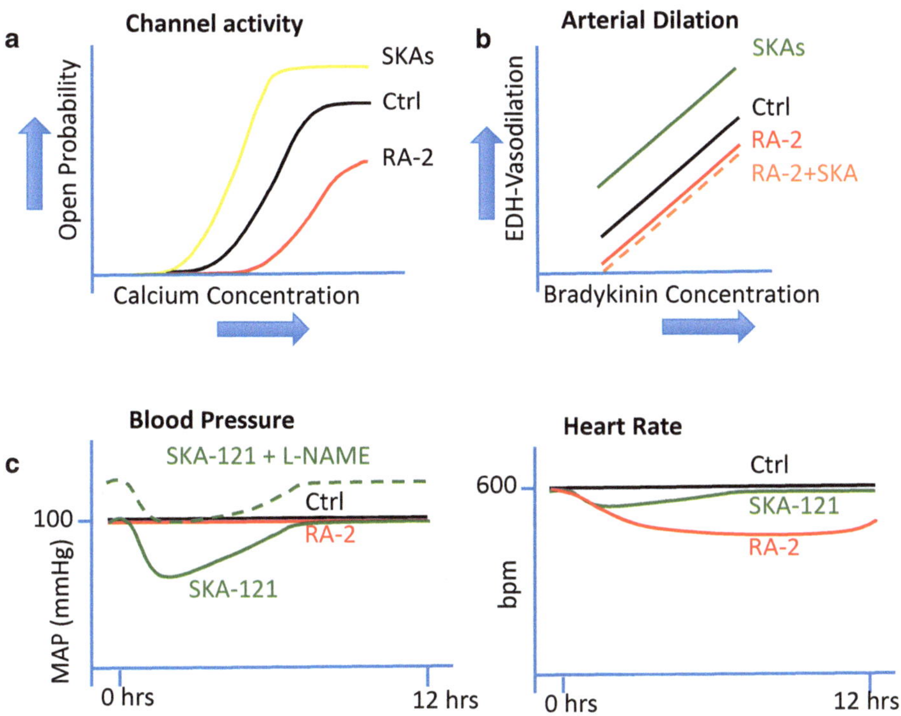

Fig. 5.5 (**a**) Schematic illustration of increase of Ca^{2+} sensitivity of $K_{Ca}3.1$-activity (open probability) by SKAs (*left* shift of the concentration-response curve). RA-2 produces a *right* shift of the curve, indicating reduced Ca^{2+} sensitivity. (**b**) Schematic illustration of SKA-induced potentiation of bradykinin-induced dilation of porcine coronary artery and antagonism by RA-2. (**c**) On *left*: Illustration of blood pressure lowering effects of SKA-121. RA-2 has no appreciable effect. On *right*: Mild bradycardia induced by SKA-121 and appreciable bradycardia evoked by RA-2

domain to S6, which can explain the molecular mechanics of channel activation ([119] for details; for review see [120]).

What would be the advantage of a positive-gating modulator compared to a more classical activator such as e.g. activators of K_{ATP} channels in cardiovascular disease? A mechanistic advantage of the $K_{Ca}2/3$ activators is presumably that they act only when there is Ca^{2+} mobilization, e.g. in form of "Ca^{2+} pulsars" in the endothelium [57] and thus initiation of endothelial function. Here, we would stimulate the active endothelium or potentiate endothelial function "when it is needed". Indeed, at present we understand well how the endothelium regulates arterial tone *in-vitro* and inhibition of NO synthesis *in-vivo* clearly elevates blood pressure while NO-donors are in the clinics since long to alleviate angina pectoris and hypertensive crisis. This fosters the pivotal role of NO in the systemic circulatory regulation. However, we do not know when endothelial EDH-vasodilator function occurs in the organisms and under physiological conditions and what are the consequences systemically or locally.

In situation of endothelial dysfunction/degeneration with loss of both channels an activator would not make sense at all while some conserved $K_{Ca}3.1/K_{Ca}2.3$ functions could be potentiated and improve thereby endothelial vasodilator function.

Systemic Effects of Positive-Gating Modulation

What do we learn from cardiovascular telemetry? In freely moving unstressed mice that have a blood pressure similar to ours (around 120/85 mmHg Systole/Diastole), acute administrations of SKA-31 (30–100 mg/kg; giving plasma levels above the EC50 for $K_{Ca}3.1$ and $K_{Ca}2$ activation) caused a rapid drop (by approx. −30 mmHg) in systolic and diastolic pressures that persisted over 1–4 h, pending on dose [111, 115]. In hypertensive mice, SKA-31 was similarly effective and lowered pressure to normotensive levels (mice treated with L-NAME or connexin(Cx)40−/− with angiotensin-II hypertension). Moreover, intra-vital microscopy on the microcirculation of cremaster skeletal muscle in anesthetized mice reveals that SKA-31 was capable to produce substantial arteriolar dilation from basal tone that did not require NO, but was abolished by genetic $K_{Ca}3.1$ deficiency [115]. This effect in resistance-size arteries explains to some extent the efficiency of SKA-31 to produce hypotension or normotension from hypertensive levels. Thus, this positive-gating modulator of $K_{Ca}3.1$ and $K_{Ca}2.3$ is active and causes the expected decrease in blood pressure.

A major concern is, however, that the pressure drop in the mice is accompanied by strong bradycardia (300 bpm at 100 mg/kg vs. 600 bpm (normal)) [115]. This is also seen in $K_{Ca}3.1$−/−, suggesting that this bradycardia is not related to activation of $K_{Ca}3.1$ but was rather caused by $K_{Ca}2$ activation in atria of the heart ensuing loss of sinus rhythm, and/or block of transmission within the conduction system. Interestingly, genetically encoded $K_{Ca}2.3$ over-expression has been shown to increase atrioventricular refractory period in young $K_{Ca}2.3$-overexpressor ($K_{Ca}2.3^{T/T}$) but decreased it in old $K_{Ca}2.3^{T/T}$ mice because of anatomical alterations related to $K_{Ca}2.3$-over-expression [75]. Moreover, genetic deficiency of the $K_{Ca}2.2$ subtype causes prolonged the PR and RR intervals [121] and caused prolongation of atrioventricular transmission, while over-expression of $K_{Ca}2.2$ had opposite effects.

Together, the data makes it likely that SKA-31 at $K_{Ca}2$-activating plasma concentrations affects the conduction system of the murine heart, which, unfortunately, may hinder a further development of this compound. Another, drawback is that SKA-31 causes severe sedation (immobility) in a $K_{Ca}3.1$-independent fashion [70], which relies likely on the bradycardia and/or the activation of central $K_{Ca}2$ channels, slowing neurotransmission, or—possibly—skeletal muscle $K_{Ca}2$, producing paresis.[2]

Interestingly, dogs that have a heart rate similar to us respond differently to SKA-31 (injected i.v.) because they show at $K_{Ca}3.1/K_{Ca}2$-activating plasma levels profound but short-lived hypotension and reflex tachycardia [113], which may point to

[2] *So far, there is no evidence that $K_{Ca}3.1$ is expressed in cardiomyocytes. In contrast, $K_{Ca}2$ channels are expressed cardiomyocytes. Moreover, $K_{Ca}3.1$ has been considered a non-neuronal channel as concluded form the absence of $K_{Ca}3.1$-mRNA in central neurons* [9, 29]. *Interestingly, $K_{Ca}3.1$ protein has recently been documented in rat brain by immunohistochemical approaches* [122]. *However, we could not clearly detect $K_{Ca}3.1$ in neuronal structures (but in the blood brain barrier) in murine brain and in human post mortem material using the current IHC approaches* [70]. *Thus, there are apparently species differences and it remains still possible that $K_{Ca}3.1$ in neurons add to cardiovascular effects and sedation described here.*

substantial species differences and request the need for thorough cardiovascular safety monitoring in large mammals.

Yet, SKA-31 and an analogue of SKA-31, SKA-19 has been found effective in the NINDS-funded anti-convulsant screening program [123]. Nonetheless, the likely disadvantageous cardiovascular profile of SKA-31 generates the need of a more selective $K_{Ca}3.1$-activator. Heike Wulff's group at UC-Davis synthesized a series of SKA-analogues [123], of which SKA-121 (5-methylnaphtho[2,1-d]oxazol-2-amine; Fig. 5.4a) has an improved selectivity profile for $K_{Ca}3.1$ (approx. 40-fold higher for $K_{Ca}3.1$ over $K_{Ca}2$). At the level of endothelium-dependent vasorelaxation, we find that SKA-121 does not act as vasorelaxant on its own in large coronary artery of the pig but potentiates the response elicited by bradykinin (Fig. 5.5a). This can be reversed by TRAM-34 [123]. Moreover, the compound does not affect endothelium-independent contraction or relaxation. In telemetry (Fig. 5.5b), SKA-121 reduced blood pressure in normotensive mice and L-NAME-treated hypertensive mice over approx. 2 h and 6 h, respectively, but did not cause blood pressure alterations in $K_{Ca}3.1$−/− mice. Importantly, heart rate was insignificantly affected in this study, suggesting substantial improvement of selectivity and cardiovascular safety of SKA-121. Neurological effects still need to be investigated in details, but sedation can still be an issue because of the higher brain/plasma concentration ratio [123]. In sum, positive gating modulators like SKA-121 are likely to have some therapeutic utilities in cardiovascular disease (hypertensive crisis, vasospasm, and central and peripheral ischemia) (Fig. 5.6).

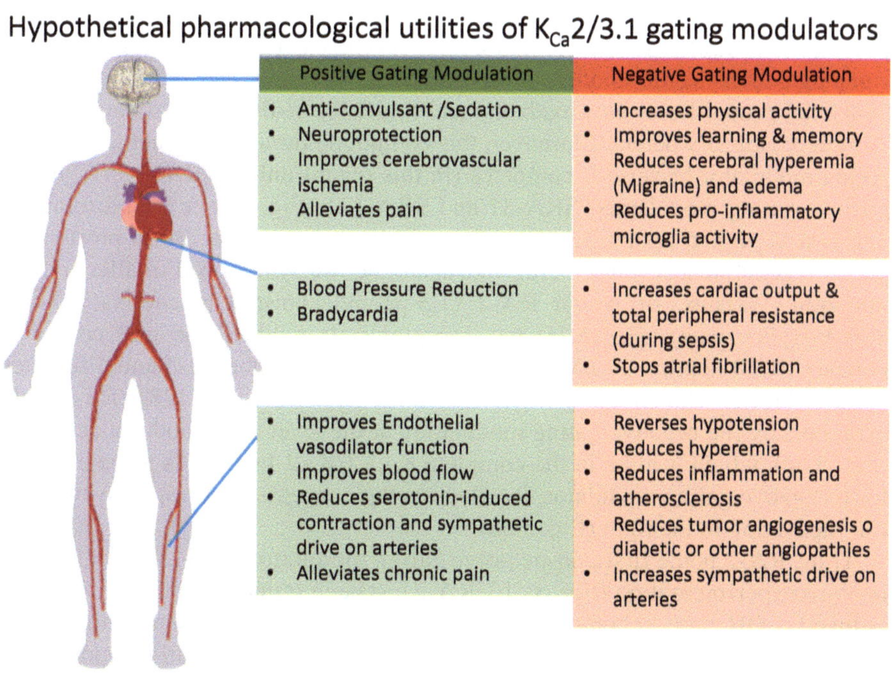

Fig. 5.6 Schematic illustration of systemic actions of the positive- and negative gating modulation of $K_{Ca}2/K_{Ca}3.1$ channels

Recent Developments: Negative-Gating Modulators

Besides the potential utility of positive-gating modulators of $K_{Ca}3.1/K_{Ca}2$ in cardiovascular pathophysiology—albeit counterintuitive at first—also negative-gating modulators of $K_{Ca}3.1/K_{Ca}2$ channel activators are of potential pharmacological value, for instance in atrial fibrillation, hypotension and sepsis, hyperemia, atherosclerosis, and restenosis after balloon catheter intervention, but perhaps also in neurological disorders associated with microglia-activation, chronic inflammation, and possible some cancer over-expressing $K_{Ca}3.1/K_{Ca}2$ channels.

The present selective $K_{Ca}3.1/K_{Ca}2$ inhibitors are mainly blockers that obstruct ion flow at the intracellular cavity below the selectivity filter (such as TRAM-34 on $K_{Ca}3.1$, Fig. 5.4c) or from the outside as classical pore blockers such as UCL-1684 (Fig. 5.4a, c), structurally mimicking the peptide blocker, apamin. The utility of UCL-1684 for in-vivo experimentation is not clear, but this very large molecule is still not drug-like. TRAM-34 has been used frequently in experimental in-vivo intervention trials and has been proven to be effective to inhibit BCI-induced neointima formation in rats and pigs [76, 77], neo-angiogenesis in Matri-gels [124], and experimental fibrosis in normal and diabetic kidneys [80, 81] and lungs [79, 125], suggesting utilities of $K_{Ca}3.1$-inhibitors for the treatment of disease characterized by pathological cell proliferation and pathological organ remodeling (for review see [8, 126]). However, TRAM-34 has been reported to be an inducer of P450 enzymes and looses selectivity for $K_{Ca}3.1$ over some other K channels at μmolar concentrations. Therefore, we have recently performed a small screening campaign focusing on another compound class (phenols and polyphenols with beneficial properties as anti-oxidants) and find 13b, tri-fluoro-benzoate ester [44, 127] as reasonably potent. Moreover, this compound turned out to act as negative-gating modulator because it competes (in this regard unlike TRAM-34) with the positive-gating modulators, SKA-31 and SKA-121 (Fig. 5.4a, c). Unfortunately, 13b is a large and lipophilic molecule and bioavailability of 13b is presumably poor. To generate a new more drug-like analogue, we synthesized smaller and less lipophilic variants, of which RA-2 (1,3-phenylenebis(methylene)bis(3-fluoro-4-hydroxybenzoate) (Fig. 5.4a) has acceptable drug likelihood and conserved potency (IC50 approx. 20 nM), a good selectivity over other K channel families, but also blocks with similar potencies all three $K_{Ca}2$-subtypes [128]. This can be explained by the negative-gating modulation at the level of calmodulin-activation (Fig. 5.4c), which is alike in the complete $K_{Ca}3.1/K_{Ca}2$ family. As to be expected for a negative gating modulator, RA-2 shifts the concentration-response curve for Ca^{2+} activation to the right (Fig. 5.5a).

Thus, RA-2 is the first pan-negative gating modulator of $K_{Ca}3.1$ and $K_{Ca}2$. In in-vitro experimentation (Fig. 5.5b), RA-2 has been found to inhibit bradykinin-induced EDH-type relaxation in porcine coronary artery and antagonizes potentiation of the bradykinin response by the positive-gating modulator, SKA-121 [128].

Systemic Effects of Negative-Gating Modulation

In respect to systemic cardiovascular regulation, RA-2 appears to be relatively safe since acute i.p. injections of up to 100 mg/kg did not produce hypertension or any disability in the mice (Fig. 5.5b). Still, RA-2 caused lasting bradycardia (drop by 150 bpm from the high levels of approx. 600 bpm, Fig. 5.5c). The simplest explanation for this bradycardia is that it mirrors baroreceptor-reflex to a higher peripheral resistance caused by $K_{Ca}3.1$ and $K_{Ca}2.3$ inhibition in small resistance-size arteries (Fig. 5.6). Other possibilities are a direct effect on the heart conduction system, caused by action potential prolongation as a result of $K_{Ca}2$ inhibition. However, we do not wish to exclude other changes in sympathetic or parasympathetic drive on the heart or central and peripheral mechanisms. Interestingly, bradycardia was absent in $K_{Ca}3.1$−/− suggesting an effect that is selectively mediated by $K_{Ca}3.1$. Besides these cardiovascular actions of RA-2, our unpublished data demonstrate a higher locomotor activity in RA-2 treated mice, which is, interestingly, also a feature of $K_{Ca}3.1$−/− mice and fosters the view of a participation of $K_{Ca}3.1$ in control of behavior and/or physical activity (Fig. 5.6).

Concluding Remarks

The current evidence derived from experimentation on gene-targeted mice and advance $K_{Ca}3.1/K_{Ca}2.3$ pharmacology assigns substantial impact of endothelial $K_{Ca}3.1$ and $K_{Ca}2.3$ in the endothelium on local arterial regulation as well as systemic cardiovascular regulation. Here, EDH and to some extent also NO are likely the downstream effectors. However, with respect to humans, we still need to be careful since the contribution of the channels to human systemic cardiovascular regulation remains unexplored although there is no doubt that human endothelium expresses these channels.

$K_{Ca}3.1$ and $K_{Ca}2.3$ are differentially regulated by cardiovascular disease at the functional and gene expression level and by other disease states. However, "loss-of-function" or "gain-of-function" (in the sense of monogenetic channelopathies) are unknown so far. Still, the $K_{Ca}2.3$-gene is linked to atrial fibrillation and $K_{Ca}2$ blockers are currently developed by the Danish Spin-off *Acesion* as novel types of atria-selective antiarrhythmic drugs.

Idiopathic forms of cardiovascular dysfunction in, particularly, human subjects go along with altered EDH and variable changes of $K_{Ca}3.1/K_{Ca}2.3$ gene expression and functions that could point at compensatory or pathogenic roles. This offers venues for endothelial selective treatment of endothelial dysfunction by activators like the positive-gating modulator, SKA-121 (Fig. 5.6). While the development of $K_{Ca}3.1/K_{Ca}2$ channel activators as antihypertensive is unlikely considering the availability of several established medications including Ca^{2+} antagonists and the—for activators—

potentially problematic tachyphylaxia, they may still be of use as alternative for the treatment of angina pectoris, local peripheral ischemia and pain, intra-surgical hypertension, and vascular protection of transplants. Not to forget that $K_{Ca}3.1/K_{Ca}2$-activators may be of advantage in early ischemic stroke and cause neuroprotection by impeding Ca^{2+}/Na^{+} overload in hypoxic neurons. Besides, such utilities in cardiovascular and cerebrovascular disease, they may also serve as targets in neurons to treat epilepsy and pain (Fig. 5.6).

In contrast, $K_{Ca}3.1$ and $K_{Ca}2$ inhibitors such as the pan-negative gating modulator of this channel class, RA-2, could have therapeutic value in situation of systemic hypotension as it occurs during sepsis, anesthesia accident, or after resurrection (Fig. 5.6). Moreover, it may have utilities for chronic inflammatory processes and proliferative angiopathies and, if considering $K_{Ca}2/3.1$ expression in brain and skeletal muscle, for the treatment of motivation loss and muscle weakness (Fig. 5.6).

Yet, these gating-modulators of $K_{Ca}3.1$ and $K_{Ca}2$ channels may be of help to shed new light on physiological and pathophysiological roles of $K_{Ca}3.1$ and $K_{Ca}2$ channels in the organisms. Considering the complex tissue expression pattern of the channels, the efficacy or undesired side effects of pharmacological manipulation remains to be tested in more detail. Still, the broad therapeutic utilities of $K_{Ca}3.1/K_{Ca}2$ gating modulation offers several attractive venues for further pharmaceutical development.

Acknowledgements The authors are supported by the Deutsche Forschungsgemeinschaft (KO1899/11-1), the Danish Heart Foundation, European Community (FP7-PEOPLE Project 321721), Department of Industry & Innovation, Government of Aragon (GIPASC-B105), and the Fondo de Investigación Sanitaria (Red HERACLES RD12/0042/0014).

References

1. Adelman JP, Maylie J, Sah P. Small-conductance Ca^{2+}-activated K+ channels: form and function. Annu Rev Physiol. 2012;74:245–69.
2. Grgic I, Kaistha BP, Hoyer J, Kohler R. Endothelial Ca+-activated K+ channels in normal and impaired EDHF-dilator responses—relevance to cardiovascular pathologies and drug discovery. Br J Pharmacol. 2009;157:509–26.
3. Feletou M. Calcium-activated potassium channels and endothelial dysfunction: therapeutic options? Br J Pharmacol. 2009;156:545–62.
4. Feletou M, Kohler R, Vanhoutte PM. Endothelium-derived vasoactive factors and hypertension: possible roles in pathogenesis and as treatment targets. Curr Hypertens Rep. 2010;12:267–75.
5. Edwards G, Feletou M, Weston AH. Endothelium-derived hyperpolarising factors and associated pathways: a synopsis. Pflugers Arch. 2010;459:863–79.
6. Feletou M, Kohler R, Vanhoutte PM. Nitric oxide: orchestrator of endothelium-dependent responses. Ann Med. 2012;44(7):694–716.
7. Köhler R, Ruth P. Endothelial dysfunction and blood pressure alterations in K^{+}-channel transgenic mice. Pflugers Arch. 2010;459:969–76.
8. Wulff H, Köhler R. Endothelial small-conductance and intermediate-conductance KCa channels: an update on their pharmacology and usefulness as cardiovascular targets. J Cardiovasc Pharmacol. 2013;61:102–12.

9. Wei AD, Gutman GA, Aldrich R, Chandy KG, Grissmer S, et al. International Union of Pharmacology. LII. Nomenclature and molecular relationships of calcium-activated potassium channels. Pharmacol Rev. 2005;57:463–72.
10. Schmitt N, Grunnet M, Olesen SP. Cardiac potassium channel subtypes: new roles in repolarization and arrhythmia. Physiol Rev. 2014;94:609–53.
11. Ellinor PT, Lunetta KL, Glazer NL, Pfeufer A, Alonso A, et al. Common variants in KCNN3 are associated with lone atrial fibrillation. Nat Genet. 2010;42:240–4.
12. Grunnet M, Bentzen BH, Sorensen US, Diness JG. Cardiac ion channels and mechanisms for protection against atrial fibrillation. Rev Physiol Biochem Pharmacol. 2012;162:1–58.
13. Yamaguchi M, Nakayama T, Fu Z, Naganuma T, Sato N, et al. Relationship between haplotypes of KCNN4 gene and susceptibility to human vascular diseases in Japanese. Med Sci Monit. 2009;15:CR389–97.
14. Köhler R. Single-nucleotide polymorphisms in vascular Ca^{2+}-activated K^+-channel genes and cardiovascular disease. Pflugers Arch. 2010;460:343–51.
15. Simms LA, Doecke JD, Roberts RL, Fowler EV, Zhao ZZ, et al. KCNN4 gene variant is associated with ileal Crohn's Disease in the Australian and New Zealand population. Am J Gastroenterol. 2010;105:2209–17.
16. Waeckel L, Bertin F, Clavreul N, Damery T, Kohler R, et al. Preserved regulation of renal perfusion pressure by small and intermediate conductance K channels in hypertensive mice with or without renal failure. Pflugers Arch. 2014;467:817–31.
17. Haddock RE, Grayson TH, Morris MJ, Howitt L, Chadha PS, et al. Diet-induced obesity impairs endothelium-derived hyperpolarization via altered potassium channel signaling mechanisms. PLoS One. 2011;6:e16423.
18. Wulff H, Castle NA. Therapeutic potential of KCa3.1 blockers: recent advances and promising trends. Expert Rev Clin Pharmacol. 2010;3:385–96.
19. Kaczorowski GJ, Knaus HG, Leonard RJ, McManus OB, Garcia ML. High-conductance calcium-activated potassium channels; structure, pharmacology, and function. J Bioenerg Biomembr. 1996;28:255–67.
20. Kaczmarek LK. Slack, slick and sodium-activated potassium channels. ISRN Neurosci. 2013;2013:354262.
21. Plüger S, Faulhaber J, Furstenau M, Lohn M, Waldschutz R, et al. Mice with disrupted BK channel beta1 subunit gene feature abnormal $Ca(^{2+})$ spark/STOC coupling and elevated blood pressure. Circ Res. 2000;87:E53–60.
22. Gollasch M, Ried C, Bychkov R, Luft FC, Haller H. K+ currents in human coronary artery vascular smooth muscle cells. Circ Res. 1996;78:676–88.
23. Hill-Eubanks DC, Werner ME, Heppner TJ, Nelson MT. Calcium signaling in smooth muscle. Cold Spring Harb Perspect Biol. 2011;3:a004549.
24. Brenner R, Perez GJ, Bonev AD, Eckman DM, Kosek JC, et al. Vasoregulation by the beta1 subunit of the calcium-activated potassium channel. Nature. 2000;407:870–6.
25. Brakemeier S, Eichler I, Knorr A, Fassheber T, Kohler R, et al. Modulation of Ca^{2+}-activated K^+ channel in renal artery endothelium in situ by nitric oxide and reactive oxygen species. Kidney Int. 2003;64:199–207.
26. Bychkov R, Burnham MP, Richards GR, Edwards G, Weston AH, et al. Characterization of a charybdotoxin-sensitive intermediate conductance Ca^{2+}-activated K^+ channel in porcine coronary endothelium: relevance to EDHF. Br J Pharmacol. 2002;137:1346–54.
27. Fernandez-Fernandez JM, Tomas M, Vazquez E, Orio P, Latorre R, et al. Gain-of-function mutation in the KCNMB1 potassium channel subunit is associated with low prevalence of diastolic hypertension. J Clin Invest. 2004;113:1032–9.
28. Tomas M, Vazquez E, Fernandez-Fernandez JM, Subirana I, Plata C, et al. Genetic variation in the KCNMA1 potassium channel alpha subunit as risk factor for severe essential hypertension and myocardial infarction. J Hypertens. 2008;26:2147–53.
29. Ishii TM, Silvia C, Hirschberg B, Bond CT, Adelman JP, et al. A human intermediate conductance calcium-activated potassium channel. Proc Natl Acad Sci U S A. 1997;94:11651–6.

30. Köhler M, Hirschberg B, Bond CT, Kinzie JM, Marrion NV, et al. Small-conductance, calcium-activated potassium channels from mammalian brain. Science. 1996;273:1709–14.
31. Alvarez J, Montero M, Garcia-Sancho J. High affinity inhibition of Ca($^{2+}$)-dependent K+ channels by cytochrome P-450 inhibitors. J Biol Chem. 1992;267:11789–93.
32. Knaus HG, McManus OB, Lee SH, Schmalhofer WA, Garcia-Calvo M, et al. Tremorgenic indole alkaloids potently inhibit smooth muscle high-conductance calcium-activated potassium channels. Biochemistry. 1994;33:5819–28.
33. Rosa JC, Galanakis D, Ganellin CR, Dunn PM, Jenkinson DH. Bis-quinolinium cyclophanes: 6,10-diaza-3(1,3),8(1,4)-dibenzena-1,5(1,4)-diquinolinacyclodecaphane (UCL 1684), the first nanomolar, non-peptidic blocker of the apamin-sensitive Ca^{2+}-activated K^+ channel. J Med Chem. 1998;41:2–5.
34. Marchenko SM, Sage SO. Calcium-activated potassium channels in the endothelium of intact rat aorta. J Physiol. 1996;492(Pt 1):53–60.
35. Van Renterghem C, Vigne P, Frelin C. A charybdotoxin-sensitive, Ca($^{2+}$)-activated K+ channel with inward rectifying properties in brain microvascular endothelial cells: properties and activation by endothelins. J Neurochem. 1995;65:1274–81.
36. Cai S, Garneau L, Sauve R. Single-channel characterization of the pharmacological properties of the K(Ca^{2+}) channel of intermediate conductance in bovine aortic endothelial cells. J Membr Biol. 1998;163:147–58.
37. Köhler R, Distler A, Hoyer J. Increased mechanosensitive currents in aortic endothelial cells from genetically hypertensive rats. J Hypertens. 1999;17:365–71.
38. Burnham MP, Bychkov R, Feletou M, Richards GR, Vanhoutte PM, et al. Characterization of an apamin-sensitive small-conductance Ca($^{2+}$)-activated K(+) channel in porcine coronary artery endothelium: relevance to EDHF. Br J Pharmacol. 2002;135:1133–43.
39. Eichler I, Wibawa J, Grgic I, Knorr A, Brakemeier S, et al. Selective blockade of endothelial Ca^{2+}-activated small- and intermediate-conductance K^+-channels suppresses EDHF-mediated vasodilation. Br J Pharmacol. 2003;138:594–601.
40. Köhler R, Degenhardt C, Kuhn M, Runkel N, Paul M, et al. Expression and function of endothelial Ca^{2+}-activated K^+ channels in human mesenteric artery: a single-cell reverse transcriptase-polymerase chain reaction and electrophysiological study in situ. Circ Res. 2000;87:496–503.
41. Chadha PS, Liu L, Rikard-Bell M, Senadheera S, Howitt L, et al. Endothelium-dependent vasodilation in human mesenteric artery is primarily mediated by myoendothelial gap junctions intermediate conductance calcium-activated K+ channel and nitric oxide. J Pharmacol Exp Ther. 2011;336:701–8.
42. Fisslthaler B, Popp R, Kiss L, Potente M, Harder DR, et al. Cytochrome P450 2C is an EDHF synthase in coronary arteries. Nature. 1999;401:493–7.
43. Ng KF, Leung SW, Man RY, Vanhoutte PM. Endothelium-derived hyperpolarizing factor mediated relaxations in pig coronary arteries do not involve Gi/o proteins. Acta Pharmacol Sin. 2008;29:1419–24.
44. Olivan-Viguera A, Valero MS, Murillo MD, Wulff H, Garcia-Otin AL, et al. Novel phenolic inhibitors of small/intermediate-conductance Ca(2)(+)-activated K(+) channels, KCa3.1 and KCa2.3. PLoS One. 2013;8:e58614.
45. Randriamboavonjy V, Kiss L, Falck JR, Busse R, Fleming I. The synthesis of 20-HETE in small porcine coronary arteries antagonizes EDHF-mediated relaxation. Cardiovasc Res. 2005;65:487–94.
46. Kacik M, Olivan-Viguera A, Kohler R. Modulation of KCa3.1 channels by eicosanoids, omega-3 fatty acids, and molecular determinants. PLoS One. 2014;9:e112081.
47. Si H, Heyken WT, Wölfle SE, Tysiac M, Schubert R, et al. Impaired endothelium-derived hyperpolarizing factor-mediated dilations and increased blood pressure in mice deficient of the intermediate-conductance Ca^{2+}-activated K^+ channel. Circ Res. 2006;99:537–44.
48. Bond CT, Sprengel R, Bissonnette JM, Kaufmann WA, Pribnow D, et al. Respiration and parturition affected by conditional overexpression of the Ca^{2+}-activated K^+ channel subunit, SK3. Science. 2000;289:1942–6.

49. Taylor MS, Bonev AD, Gross TP, Eckman DM, Brayden JE, et al. Altered expression of small-conductance Ca^{2+}-activated K+ (SK3) channels modulates arterial tone and blood pressure. Circ Res. 2003;93:124–31.
50. Brähler S, Kaistha A, Schmidt VJ, Wolfle SE, Busch C, et al. Genetic deficit of SK3 and IK1 channels disrupts the endothelium-derived hyperpolarizing factor vasodilator pathway and causes hypertension. Circulation. 2009;119:2323–32.
51. Wolfle SE, Schmidt VJ, Hoyer J, Kohler R, de Wit C. Prominent role of KCa3.1 in endothelium-derived hyperpolarizing factor-type dilations and conducted responses in the microcirculation in vivo. Cardiovasc Res. 2009;82:476–83.
52. Milkau M, Kohler R, de Wit C. Crucial importance of the endothelial K+ channel SK3 and connexin40 in arteriolar dilations during skeletal muscle contraction. FASEB J. 2010; 24:3572–9.
53. Hasenau AL, Nielsen G, Morisseau C, Hammock BD, Wulff H, et al. Improvement of endothelium-dependent vasodilations by SKA-31 and SKA-20, activators of small- and intermediate-conductance Ca^{2+}-activated K+-channels. Acta Physiol (Oxf). 2011;203:117–26.
54. Köhler R, Heyken WT, Heinau P, Schubert R, Si H, et al. Evidence for a functional role of endothelial transient receptor potential V4 in shear stress-induced vasodilatation. Arterioscler Thromb Vasc Biol. 2006;26:1495–502.
55. Qian X, Francis M, Kohler R, Solodushko V, Lin M, et al. Positive feedback regulation of agonist-stimulated endothelial Ca^{2+} dynamics by KCa3.1 channels in mouse mesenteric arteries. Arterioscler Thromb Vasc Biol. 2014;34:127–35.
56. Watanabe H, Vriens J, Prenen J, Droogmans G, Voets T, et al. Anandamide and arachidonic acid use epoxyeicosatrienoic acids to activate TRPV4 channels. Nature. 2003;424:434–8.
57. Ledoux J, Taylor MS, Bonev AD, Hannah RM, Solodushko V, et al. Functional architecture of inositol 1,4,5-trisphosphate signaling in restricted spaces of myoendothelial projections. Proc Natl Acad Sci U S A. 2008;105:9627–32.
58. Hill CE. Tudor Griffith, gap junctions and conducted vasodilatation: electromechanical coupling back in the limelight. J Cardiovasc Pharmacol. 2013;61:93–101.
59. de Wit C, Griffith TM. Connexins and gap junctions in the EDHF phenomenon and conducted vasomotor responses. Pflugers Arch. 2010;459:897–914.
60. Billaud M, Lohman AW, Johnstone SR, Biwer LA, Mutchler S, et al. Regulation of cellular communication by signaling microdomains in the blood vessel wall. Pharmacol Rev. 2014;66:513–69.
61. Sandow SL, Neylon CB, Chen MX, Garland CJ. Spatial separation of endothelial small- and intermediate-conductance calcium-activated potassium channels (K_{Ca}) and connexins: possible relationship to vasodilator function? J Anat. 2006;209:689–98.
62. Ma X, Du J, Zhang P, Deng J, Liu J, et al. Functional role of TRPV4-KCa2.3 signaling in vascular endothelial cells in normal and streptozotocin-induced diabetic rats. Hypertension. 2013;62:134–9.
63. Berrout J, Mamenko M, Zaika OL, Chen L, Zang W, et al. Emerging role of the calcium-activated, small conductance, SK3 K+ channel in distal tubule function: regulation by TRPV4. PLoS One. 2014;9:e95149.
64. Sukumaran SV, Singh TU, Parida S, Narasimha Reddy CE, Thangamalai R, et al. TRPV4 channel activation leads to endothelium-dependent relaxation mediated by nitric oxide and endothelium-derived hyperpolarizing factor in rat pulmonary artery. Pharmacol Res. 2013;78:18–27.
65. Hoyer J, Kohler R, Distler A. Mechanosensitive Ca^{2+} oscillations and STOC activation in endothelial cells. FASEB J. 1998;12:359–66.
66. Sonkusare SK, Bonev AD, Ledoux J, Liedtke W, Kotlikoff MI, et al. Elementary Ca^{2+} signals through endothelial TRPV4 channels regulate vascular function. Science. 2012; 336:597–601.
67. Hartmannsgruber V, Heyken WT, Kacik M, Kaistha A, Grgic I, et al. Arterial response to shear stress critically depends on endothelial TRPV4 expression. PLoS One. 2007;2:e827.

68. Saliez J, Bouzin C, Rath G, Ghisdal P, Desjardins F, et al. Role of caveolar compartmentation in endothelium-derived hyperpolarizing factor-mediated relaxation: Ca^{2+} signals and gap junction function are regulated by caveolin in endothelial cells. Circulation. 2008;117:1065–74.
69. Dominguez Rieg JA, Burt JM, Ruth P, Rieg T. P2Y receptor activation decreases blood pressure via intermediate conductance potassium channels and connexin 37. Acta Physiol (Oxf). 2014;213:628–41.
70. Lambertsen KL, Gramsbergen JB, Sivasaravanaparan M, Ditzel N, Sevelsted-Moller LM, et al. Genetic KCa3.1-deficiency produces locomotor hyperactivity and alterations in cerebral monoamine levels. PLoS One. 2012;7:e47744.
71. Liang Z, Chen L, McClafferty H, Lukowski R, Macgregor D, et al. Control of HPA stress axis activity by the intermediate conductance calcium-activated potassium channel, SK4. J Physiol. 2011;589:5965–86.
72. Grgic I, Kaistha BP, Paschen S, Kaistha A, Busch C, et al. Disruption of the Gardos channel (KCa3.1) in mice causes subtle erythrocyte macrocytosis and progressive splenomegaly. Pflugers Arch. 2009;458:291–302.
73. Rada CC, Pierce SL, Nuno DW, Zimmerman K, Lamping KG, et al. Overexpression of the SK3 channel alters vascular remodeling during pregnancy, leading to fetal demise. Am J Physiol Endocrinol Metab. 2012;303:E825–31.
74. Wandall-Frostholm C, Skaarup LM, Sadda V, Nielsen G, Hedegaard ER, et al. Pulmonary hypertension in wild type mice and animals with genetic deficit in KCa2.3 and KCa3.1 channels. PLoS One. 2014;9:e97687.
75. Mahida S, Mills RW, Tucker NR, Simonson B, Macri V, et al. Overexpression of KCNN3 results in sudden cardiac death. Cardiovasc Res. 2014;101:326–34.
76. Köhler R, Wulff H, Eichler I, Kneifel M, Neumann D, et al. Blockade of the intermediate-conductance calcium-activated potassium channel as a new therapeutic strategy for restenosis. Circulation. 2003;108:1119–25.
77. Tharp DL, Wamhoff BR, Wulff H, Raman G, Cheong A, et al. Local delivery of the KCa3.1 blocker, TRAM-34, prevents acute angioplasty-induced coronary smooth muscle phenotypic modulation and limits stenosis. Arterioscler Thromb Vasc Biol. 2008;28:1084–9.
78. Toyama K, Wulff H, Chandy KG, Azam P, Raman G, et al. The intermediate-conductance calcium-activated potassium channel KCa3.1 contributes to atherogenesis in mice and humans. J Clin Invest. 2008;118:3025–37.
79. Roach KM, Wulff H, Feghali-Bostwick C, Amrani Y, Bradding P. Increased constitutive inverted question markSMA and Smad2/3 expression in idiopathic pulmonary fibrosis myofibroblasts is K Ca 31-dependent. Respir Res. 2014;15:155.
80. Grgic I, Kiss E, Kaistha BP, Busch C, Kloss M, et al. Renal fibrosis is attenuated by targeted disruption of KCa3.1 potassium channels. Proc Natl Acad Sci U S A. 2009;106:14518–23.
81. Huang C, Shen S, Ma Q, Gill A, Pollock CA, et al. KCa3.1 mediates activation of fibroblasts in diabetic renal interstitial fibrosis. Nephrol Dial Transplant. 2014;29:313–24.
82. Freise C, Heldwein S, Erben U, Hoyer J, Kohler R, et al. K-channel inhibition reduces portal perfusion pressure in fibrotic rats and fibrosis associated characteristics of hepatic stellate cells. Liver Int. 2014;35:1244–52.
83. Chen YJ, Lam J, Gregory CR, Schrepfer S, Wulff H. The Ca(2)(+)-activated K(+) channel KCa3.1 as a potential new target for the prevention of allograft vasculopathy. PLoS One. 2013;8:e81006.
84. Grgic I, Wulff H, Eichler I, Flothmann C, Kohler R, et al. Blockade of T-lymphocyte KCa3.1 and Kv1.3 channels as novel immunosuppression strategy to prevent kidney allograft rejection. Transplant Proc. 2009;41:2601–6.
85. Hua X, Deuse T, Chen YJ, Wulff H, Stubbendorff M, et al. The potassium channel KCa3.1 as new therapeutic target for the prevention of obliterative airway disease. Transplantation. 2013;95:285–92.
86. Si H, Grgic I, Heyken WT, Maier T, Hoyer J, et al. Mitogenic modulation of Ca^{2+}-activated K+ channels in proliferating A7r5 vascular smooth muscle cells. Br J Pharmacol. 2006;148:909–17.

87. Cheong A, Bingham AJ, Li J, Kumar B, Sukumar P, et al. Downregulated REST transcription factor is a switch enabling critical potassium channel expression and cell proliferation. Mol Cell. 2005;20:45–52.
88. Brakemeier S, Kersten A, Eichler I, Grgic I, Zakrzewicz A, et al. Shear stress-induced up-regulation of the intermediate-conductance Ca($^{2+}$)-activated K(+) channel in human endothelium. Cardiovasc Res. 2003;60:488–96.
89. Takai J, Santu A, Zheng H, Koh SD, Ohta M, et al. Laminar shear stress upregulates endothelial Ca(2)(+)-activated K(+) channels KCa2.3 and KCa3.1 via a Ca(2)(+)/calmodulin-dependent protein kinase kinase/Akt/p300 cascade. Am J Physiol Heart Circ Physiol. 2013;305:H484–93.
90. Köhler R, Brakemeier S, Kuhn M, Behrens C, Real R, et al. Impaired hyperpolarization in regenerated endothelium after balloon catheter injury. Circ Res. 2001;89:174–9.
91. Köhler R, Eichler I, Schonfelder H, Grgic I, Heinau P, et al. Impaired EDHF-mediated vasodilation and function of endothelial Ca-activated K channels in uremic rats. Kidney Int. 2005;67:2280–7.
92. Chennupati R, Lamers WH, Koehler SE, De Mey JG. Endothelium-dependent hyperpolarization-related relaxations diminish with age in murine saphenous arteries of both sexes. Br J Pharmacol. 2013;169:1486–99.
93. Yap FC, Taylor MS, Lin MT. Ovariectomy-induced reductions in endothelial SK3 channel activity and endothelium-dependent vasorelaxation in murine mesenteric arteries. PLoS One. 2014;9:e104686.
94. Climent B, Moreno L, Martinez P, Contreras C, Sanchez A, et al. Upregulation of SK3 and IK1 channels contributes to the enhanced endothelial calcium signaling and the preserved coronary relaxation in obese Zucker rats. PLoS One. 2014;9:e109432.
95. Chadha PS, Haddock RE, Howitt L, Morris MJ, Murphy TV, et al. Obesity up-regulates intermediate conductance calcium-activated potassium channels and myoendothelial gap junctions to maintain endothelial vasodilator function. J Pharmacol Exp Ther. 2010;335:284–93.
96. Chantome A, Potier-Cartereau M, Clarysse L, Fromont G, Marionneau-Lambot S, et al. Pivotal role of the lipid Raft SK3-Orai1 complex in human cancer cell migration and bone metastases. Cancer Res. 2013;73:4852–61.
97. Turner KL, Honasoge A, Robert SM, McFerrin MM, Sontheimer H. A proinvasive role for the Ca($^{2+}$)-activated K(+) channel KCa3.1 in malignant glioma. Glia. 2014;62:971–81.
98. Grossinger EM, Weiss L, Zierler S, Rebhandl S, Krenn PW, et al. Targeting proliferation of chronic lymphocytic leukemia (CLL) cells through KCa3.1 blockade. Leukemia. 2014;28:954–8.
99. D'Alessandro G, Catalano M, Sciaccaluga M, Chece G, Cipriani R, et al. KCa3.1 channels are involved in the infiltrative behavior of glioblastoma in vivo. Cell Death Dis. 2013;4:e773.
100. Ouadid-Ahidouch H, Ahidouch A. K(+) channels and cell cycle progression in tumor cells. Front Physiol. 2013;4:220.
101. Huang X, Jan LY. Targeting potassium channels in cancer. J Cell Biol. 2014;206:151–62.
102. Koch Hansen L, Sevelsted-Moller L, Rabjerg M, Larsen D, Hansen TP, et al. Expression of T-cell KV1.3 potassium channel correlates with pro-inflammatory cytokines and disease activity in ulcerative colitis. J Crohns Colitis. 2014;8:1378–91.
103. Chen YJ, Raman G, Bodendiek S, O'Donnell ME, Wulff H. The KCa3.1 blocker TRAM-34 reduces infarction and neurological deficit in a rat model of ischemia/reperfusion stroke. J Cereb Blood Flow Metab. 2011;31:2363–74.
104. Wulff H, Miller MJ, Hansel W, Grissmer S, Cahalan MD, et al. Design of a potent and selective inhibitor of the intermediate-conductance Ca^{2+}-activated K^+ channel, IKCa1: a potential immunosuppressant. Proc Natl Acad Sci U S A. 2000;97:8151–6.
105. Chen YJ, Wallace BK, Yuen N, Jenkins DP, Wulff H, et al. Blood-brain barrier KCa3.1 channels: evidence for a role in brain na uptake and edema in ischemic stroke. Stroke. 2015;46(1):237–44.
106. Mauler F, Hinz V, Horvath E, Schuhmacher J, Hofmann HA, et al. Selective intermediate-/small-conductance calcium-activated potassium channel (KCNN4) blockers are potent and

effective therapeutics in experimental brain oedema and traumatic brain injury caused by acute subdural haematoma. Eur J Neurosci. 2004;20:1761–8.
107. Ataga KI, Reid M, Ballas SK, Yasin Z, Bigelow C, et al. Improvements in haemolysis and indicators of erythrocyte survival do not correlate with acute vaso-occlusive crises in patients with sickle cell disease: a phase III randomized, placebo-controlled, double-blind study of the Gardos channel blocker senicapoc (ICA-17043). Br J Haematol. 2011;153:92–104.
108. Bouhy D, Ghasemlou N, Lively S, Redensek A, Rathore KI, et al. Inhibition of the Ca^{2+}-dependent K+ channel, KCNN4/KCa3.1, improves tissue protection and locomotor recovery after spinal cord injury. J Neurosci. 2011;31:16298–308.
109. Schlichter LC, Kaushal V, Moxon-Emre I, Sivagnanam V, Vincent C. The Ca^{2+} activated SK3 channel is expressed in microglia in the rat striatum and contributes to microglia-mediated neurotoxicity in vitro. J Neuroinflammation. 2010;7:4.
110. Maezawa I, Zimin PI, Wulff H, Jin LW. Amyloid-beta protein oligomer at low nanomolar concentrations activates microglia and induces microglial neurotoxicity. J Biol Chem. 2011;286:3693–706.
111. Sankaranarayanan A, Raman G, Busch C, Schultz T, Zimin PI, et al. Naphtho[1,2-d]thiazol-2-ylamine (SKA-31), a new activator of KCa2 and KCa3.1 potassium channels, potentiates the endothelium-derived hyperpolarizing factor response and lowers blood pressure. Mol Pharmacol. 2009;75:281–95.
112. Mishra RC, Belke D, Wulff H, Braun AP. SKA-31, a novel activator of SK(Ca) and IK(Ca) channels, increases coronary flow in male and female rat hearts. Cardiovasc Res. 2013;97:339–48.
113. Damkjaer M, Nielsen G, Bodendiek S, Staehr M, Gramsbergen JB, et al. Pharmacological activation of KCa3.1/KCa2.3 channels produces endothelial hyperpolarization and lowers blood pressure in conscious dogs. Br J Pharmacol. 2012;165:223–34.
114. Coleman N, Brown BM, Olivan-Viguera A, Singh V, Olmstead MM, et al. New positive Ca^{2+}-activated K+ channel gating modulators with selectivity for KCa3.1. Mol Pharmacol. 2014;86:342–57.
115. Radtke J, Schmidt K, Wulff H, Kohler R, de Wit C. Activation of KCa3.1 by SKA-31 induces arteriolar dilatation and lowers blood pressure in normo- and hypertensive connexin40-deficient mice. Br J Pharmacol. 2013;170:293–303.
116. Waeckel L, Badier-Commander C, Damery T, Kohler R, Sansilvestri-Morel P, et al. Vascular dysfunctions in the isolated aorta of double-transgenic hypertensive mice developing aortic aneurysm. Pflugers Arch. 2014;467:1945–63.
117. Soder RP, Parajuli SP, Hristov KL, Rovner ES, Petkov GV. SK channel-selective opening by SKA-31 induces hyperpolarization and decreases contractility in human urinary bladder smooth muscle. Am J Physiol Regul Integr Comp Physiol. 2013;304:R155–63.
118. Zhang M, Pascal JM, Schumann M, Armen RS, Zhang JF. Identification of the functional binding pocket for compounds targeting small-conductance Ca(2)(+)-activated potassium channels. Nat Commun. 2012;3:1021.
119. Zhang M, Pascal JM, Zhang JF. Unstructured to structured transition of an intrinsically disordered protein peptide in coupling Ca(2)(+)-sensing and SK channel activation. Proc Natl Acad Sci U S A. 2013;110:4828–33.
120. Cui M, Qin G, Yu K, Bowers MS, Zhang M. Targeting the small- and intermediate-conductance Ca-activated potassium channels: the drug-binding pocket at the channel/calmodulin interface. Neurosignals. 2014;22:65–78.
121. Zhang Q, Timofeyev V, Lu L, Li N, Singapuri A, et al. Functional roles of a Ca^{2+}-activated K+ channel in atrioventricular nodes. Circ Res. 2008;102:465–71.
122. Engbers JD, Anderson D, Asmara H, Rehak R, Mehaffey WH, et al. Intermediate conductance calcium-activated potassium channels modulate summation of parallel fiber input in cerebellar Purkinje cells. Proc Natl Acad Sci U S A. 2012;109:2601–6.
123. Coleman N, Nguyen HM, Cao Z, Brown BM, Jenkins DP, et al. The riluzole derivative 2-amino-6-trifluoromethylthio-benzothiazole (SKA-19), a mixed K2 activator and Na blocker, is a potent novel anticonvulsant. Neurotherapeutics. 2014;12:234–49.

124. Grgic I, Eichler I, Heinau P, Si H, Brakemeier S, et al. Selective blockade of the intermediate-conductance Ca^{2+}-activated K^+ channel suppresses proliferation of microvascular and macrovascular endothelial cells and angiogenesis in vivo. Arterioscler Thromb Vasc Biol. 2005;25:704–9.
125. Van Der Velden J, Sum G, Barker D, Koumoundouros E, Barcham G, et al. K(Ca)3.1 channel-blockade attenuates airway pathophysiology in a sheep model of chronic asthma. PLoS One. 2013;8:e66886.
126. Köhler R, Kaistha BP, Wulff H. Vascular KCa-channels as therapeutic targets in hypertension and restenosis disease. Expert Opin Ther Targets. 2010;14:143–55.
127. Lamoral-Theys D, Pottier L, Kerff F, Dufrasne F, Proutiere F, et al. Simple di- and trivanillates exhibit cytostatic properties toward cancer cells resistant to pro-apoptotic stimuli. Bioorg Med Chem. 2010;18:3823–33.
128. Olivan-Viguera A, Valero MS, Coleman N, Brown BM, Laria C, et al. A novel pan-negative-gating modulator of KCa2/3 channels, the fluoro-di-benzoate, RA-2, inhibits EDH-type relaxation in coronary artery and produces bradycardia in vivo. Mol Pharmacol. 2015;87:1–12.

Part II
Ion Channel Regulation of Vascular Tone and Blood Flow. Changes with Hypertension: Smooth Muscle Cell

Chapter 6
Regional Variation in Arterial Myogenic Responsiveness: Links to Potassium Channel Diversity/Function

Michael A. Hill, Yan Yang, Zahra Nourian, Barry D. Kyle, Kwangseok Hong, and Andrew P. Braun

Abstract Regional variation in small artery myogenic responsiveness is associated with differences in relationships amongst intraluminal pressure, smooth muscle cell (SMC) membrane potential (Em) and vessel diameter. For example, under *in vitro* conditions, small arteries from cremaster muscle show a steeper relationship between Em and myogenic contraction compared with cerebral arteries. To explain this difference, we hypothesized that the function/regulation of the large conductance, Ca^{2+}-activated, K^+ channel (BK_{Ca}) differs between these vascular beds. This was based on previous observations by Nelson and colleagues that BK_{Ca}, activated by sarcoplasmic reticulum (SR)-generated Ca^{2+} sparks, exerts a hyperpolarizing influence that opposes myogenic constriction. To test this, studies were performed using Ca^{2+} imaging, vessel myography, isolated cell electrophysiology and molecular biology techniques on small resistance arteries from the cerebral and cremaster muscle vasculatures. While BK_{Ca} in SMCs of both small arteries showed a similar conductance and voltage sensitivity, Ca^{2+} sensitivity was 2–3-fold greater in cerebral SMCs. Single channel open times were greater in cerebral SMCs compared with those of cremaster SMCs. Conversely, closed times were significantly shorter in cerebral SMCs. In addition to variation in biophysical characteristics, β_1-BK_{Ca} subunit expression was decreased in cremaster SMCs. Further, siRNA-induced knockdown of the β1 subunit of the BK_{Ca} holo-channel shifted gating behavior of cerebral BK_{Ca} channels to resemble that observed in cremaster SMCs. Collectively, the data indicate that while BK_{Ca} is present in both vascular preparations expression levels and modes of regulation differ. In particular, BK_{Ca} in small cerebral arteries is configured to show a higher Ca^{2+} sensitivity resulting in greater opening at

M.A. Hill, Ph.D. (✉) • Y. Yang • Z. Nourian • K. Hong
Dalton Cardiovascular Research Center and Department of Medical Pharmacology and Physiology, University of Missouri, Columbia, MO 65211, USA
e-mail: hillmi@missouri.edu

B.D. Kyle • A.P. Braun
Department of Physiology and Pharmacology, Cumming School of Medicine, University of Calgary, Calgary, AB, Canada

I. Levitan, A.M. Dopico (eds.), *Vascular Ion Channels in Physiology and Disease*, DOI 10.1007/978-3-319-29635-7_6

physiological levels of membrane potential. Heterogeneity in SMC ion channel function is not limited to BK_{Ca} as vascular bed differences are also apparent for other K^+ channels including the voltage-gated Kv and Kv7 families of channels. With respect to the latter, Kv7 channels appear to play a greater role in cerebral vasculature compared to the coronary circulation. From a physiological perspective it is suggested that differences in local ion channel function allow for regional differences in the regulation of myogenic tone and hence the control of tissue hemodynamics.

Keywords Arterioles • Vascular smooth muscle • Signal transduction • Ion channels • Mechanotransduction

The ability of arterioles to constrict to acute increases in intraluminal pressure has been recognized for well over 100 years and is commonly referred to as the myogenic response [1–4]. The physiological significance of this mechanically-induced contraction relates to its role in blood flow autoregulation, setting of basal vascular tone and its contribution to the regulation of capillary pressure and prevention of pressure-induced damage [2].

Although the basic pathways underlying myogenic contraction are known to lie within the vascular smooth muscle cells (VSMCs) of the arteriolar wall, there remains doubt as to the precise molecular mechanisms by which the pressure-mediated mechanical stimulus is sensed and ultimately converted to an activation of the contractile proteins. Substantial evidence exists for a pivotal role of membrane depolarization, opening of voltage-gated Ca^{2+} channels, increased intracellular Ca^{2+} (Ca^{2+}_i) and Ca^{2+}-dependent phosphorylation of the regulatory myosin light chain [5–11]. These events are supported by a number of other protein kinase-based mechanisms that involve pathways leading to alterations in Ca^{2+} sensitivity and the rearrangement of cytoskeletal elements [8, 12–14].

While the exact event(s) underlying the pressure-induced regulation of membrane potential are uncertain they appear to involve a number of different ion channels (including stretch/mechanogated channels, various TRP channels and possibly TMEM16A/ANO1 chloride channels) [15–20]. Activation of these channels occurs either by direct mechanical perturbation or in response to integrin and/or second messenger-mediated mechanisms [19, 21, 22]. In addition to contributing to the initial pressure-induced depolarization, various ion channels also underlie movement of Ca^{2+} from the extracellular space into VSMCs, release of Ca^{2+} from the intracellular stores [2, 4] and to repolarization both following restoration of the initial pressure level and as a feedback mechanism controlling the extent of myogenic constriction [23–25]. It has also become evident that considerable complexity surrounds ion channel structure and regulation. Rather than being simple pores, ion channels and their gating behavior are regulated through their expression levels, existence of isoforms, splice variants and regulatory subunits, post-translational modification, dynamic trafficking to and from the cell surface, and spatial location within signaling microdomains [26–32].

This chapter will present evidence suggesting that heterogeneity in K+ channel expression and function may contribute to differences in how myogenic reactivity is controlled in different vascular beds. An underlying assumption is that this regional heterogeneity allows fine-tuning of the basic myogenic contractile mechanism to specifically serve local tissue requirements. A particular example of this is shown in small arteries from rat skeletal muscle (cremaster muscle) where there is a steeper relationship between Em and myogenic contraction compared with that in cerebral vessels (Fig. 6.1). It was suggested that such steeper relationship allows maintenance of a high peripheral resistance in resting skeletal muscle vasculature while at the same time maintaining adequate cerebral perfusion [11, 33]. An additional

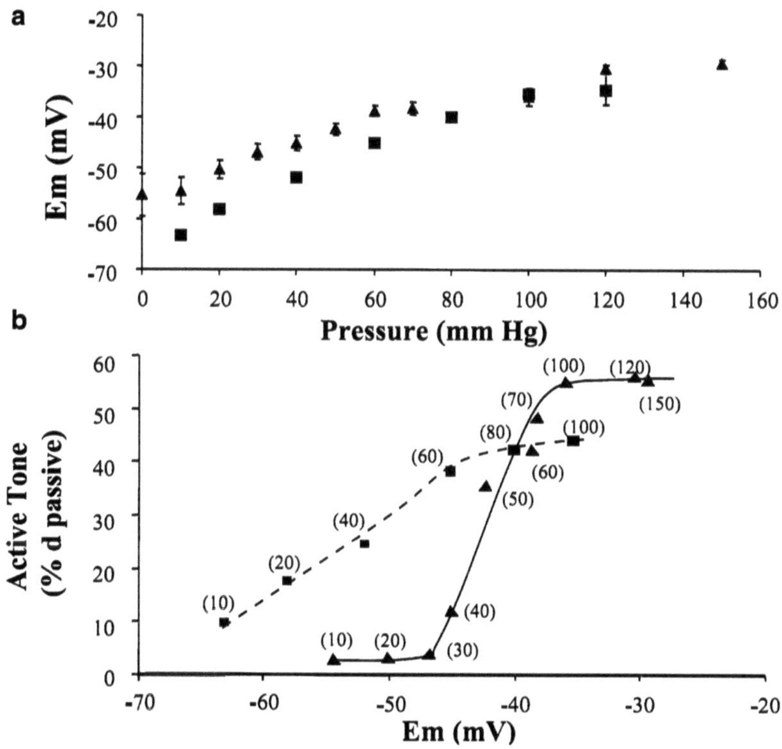

Fig. 6.1 Small arteries from skeletal muscle demonstrate a steeper Em-myogenic tone relationship compared with cerebral vessels. Comparison of data from cerebral vessels (*squares*) [data replotted from Knot et al. [24]] with skeletal muscle arterioles (*triangles*) from Kotecha and Hill[11] shows an upward shift in the pressure-Em relationship for skeletal muscle arterioles with largest differences between data sets evident at pressures below approx. 80 mmHg (**a**). Above this pressure, cerebral vessels (approx. 200 μm diameter) exhibit a lesser extent of myogenic constriction compared with skeletal muscle arterioles (approx. 150 μm diameter) (**b**). Note that for both vessel sets, active tone is plotted relative to the passive diameter at each pressure. Data are normalized as cerebral vessels are slightly larger than cremaster muscle vessels. *Numbers* in *parentheses* indicate intraluminal pressures (in mmHg). From Kotecha and Hill [11]. *Em* membrane potential

example of ion channel diversity affecting myogenic reactivity can be seen between branch orders within a vascular bed. Chen and colleagues thus showed that first order mesenteric arteries from rats express higher levels of the voltage-gated K^+ channel (Kv) than do the smaller, more myogenically reactive third order vessels [34]. Further, *in vitro* treatment of the first order vessels, for 48 h, with a targeted oligonucleotide decreased Kv channel expression at the protein level and increased myogenic reactivity. Similarly, Loutzenhiser and colleagues have shown variation in the expression and function of ion channels, including K_{ir} and voltage-operated Ca^{2+} channels (VOCC), in the renal vasculature of the rat [35, 36]. Thus, angiotensin-mediated constriction of the afferent arterioles was shown to be dependent on Ca^{2+} entry via VOCCs while that of the efferent arteriole showed a greater dependence on Ca^{2+} release from intracellular stores and subsequent store-operated Ca^{2+} entry [36]. In the interlobular artery the expression and function of K_{ir} increases along the length (proximal to distal) of the vessel. Interestingly, associated with this pattern of channel expression is a decrease in vessel diameter and increases in myogenic reactivity and vascular resistance [35].

Ion channel diversity may also occur in regard to intracellular channels. Westcott et al. have shown variation in the expression of SR ryanodine receptors (RyR; an intracellular Ca^{2+} release channel) between mouse (C57Bl/6 strain) feed arteries (arteries entering a tissue such as a specific skeletal muscle; diameter approximately 190 μm) and smaller intramuscular arterioles (diameter approximately 35 μm) [37, 38]. Thus, the feed arteries demonstrated significantly greater expression of the RyR2 isoform than did the smaller arterioles. From a functional perspective RyR contributed to Ca^{2+} signals and myogenic tone in the larger vessels but not in the smaller arterioles. Both vessels exhibited similar complements of inositol trisphosphate receptors (IP_3R1 dominant in both).

The following sections draw heavily on our comparisons of BK_{Ca} expression and function between cerebral and skeletal muscle circulations. Particular emphasis has been placed at the level of small arteries. It is, however, acknowledged that considerable heterogeneity may also exist within the microvascular bed of any tissue. Supporting this latter point, and as outlined above, Westcott et al. have demonstrated ion channel diversity within mouse skeletal muscle microvasculature.

Large Conductance, Ca^{2+}-Activated K^+ Channels (BK_{Ca})

The BK_{Ca} channel is widely expressed in the body and vascular smooth muscle BK_{Ca} channels exhibit a single channel conductance of approximately 240 pS under conditions of symmetric 150 mM KCl [33, 39–41]. Its large conductance, along with its relatively high level of cellular expression, makes BK_{Ca} a dominant K^+ channel. In excitable cells, including VSMCs, BK_{Ca} plays a significant role in the regulation of membrane potential. In doing so, BK_{Ca} contributes to the regulation of the open state of VOCCs, which are major contributors to Ca^{2+} influx, intracellular Ca^{2+} handling, contractile activity and thus artery/arteriolar tone [41]. In addition, activation of BK_{Ca} by both depolarization and increased intracellular Ca^{2+} results in hyperpolarization that conceivably acts to limit agonist and mechanically-induced vasoconstriction in small arteries and arterioles [25].

An important question relates to whether the BK_{Ca} channel can be regulated locally or is simply ubiquitously expressed throughout the vasculature and regulated on a more global level. While channel opening is affected by membrane potential and cytosolic free Ca^{2+} the function of BK_{Ca} is regulated at a variety of other levels providing multiple mechanisms to fine tune channel activity in a differential manner to meet local needs. As outlined above these levels of regulation include differences in expression [33, 42, 43], the existence of splice variants [28], subunit composition [33, 44, 45], modulation by phosphorylation and small molecules [46], trafficking of subunits [27] and spatial distribution within the VSMC [47, 48].

In support of local differences in channel activity, comparison of VSMCs from cerebral and cremaster muscle arteries shows a higher expression for β1:α BK_{Ca} in the cerebral vasculature, both at the mRNA and protein levels [33, 43]. Further, using a cell surface biotinylation assay the α subunit of BK_{Ca} was expressed at significantly higher levels in cerebral vessel SMCs compared with that for cremaster VSMCs [43]. Coincident with this difference is a greater whole cell, iberiotoxin (a highly selective inhibitor of BK_{Ca})-sensitive, K$^+$ conductance in cerebral VSMCs. Importantly, this electrophysiological difference persists when K$^+$ current data are normalized for cellular capacitance indicating that the results cannot be explained by a difference in cell size between the two sites.

Molecular Heterogeneity BK_{Ca} Subunits

The functional BK_{Ca} channel is comprised of four α-subunits that assemble around a central axis forming a K$^+$-selective conduction pore (Fig. 6.2). Each α-subunit contains seven transmembrane-spanning segments, denoted S0–S6, with an extracellular amino terminus and a long intracellular carboxyl terminus. The S1 to S6 membrane-spanning segments are a common structural feature of voltage-gated K$^+$ channels, in general, while the S0 segment located at the amino terminus is found only in BK_{Ca} (Slo1) and the related H$^+$-gated large conductance K$^+$ channel encoded by Slo3 [49–51]. The latter are not, however, gated by physiological calcium levels. The S0 segment of BK_{Ca} appears to be important for modulation of the channel by the associated β1 subunit (described below; [52]). Contained within the membrane spanning segments are the voltage sensor domain (S1–S4) and conduction pore (S5–S6); these assignments are based largely on recent crystal structure data from closely related voltage-gated Kv channels [53, 54] and are supported by a low resolution (~20 Å) structure of the holo-channel in a cell membrane environment determined by electron cryo-microscopy [55]. The large intracellular domain accounts for approximately two-thirds of the α-subunit amino acid sequence and contains a number of known regulatory sites that are modulated by Ca^{2+}, phosphorylation and

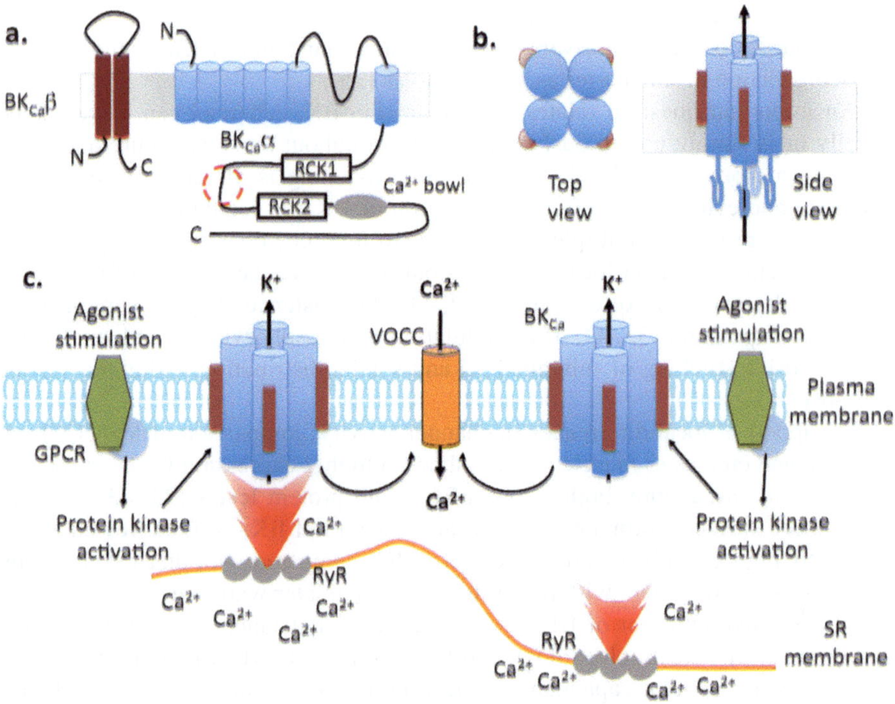

Fig. 6.2 Structure and regulation of BK$_{Ca}$. (**a**) Smooth muscle cell BK$_{Ca}$ holo-channel consists of α pore-forming subunits and the regulatory β1 subunit. The α subunit is characterized by seven membrane spanning segments (S0–S6); a pore region between S5 and S6; a long cytoplasmic tail containing a number of regulatory sites; and an extracellular amino terminus. The *red circle* between RCK1 and two domains indicates the C2 region known to exhibit several splice variants including STREX (see text for more detail). The β1 subunit has two membrane spanning segments and intracellular amino and carboxy termini. (**b**) The functional smooth muscle cell channel typically consists of four α and four β subunits. (**c**) BK$_{Ca}$ is regulated by Ca^{2+} released from the SR in the form of Ca^{2+} sparks and a number of intracellular signaling molecules including phosphorylation events mediated by various activated kinases. *BK$_{Ca}$* large conductance Ca^{2+}-activated K^+ channel, *RCK* regulator of K^+ conductance, *RyR* ryanodine receptor, *VOCC* voltage operated Ca^{2+} channel

a variety of small molecules (including lipids, reactive oxygen species, CO_2 and H^+) [46, 56–58]. Two RCK (regulators of K^+ conductance) domains have been described in the large, intracellular C-terminal domain of each BK$_{Ca}$ alpha subunit and contain functionally relevant, high affinity Ca^{2+} binding sites (Kd 1–10 μM). The Ca^{2+} bowl region, denoted by its density of acidic amino acids, flanks the C-terminal end of RCK2 and appears to be responsible for much of the Ca^{2+} sensitivity exhibited by BK$_{Ca}$ channels. Occupancy of these sites in the presence of micromolar levels of cytosolic Ca^{2+} promotes opening of the channel's C-terminal activation gate through an allosteric mechanism, leading to greater open probability. Electrophysiologically, this process is typically observed as a left-ward shift of the BK$_{Ca}$ conductance-voltage (G-V) relationship to more negative voltages [59]. The proximal RCK1 domain also contains a low affinity (mM range) binding site for intracellular Mg^{2+}; occupancy of this site by Mg^{2+} increases BK_{Ca} open probability and shifts the channel's G-V curve to more negative potentials. This effect of Mg^{2+} on channel gating appears to occur via an electrostatic interaction between the Mg^{2+} binding site in RCK1 and the intracellular face of the voltage sensor domain [59]; such coupling could allow for modulation of voltage-dependent gating by regulatory events within the C-terminal region. In the intact channel (containing four α-subunits) the 8 RCK domains form a 'Ca^{2+}-dependent gating ring' the mobility of which is involved in regulating channel opening [60]. Recent crystallographic structures of the intracellular gating ring apparatus in both the closed (Ca^{2+}-free) and open (Ca^{2+}-bound) configurations confirm the critical involvement of this structure in the regulation of the ion conduction pore [61, 62].

In VSMCs each α-subunit is often associated with a β1 subunit that, while not being absolutely necessary for channel activity, enhances Ca^{2+} sensitivity and increases channel opening probability [61]. The β1 subunit has also been suggested to influence trafficking of BK_{Ca} and hence its distribution between the membrane and intracellular compartments [63]. More recently, an additional γ subunit, LRRC26 (a leucine-rich repeating-containing peptide), has been demonstrated for BK_{Ca} channels [44, 64] and shown to markedly increase voltage sensitivity, thus allowing the channel to open at more hyperpolarized (and perhaps physiological) membrane potentials. A schematic diagram illustrating the overall structure of the BK_{Ca} channel and the association of its subunits is shown in Fig. 6.2.

The BK_{Ca} α-subunit is encoded by a single gene (also referred to as Slo 1 or KCNMA1) that contains 27 distinct exons [65]. Alternate splicing of BK_{Ca} α-subunit pre-mRNA allows considerable diversity of the ultimately translated protein and, to date, more than ten splice sites have been described for mammalian BK_{Ca} α subunit mRNA [66]. Particular focus has been placed on spliced variants involving the site designated C2 as at least five variants have been described at this position and this region lies close to C terminal regulatory domains including RCK 1 and 2 and the so-called "Ca^{2+} bowl" located at the distal end of RCK2 [67]. Diversity at the functional protein level results from the splice variants showing differences in voltage and Ca^{2+} sensitivity, regulation by phosphorylation and cellular localization/trafficking (including to the plasma and ER membranes) as described earlier.

Thus the existence of splice variants of αBK_{Ca} raises the possibility that these may impact channel function and that this could occur at a tissue-specific level. One splice variant that has received considerable attention is STREX (STRess axis regulated EXon) that encodes for a 59 amino acid insertion (relative to the ZERO variant that lacks the insertion) located between RCK1 and RCK2 in the C terminus of the α subunit. Importantly the STREX insert is of functional significance as STREX-containing channels open at more negative potentials compared with the ZERO form [68]. Further, Shipston and colleagues, in particular, have shown that the STREX insert changes the response of BK_{Ca} channels to cAMP signaling relative to the variant lacking this exon (ZERO) [69]. Thus, in BK_{Ca} channels heterologously expressed in HEK293 cells, the STREX variant channel is inhibited by the cAMP/protein kinase A (PKA) signaling pathway, while in the ZERO form cAMP/PKA signaling is associated with increased channel activity and larger whole cell currents [68].

Studies in non-vascular tissues have examined the role of the STREX insert in the inherent mechanosensitivity demonstrated by BK_{Ca} [70–73]. In channels isolated from chick ventricular myocytes and heterologously expressed in CHO cells (referred to as stretch activated BK_{Ca} channel, $SAK_{Ca}C$), it has been suggested that the STREX sequence is critical for mechanosensitivity [70–73]. The mechanism by which STREX confers mechanosensitivity has been suggested to relate to the length of the linker region joining S6 and the RCK1 domain (see Fig. 6.2), but importantly occurs independently of intracellular Ca^{2+} [70–73]. In apparent contrast to these observations, in studies of freshly isolated murine colonic SMCs, Wang et al. demonstrated mechanosensitivity of native BK_{Ca} channels in freshly isolated murine colonic SMCs, in which only the ZERO variant was expressed and not STREX (as determined by end-point RT-PCR) [70–73]. Based on these observations, it was therefore concluded that the presence of the STREX insert was not an absolute requirement for BK_{Ca} mechanosensitivity.

In our own studies we have examined the relative mRNA expression levels of three splice variants within the C-terminus of the BK_{Ca} α-subunit; ZERO, STREX and SS4. ZERO is the predominant isoform of these variants in small arteries from both cremaster muscle and cerebral circulations of the rat although the STREX insert is significantly more highly expressed in the vessels from cremaster muscle (approximately 29 compared with 16 % of total αBK_{Ca} α-subunit) [43]. Sequencing of PCR products generated from isolated VSMCs confirmed that ZERO and STREX were identical in both vessel types [43]. While at this point in time, the functional significance of these differences in expression levels is unclear, particularly as they relate to the generation of differing levels of myogenic tone, observed differences in ZERO/STREX expression did not alter the distribution of BK_{Ca} alpha subunits between the membrane and the cytosol. [43]. Interestingly, the vascular preparations contain a mixed population of STREX and ZERO BK_{Ca} α-subunits raising the possibility of there being both homomeric and heteromeric holo-channels. Studies are required to recapitulate these differences in expression systems and in fully functional arterioles, where levels of these variants can be manipulated.

In regard to modulation of BK_{Ca} function by β accessory subunits, four genes have been identified and encode for the distinct proteins referred to as β1, β2, β3 and β4. As stated above, arterial VSMCs typically express the β1 subunit. At the mRNA level both cerebral and cremaster VSMCs express predominantly the β1 subunit although small amounts of β4 subunit can be detected. At the protein level the β1 subunit is seen in both vasculatures although significantly greater levels were seen in small cerebral arteries [33]. Despite the presence of small levels of β4 subunit mRNA measured using real time PCR, Western blotting analysis does not show measurable quantities of the protein in homogenates of cerebral arteries. Further, no evidence for β2 or β3 subunit mRNA expression was obtained in either vasculatures (unpublished observations).

A cautionary note is that quantitative mRNA and, in particular, protein expression data are often necessarily derived from homogenates of whole microvessels and not purely from VSMCs. As native arterial endothelial cells are generally not thought to contain significant levels of BK_{Ca} and they represent a relatively small component of the vessel wall, it is assumed that the bulk of the measured signal does, in fact, come from the VSMCs. This is, however, currently a matter of interest and debate, in part, because endothelial cell data have largely been derived from large vessels and expression of BK_{Ca} channels may be up-regulated when endothelial cells are cultured or in pathophysiological states [74, 75]. Importantly though, vessel homogenates contain other cell types (for example neurons, fibroblasts and immune cells), which are resident in the adventitial layer of the vessel wall. More stringent studies are required, therefore, using purified VSMCs and appropriate markers for non-smooth muscle cell contamination.

From a functional perspective the importance of the BK_{Ca} β1 subunit to arteriolar function was first demonstrated in the $\beta1^{-/-}$ mouse [76, 77]. Cerebral blood vessels from the homozygous global knockout animal showed increased arterial tone in the face of decreased coupling between BK_{Ca} and SR-derived Ca^{2+} sparks. The decrease in coupling between the Ca^{2+} sparks and BK_{Ca} was further shown to be a result of a decrease in the Ca^{2+} sensitivity of the channel resulting from the absence of the regulatory β1 subunit.

The BK_{Ca} $\beta1^{-/-}$ mouse has been reported to be either mildly hypertensive [76, 77] or not hypertensive at all [78]. This, in part, may relate to differences in methodologies for assessing blood pressure, namely direct cannulation, tail cuff, or telemetry. Interestingly, 24 h monitoring by telemetry (in the absence of anesthesia and stress due to handling) showed a mild increase in systemic blood pressure at night, but statistically unchanged when considered over the 24 h period [78]. It would be expected that, if BK_{Ca} activity were to oppose myogenic tone in all vascular beds, a higher blood pressure would be seen in the $\beta1^{-/-}$ mouse. Vascular heterogeneity in this feedback mechanism could conceivably, however, obscure such a relationship. Alternatively in the whole animal, compensatory mechanisms may develop in the knock-out model.

The complex molecular composition of BK_{Ca} also raises a number of questions as to whether channel subunit stoichiometry varies between cells or between channels within a given cell? For example in VSMCs is there always a 1:1 relationship between the α and β subunits in VSMCs and do heteromultimeric channels exist between the various α subunit splice variants (for example STREX and ZERO)? Are these relationships fixed or do they vary on a more dynamic basis, particularly given the recent work of Leo and colleagues showing stimulus-dependent trafficking of the β1 subunit? Substantial evidence exists for variation in β subunit expression in both physiological (for example pregnancy where both α and β subunits decrease prior to parturition [79]) and pathophysiological states (diabetes [80], hypertension [81] and increased vascular resistance associated with heart failure [82]). The existence of such variation raises additional questions as to their potential effects on electrophysiological properties and regulation by post-translational mechanisms (phosphorylation, palmitoylation, fatty acids, steroid hormones, redox state) and cellular location (trafficking).

BK_{Ca} Regulation by Voltage and Ca^{2+}

BK_{Ca} is a member of the KV superfamily as indicated by similarities in amino acid sequence and biophysical properties [83–85]. Together with Slo 2 and Slo3, the mammalian Slo1 gene and its orthologs encode for potassium channels of large conductance [50, 51]. These channels share structural similarities with Slo 1 and 3 having an S0 membrane spanning segment at its amino terminus while Slo 1, 2 and 3 all exhibit a long intracellular carboxy terminus containing two RCK domains [50, 51]. However BK_{Ca} is distinct from other members of the Kv family of channels due to its regulation by intracellular Ca^{2+} levels. A number of Ca^{2+} (and divalent cation) binding sites have been identified, particularly in the intracellular C-terminus of the α-subunit. Further, the presence of the β1 subunit facilitates Ca^{2+} sensitivity in VSMCs.

In comparative studies, BK_{Ca} of cerebral and cremaster muscle VSMCs showed similar responsiveness to applied voltage steps. Thus, Boltzmann-type analysis of single channel openings activated by negative and positive voltage pulses showed similar Boltzmann constants (Z) over a $[Ca^{2+}]$ range of 0.05–100 μM [45]. In contrast to voltage characteristics differences in BK_{Ca} Ca^{2+} sensitivity were observed such that skeletal muscle VSMCs required a 2–3-fold increase in cytosolic Ca^{2+} concentration compared with cerebral VSMCs to affect half maximal channel opening. Alternatively the differences in Ca^{2+} sensitivity can be compensated for by an additional depolarization of approximately 30 mV. As membrane potential is approximately −35 to −45 mV in both vessels (as measured under *in vitro* pressurized conditions [11, 24]) it is unlikely that voltage alone can compensate for the difference in Ca^{2+} sensitivity between skeletal muscle and cerebral BK_{Ca} channels. Alternatively, an argument can be made that in the presence of additional stimuli, particularly those resulting in the activation of PKA and PKG, may cause a leftward shift in the open probability-voltage relationship thereby allowing BK_{Ca} channels to open at more negative membrane potentials [39, 86].

Single channel studies have further showed that cremaster VSMC BK_{Ca} channels exhibit shorter opening times and longer closed times compared with BK_{Ca} of cerebral vessels. Based on studies of expression systems [87] these differences could relate to variation in the number of β1 subunits within the holo-channel. These biophysical observations are also consistent with the molecular data showing decreased BK_{Ca} β1 subunit expression in cremaster VSMCs.

Supporting variation in β1 subunit expression between cremaster and cerebral VSMCs, acute (48 h) *in vitro* knockdown of cerebral VSMC β1 BK_{Ca} using specific siRNA resulted in decreased expression of the β1 subunit, attenuated whole cell K^+ currents and decreased responsiveness of BK_{Ca} channels to estrogen [45]. Further, knockdown of the β1 subunit decreased single channel open time and increased closed time while also decreasing the amplitude and frequency of BK_{Ca}-dependent STOCs in whole cell recordings. From these data we concluded that following the decrease in BK_{Ca} β1 subunit expression the electrophysiological properties of cerebral VSMC BK_{Ca} activity more closely resembled those of cremaster muscle. This observation was taken to be further evidence that the cerebral VSMCs channel contains a higher relative proportion of functional β1 subunit.

While the modulatory roles of BK_{Ca} β-subunits in channel activity are well-recognized, recent data have revealed a novel set of γ-subunits that can also co-assemble with BK_{Ca} α-subunits and have a positive regulatory effect on channel activity [64, 88, 89]. These γ-subunits belong to a family of leucine-rich repeat containing (LRRC) proteins and Evanson et al. [44] have shown that the LRRC26 variant is endogenously expressed in cerebral VSMCs and enhances the Ca^{2+} and voltage-dependent gating of BK_{Ca} channels.

BK_{Ca} Regulation by Phosphorylation

As many vasoactive stimuli (including vasoactive hormones and locally acting factors including NO) result in the production of cyclic nucleotides and protein kinase activation (including protein kinases A, G and C), regional differences in phosphorylation could conceivably result in tissue-specific differences in BK_{Ca} activity. Using proteomics and radiolabeling approaches, numerous potential phosphorylation sites have been identified in BK_{Ca} [90, 91]. In VSMCs stimuli elevating cAMP and cGMP typically activate BK_{Ca} and lead to vasodilatation while PKC activation often causes channel closure and vasoconstriction [86]. It should be noted, however, that PKC effects may be isoform specific as evidence has also been presented for it activating BK_{Ca} [92]. Further variation may exist in these relationships, however, as the expression of different splice variants (for example the STREX insert referred to earlier) can markedly change modulation of BK_{Ca} by cyclic nucleotides and phosphorylation.

An important question is how, at a mechanistic level, do cyclic nucleotides and protein kinases alter conductivity of BK_{Ca}? In regard to the action of NO/cGMP/PKG signaling in VSMCs, Kyle, Braun and colleagues have identified, using biochemical, cellular and electrophysiological techniques, three specific serine phosphorylation sites (i.e. Ser691, Ser873 and Ser1111–1113) in the C-terminus of BK_{Ca} [91]. The data strongly suggested that maximal augmentation of BK_{Ca} current by NO/cGMP signaling in A7r5 smooth muscle cells requires phosphorylation of all three Ser locations and that disruption of any one of the three sites by Ser to Ala substitution could prevent NO/cGMP stimulation of BK_{Ca} current. In both A7r5 cells and rat cerebral VSMCs, BK_{Ca} channels co-localize with PKGIα, as determined by proximity ligation assay analysis, and this co-distribution can be further augmented in the presence of a NO donor (Fig. 6.3). In rat brain, BK_{Ca} channels appear to be constitutively phosphorylated at multiple Ser and Thr sites [90] and such modifications may help to stabilize the channel's tertiary structure or create interacting sites for auxiliary proteins. The extent to which BK_{Ca} channels in other tissues or following heterologous expression undergo constitutive phosphorylation is presently unknown, along with the specific protein kinases responsible for these modifications.

BK_{Ca} current is also augmented by cAMP/PKA signaling and Ser873 has been implicated as a putative phosphorylation site in the channel targeted by PKA [93].

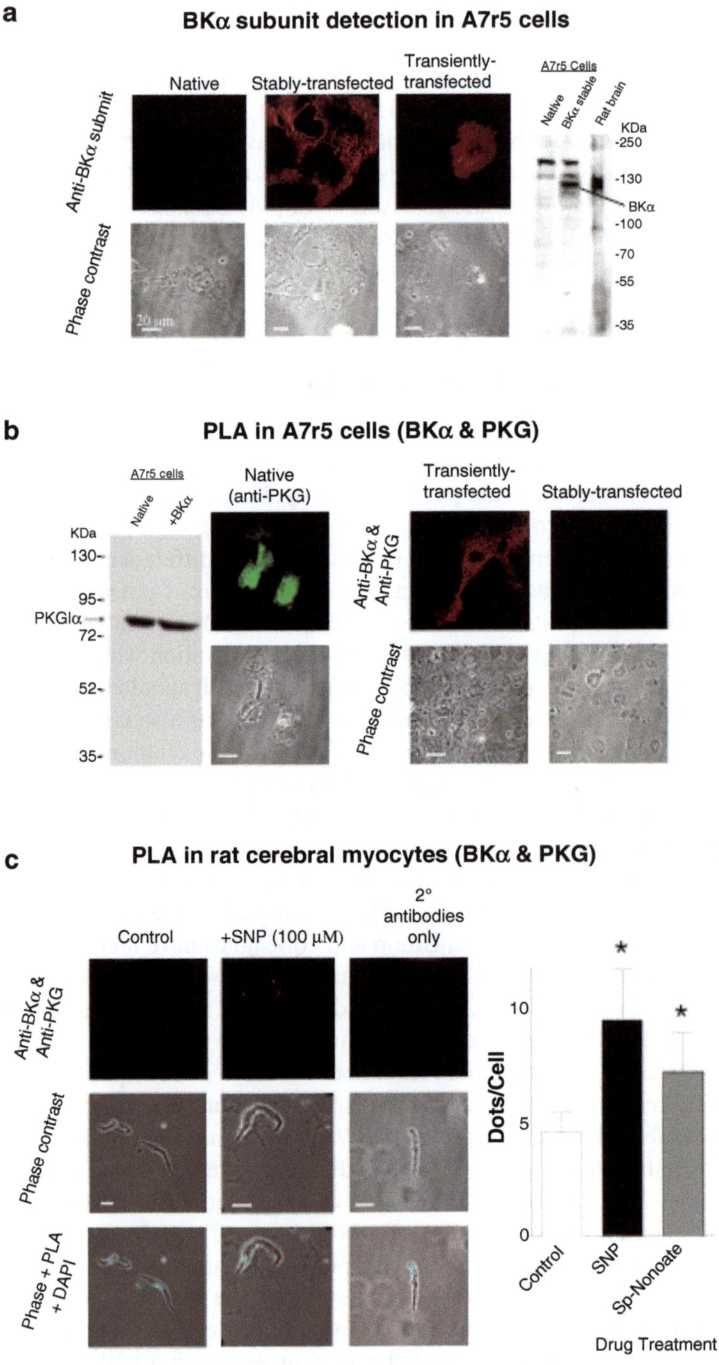

Fig. 6.3 Immuno-detection of BK_{Ca} channels and PKGIα in VSMCs. Native A7r5 aortic smooth muscle cells exhibit very low levels of BK_{Ca} channel expression, as revealed by immunostaining of the BK_{Ca} α subunit, however, channels are more readily detected in these cells following transient or stable expression of a cloned mammalian BK_{Ca} α subunit. The scale bar shown in each phase contrast micrograph represents 20 μm and also applies to the accompanying

In STREX-containing BK_{Ca} channels, PKA signaling has the opposite effect and decreases current amplitude by modifying a serine residue within the STREX insert [69]. Functionally, STREX phosphorylation by PKA is reported to have a dominant-negative effect on BK_{Ca} channel regulation and is able to override any potential stimulatory effect of PKA-mediated phosphorylation at additional sites (e.g. Ser873). Moreover, the presence of only a single STREX-containing α-subunit within a BK_{Ca} tetrameric channel is sufficient to change the effect of PKA-mediated phosphorylation on BK_{Ca} gating from positive to negative. This inhibitory effect of PKA signaling on STREX-containing BK_{Ca} channels also appears to require fatty acid modification (i.e. palmitoylation) of the cysteine-rich STREX insert [29, 30]. Additional aspects of BK_{Ca} channel phosphorylation are described by Schubert and Nelson [86] and Kyle and Braun [46].

Studies from our laboratory have shown that VSMCs from cremaster arterioles exhibit less activation of BK_{Ca} in response to forskolin (activator of adenylyl cyclase) as compared with cells isolated from small cerebral arterioles (Yang et al., unpublished data; Fig. 6.4). These observations at the whole cell level may thus reflect differences in BK_{Ca} channel density/expression as mentioned earlier or regulatory differences at the single channel level. Interestingly, the BK_{Ca} response to forskolin stimulation was blocked by a PKG inhibitor (KT5823), suggesting a 'cross over' between PKA and PKG-dependent pathways as reported earlier by Barman and colleagues [94]. The exact mechanism(s) by which evoked cAMP production promotes PKG activity (e.g. direct binding and stimulation of PKG, indirect elevation of cellular cGMP via reduced breakdown, etc.) has not been fully elucidated.

Additional post-translational enhancement of BK_{Ca} current occurs through direct tyrosine phosphorylation of the channel by c-Src and Pyk2 tyrosine

Fig. 6.3 (continued) fluorescent image immediately above. Western blotting demonstrates that the expressed BK_{Ca} α subunit (~130 kDa MW) is comparable to that seen in rat brain (panel **a**). As shown in panel **b**, endogenous PKGIα is readily detected in A7r5 cells by either western blotting or immunocytochemistry. Immunostaining of A7r5 cells using the PLA technique (*right-hand* images) demonstrates co-localization (<40 nm) of BK_{Ca} channels with endogenous PKGIα, as revealed by the appearance of distinct *red fluorescent dots* (panel **b**). Scale bar shown in each phase contrast micrograph represents 20 μm. BK_{Ca} channels and PKGIα co-localization can also be readily detected by PLA staining in acutely isolated rat cerebral myocytes (panel **c**). The *third row* of images shows merged detection of BKα+PKGIα by PLA (*red* fluorescence), along with DAPI staining for cell nuclei (*blue* fluorescence), superimposed on phase contrast micrographs of cerebral myocytes.The scale bar depicted in each phase contrast micrograph represents 20 μm and applies to the accompanying images within the column. As quantified in the histogram, the number of *red dots* was observed to increase following treatment with an NO donor, either SNP (100 μM) or sp nonoate (5 μM), suggesting that NO promotes BK_{Ca}-PKG co-localization. *Red dots* are not observed in the absence of primary antibody versus BK_{Ca} α subunit or PKGIα. Data are presented as mean ± SEM, n = 10–15 cells under each condition. *Asterisk* indicates a statistically significant difference compared with control, as determined by one-way ANOVA and a Dunnett's post-hoc test; $P < 0.05$. *BK_{Ca}* large conductance Ca^{2+}-activated K^+ channel, *PKGIα* protein kinase G type Iα, *NO* nitric oxide, *PLA* proximity ligation assay, *SNP* sodium nitroprusside, *sp nonoate* spermine nonoate

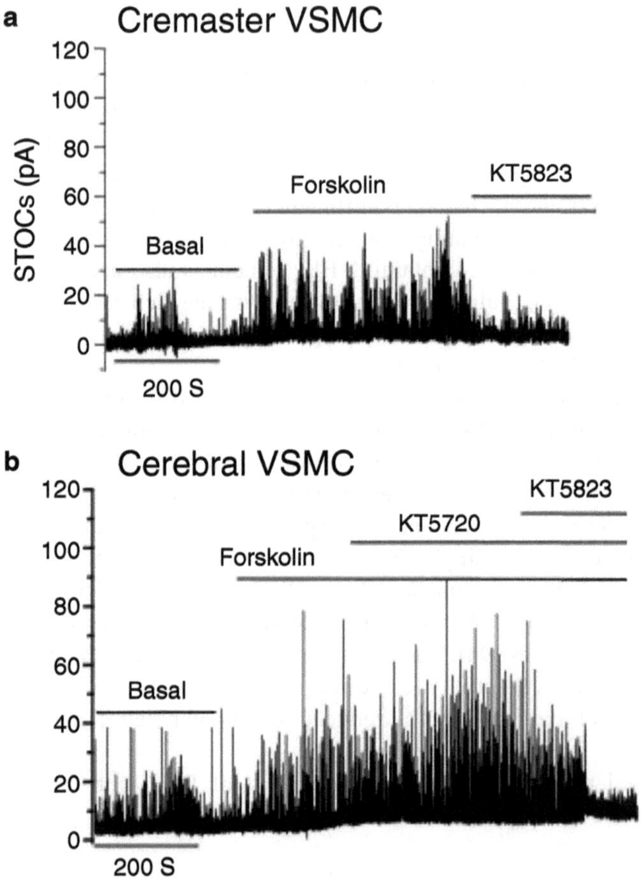

Fig. 6.4 Forskolin increases BK_{Ca}-mediated STOC activity in a PKG-dependent manner. Panels **a** and **b** show representative traces depicting STOC activity following forskolin (30 μM) stimulation in cremaster and cerebral VSMCs, respectively. STOCs were inhibited by the PKG inhibitor KT5823 (300 nM) but not the PKA inhibitor KT5720 (300 nM). Note the higher amplitude and frequency of STOCs consistent with earlier studies [33]. *PKA* protein kinase A, *PKG* protein kinase G, *STOC* spontaneous transient outward current, *VSMC* vascular smooth muscle cells

kinases [95, 96]. These kinases have been implicated in integrin-based signaling and the subsequent modulation of BK_{Ca} current [97, 98], thereby providing a mechanism by which ion channel conductance and membrane potential can be influenced by the extracellular environment and mechanical forces.

Whole cell BK_{Ca} currents (and STOCs) in VSMCs isolated from both cremaster and cerebral small arteries are increased by the PP1 and PP2 phosphatase inhibitor, okadaic acid, and decreased by the application of alkaline phosphatase

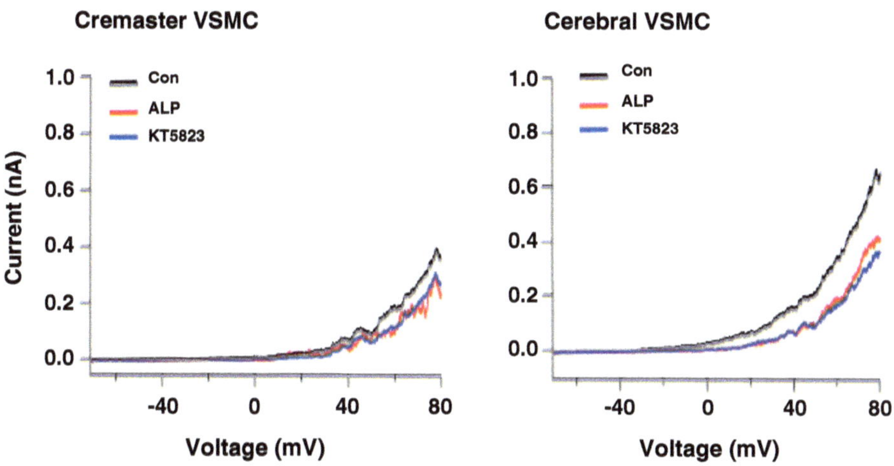

Fig. 6.5 Alkaline phosphatase decreases whole cell currents to a greater extent in cerebral VSMCs compared with cremaster VSMCs. Electrophysiological recordings were conducted in whole cell mode with voltage ramps applied. Alkaline phosphatase (ALP; 350 U/ml) was delivered intracellularly via the patch pipette while the PKG inhibitor, KT5823 (300 nM), was added to the bath. At 70 mV AP decreased current by 49.3 ± 0.9 % in cerebral VSMCs and 28.6 ± 2.1 % in cremaster VSMCs ($n = 13$ cells per group; $p < 0.05$). Following treatment with ALP VSMCs from both vessel types no longer showed significant inhibition by KT5823 consistent with a basal level of BK_{Ca} channel phosphorylation and PKG activity. *ALP* alkaline phosphatase, *BK_{Ca}* large conductance Ca^{2+}-activated K^+ channel, *PKG* protein kinase G, *VSMC* vascular smooth muscle cells

through a patch clamp pipette suggesting that channel activity can also be regulated by dephosphorylation (unpublished observations; Fig. 6.5). Interestingly the effect of alkaline phosphatase on BK_{Ca} current appeared greater in cerebral VSMCs suggesting a higher level of basal phosphorylation. This suggestion is line with recent proteomic data demonstrating that BK_{Ca} channels in rat brain exhibit a high level of constitutive phosphorylation [90]. Further consistent with a basal level of phosphorylation our studies have shown that the PKG inhibitor, KT5823, reduces baseline levels of BK_{Ca}-mediated STOCs (both in terms of amplitude and frequency). This effect occurred similarly in cremaster and cerebral VSMCs, although the absolute effect was greater in cerebral vessels due to their overall higher STOC activity.

BK_{Ca} Regulation by Its Local Environment

BK_{Ca} channel function is likely influenced by its local environment. For example, its regulation in VSMCs by elevated $[Ca^{2+}]_i$ is facilitated by close proximity to the SR and focal Ca^{2+} release. This arrangement is necessary for the channel to have access to local [Ca^{2+}]s in the range of 5–10 μM that are required to activate BK_{Ca} (at physiological membrane potentials) while bulk cytosolic [Ca^{2+}] is typically <250 nM in myogenically active arteries [10] and only achieves levels of approximately 1 μM during maximal agonist stimulation. In addition to spatial considerations, BK_{Ca} is known to interact with ECM proteins [97, 98] and is also influenced by its local lipid environment [58]. As an example of the former fibronectin potentiates BK_{Ca} current at the single channel level through a mechanism involving integrin activation ($\alpha_5\beta_1$), the tyrosine kinase c-SRC and subsequent modulation of BK_{Ca} Ca^{2+} sensitivity [97, 99].

As stated above, regulation of BK_{Ca} by intracellular Ca^{2+} has been estimated to require local Ca^{2+} concentrations in the order of 5–10 μM [47, 100]. As such Ca^{2+} levels, on a global cellular basis, would cause maximal VSMC contraction and likely activate Ca^{2+}-dependent proteases and cell death pathways, a spatially restricted increase in Ca^{2+} would be required to activate BK_{Ca} channels, thereby preserving cell viability. In this regard, Nelson and colleagues first presented evidence in VSMCs for Ca^{2+} sparks as a regulatory process for BK_{Ca} activity. Sparks are spatially-restricted Ca^{2+} release events from the SR that utilize a ryanodine-sensitive release mechanism [25, 101].

Recent studies have shown, however, that while a number of vascular beds exhibit Ca^{2+} spark- dependent regulation of BK_{Ca} [25, 100, 102, 103], this situation may not always be the case [37, 42]. In addition other studies have suggested that in contrast to promoting hyperpolarization and relaxation, Ca^{2+} sparks may promote contraction, perhaps through summation to more global Ca^{2+} events such as Ca^{2+} waves [103].

In studies comparing VSMCs from cremaster muscle arterioles with those from cerebral vessels Ca^{2+} sparks were only observed in the latter. It would appear that this result was unlikely to simply reflect methodological differences as both cell preparations were studied using the same spinning disk confocal microscope system. That being said, both sets of VSMCs did exhibit an accepted electrophysiological correlate or readout of Ca^{2+} sparks (i.e. the presence of STOCs) and, further, the STOCs observed in cremaster VSMCs were attenuated by both ryanodine (at concentrations that would be expected to deplete SR Ca^{2+} stores) and by the BK_{Ca} channel inhibitor, iberiotoxin. These latter observations strongly suggest that communication does occur between ryanodine-sensitive Ca^{2+} stores in the SR and plasmalemmal BK_{Ca} channels in both cremaster and cerebral VSMCs.

As an additional example of BK_{Ca} regulation via the local cellular environment, BK_{Ca} channels are known to organize into macromolecular complexes consisting of various receptors, scaffolding proteins and specific membrane domains. For example, β_2-adrenoceptors, together with BK_{Ca}, associate with A-kinase anchoring proteins (AKAPs), PKA and voltage-gated Ca^{2+} channels [104]. Similarly, angiotensin II receptors (AT1R) have been reported to complex with BK_{Ca} through caveolin 1 [105]. Further, BK_{Ca} associates with specific lipids [58] providing a further possible mechanism for the targeting/trafficking of BK_{Ca} channels to spatially restricted domains/signaling complexes.

Conclusion

The activity of BK_{Ca} shows considerable complexity due to regulation of the channel at multiple levels. This complexity likely contributes to tissue diversity in channel function and its contribution to vasoregulation, particularly as this relates to the control of myogenic tone. Thus, in comparing BK_{Ca} function in cremaster and cerebral VSMCs, a number of differences are apparent. Overall, mRNA and protein expression levels are less in cremaster VSMCs, subtle differences exist with regard to splice variant expression, channels differ in their activation by Ca^{2+} (both in terms of channel Ca^{2+} sensitivity and reliance on Ca^{2+} sparks) and phosphorylation status of the channel appears to be relatively higher in cerebral VSMCs. Similarities and differences in their cellular spatial localization are yet to be examined in any detail. An obvious question is whether these regional differences in the BK_{Ca} channel properties are of physiological significance. Given the importance of this channel in the regulation of membrane potential and intracellular Ca^{2+} handling, along with vascular bed-specific differences in the relationships between membrane potential and myogenic tone, it is tempting to speculate that differences in BK_{Ca} contribute to the local control of vessel diameter, and hence blood flow, in resistance vessels. Thus, variation in the molecular make-up of the channel along with differences in post-translational regulation would allow the same basic ion channel to contribute to differing degrees to VSM function within a given vascular bed.

An additional question in understanding the relevance of K^+ channel distribution to vasoregulatory mechanisms relates to why BK_{Ca} is predominantly expressed in VSMCs while intermediate and small conductance, Ca^{2+}-activated, K^+ channels (IK_{Ca} and SK_{Ca}) are primarily located in endothelial cells. As endothelial cells do not typically exhibit sparks, it is conceivable that they express channels with higher Ca^{2+} sensitivity as a reflection of their more modest Ca^{2+} dynamics. Activation of IK_{Ca} and SK_{Ca} in the endothelium tends to support the vasodilatory effects of Ca^{2+}-mobilizing stimuli and thus these channels function closely with other pathways, including eNOS, towards a common goal (i.e. vasodilation). Further, the higher Ca^{2+} affinity of IK_{Ca} and SK_{Ca} will provide a prolonged signal to support vasodilation. Thus, endothelial cells do not appear to require two distinct Ca^{2+} operated mechanisms, while in VSMCs, Ca^{2+} evokes contraction at sub-micromolar levels and sparks evoke relaxation by locally increasing Ca^{2+} to higher levels (approximately 10 μM). As a consequence there is a dichotomy in Ca^{2+} signaling that requires two distinct pathways and sensors. The lower affinity Ca^{2+} sensor of BK_{Ca}, together with its spatial localization, will limit its activation/opening thereby preventing the channel from interfering with evoked contraction occurring at sub-micromolar Ca^{2+} levels. However, the much higher conductance will provide a rapid and robust Vm change to oppose voltage-gated Ca^{2+} channel activity when needed. As these mechanisms may also vary between tissues, understanding of the complexity of K^+ channel distribution, molecular structure and function requires considerable further work so that these properties can be exploited in the development of vascular bed-specific therapeutics.

Acknowledgements Original experimental work described in this paper was supported by NIH grant HL92241 to M.A.H.; A.P.B. is supported by grants from the Canadian Institutes of Health Research (MOP 97091) and the Natural Sciences and Engineering Research Council (RGPIN/312240).

References

1. Bayliss WM. On the local reactions of the arterial wall to changes of internal pressure. J Physiol. 1902;28:220–31.
2. Davis MJ, Hill MA, Kuo L. Local regulation of microvascular perfusion. In: Tuma RF, Duran WN, Ley K, editors. Handbook of physiology microcirculation. 2nd ed. San Diego: Academic Press; 2008.
3. Johnson PC. The myogenic response. In: Bohr DF, Somlyo AP, Sparks Jr HV, editors. Handbook of physiology, the cardiovascular system, vascular smooth muscle. New York: Wiley-Blackwell; 1980. p. 409–42.
4. Davis MJ, Hill MA. Signaling mechanisms underlying the vascular myogenic response. Physiol Rev. 1999;79:387–423.
5. Johnson RP, El-Yazbi AF, Takeya K, Walsh EJ, Walsh MP, et al. Ca2+ sensitization owing to Rho kinase-dependent phosphorylation of MYPT1-T855 contributes to myogenic control of arterial diameter. J Physiol. 2009;587:2537–53.
6. Knot HJ, Nelson MT. Regulation of arterial diameter and wall [Ca2+] in cerebral arteries of rat by membrane potential and intravascular pressure. J Physiol. 1998;508(Pt 1):199–209.
7. Meininger GA, Zawieja DC, Falcone JC, Hill MA, Davey JP. Calcium measurement in isolated arterioles during myogenic and agonist stimulation. Am J Physiol. 1991;261:H950–9.
8. Moreno-Dominguez A, Colinas O, El-Yazbi A, Walsh E, Hill MA, et al. Calcium sensitization due to myosin light chain phosphatase inhibition and cytoskeletal reorganization in the myogenic response of skeletal muscle resistance arteries. J Physiol. 2013;591(Pt 5):1235–50.
9. Osol G, Brekke JF, McElroy-Yaggy K, Gokina NI. Myogenic tone, reactivity, and forced dilatation: a three-phase model of in vitro arterial myogenic behavior. Am J Physiol Heart Circ Physiol. 2002;283:H2260–7.
10. Zou H, Ratz PH, Hill MA. Role of myosin phosphorylation and [Ca2+]i in myogenic reactivity and arteriolar tone. Am J Physiol. 1995;269:H1590–6.
11. Kotecha N, Hill MA. Myogenic contraction in rat skeletal muscle arterioles: smooth muscle membrane potential and Ca(2+) signaling. Am J Physiol Heart Circ Physiol. 2005;289:H1326–34.
12. Moreno-Dominguez A, El-Yazbi AF, Zhu HL, Colinas O, Zhong XZ, et al. Cytoskeletal reorganization evoked by Rho-associated kinase- and protein kinase C-catalyzed phosphorylation of cofilin and heat shock protein 27, respectively, contributes to myogenic constriction of rat cerebral arteries. J Biol Chem. 2014;289:20939–52.
13. El-Yazbi AF, Johnson RP, Walsh EJ, Takeya K, Walsh MP, et al. Pressure-dependent contribution of Rho kinase-mediated calcium sensitization in serotonin-evoked vasoconstriction of rat cerebral arteries. J Physiol. 2010;588:1747–62.
14. Walsh MP, Cole WC. The role of actin filament dynamics in the myogenic response of cerebral resistance arteries. J Cereb Blood Flow Metab. 2013;33:1–12.
15. Davis MJ, Donovitz JA, Hood JD. Stretch-activated single-channel and whole cell currents in vascular smooth muscle cells. Am J Physiol. 1992;262:C1083–8.
16. Earley S, Brayden JE. Transient receptor potential channels and vascular function. Clin Sci (Lond). 2010;119:19–36.
17. Earley S, Waldron BJ, Brayden JE. Critical role for transient receptor potential channel TRPM4 in myogenic constriction of cerebral arteries. Circ Res. 2004;95:922–9.
18. Welsh DG, Morielli AD, Nelson MT, Brayden JE. Transient receptor potential channels regulate myogenic tone of resistance arteries. Circ Res. 2002;90:248–50.

19. Gonzales AL, Yang Y, Sullivan MN, Sanders L, Dabertrand F, et al. A PLCgamma1-dependent, force-sensitive signaling network in the myogenic constriction of cerebral arteries. Sci Signal. 2014;7:ra49.
20. Bulley S, Neeb ZP, Burris SK, Bannister JP, Thomas-Gatewood CM, et al. TMEM16A/ANO1 channels contribute to the myogenic response in cerebral arteries. Circ Res. 2012;111:1027–36.
21. Ju M, Shi J, Saleh SN, Albert AP, Large WA. Ins(1,4,5)P3 interacts with PIP2 to regulate activation of TRPC6/C7 channels by diacylglycerol in native vascular myocytes. J Physiol. 2010;588:1419–33.
22. Large WA, Saleh SN, Albert AP. Role of phosphoinositol 4,5-bisphosphate and diacylglycerol in regulating native TRPC channel proteins in vascular smooth muscle. Cell Calcium. 2009;45:574–82.
23. Brayden JE, Nelson MT. Regulation of arterial tone by activation of calcium-dependent potassium channels. Science. 1992;256:532–5.
24. Knot HJ, Nelson MT. Regulation of membrane potential and diameter by voltage-dependent K+ channels in rabbit myogenic cerebral arteries. Am J Physiol. 1995;269:H348–55.
25. Nelson MT, Cheng H, Rubart M, Santana LF, Bonev AD, et al. Relaxation of arterial smooth muscle by calcium sparks. Science. 1995;270:633–7.
26. Hill MA, Yang Y, Ella SR, Davis MJ, Braun AP. Large conductance, Ca(2+)-activated K(+) channels (BK(Ca)) and arteriolar myogenic signaling. FEBS Lett. 2010;584:2033–42.
27. Leo MD, Bannister JP, Narayanan D, Nair A, Grubbs JE, et al. Dynamic regulation of beta1 subunit trafficking controls vascular contractility. Proc Natl Acad Sci U S A. 2014;111:2361–6.
28. Shipston MJ. Alternative splicing of potassium channels: a dynamic switch of cellular excitability. Trends Cell Biol. 2001;11:353–8.
29. Shipston MJ. S-acylation dependent post-translational cross-talk regulates large conductance calcium- and voltage-activated potassium (BK) channels. Front Physiol. 2014;5:281.
30. Shipston MJ. Ion channel regulation by protein S-acylation. J Gen Physiol. 2014;143:659–78.
31. Shao D, Okuse K, Djamgoz MB. Protein-protein interactions involving voltage-gated sodium channels: post-translational regulation, intracellular trafficking and functional expression. Int J Biochem Cell Biol. 2009;41:1471–81.
32. Schulz DJ, Temporal S, Barry DM, Garcia ML. Mechanisms of voltage-gated ion channel regulation: from gene expression to localization. Cell Mol Life Sci. 2008;65:2215–31.
33. Yang Y, Murphy TV, Ella SR, Grayson TH, Haddock R, et al. Heterogeneity in function of small artery smooth muscle BKCa: involvement of the beta1-subunit. J Physiol. 2009;587:3025–44.
34. Chen TT, Luykenaar KD, Walsh EJ, Walsh MP, Cole WC. Key role of Kv1 channels in vasoregulation. Circ Res. 2006;99:53–60.
35. Chilton L, Loutzenhiser K, Morales E, Breaks J, Kargacin GJ, et al. Inward rectifier K(+) currents and Kir2.1 expression in renal afferent and efferent arterioles. J Am Soc Nephrol. 2008;19:69–76.
36. Loutzenhiser R, Epstein M. Renal microvascular actions of calcium antagonists. J Am Soc Nephrol. 1990;1:S3–12.
37. Westcott EB, Jackson WF. Heterogeneous function of ryanodine receptors, but not IP3 receptors, in hamster cremaster muscle feed arteries and arterioles. Am J Physiol Heart Circ Physiol. 2011;300:H1616–30.
38. Westcott EB, Goodwin EL, Segal SS, Jackson WF. Function and expression of ryanodine receptors and inositol 1,4,5-trisphosphate receptors in smooth muscle cells of murine feed arteries and arterioles. J Physiol. 2012;590:1849–69.
39. Jackson WF, Blair KL. Characterization and function of Ca(2+)-activated K+ channels in arteriolar muscle cells. Am J Physiol. 1998;274:H27–34.
40. Ledoux J, Werner ME, Brayden JE, Nelson MT. Calcium-activated potassium channels and the regulation of vascular tone. Physiology (Bethesda). 2006;21:69–78.
41. Nelson MT, Quayle JM. Physiological roles and properties of potassium channels in arterial smooth muscle. Am J Physiol. 1995;268:C799–822.

42. Yang Y, Ella SR, Braun AP, Korhuis RJ, Davis MJ, et al. Manipulation of arteriolar BKCa using subunit directed siRNA (Abstract). FASEB J 2010.
43. Nourian Z, Li M, Leo MD, Jaggar JH, Braun AP, et al. Large conductance Ca2+-activated K+ channel (BKCa) alpha subunit splice variants in resistance arteries from rat cerebral and skeletal muscle vasculature. PLoS One. 2014;9(6):e98863.
44. Evanson KW, Bannister JP, Leo MD, Jaggar JH. LRRC26 is a functional BK channel auxiliary gamma subunit in arterial smooth muscle cells. Circ Res. 2014;115:423–31.
45. Yang Y, Sohma Y, Nourian Z, Ella SR, Li M, et al. Mechanisms underlying regional differences in the Ca2+ sensitivity of BK(Ca) current in arteriolar smooth muscle. J Physiol. 2013;591:1277–93.
46. Kyle BD, Braun AP. The regulation of BK channel activity by pre- and post-translational modifications. Front Physiol. 2014;5:316.
47. Zhuge R, Fogarty KE, Tuft RA, Walsh Jr JV. Spontaneous transient outward currents arise from microdomains where BK channels are exposed to a mean Ca(2+) concentration on the order of 10 microM during a Ca(2+) spark. J Gen Physiol. 2002;120:15–27.
48. Fakler B, Adelman JP. Control of K(Ca) channels by calcium nano/microdomains. Neuron. 2008;59:873–81.
49. Wu RS, Marx SO. The BK potassium channel in the vascular smooth muscle and kidney: alpha- and beta-subunits. Kidney Int. 2010;78:963–74.
50. Salkoff L, Butler A, Ferreira G, Santi C, Wei A. High-conductance potassium channels of the SLO family. Nat Rev Neurosci. 2006;7:921–31.
51. Liu J, Bukiya AN, Kuntamallappanavar G, Singh AK, Dopico AM. Distinct sensitivity of slo1 channel proteins to ethanol. Mol Pharmacol. 2013;83:235–44.
52. Morrow JP, Zakharov SI, Liu G, Yang L, Sok AJ, et al. Defining the BK channel domains required for beta1-subunit modulation. Proc Natl Acad Sci U S A. 2006;103:5096–101.
53. Doyle DA, Morais Cabral J, Pfuetzner RA, Kuo A, Gulbis JM, et al. The structure of the potassium channel: molecular basis of K+ conduction and selectivity. Science. 1998;280:69–77.
54. Jiang Y, Lee A, Chen J, Ruta V, Cadene M, et al. X-ray structure of a voltage-dependent K^+ channel. Nature. 2003;423:33–41.
55. Wang L, Sigworth FJ. Structure of the BK potassium channel in a lipid membrane from electron cryomicroscopy. Nature. 2009;461:292–5.
56. Hou S, Heinemann SH, Hoshi T. Modulation of BKCa channel gating by endogenous signaling molecules. Physiology (Bethesda). 2009;24:26–35.
57. Dick GM, Sanders KM. (Xeno)estrogen sensitivity of smooth muscle BK channels conferred by the regulatory beta1 subunit: a study of beta1 knockout mice. J Biol Chem. 2001;276:44835–40.
58. Dopico AM, Bukiya AN. Lipid regulation of BK channel function. Front Physiol. 2014;5:312.
59. Cui J, Yang H, Lee US. Molecular mechanisms of BK channel activation. Cell Mol Life Sci. 2009;66:852–75.
60. Yuan P, Leonetti MD, Pico AR, Hsiung Y, MacKinnon R. Structure of the human BK channel Ca2+-activation apparatus at 3.0 A resolution. Science. 2010;329:182–6.
61. McManus OB, Helms LM, Pallanck L, Ganetzky B, Swanson R, et al. Functional role of the beta subunit of high conductance calcium-activated potassium channels. Neuron. 1995;14:645–50.
62. Wu Y, Yang Y, Ye S, Jiang Y. Structure of the gating ring from the large-conductance $Ca(2^+)$-gated K(+) channel. Nature. 2010;466:393–7.
63. Toro B, Cox N, Wilson RJ, Garrido-Sanabria E, Stefani E, et al. KCNMB1 regulates surface expression of a voltage and Ca2+-activated K+ channel via endocytic trafficking signals. Neuroscience. 2006;142:661–9.
64. Yan J, Aldrich RW. LRRC26 auxiliary protein allows BK channel activation at resting voltage without calcium. Nature. 2010;466:513–6.
65. Tseng-Crank J, Foster CD, Krause JD, Mertz R, Godinot N, et al. Cloning, expression, and distribution of functionally distinct Ca(2+)-activated K+ channel isoforms from human brain. Neuron. 1994;13:1315–30.

66. Poulsen AN, Wulf H, Hay-Schmidt A, Jansen-Olesen I, Olesen J, et al. Differential expression of BK channel isoforms and beta-subunits in rat neuro-vascular tissues. Biochim Biophys Acta. 2009;1788:380–9.
67. Chen L, Tian L, MacDonald SH, McClafferty H, Hammond MS, et al. Functionally diverse complement of large conductance calcium- and voltage-activated potassium channel (BK) alpha-subunits generated from a single site of splicing. J Biol Chem. 2005;280:33599–609.
68. Tian L, Coghill LS, McClafferty H, MacDonald SH, Antoni FA, et al. Distinct stoichiometry of BKCa channel tetramer phosphorylation specifies channel activation and inhibition by cAMP-dependent protein kinase. Proc Natl Acad Sci U S A. 2004;101:11897–902.
69. Tian L, Duncan RR, Hammond MS, Coghill LS, Wen H, et al. Alternative splicing switches potassium channel sensitivity to protein phosphorylation. J Biol Chem. 2001;276:7717–20.
70. Naruse K, Tang QY, Sokabe M. Stress-Axis Regulated Exon (STREX) in the C terminus of BK(Ca) channels is responsible for the stretch sensitivity. Biochem Biophys Res Commun. 2009;385:634–9.
71. Tang QY, Qi Z, Naruse K, Sokabe M. Characterization of a functionally expressed stretch-activated BKca channel cloned from chick ventricular myocytes. J Membr Biol. 2003;196:185–200.
72. Wang W, Huang H, Hou D, Liu P, Wei H, et al. Mechanosensitivity of STREX-lacking BKCa channels in the colonic smooth muscle of the mouse. Am J Physiol Gastrointest Liver Physiol. 2010;299:G1231–40.
73. Zhao HC, Agula H, Zhang W, Wang F, Sokabe M, et al. Membrane stretch and cytoplasmic Ca2+ independently modulate stretch-activated BK channel activity. J Biomech. 2010;43:3015–9.
74. Sandow SL, Grayson TH. Limits of isolation and culture: intact vascular endothelium and BKCa. Am J Physiol Heart Circ Physiol. 2009;297:H1–7.
75. Riddle MA, Hughes JM, Walker BR. Role of caveolin-1 in endothelial BKCa channel regulation of vasoreactivity. Am J Physiol Cell Physiol. 2011;301:C1404–14.
76. Brenner R, Perez GJ, Bonev AD, Eckman DM, Kosek JC, et al. Vasoregulation by the beta1 subunit of the calcium-activated potassium channel. Nature. 2000;407:870–6.
77. Pluger S, Faulhaber J, Furstenau M, Lohn M, Waldschutz R, et al. Mice with disrupted BK channel beta1 subunit gene feature abnormal Ca(2+) spark/STOC coupling and elevated blood pressure. Circ Res. 2000;87:E53–60.
78. Xu H, Garver H, Galligan JJ, Fink GD. Large-conductance Ca2+-activated K+ channel beta1-subunit knockout mice are not hypertensive. Am J Physiol Heart Circ Physiol. 2011;300:H476–85.
79. Matharoo-Ball B, Ashford ML, Arulkumaran S, Khan RN. Down-regulation of the alpha- and beta-subunits of the calcium-activated potassium channel in human myometrium with parturition. Biol Reprod. 2003;68:2135–41.
80. McGahon MK, Dash DP, Arora A, Wall N, Dawicki J, et al. Diabetes downregulates large-conductance $Ca2^{+}$-activated potassium beta 1 channel subunit in retinal arteriolar smooth muscle. Circ Res. 2007;100:703–11.
81. Amberg GC, Santana LF. Downregulation of the BK channel beta1 subunit in genetic hypertension. Circ Res. 2003;93:965–71.
82. Wan E, Kushner JS, Zakharov S, Nui XW, Chudasama N, et al. Reduced vascular smooth muscle BK channel current underlies heart failure-induced vasoconstriction in mice. FASEB J. 2013;27:1859–67.
83. Cui J, Cox DH, Aldrich RW. Intrinsic voltage dependence and Ca2+ regulation of mslo large conductance Ca-activated K^{+} channels. J Gen Physiol. 1997;109:647–73.
84. Horrigan FT, Aldrich RW. Allosteric voltage gating of potassium channels II. Mslo channel gating charge movement in the absence of Ca(2+). J Gen Physiol. 1999;114:305–36.
85. Stefani E, Ottolia M, Noceti F, Olcese R, Wallner M, et al. Voltage-controlled gating in a large conductance Ca2+-sensitive K+ channel (hslo). Proc Natl Acad Sci U S A. 1997;94:5427–31.
86. Schubert R, Nelson MT. Protein kinases: tuners of the BKCa channel in smooth muscle. Trends Pharmacol Sci. 2001;22:505–12.

87. Nimigean CM, Magleby KL. The beta subunit increases the Ca2+ sensitivity of large conductance Ca2+-activated potassium channels by retaining the gating in the bursting states. J Gen Physiol. 1999;113:425–40.
88. Gonzalez-Perez V, Xia XM, Lingle CJ. Functional regulation of BK potassium channels by gamma1 auxiliary subunits. Proc Natl Acad Sci U S A. 2014;111:4868–73.
89. Zhang J, Yan J. Regulation of BK channels by auxiliary gamma subunits. Front Physiol. 2014;5:401.
90. Yan J, Olsen JV, Park KS, Li W, Bildl W, et al. Profiling the phospho-status of the BKCa channel alpha subunit in rat brain reveals unexpected patterns and complexity. Mol Cell Proteomics. 2008;7:2188–98.
91. Kyle BD, Hurst S, Swayze RD, Sheng J, Braun AP. Specific phosphorylation sites underlie the stimulation of a large conductance, Ca(2+)-activated K(+) channel by cGMP-dependent protein kinase. FASEB J. 2013;27:2027–38.
92. Barman SA, Zhu S, White RE. PKC activates BKCa channels in rat pulmonary arterial smooth muscle via cGMP-dependent protein kinase. Am J Physiol Lung Cell Mol Physiol. 2004;286:L1275–81.
93. Nara M, Dhulipala PD, Wang YX, Kotlikoff MI. Reconstitution of beta-adrenergic modulation of large conductance, calcium-activated potassium (maxi-K) channels in Xenopus oocytes. Identification of the camp-dependent protein kinase phosphorylation site. J Biol Chem. 1998;273:14920–4.
94. Barman SA, Zhu S, Han G, White RE. cAMP activates BKCa channels in pulmonary arterial smooth muscle via cGMP-dependent protein kinase. Am J Physiol Lung Cell Mol Physiol. 2003;284:L1004–11.
95. Ling S, Sheng JZ, Braun AP. The calcium-dependent activity of large-conductance, calcium-activated K+ channels is enhanced by Pyk2- and Hck-induced tyrosine phosphorylation. Am J Physiol Cell Physiol. 2004;287:C698–706.
96. Ling S, Woronuk G, Lev S, Braun AP. Enhanced activity of a large conductance, calcium-sensitive K+ channel in the presence of Src tyrosine kinase. J Biol Chem. 2000;275: 30683–9.
97. Wu X, Yang Y, Gui P, Sohma Y, Meininger GA, et al. Potentiation of large conductance, Ca2+-activated K^+ (BK) channels by alpha5beta1 integrin activation in arteriolar smooth muscle. J Physiol. 2008;586:1699–713.
98. Yang Y, Wu X, Gui P, Wu J, Sheng JZ, et al. Alpha5beta1 integrin engagement increases large conductance, $Ca2^+$-activated K+ channel current and $Ca2^+$ sensitivity through c-src-mediated channel phosphorylation. J Biol Chem. 2010;285:131–41.
99. Yang Y, Wu X, Gui P, Wu J, Sheng JZ, et al. {alpha}5{beta}1 integrin engagement increases BK channel current and Ca2+ sensitivity through c-Src mediated channel phosphorylation. J Biol Chem. 2009;285:131–41.
100. Perez GJ, Bonev AD, Nelson MT. Micromolar Ca(2+) from sparks activates $Ca(2^+)$-sensitive K(+) channels in rat cerebral artery smooth muscle. Am J Physiol Cell Physiol. 2001;281:C1769–75.
101. Jaggar JH, Wellman GC, Heppner TJ, Porter VA, Perez GJ, et al. $Ca2^+$ channels, ryanodine receptors and Ca(2+)-activated K+ channels: a functional unit for regulating arterial tone. Acta Physiol Scand. 1998;164:577–87.
102. Zhuge R, Fogarty KE, Baker SP, McCarron JG, Tuft RA, et al. $Ca(2^+)$ spark sites in smooth muscle cells are numerous and differ in number of ryanodine receptors, large-conductance K(+) channels, and coupling ratio between them. Am J Physiol Cell Physiol. 2004;287:C1577–88.
103. Kur J, Bankhead P, Scholfield CN, Curtis TM, McGeown JG. $Ca(2^+)$ -sparks promote myogenic tone in retinal arterioles. Br J Pharmacol. 2012;168:1675–86.
104. Liu G, Shi J, Yang L, Cao L, Park SM, et al. Assembly of a Ca2+-dependent BK channel signaling complex by binding to beta2 adrenergic receptor. EMBO J. 2004;23:2196–205.
105. Lu T, Zhang DM, Wang XL, He T, Wang RX, et al. Regulation of coronary arterial BK channels by caveolae-mediated angiotensin II signaling in diabetes mellitus. Circ Res. 2010; 106:1164–73.

Chapter 7
Ion Channel Trafficking and Control of Arterial Contractility

M. Dennis Leo and Jonathan H. Jaggar

Abstract Ion channels control many cellular processes, including neuronal excitability and arterial contractility. Vascular smooth muscle cell (myocyte) plasma membrane ion channels regulate membrane potential and extracellular calcium (Ca^{2+}) influx, which alters regional blood flow and systemic blood pressure. Ion channels can be homomers or heteromers of several pore-forming subunits and may associate with auxiliary subunits that can modify channel properties, including activity and surface expression. Recent studies have investigated pathways that control the surface expression of some plasma membrane ion channels in myocytes. These findings have also found that vasoregulatory stimuli can control the number of plasma membrane ion channels to regulate contractility. Here, we review current literature describing mechanisms of trafficking of voltage-gated K^+ (Kv), voltage-gated Ca^{2+} ($Ca_v1.2$), large-conductance Ca^{2+}-activated potassium (BK_{Ca}) and transient receptor potential (TRP) channels in arterial myocytes. These studies indicate that regulated ion channel trafficking is a functional mechanism to control vascular contractility.

Keywords Ion channel • Accessory subunits • Protein trafficking

Introduction

Vascular smooth muscle (myocyte) ion channels regulate membrane potential and contractility [1–3]. Current (*I*) generated by a membrane ion channel population is the product of the number of channels (*N*), single channel open probability (P_o) and single channel current (*i*), such that $I = N \cdot P_o \cdot i$. Vasoregulatory stimuli can modify myocyte ion channel function via transcriptional or post-translational modification and by directly interacting with these proteins or auxiliary subunits. Myocyte ion

M.D. Leo • J.H. Jaggar (✉)
Department of Physiology, University of Tennessee Health Science Center, Memphis, TN 38163, USA
e-mail: jjaggar@uthsc.edu

I. Levitan, A.M. Dopico (eds.), *Vascular Ion Channels in Physiology and Disease*, DOI 10.1007/978-3-319-29635-7_7

channel activity (NP_o) can also be modulated by mechanisms that regulate the number of functional plasma membrane ion channels. This chapter will review studies that have investigated trafficking pathways of several ion channels that are expressed in vascular myocytes. If available, functional significance of these trafficking mechanisms in the vasculature is provided.

L-Type Voltage-Gated Ca^{2+} ($Ca_V1.2$) Channels

Voltage-gated Ca^{2+} channels are heteromeric complexes composed of a pore-forming α_{1C} ($Ca_V1.2$) and auxiliary $\alpha_2\delta$ and β subunits (Fig. 7.1). $Ca_V1.2$ is encoded by the CACNA1C gene which has 55 exons in humans and 52 exons in rodents [4, 5]. $Ca_V1.2$ is a high voltage-activated, L-type Ca^{2+} channel that is sensitive to several blockers, including antihypertensive dihydropyridines [3]. Membrane

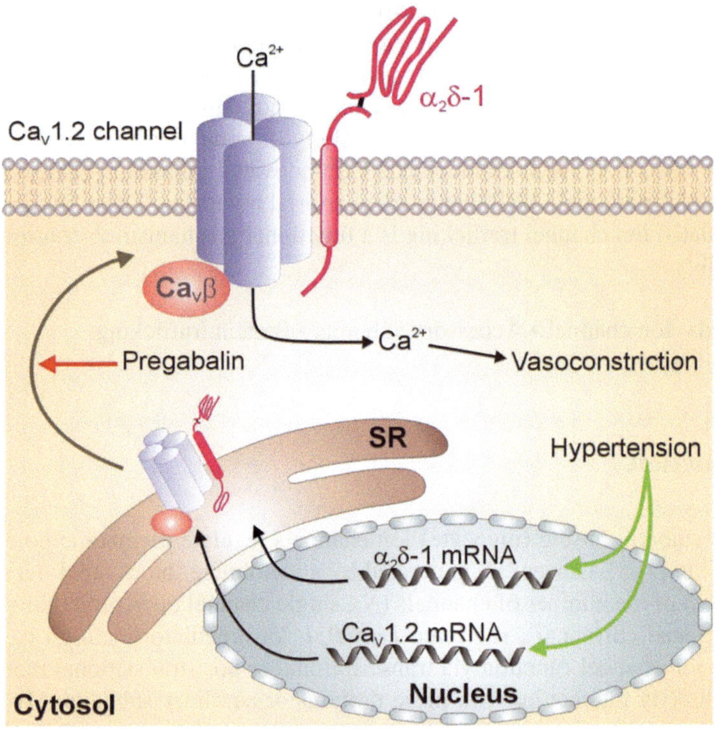

Fig. 7.1 $Ca_V1.2$ trafficking in arterial myocytes. $\alpha_2\delta$-1 is required for functional surface expression of $Ca_V1.2$ subunits. Pregabalin, a $\alpha_2\delta$-1 ligand, inhibits surface trafficking of $Ca_V1.2$ channels, leading to vasodilation. $\alpha_2\delta$-1 and $Ca_V1.2$ expression are both elevated in genetic hypertension, which promotes surface trafficking of higher numbers of $Ca_V1.2$ channels, leading to an increase in current density and vasoconstriction. *Green arrow* indicates activation. *Red arrow* indicates inhibition

depolarization activates $Ca_V1.2$ channels in arterial myocytes, leading to Ca^{2+} influx, an elevation in global intracellular Ca^{2+} concentration, activation of Ca^{2+}/calmodulin-dependent myosin light chain kinase and vasoconstriction [6, 7].

$Ca_V1.2$ channel surface expression and function is regulated by $\alpha_2\delta$ and β subunits, of which there are multiple isoforms [3]. Four different $\alpha_2\delta$ subunits ($\alpha_2\delta$-1-4) have been identified that arise from alternative splicing of a single gene (CACNA2D1). Post-translational cleavage of $\alpha_2\delta$ results in a glycosylated extracellular α_2 and a smaller membrane-spanning δ subunit, which then re-associate via a disulfide bond to form a functional protein [8–10]. Structurally, $\alpha_2\delta$ subunits are type I transmembrane (TM)-spanning proteins [8, 11]. Arterial myocytes express only $\alpha_2\delta$-1 [12]. $\alpha_2\delta$-1 knockdown reduced $Ca_V1.2$ surface expression and decreased $[Ca^{2+}]_i$ in cerebral artery myocytes, indicating that $\alpha_2\delta$-1 is an important stimulator of $Ca_V1.2$ surface trafficking [12]. Pregabalin (Lyrica™), a $\alpha_2\delta$-1/2 ligand, is an antiallodynic and antihyperalgesic prescribed for the treatment of neuropathic pain in humans [13]. Pregabalin decreased surface expression of both $\alpha_2\delta$-1 and $Ca_V1.2$ channels without altering total protein in rat cerebral arteries [12] (Fig. 7.1). These data were consistent with those obtained in DRG neurons with gabapentin, another gabapentinoid compound, which reduced $\alpha_2\delta$-2 surface expression by preventing Rab11-dependent recycling to the plasma membrane [14]. Pregabalin is also a weak $Ca_V1.2$ channel pore blocker that does not directly affect channel voltage-dependence in cerebral artery myocytes [12]. Pregabalin caused concentration-dependent relaxation of pressurized (60 mmHg) rat cerebral arteries via a mechanism involving both inhibition of $Ca_V1.2$ anterograde trafficking and surface channel pore block [12].

Transcriptional upregulation of both $\alpha_2\delta$-1 and $Ca_V1.2$ occurs in cerebral artery myocytes of the spontaneously hypertensive rat (SHR), a genetic model of hypertension [15] (Fig. 7.1). $\alpha_2\delta$-1 and $Ca_V1.2$ surface protein expression were also higher in SHR arteries, when compared with normotensive control arteries [15]. Pregabalin was more effective at inhibiting surface expression of $\alpha_2\delta$-1 and $Ca_V1.2$ in SHR cerebral arteries than in normotensive controls [15] (Fig. 7.1). Conceivably, targeting of vascular $\alpha_2\delta$-1 may represent a novel therapeutic approach for inducing vasodilation during hypertension.

Association with auxiliary β subunits can modify the biophysical properties and promote surface trafficking of $Ca_V1.2$ channels. Binding of recombinant β subunits to the $Ca_V1.2$ I–II loop produces a C-terminus-dependent conformational rearrangement that weakens the ER retention signal, leading to surface trafficking [26]. It is possible, but not shown, that a similar mechanism exists in vascular myocytes. A decrease in whole-cell $Ca_V1.2$ currents, but an increase in local Ca^{2+} sparklet activity was observed in mesenteric artery myocytes from a mouse model of genetic hypertension [27]. This was associated with a shift in channel composition from $\alpha_1c/\alpha_2\delta/\beta_3$ to $\alpha_1c/\alpha_2\delta/\beta_2$ in myocytes [27].

Alternative splicing of $Ca_V1.2$ gives rise to several variants with different biophysical properties. 5′ RACE-PCR revealed the expression of a novel exon 1c (e1c) which encodes a cysteine-rich $Ca_V1.2$ N-terminus [16]. E1c was the principal smooth muscle $Ca_V1.2$ variant, with a small proportion of e1b-containing $Ca_V1.2$ in rat cerebral artery myocytes [16]. Knockdown of $Ca_V1.2$ channels containing the e1c-encoded N-terminus reduced $Ca_V1.2$ protein and myogenic tone more than e1b-specific knockdown in cerebral arteries [16]. These data indicated that $Ca_V1.2e1c$ was the principal functional $Ca_V1.2$ variant in arterial myocytes [17]. $\alpha_2\delta$-1 more effectively surface-trafficked $Ca_V1.2e1c$ than $Ca_V1.2e1b$ channels in cerebral artery myocytes, and evidence was provided to demonstrate that this association regulated contractility [17]. The $Ca_V1.2$ N-terminus also contains a α-CaMKII binding site that regulates surface expression of recombinant channels [18]. The amino acid sequence CISI located in the $Ca_V1.2$ distal N-terminus mediated this function in an heterologous expression system [18].

$Ca_V1.2$ α subunits are primarily plasma membrane-localized, with a small proportion of total protein intracellular, in arterial myocytes [12]. Western blotting detected two distinct $Ca_V1.2$ bands, one at ~240 kDa and a second band at ~190 kDa in arterial myocytes [12, 17, 19, 20]. This difference occurs due to post-translational modification that cleaves ~300 amino acids of the distal $Ca_V1.2$ C-terminus [20, 21]. *In vitro*, proteolytic enzymes, including calpain and chymotrypsin cleave the C-terminal fragment from full-length $Ca_V1.2$ protein [22, 23]. Enzymes responsible for this modification *in vivo* have not yet been identified. A proline-rich domain in the C-terminal fragment aids in its re-association with $Ca_V1.2$ [23]. Re-association of the fragment with the truncated $Ca_V1.2$ protein right-shifts voltage-dependent activation, including in arterial myocytes [19, 24]. Overexpression of the CCT fragment decreased $Ca_V1.2$ mRNA and total protein, reducing myocyte $Ca_V1.2$ currents and myogenic tone in pressurized (60 mmHg) cerebral arteries [19]. These data indicated that CCT acts as a bimodal vasodilator, reducing both $Ca_V1.2$ voltage-sensitivity and transcription, and thus surface expression, in cerebral artery myocytes. $Ca_V1.2$ gene expression is also modulated by mitochondria in arterial myocytes [25]. IP_3R-mediated sarcoplasmic reticulum (SR) Ca^{2+} release elevates the Ca^{2+} concentration within spatially localized mitochondrial Ca^{2+} ($[Ca^{2+}]_{mito}$), which stimulates mitochondrial ROS production [25]. Endothelin-1, a vasoconstrictor, activates this pathway, leading to mitochondrial ROS-dependent activation of NF-κB, a transcription factor that stimulates $Ca_V1.2$ transcription, leading to vasoconstriction [25].

Voltage-Gated K^+ (Kv) Channels

Kv channels are a family of ~40 six transmembrane domain proteins that are subdivided into 12 types (Kv1–12) [28]. Kv channels contain a gated pore domain that confers ion specificity and a voltage-sensing domain (VSD) that responds to changes in membrane potential. Membrane depolarization causes translocation of the positively-charged S4 segment of the VSD to the extracellular surface of the plasma membrane, which elevates P_o [29, 30]. Kv channels are major contributors to resting membrane potential in vascular myocytes [31]. Kv channel activation induces a shift to more negative membrane potentials, which reduces voltage-dependent Ca^{2+}

channel activity in arterial myocytes, leading to myocyte relaxation and eventual vasodilation [32].

Vascular myocytes express several Kv family members, including subtypes of Kv1-Kv4, Kv7, Kv9 and Kv11 [33–37]. Kv channels can form homo- or heterotetramers, which contributes to a vast array of Kv current phenotypes [38]. Amino acids located in the channel pore of Kv1, but not Kv2-4, regulate cell surface expression [39]. A consensus amino acid sequence 'VXXSL' in the C-terminal region of Kv1 channels regulated recombinant protein cell surface expression [40]. Caveolin-1 interacts with recombinant Kv1.5 to target channels to plasma membrane lipid raft microdomains [41]. Kv1.5 co-localized with caveolin-1 and 5-HT_{2A} receptors in rat pulmonary artery myocytes with 5-HT-induced Kv1.5 endocytosis suggested to contribute to the pathogenesis of pulmonary hypertension [42]. Postsynaptic density-95 (PSD95) is a scaffolding protein that anchors Kv1 channels to the plasma membrane in cerebral artery myocytes [43]. Kv1.2 contains a high affinity binding site for PSD95, with evidence that PSD95 interacts with Kv1.5 directly via its low affinity binding site [43]. Transfection of a peptide corresponding to the Kv1.2 PSD95 binding motif suppressed Kv1.5 currents, leading to myocyte depolarization and vasoconstriction in rat cerebral arteries [44]. Rab4 and Rab11 dominant-negative mutants decreased Kv1.5 surface expression in atrial cardiomyocytes [45]. Recombinant Kv1.5 was also found to associate with Rab4, Rab5, Rab7 and Rab11 in H9c2 and HEK293 cells [46]. Oxyhemoglobin, a blood component mediating the pathogenesis of cerebral vasospasm following subarachnoid hemorrhage, constricts cerebral arteries [47]. Oxyhemoglobin-induced tyrosine kinase activation reduced plasma membrane Kv1.5 expression in rabbit cerebral arteries [48]. Oxyhemoglobin-induced Kv1.5 inhibition may therefore contribute to cerebral artery constriction following subarachnoid hemorrhage.

Kv2.1 is expressed in cerebral [49] and mesenteric arteries [33]. Heteromultimeric Kv2.1/Kv9.3 channels contribute to opposition of the myogenic response in rat middle cerebral arteries [50]. In an heterologous expression system, interaction between the N- and C-termini of Kv2.1 regulated cell surface expression [51]. Src-dependent tyrosine phosphorylation reduced surface expression of recombinant Kv3.1 channels [52]. Kv3.1 also possesses multiple zinc-binding sites that affect channel activity and localization in neurons, a property not yet described in arterial myocytes [53].

Large-Conductance Ca^{2+}-Activated Potassium (BK_{Ca}, BK, *Slo1*, $K_{Ca}1.1$) Channels

BK channels have a high single channel conductance, can be activated by both intracellular Ca^{2+} and membrane depolarization, and are expressed in many cell types [2, 54]. In vascular myocytes, BK channel activation leads to a shift to more negative membrane potentials, which reduces voltage-dependent Ca^{2+} channel activity, leading to myocyte relaxation and eventual vasodilation [55, 56]. A functional BK channel consists of four pore-forming BK α-subunits [57]. Each α subunit is a seven transmembrane domain protein with an extracellular N-terminus and a long C-terminus. Accessory β subunits contain two transmembrane proteins with intracellular N- and C-termini, of which four have been identified (β1–4) that exhibit tissue-specific expression [2, 54]. β1 is expressed in smooth muscle cells, including those in the vasculature [2]. Association of β1 with BKα increases apparent Ca^{2+}-sensitivity and slows channel activation and deactivation [54]. Recently, γ subunits have also been identified, including in vascular myocytes [58, 59]. Association of BK channels with γ subunits elevates voltage-sensitivity, leading to activation and vasodilation [58].

BK channels are encoded by the KCNMA1 gene, which is located on the long (q) arm of chromosome 10 at position 22.3. BKα has at least ten different splice variants that exhibit varying degrees of expression in different tissues [60]. The channel has been shown to undergo extensive splicing that can affect surface expression [61]. Recent reports also indicate functions for BK channels located on the inner mitochondrial and outer nuclear membranes [62]. A BK channel N-terminal variant with a splice site located between the S0 and S1 transmembrane domains was first identified in rat myometrium and brain [63]. An endoplasmic reticulum (ER) retention sequence prevents this variant from trafficking to the plasma membrane. Although not exclusive to vascular myocytes, two variants of BKα are commonly described, a 'ZERO' variant, and a STRess axis regulated EXon (STREX) variant, first identified in chromaffin cells as a stress adaptive response to hypophysectomy [64]. The STREX exon is a 58 amino acid insertion between RCK1 and RCK2 in the BKα C-terminus. BK channels with this insertion exhibit increased apparent Ca^{2+}- and voltage-sensitivity and are inhibited by protein kinase A (PKA), in contrast to ZERO variant channels which are activated by PKA [65]. The recombinant BKα-STREX variant demonstrated higher membrane expression than the ZERO variant when overexpressed in CHO-K1 cells [66]. Only a small proportion of BK channel mRNA contained the STREX insert in rat whole middle cerebral and cremaster arteries [66]. This study also raised the possibility that BK splice variants may differ between arteries of the cerebral and peripheral circulations with the latter expressing more of the STREX variant [66]. Another study found that BKα clones obtained from pure rat cerebral artery myocytes did not contain the STREX variant [67]. *Slo1* exons 18 and 19 encode part of the RCK1-RCK2 linker, a region that contains multiple trafficking motifs required for cell surface expression of recombinant protein [68]. Exon 18- and 19-encoded amino acid sequences are present in myocyte BK channels, but physiological function is unclear. A novel C-terminal deficient BK channel splice variant truncated at Gly781 was cloned from the rabbit medullary thick ascending limb (mTAL) cell line [69]. Overexpression studies in mTAL and HEK293 cells revealed that the protein is predominantly intracellular and is not trafficked to the plasma membrane [69]. S-palmitoylation of the S0-S1 segment also stimulates cell surface expression of recombinant BKα [70, 71]. BK channels are stretch-sensitive, a property that may involve close association with structural proteins, including actin [72].

The distal end of the recombinant BK channel C-terminal tail contains an actin binding domain that mediates this interaction and is suggested to play a key role in membrane trafficking [73]. BK channels also interact with caveolin-1 through a caveolin-binding motif [74]. Although this association does not appear to regulate channel activity, it serves to anchor the channel to the plasma membrane and thereby plays an indirect role in surface expression [74].

BK β1 subunits are expressed in vascular myocytes [2]. Association of β1 with BK channels elevates apparent Ca^{2+}-sensitivity and enhances coupling of Ca^{2+} sparks to BK channels in arterial myocytes, leading to vasodilation [2]. β1 attenuates cell surface expression of recombinant BK channels when expressed in HEK293 cells [75]. This effect was attributed to an endocytic signal present in the β1 C-terminus. In contrast, β1 co-expression stimulated surface trafficking of a "VEDEC" variant of recombinant BK channels in HEK293 cells [76]. The amino acid sequence VEDEC located at the BK channel C-terminal end prevents exit from the ER [76]. A recent article from our group revealed that although most BKα subunits are plasma membrane-resident, only a small fraction of β1 subunits are located at the cell surface in cerebral and mesenteric artery myocytes, including human cerebral artery myocytes [77]. Nitric oxide (NO), a principal endothelium-derived vasodilator, stimulated rapid (<1 min) PKG-mediated surface trafficking of intracellular β1 subunits, which associated with BKα to increase apparent Ca^{2+}-sensitivity, thereby elevating open probability [77] (Figs. 7.2 and 7.3). Concanavalin A,

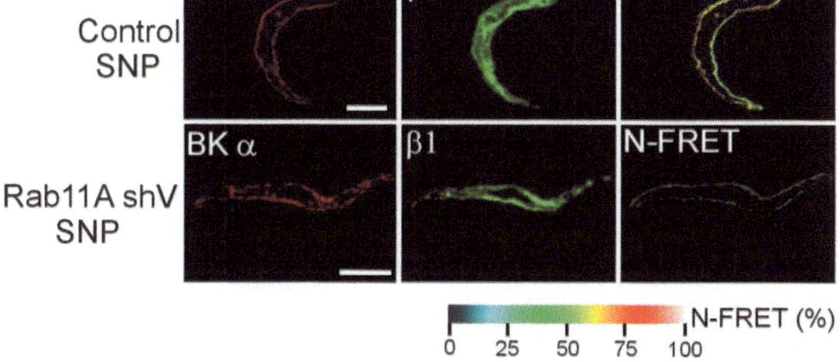

Fig. 7.2 BK β1 and Rab11A are located in close spatial proximity in arterial myocytes. (**a**) Immunofluorescence and immunoFRET images of Rab11A and BK β1 in the same arterial myocyte indicates close spatial proximity of these proteins. (**b**) Immunofluorescence and N-FRET images for BKα and β1 proteins in cells treated with control or Rab11A shV in the presence of sodium nitroprusside (SNP), a NO donor. Scale bar, 10 μm. Modified from [77]

an endocytosis inhibitor, increased surface β1 protein, indicating that β1 is constantly recycled and that NO shifts this equilibrium in favor of β1 surface expression [77]. Iberiotoxin, a selective BK channel inhibitor, Rab11A knockdown and brefeldin A, a trafficking inhibitor, each reduced NO-induced cerebral vasodilation ~60 %, indicating that stimulated β1 surface trafficking is a principal mechanism by which NO promotes vasodilation in cerebral arteries [77]. This study also provided the first evidence that different native ion channel subunits can traffic independently and that physiological stimuli can regulate mobility of an auxiliary subunits to acutely control functional surface ion channel activity (Fig. 7.3).

Rab GTPases are a large family of small Ras-like GTPases, with more than 60 identified members [78]. Rab proteins cycle between a GDP-bound 'inactive' state and a GTP-bound 'active' form. They are present on many organelles and modulate vesicular protein transport between different cellular compartments [78, 79]. In arterial myocytes, intracellular β1 was associated with rab11A-positive recycling endosomes, which controlled surface expression [77] (Figs. 7.2 and 7.3). Rab11B was not involved in either BKα or β1 trafficking in arterial myocytes [77], but this GTPase regulated recombinant BK channel surface expression in CHO cells [80]. These studies suggest that differences may exist in ion channel trafficking pathways in native tissues and heterologous expression systems.

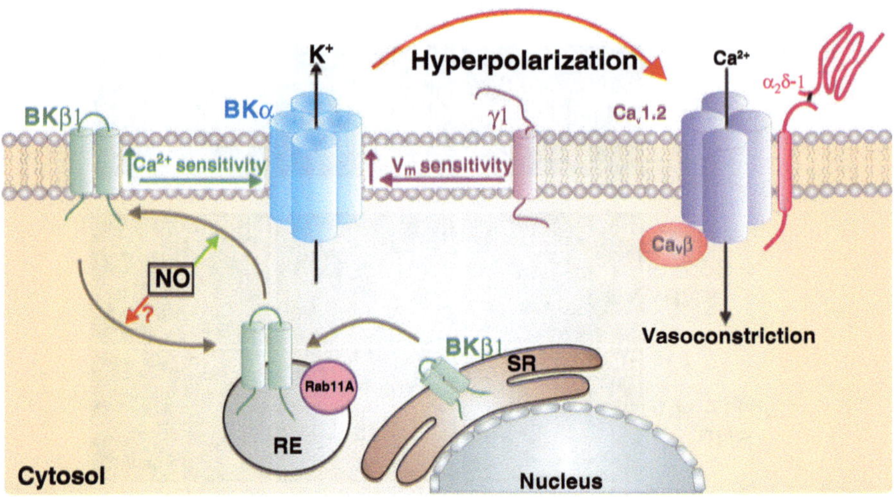

Fig. 7.3 Schematic representation of BK channel regulation by β1 surface trafficking and γ1 subunits in arterial myocytes. Nitric oxide (NO) stimulates rapid (<1 min) surface expression of β1 subunits located within Rab11A-positive recycling endosomes (RE) which then associate with BKα to increase apparent Ca^{2+}-sensitivity, leading to a shift to more negative membrane potentials and vasodilation. γ1 subunits are surface expressed and elevate BK channel voltage-sensitivity, leading to activation and vasodilation. *SR* sarcoplasmic reticulum. *Green arrow* indicates activation. *Red arrow* indicates inhibition

Transient Receptor Potential (TRP) Channels

TRP proteins are a large family of non-selective cation channels that can respond to a wide variety of stimuli, including temperature, pressure and vasoregulatory agents [81]. TRP proteins are divided into several families that include canonical (TRPC), vanniloid (TRPV), polycystin (TRPP) and melastatin (TRPM) [81]. Mechanisms of recombinant TRP channel trafficking have been described, but few studies have investigated TRP channel trafficking in arterial myocytes. As similar pathways may also exist in native arterial myocytes, data obtained in recombinant expression systems and non-vascular cells will be discussed here.

TRPC channels contain seven family members (TRPC1–7), several of which are expressed in vascular myocytes [81]. Plasma membrane insertion and retention of recombinant TRPC1 is dependent on its interaction with STIM1 and Orai1 [82]. Orai1-mediated local Ca^{2+} entry stimulated plasma membrane insertion of TRPC1 channels in human submandibular gland cells [83]. TRPC3 is primarily located in the plasma membrane of arterial myocytes [84]. A TRPC3 C-terminal calmodulin and IP3R binding (CIRB) domain can directly interact with IP_3 receptor-1 (IP_3R1) [85]. Endothelin-1, a vasoconstrictor, stimulated physical coupling between TRPC3 and IP_3R1, leading to channel activation in myocytes and cerebral vasoconstriction [85]. TRPC3 protein expression was elevated in SHR mesenteric artery myocytes, leading to enhanced coupling with IP_3R1 and vasoconstriction associated with hypertension [84]. TRPC3 associates with Rab7 in rat brain [86]. Rab7 is implicated in endocytosis and lysosome formation, indicating that TRPC3 expression is likely regulated by Rab7-mediated endocytosis and degradation [87]. In hippocampal neurons, TRPC3 associated with VAMP2, a SNARE complex protein that is involved in plasma membrane vesicle fusion [88]. β_2 adrenergic receptor stimulation of a $G\alpha_s$-cAMP pathway potentiated plasma membrane TRPC5 channel trafficking in HEK293 cells [89]. TRPC5 activation has not been described to induce vasoconstriction, although TRPC5 activity was important for motility of human saphenous vein myocytes [90]. Experiments in HEK293 cells revealed a single conserved cysteine residue at position 553 in TRPC5 to control multimer formation and plasma membrane trafficking [91]. Inhibition of PI3 kinase decreased TRPC6 plasma membrane localization in A7r5 cells, a smooth muscle cell line derived from rat thoracic aorta [92]. Recombinant TRP channels also interact with cytoskeletal proteins, including actin and caveolin-1, that can affect cell surface expression [93, 94].

TRPV proteins consist of six members of which TRPV1, V2 and V4 have been identified in vascular myocytes [95]. TRPV4 is expressed in rat cerebral and mouse mesenteric artery myocytes [96]. TRPV4 activation stimulates Ca^{2+} sparks and BK channel activity, leading to a shift to more negative myocyte membrane potentials and vasodilation [95, 97]. Mutation analysis of recombinant TRPV4 revealed C-terminal residues 838–857 to be critical for protein folding, maturation and ER exit [98]. Intracellular Ca^{2+} store depletion stimulated plasma membrane insertion of TRPV4-TRPC1 heteromeric channels in HEK293 and human endothelial cells [99].

TRPP family members TRPP1 and TRPP2 are expressed in vascular myocytes [100, 101]. TRPP1 is proposed to be a mechanical sensor that regulates activity of the TRPP2 ion channel [102]. TRPP2 was the predominant TRPP isoform and was primarily located in the plasma membrane in rat cerebral artery myocytes [100]. TRPP2 knockdown decreased swelling-induced cation currents (I_{Cat}) in rat cerebral artery myocytes [100] but elevated stretch-induced I_{Cat} in murine mesenteric artery myocytes [101]. These data suggest that TRPP2 may differentially regulate I_{Cat} in different species and/or vascular beds.

All eight TRPM subfamily members are expressed in vascular myocytes [81]. TRPM4 has two splice variants, TRPM4a and 4b [103]. The function of TRPM4a is unclear [103]. TRPM4b, which is typically referred to as TRPM4, is selective for monovalent cations and requires high $[Ca^{2+}]_i$ for activation [95]. TRPM4 knockdown decreased intravascular pressure-induced smooth muscle depolarization and constriction in rat cerebral arteries [104]. TRPM4 overexpression in A7r5 cells indicated channels to be dynamic and present in two distinct compartments, on the plasma membrane and in small, mobile intracellular vesicles [104]. Surface biotinylation experiments showed that approximately half of total TRPM4 was plasma membrane-localized in rat cerebral arteries [104]. PKC activation stimulated rapid (~10 min) surface trafficking of TRPM4 in rat cerebral arteries [104]. Pressure-induced PKC activation was proposed to stimulate surface trafficking of TRPM4, leading to arterial myocyte depolarization [104]. TRPM7 is a mechano-sensitive channel, which regulates Mg^{2+} homeostasis in vascular myocytes [105]. Extracellular fluid flow stimulated rapid surface expression of TRPM7 in A7r5 cells and may contribute to a cellular response to vessel wall injury [106].

TRP channels are activated by a wide range of mechanical and chemical stimuli to alter arterial contractility. Evidence also suggests that many vasoregulatory stimuli alter trafficking of TRP channels to modulate activity. Therefore, future studies should investigate trafficking of these channels to better understand functions of these proteins in arterial myocytes.

Conclusion

Arterial contractility is determined by the coordinated activity of ion channels that stimulate vasoconstriction or dilation. Recent studies have begun to identify trafficking pathways that control the number of functional surface ion channel proteins in arterial myocytes, but much remains unclear. Mechanisms that control the surface expression of ion channels, functional significance and pathological alterations in arterial myocytes are poorly understood. Future studies should be designed to uncover ion channel trafficking pathways and regulatory mechanisms in the vasculature.

References

1. Ko EA, Han J, Jung ID, Park WS. Physiological roles of K^+ channels in vascular smooth muscle cells. J Smooth Muscle Res. 2008;44(2):65–81.
2. Wu RS, Marx SO. The BK potassium channel in the vascular smooth muscle and kidney: alpha- and beta-subunits. Kidney Int. 2010;78(10):963–74.
3. Hofmann F, Flockerzi V, Kahl S, Wegener JW. L-type $Ca_V1.2$ calcium channels: from in vitro findings to in vivo function. Physiol Rev. 2014;94(1):303–26.
4. Soldatov NM. Genomic structure of human L-type Ca^{2+} channel. Genomics. 1994;22(1):77–87.
5. Tang ZZ, Liang MC, Lu S, Yu D, Yu CY, Yue DT, Soong TW. Transcript scanning reveals novel and extensive splice variations in human l-type voltage-gated calcium channel, $Ca_V1.2$ α_1 subunit. J Biol Chem. 2004;279(43):44335–43.
6. Harder DR. Pressure-dependent membrane depolarization in cat middle cerebral artery. Circ Res. 1984;55(2):197–202.
7. Knot HJ, Nelson MT. Regulation of arterial diameter and wall $[Ca^{2+}]$ in cerebral arteries of rat by membrane potential and intravascular pressure. J Physiol. 1998;508(Pt 1):199–209.
8. Davies A, Hendrich J, Van Minh AT, Wratten J, Douglas L, Dolphin AC. Functional biology of the $\alpha_2\delta$ subunits of voltage-gated calcium channels. Trends Pharmacol Sci. 2007;28(5):220–8.
9. Cole RL, Lechner SM, Williams ME, Prodanovich P, Bleicher L, Varney MA, Gu G. Differential distribution of voltage-gated calcium channel alpha-2 delta ($\alpha_2\delta$) subunit mRNA-containing cells in the rat central nervous system and the dorsal root ganglia. J Comp Neurol. 2005;491(3):246–69.
10. Andrade A, Sandoval A, Oviedo N, De WM, Elias D, Felix R. Proteolytic cleavage of the voltage-gated Ca^{2+} channel a_2d subunit: structural and functional features. Eur J Neurosci. 2007;25(6):1705–10.
11. Robinson P, Etheridge S, Song L, Shah R, Fitzgerald EM, Jones OT. Targeting of voltage-gated calcium channel alpha2delta-1 subunit to lipid rafts is independent from a GPI-anchoring motif. PLoS One. 2011;6(6):e19802.
12. Bannister JP, Adebiyi A, Zhao G, Narayanan D, Thomas CM, Feng JY, Jaggar JH. Smooth muscle cell $\alpha_2\delta$-1 subunits are essential for vasoregulation by $Ca_V1.2$ channels. Circ Res. 2009;105(10):948–55.
13. Toth C. Pregabalin: latest safety evidence and clinical implications for the management of neuropathic pain. Ther Adv Drug Saf. 2014;5(1):38–56.
14. Tran-Van-Minh A, Dolphin AC. The alpha2delta ligand gabapentin inhibits the Rab11-dependent recycling of the calcium channel subunit alpha2delta-2. J Neurosci. 2010;30(38):12856–67.
15. Bannister JP, Bulley S, Narayanan D, Thomas-Gatewood C, Luzny P, Pachuau J, Jaggar JH. Transcriptional upregulation of alpha2delta-1 elevates arterial smooth muscle cell voltage-dependent Ca^{2+} channel surface expression and cerebrovascular constriction in genetic hypertension. Hypertension. 2012;60(4):1006–15.
16. Cheng X, Liu J, Asuncion-Chin M, Blaskova E, Bannister JP, Dopico AM, Jaggar JH. A novel $Ca_V1.2$ N terminus expressed in smooth muscle cells of resistance size arteries modifies channel regulation by auxiliary subunits. J Biol Chem. 2007;282(40):29211–21.
17. Bannister JP, Thomas-Gatewood CM, Neeb ZP, Adebiyi A, Cheng X, Jaggar JH. $Ca_V1.2$ channel N-terminal splice variants modulate functional surface expression in resistance size artery smooth muscle cells. J Biol Chem. 2011;286(17):15058–66.
18. Simms BA, Souza IA, Rehak R, Zamponi GW. The $Ca_V1.2$ N terminus contains a CaM kinase site that modulates channel trafficking and function. Pflugers Arch. 2015;467(4):677–86.

19. Bannister JP, Leo MD, Narayanan D, Jangsangthong W, Nair A, Evanson KW, Pachuau J, Gabrick KS, Boop FA, Jaggar JH. The voltage-dependent L-type Ca^{2+} ($Ca_V1.2$) channel C-terminus fragment is a bi-modal vasodilator. J Physiol. 2013;591(Pt 12):2987–98.
20. Gomez-Ospina N, Tsuruta F, Barreto-Chang O, Hu L, Dolmetsch R. The C terminus of the L-type voltage-gated calcium channel $Ca_V1.2$ encodes a transcription factor. Cell. 2006;127(3):591–606.
21. Schroder E, Byse M, Satin J. L-type calcium channel C terminus autoregulates transcription. Circ Res. 2009;104(12):1373–81.
22. De Jongh KS, Colvin AA, Wang KK, Catterall WA. Differential proteolysis of the full-length form of the L-type calcium channel alpha 1 subunit by calpain. J Neurochem. 1994;63(4):1558–64.
23. Gerhardstein BL, Gao T, Bunemann M, Puri TS, Adair A, Ma H, Hosey MM. Proteolytic processing of the C terminus of the a_{1C} subunit of L-type calcium channels and the role of a proline-rich domain in membrane tethering of proteolytic fragments. J Biol Chem. 2000;275(12):8556–63.
24. Hulme JT, Yarov-Yarovoy V, Lin TW, Scheuer T, Catterall WA. Autoinhibitory control of the $Ca_V1.2$ channel by its proteolytically processed distal C-terminal domain. J Physiol. 2006;576(Pt 1):87–102.
25. Narayanan D, Xi Q, Pfeffer LM, Jaggar JH. Mitochondria control functional $Ca_V1.2$ expression in smooth muscle cells of cerebral arteries. Circ Res. 2010;107(5):631–41.
26. Fang K, Colecraft HM. Mechanism of auxiliary beta-subunit-mediated membrane targeting of L-type ($Ca_V1.2$) channels. J Physiol. 2011;589(Pt 18):4437–55.
27. Tajada S, Cidad P, Colinas O, Santana LF, Lopez-Lopez JR, Perez-Garcia MT. Down-regulation of $Ca_V1.2$ channels during hypertension: how fewer $Ca_V1.2$ channels allow more Ca^{2+} into hypertensive arterial smooth muscle. J Physiol. 2013;591(Pt 24):6175–91.
28. Gutman GA, Chandy KG, Grissmer S, Lazdunski M, McKinnon D, Pardo LA, Robertson GA, Rudy B, Sanguinetti MC, Stuhmer W, Wang X. International Union of Pharmacology. LIII. Nomenclature and molecular relationships of voltage-gated potassium channels. Pharmacol Rev. 2005;57(4):473–508.
29. Jiang Y, Lee A, Chen J, Ruta V, Cadene M, Chait BT, MacKinnon R. X-ray structure of a voltage-dependent K^+ channel. Nature. 2003;423(6935):33–41.
30. Ghosh S, Nunziato DA, Pitt GS. KCNQ1 assembly and function is blocked by long-QT syndrome mutations that disrupt interaction with calmodulin. Circ Res. 2006;98(8):1048–54.
31. Knot HJ, Nelson MT. Regulation of membrane potential and diameter by voltage-dependent K^+ channels in rabbit myogenic cerebral arteries. Am J Physiol. 1995;269(1 Pt 2):H348–55.
32. Korovkina VP, England SK. Molecular diversity of vascular potassium channel isoforms. Clin Exp Pharmacol Physiol. 2002;29(4):317–23.
33. Lu Y, Hanna ST, Tang G, Wang R. Contributions of Kv1.2, Kv1.5 and Kv2.1 subunits to the native delayed rectifier K^+ current in rat mesenteric artery smooth muscle cells. Life Sci. 2002;71(12):1465–73.
34. Nelson MT, Quayle JM. Physiological roles and properties of potassium channels in arterial smooth muscle. Am J Physiol. 1995;268(4 Pt 1):C799–822.
35. Yeung SY, Pucovsky V, Moffatt JD, Saldanha L, Schwake M, Ohya S, Greenwood IA. Molecular expression and pharmacological identification of a role for Kv7 channels in murine vascular reactivity. Br J Pharmacol. 2007;151(6):758–70.
36. Miguel-Velado E, Moreno-Dominguez A, Colinas O, Cidad P, Heras M, Perez-Garcia MT, Lopez-Lopez JR. Contribution of Kv channels to phenotypic remodeling of human uterine artery smooth muscle cells. Circ Res. 2005;97(12):1280–7.
37. Vandenberg JI, Perry MD, Perrin MJ, Mann SA, Ke Y, Hill AP. hERG K^+ channels: structure, function, and clinical significance. Physiol Rev. 2012;92(3):1393–478.
38. Ko EA, Park WS, Firth AL, Kim N, Yuan JX, Han J. Pathophysiology of voltage-gated K+ channels in vascular smooth muscle cells: modulation by protein kinases. Prog Biophys Mol Biol. 2010;103(1):95–101.

39. Zhu J, Gomez B, Watanabe I, Thornhill WB. Amino acids in the pore region of Kv1 potassium channels dictate cell-surface protein levels: a possible trafficking code in the Kv1 subfamily. Biochem J. 2005;388(Pt 1):355–62.
40. Li D, Takimoto K, Levitan ES. Surface expression of Kv1 channels is governed by a C-terminal motif. J Biol Chem. 2000;275(16):11597–602.
41. McEwen DP, Li Q, Jackson S, Jenkins PM, Martens JR. Caveolin regulates Kv1.5 trafficking to cholesterol-rich membrane microdomains. Mol Pharmacol. 2008;73(3):678–85.
42. Cogolludo A, Moreno L, Lodi F, Frazziano G, Cobeno L, Tamargo J, Perez-Vizcaino F. Serotonin inhibits voltage-gated K^+ currents in pulmonary artery smooth muscle cells: role of 5-HT2A receptors, caveolin-1, and Kv1.5 channel internalization. Circ Res. 2006;98(7):931–8.
43. Joseph BK, Thakali KM, Pathan AR, Kang E, Rusch NJ, Rhee SW. Postsynaptic density-95 scaffolding of Shaker-type K^+ channels in smooth muscle cells regulates the diameter of cerebral arteries. J Physiol. 2011;589(Pt 21):5143–52.
44. Moore CL, Nelson PL, Parelkar NK, Rusch NJ, Rhee SW. Protein kinase A-phosphorylated Kv1 channels in PSD95 signaling complex contribute to the resting membrane potential and diameter of cerebral arteries. Circ Res. 2014;114(8):1258–67.
45. McEwen DP, Schumacher SM, Li Q, Benson MD, Iniguez-Lluhi JA, Van Genderen KM, Martens JR. Rab-GTPase-dependent endocytic recycling of Kv1.5 in atrial myocytes. J Biol Chem. 2007;282(40):29612–20.
46. Zadeh AD, Xu H, Loewen ME, Noble GP, Steele DF, Fedida D. Internalized Kv1.5 traffics via Rab-dependent pathways. J Physiol. 2008;586(Pt 20):4793–813.
47. Dietrich HH, Dacey Jr RG. Molecular keys to the problems of cerebral vasospasm. Neurosurgery. 2000;46(3):517–30.
48. Ishiguro M, Morielli AD, Zvarova K, Tranmer BI, Penar PL, Wellman GC. Oxyhemoglobin-induced suppression of voltage-dependent K^+ channels in cerebral arteries by enhanced tyrosine kinase activity. Circ Res. 2006;99(11):1252–60.
49. Amberg GC, Rossow CF, Navedo MF, Santana LF. NFATc3 regulates Kv2.1 expression in arterial smooth muscle. J Biol Chem. 2004;279(45):47326–34.
50. Zhong XZ, Abd-Elrahman KS, Liao CH, El-Yazbi AF, Walsh EJ, Walsh MP, Cole WC. Stromatoxin-sensitive, heteromultimeric Kv2.1/Kv9.3 channels contribute to myogenic control of cerebral arterial diameter. J Physiol. 2010;588(Pt 22):4519–37.
51. Mohapatra DP, Siino DF, Trimmer JS. Interdomain cytoplasmic interactions govern the intracellular trafficking, gating, and modulation of the Kv2.1 channel. J Neurosci. 2008;28(19):4982–94.
52. Bae SH, Kim DH, Shin SK, Choi JS, Park KS. Src regulates membrane trafficking of the Kv3.1b channel. FEBS Lett. 2014;588(1):86–91.
53. Gu Y, Barry J, Gu C. Kv3 channel assembly, trafficking and activity are regulated by zinc through different binding sites. J Physiol. 2013;591(Pt 10):2491–507.
54. Torres YP, Granados ST, Latorre R. Pharmacological consequences of the coexpression of BK channel alpha and auxiliary beta subunits. Front Physiol. 2014;5:383.
55. Nelson MT, Brayden JE. Regulation of arterial tone by calcium-dependent K^+ channels and ATP-sensitive K^+ channels. Cardiovasc Drugs Ther. 1993;7 Suppl 3:605–10.
56. Jaggar JH, Wellman GC, Heppner TJ, Porter VA, Perez GJ, Gollasch M, Kleppisch T, Rubart M, Stevenson AS, Lederer WJ, Knot HJ, Bonev AD, Nelson MT. Ca^{2+} channels, ryanodine receptors and Ca^{2+}-activated K^+ channels: a functional unit for regulating arterial tone. Acta Physiol Scand. 1998;164(4):577–87.
57. Toro L, Li M, Zhang Z, Singh H, Wu Y, Stefani E. MaxiK channel and cell signalling. Pflugers Arch. 2014;466(5):875–86.
58. Evanson KW, Bannister JP, Leo MD, Jaggar JH. LRRC26 is a functional BK channel auxiliary gamma subunit in arterial smooth muscle cells. Circ Res. 2014;115(4):423–31.
59. Yan J, Aldrich RW. LRRC26 auxiliary protein allows BK channel activation at resting voltage without calcium. Nature. 2010;466(7305):513–6.

60. Poulsen AN, Wulf H, Hay-Schmidt A, Jansen-Olesen I, Olesen J, Klaerke DA. Differential expression of BK channel isoforms and beta-subunits in rat neuro-vascular tissues. Biochim Biophys Acta. 2009;1788(2):380–9.
61. Kyle BD, Braun AP. The regulation of BK channel activity by pre- and post-translational modifications. Front Physiol. 2014;5:316.
62. Singh H, Stefani E, Toro L. Intracellular BK_{Ca} (iBK_{Ca}) channels. J Physiol. 2012;590(Pt 23):5937–47.
63. Zarei MM, Zhu N, Alioua A, Eghbali M, Stefani E, Toro L. A novel MaxiK splice variant exhibits dominant-negative properties for surface expression. J Biol Chem. 2001;276(19):16232–9.
64. Xie J, McCobb DP. Control of alternative splicing of potassium channels by stress hormones. Science. 1998;280(5362):443–6.
65. Tian L, Duncan RR, Hammond MS, Coghill LS, Wen H, Rusinova R, Clark AG, Levitan IB, Shipston MJ. Alternative splicing switches potassium channel sensitivity to protein phosphorylation. J Biol Chem. 2001;276(11):7717–20.
66. Nourian Z, Li M, Leo MD, Jaggar JH, Braun AP, Hill MA. Large conductance Ca^{2+}-activated K^+ channel (BK_{Ca}) alpha-subunit splice variants in resistance arteries from rat cerebral and skeletal muscle vasculature. PLoS One. 2014;9(6):e98863.
67. Jaggar JH, Li A, Parfenova H, Liu J, Umstot ES, Dopico AM, Leffler CW. Heme is a carbon monoxide receptor for large-conductance Ca^{2+}-activated K^+ channels. Circ Res. 2005;97(8):805–12.
68. Chen L, Jeffries O, Rowe IC, Liang Z, Knaus HG, Ruth P, Shipston MJ. Membrane trafficking of large conductance calcium-activated potassium channels is regulated by alternative splicing of a transplantable, acidic trafficking motif in the RCK1-RCK2 linker. J Biol Chem. 2010;285(30):23265–75.
69. Wang SX, Ikeda M, Guggino WB. The cytoplasmic tail of large conductance, voltage- and Ca^{2+}-activated K^+ (MaxiK) channel is necessary for its cell surface expression. J Biol Chem. 2003;278(4):2713–22.
70. Tian L, McClafferty H, Knaus HG, Ruth P, Shipston MJ. Distinct acyl protein transferases and thioesterases control surface expression of calcium-activated potassium channels. J Biol Chem. 2012;287(18):14718–25.
71. Shipston MJ. Regulation of large conductance calcium- and voltage-activated potassium (BK) channels by S-palmitoylation. Biochem Soc Trans. 2013;41(1):67–71.
72. Piao L, Ho WK, Earm YE. Actin filaments regulate the stretch sensitivity of large-conductance, Ca^{2+}-activated K^+ channels in coronary artery smooth muscle cells. Pflugers Arch. 2003;446(5):523–8.
73. Zou S, Jha S, Kim EY, Dryer SE. A novel actin-binding domain on Slo1 calcium-activated potassium channels is necessary for their expression in the plasma membrane. Mol Pharmacol. 2008;73(2):359–68.
74. Alioua A, Lu R, Kumar Y, Eghbali M, Kundu P, Toro L, Stefani E. Slo1 caveolin-binding motif, a mechanism of caveolin-1-Slo1 interaction regulating Slo1 surface expression. J Biol Chem. 2008;283(8):4808–17.
75. Toro B, Cox N, Wilson RJ, Garrido-Sanabria E, Stefani E, Toro L, Zarei MM. KCNMB1 regulates surface expression of a voltage and Ca^{2+}-activated K^+ channel via endocytic trafficking signals. Neuroscience. 2006;142(3):661–9.
76. Kim EY, Zou S, Ridgway LD, Dryer SE. Beta1-subunits increase surface expression of a large-conductance Ca^{2+}-activated K^+ channel isoform. J Neurophysiol. 2007;97(5):3508–16.
77. Leo MD, Bannister JP, Narayanan D, Nair A, Grubbs JE, Gabrick KS, Boop FA, Jaggar JH. Dynamic regulation of beta1 subunit trafficking controls vascular contractility. Proc Natl Acad Sci U S A. 2014;111(6):2361–6.
78. Zerial M, McBride H. Rab proteins as membrane organizers. Nat Rev Mol Cell Biol. 2001;2(2):107–17.
79. Loirand G, Sauzeau V, Pacaud P. Small G proteins in the cardiovascular system: physiological and pathological aspects. Physiol Rev. 2013;93(4):1659–720.

80. Sokolowski S, Harvey M, Sakai Y, Jordan A, Sokolowski B. The large conductance calcium-activated K^+ channel interacts with the small GTPase Rab11b. Biochem Biophys Res Commun. 2012;426(2):221–5.
81. Earley S, Brayden JE. Transient receptor potential channels and vascular function. Clin Sci (Lond). 2010;119(1):19–36.
82. Pani B, Ong HL, Brazer SC, Liu X, Rauser K, Singh BB, Ambudkar IS. Activation of TRPC1 by STIM1 in ER-PM microdomains involves release of the channel from its scaffold caveolin-1. Proc Natl Acad Sci U S A. 2009;106(47):20087–92.
83. Cheng KT, Liu X, Ong HL, Swaim W, Ambudkar IS. Local Ca^{2+} entry via Orai1 regulates plasma membrane recruitment of TRPC1 and controls cytosolic Ca^{2+} signals required for specific cell functions. PLoS Biol. 2011;9(3):e1001025.
84. Adebiyi A, Thomas-Gatewood CM, Leo MD, Kidd MW, Neeb ZP, Jaggar JH. An elevation in physical coupling of type 1 inositol 1,4,5-trisphosphate (IP_3) receptors to transient receptor potential 3 (TRPC3) channels constricts mesenteric arteries in genetic hypertension. Hypertension. 2012;60(5):1213–9.
85. Adebiyi A, Zhao G, Narayanan D, Thomas-Gatewood CM, Bannister JP, Jaggar JH. Isoform-selective physical coupling of TRPC3 channels to IP_3 receptors in smooth muscle cells regulates arterial contractility. Circ Res. 2010;106(10):1603–12.
86. Lockwich T, Pant J, Makusky A, Jankowska-Stephens E, Kowalak JA, Markey SP, Ambudkar IS. Analysis of TRPC3-interacting proteins by tandem mass spectrometry. J Proteome Res. 2008;7(3):979–89.
87. Bucci C, Thomsen P, Nicoziani P, McCarthy J, van Deurs B. Rab7: a key to lysosome biogenesis. Mol Biol Cell. 2000;11(2):467–80.
88. Singh BB, Lockwich TP, Bandyopadhyay BC, Liu X, Bollimuntha S, Brazer SC, Combs C, Das S, Leenders AG, Sheng ZH, Knepper MA, Ambudkar SV, Ambudkar IS. VAMP2-dependent exocytosis regulates plasma membrane insertion of TRPC3 channels and contributes to agonist-stimulated Ca^{2+} influx. Mol Cell. 2004;15(4):635–46.
89. Hong C, Kim J, Jeon JP, Wie J, Kwak M, Ha K, Kim H, Myeong J, Kim SY, Jeon JH, So I. Gs cascade regulates canonical transient receptor potential 5 (TRPC5) through cAMP mediated intracellular Ca^{2+} release and ion channel trafficking. Biochem Biophys Res Commun. 2012;421(1):105–11.
90. Xu SZ, Muraki K, Zeng F, Li J, Sukumar P, Shah S, Dedman AM, Flemming PK, McHugh D, Naylor J, Cheong A, Bateson AN, Munsch CM, Porter KE, Beech DJ. A sphingosine-1-phosphate-activated calcium channel controlling vascular smooth muscle cell motility. Circ Res. 2006;98(11):1381–9.
91. Hong C, Kwak M, Myeong J, Ha K, Wie J, Jeon JH, So I. Extracellular disulfide bridges stabilize TRPC5 dimerization, trafficking, and activity. Pflugers Arch. 2014;27.
92. Monet M, Francoeur N, Boulay G. Involvement of phosphoinositide 3-kinase and PTEN protein in mechanism of activation of TRPC6 protein in vascular smooth muscle cells. J Biol Chem. 2012;287(21):17672–81.
93. Lockwich T, Singh BB, Liu X, Ambudkar IS. Stabilization of cortical actin induces internalization of transient receptor potential 3 (Trp3)-associated caveolar Ca^{2+} signaling complex and loss of Ca^{2+} influx without disruption of Trp3-inositol trisphosphate receptor association. J Biol Chem. 2001;276(45):42401–8.
94. Lockwich TP, Liu X, Singh BB, Jadlowiec J, Weiland S, Ambudkar IS. Assembly of Trp1 in a signaling complex associated with caveolin-scaffolding lipid raft domains. J Biol Chem. 2000;275(16):11934–42.
95. Earley S. Vanilloid and melastatin transient receptor potential channels in vascular smooth muscle. Microcirculation. 2010;17(4):237–49.
96. Earley S, Pauyo T, Drapp R, Tavares MJ, Liedtke W, Brayden JE. TRPV4-dependent dilation of peripheral resistance arteries influences arterial pressure. Am J Physiol Heart Circ Physiol. 2009;297(3):H1096–102.
97. Filosa JA, Yao X, Rath G. TRPV4 and the regulation of vascular tone. J Cardiovasc Pharmacol. 2013;61(2):113–9.

98. Lei L, Cao X, Yang F, Shi DJ, Tang YQ, Zheng J, Wang K. A TRPV4 channel C-terminal folding recognition domain critical for trafficking and function. J Biol Chem. 2013;288(15):10427–39.
99. Ma X, Cao J, Luo J, Nilius B, Huang Y, Ambudkar IS, Yao X. Depletion of intracellular Ca^{2+} stores stimulates the translocation of vanilloid transient receptor potential 4-c1 heteromeric channels to the plasma membrane. Arterioscler Thromb Vasc Biol. 2010;30(11):2249–55.
100. Narayanan D, Bulley S, Leo MD, Burris SK, Gabrick KS, Boop FA, Jaggar JH. Smooth muscle cell transient receptor potential polycystin-2 (TRPP2) channels contribute to the myogenic response in cerebral arteries. J Physiol. 2013;591(Pt 20):5031–46.
101. Sharif-Naeini R, Folgering JH, Bichet D, Duprat F, Lauritzen I, Arhatte M, Jodar M, Dedman A, Chatelain FC, Schulte U, Retailleau K, Loufrani L, Patel A, Sachs F, Delmas P, Peters DJ, Honore E. Polycystin-1 and -2 dosage regulates pressure sensing. Cell. 2009;139(3):587–96.
102. Nauli SM, Alenghat FJ, Luo Y, Williams E, Vassilev P, Li X, Elia AE, Lu W, Brown EM, Quinn SJ, Ingber DE, Zhou J. Polycystins 1 and 2 mediate mechanosensation in the primary cilium of kidney cells. Nat Genet. 2003;33(2):129–37.
103. Earley S. TRPM4 channels in smooth muscle function. Pflugers Arch. 2013;465(9):1223–31.
104. Crnich R, Amberg GC, Leo MD, Gonzales AL, Tamkun MM, Jaggar JH, Earley S. Vasoconstriction resulting from dynamic membrane trafficking of TRPM4 in vascular smooth muscle cells. Am J Physiol Cell Physiol. 2010;299(3):C682–94.
105. He Y, Yao G, Savoia C, Touyz RM. Transient receptor potential melastatin 7 ion channels regulate magnesium homeostasis in vascular smooth muscle cells: role of angiotensin II. Circ Res. 2005;96(2):207–15.
106. Oancea E, Wolfe JT, Clapham DE. Functional TRPM7 channels accumulate at the plasma membrane in response to fluid flow. Circ Res. 2006;98(2):245–53.

Chapter 8
Abnormalities of Vascular Ion Channels During Hypertension

Anup K. Srivastava, Lee Ann MacMillan-Crow, Sung W. Rhee, and Nancy J. Rusch

Abstract Homeostasis of vascular tone in small arteries and arterioles relies on the balance between Ca^{2+}-dependent activation of vascular smooth muscle cells (VSMCs) and the vasodilator influences that counteract it. During the development of hypertension, the homeostasis of vascular tone is disrupted as pathophysiological events impinging on VSMCs promote vasoconstriction, rises in peripheral vascular resistance, and elevation of blood pressure. One critical process that contributes to hypertension is the "remodeling" of ion channels in the plasma membrane of VSMCs, which increases cell excitability. Most studies suggest that elevation of blood pressure is associated with a profound loss of voltage-gated K^+ channels encoded by several gene families. The loss of K^+ channels causes depolarization of VSMCs, opening of voltage-sensitive Ca^{2+} channels, Ca^{2+} influx and accentuated vasoconstriction. In tandem, even short-term elevations of blood pressure promote the expression of voltage-sensitive Ca^{2+} channels in VSMCs, effectively increasing the number of pathways for Ca^{2+} influx and Ca^{2+}-dependent activation of contractile proteins. Drawing on findings from many laboratories, this chapter will review evidence for the induction of abnormalities of vascular K^+ and Ca^{2+} channels during hypertension and discuss mechanisms that may mediate this process. New areas of investigation that may spur us to reconsider current dogma or provide important clues to the process of ion channel remodeling during hypertension also will be discussed.

Keywords Ion channels • Potassium channels • Calcium channels • Membrane potential • Vascular smooth muscle • Hypertension • Pulmonary hypertension • Vascular reactivity • Electrophysiology

A.K. Srivastava • L.A. MacMillan-Crow • S.W. Rhee • N.J. Rusch, Ph.D. (✉)
Department of Pharmacology and Toxicology, University of Arkansas for Medical Sciences, 4301 W. Markham Street, Little Rock, AR 72205-7199, USA
e-mail: nrusch@uams.edu

I. Levitan, A.M. Dopico (eds.), *Vascular Ion Channels in Physiology and Disease*, DOI 10.1007/978-3-319-29635-7_8

Introduction

The diameter of small arteries and arterioles is positively regulated by the concentration of free cytosolic calcium ($[Ca^{2+}_i]$) in the vascular smooth muscle cells (VSMCs). Many Ca^{2+}-permeable ion channels contribute to rises in $[Ca^{2+}_i]$ during excitation of VSMCs, and thereby contribute to vascular tone and blood pressure. Some of these ion channels include the L-type Ca^{2+} channels that mediate voltage-dependent Ca^{2+} influx, inositol triphosphate receptors (IP_3R) that mediate release of Ca^{2+} from the sarcoplasmic reticulum, and receptor-operated channels that mediate agonist-induced cation influx. The Ca^{2+} mobilization in VSMCs that mediates contraction is counteracted under normal conditions by Ca^{2+} uptake mechanisms and by the opening of K^+ channels, which mediate K^+ efflux, hyperpolarization of the plasma membrane, closure of voltage-gated Ca^{2+} channels, reduced $[Ca^{2+}_i]$, and ultimately, VSMC relaxation and arterial dilation. It has been recognized for more than two decades that the delicate balance between the ionic pathways that positively and negatively regulate arterial tone is disrupted during hypertension and that "remodeling of vascular ion channels" leads to pathogenic vasoconstriction [1–3]. Thus, an increased peripheral vascular resistance that relies on augmented vascular tone is a hallmark finding of experimental and human forms of hypertension. Primary remodeling events of vascular ion channels during hypertension include a loss of K^+ channels, resulting in VSMC depolarization and excitation. Concurrently, several types of Ca^{2+}-permeable ion channels become over-abundant early in the pathogenesis of hypertension. In particular, the long-lasting ("L-type") Ca^{2+} channels in the plasma membrane of VSMCs, which mediate the sustained voltage-dependent Ca^{2+} influx required for tonic vascular contraction, upregulate during the development of hypertension (Fig. 8.1). This chapter will review our current knowledge of vascular ion channel remodeling and its contribution to elevation of blood pressure during chronic hypertension with a focus on Ca^{2+} and K^+ channel abnormalities in VSMCs. We will evaluate the growing knowledge of cellular mechanisms underlying vascular ion channel remodeling, and comment on future investigations that may help us to more fully understand the cellular processes by which ion channel dysregulation results in abnormal vascular tone and hypertension.

Depolarization of VSMCs as a Defect of Hypertension

A less negative resting membrane potential (E_m) associated with a heightened excitability of VSMCs appears to be a fundamental defect of hypertension, which is shared by many forms of the disease [4–10]. The abnormal depolarization of VSMCs during hypertension was discovered more than 30 years ago, when microelectrode measurements of resting E_m in isolated and *in-situ* arteries from several rat models of hypertension revealed that VSMCs of hypertensive animals were

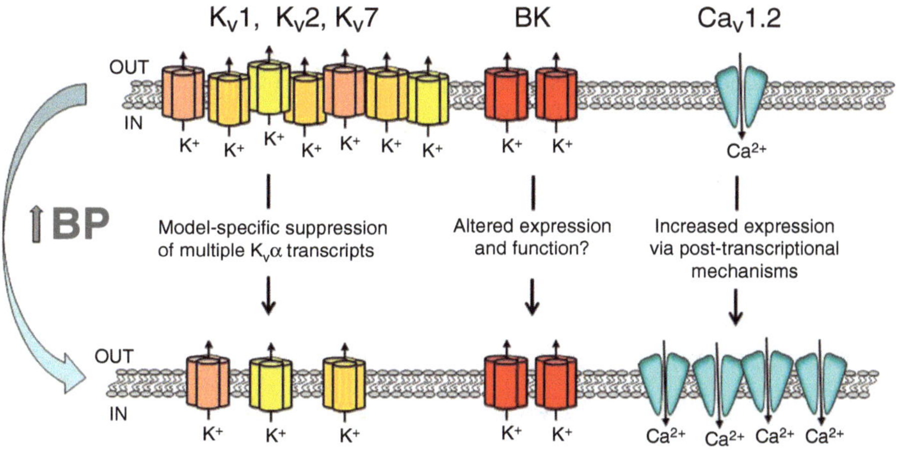

Fig. 8.1 Depiction of changes in K^+ channel and Ca^{2+} channel expression in plasma membranes of VSMCs during hypertension. The expression levels of three families of voltage-gated K^+ (K_v) channels including *Shaker*-type K_v1, *Shab*-like K_v2 and *KCNQ*-encoded K_v7 channels are reduced in VSMCs exposed to elevated blood pressure, a change associated with loss of the corresponding transcripts. The fate in VSMCs of the high-conductance, Ca^{2+}-sensitive K^+ channel during hypertension is unclear, and both increased and decreased expression and function of BK channels have been reported in arteries of hypertensive animals. In contrast, the expression and function of voltage-sensitive "L-type" Ca^{2+} ($Ca_v1.2$) channels are consistently reported to be increased in VSMCs during hypertension, resulting in enhanced Ca^{2+}-dependent vascular tone

depolarized compared to those of age-matched controls. For example, *in-situ* mesenteric arteries of spontaneously hypertensive rats (SHR) showed a resting E_m of −38 mV compared to −43 mV in similar arteries of normotensive Wistar-Kyoto (WKY) rats [6]. Similarly, *in-situ* mesenteric and gracilis arteries from rats with salt-induced hypertension exhibited a resting E_m of −40 mV compared to −44 mV in arteries of control rats [7]. Two endogenous influences associated with hypertension, namely enhanced sympathetic neural output and increased intraluminal pressure, were observed to promote depolarization of native VSMCs, resulting in the opening of voltage-gated Ca^{2+} channels, Ca^{2+} influx and vasoconstriction [5, 8]. However, even VSMCs in isolated arterial segments of hypertensive rats, which were free of variable native influences, exhibited a depolarized resting E_m compared to arteries of normotensive animals [9, 10]. For example, average resting E_m values were −43 mV and −55 mV in isolated cerebral arteries from SHR and WKY rats, respectively [11]. Similarly, the resting E_m of VSMCs is less negative in isolated renal and mesenteric arteries from hypertensive rats compared to normotensive controls [9, 10].

Since the resting E_m of VSMCs primarily relies on a high resting K^+ permeability of the plasma membrane, which allows K^+ efflux to establish a negative transmembrane potential, the abnormal depolarization of VSMCs during hypertension was attributed to an inactivation or a loss of plasma membrane-delineated K^+ channels. It was recognized later that VSMCs express a preponderance of voltage-gated K^+

(K_V) channels and high-conductance Ca^{2+}-activated K^+ (BK) channels [1–3, 12–14]. Thus, subsequent studies of vascular K^+ channel abnormalities in hypertension focused on identifying changes in the expression and properties of these two gene families of K^+ channels. The following sections of this chapter will review these findings, which provide compelling evidence of K^+ channel remodeling in VSMCs as a prominent feature of hypertension.

Loss of Voltage-Gated (K_v) Channels Implicated in VSMC Depolarization

The critical event associated with depolarization of VSMCs during chronic hypertension is the inactivation and/or loss of several gene families of K_v channels, including *Shaker*-like (K_v1), *Shab*-like (K_v2) and *KCNQ* (K_v7). These families of K^+ channels, and others not discussed here [12–14], collectively contribute to the negative resting E_m that suppresses the excitability of VSMCs. A relatively small body of literature implicates K_v channel inactivation in VSMCs as an important mechanism of membrane depolarization and vascular activation during hypertension. Patch-clamp studies have directly demonstrated that elevated [$Ca^{2+}{}_i$] can inactivate K_v channels in VSMCs [15, 16], but it has not been concluded that this event occurs during hypertension. However, it is hypothesized that a transient rise in blood pressure, which is associated with stretch-induced depolarization and increased [$Ca^{2+}{}_i$] in VSMCs, may inactivate K_v channels, resulting in sustained membrane depolarization and vasoconstriction [16]. Similarly, it is also possible that other signaling molecules modified in hypertension, inhibit K_v channel function. For example, PKC signaling associated with vascular activation also inhibits K_V channels in VSMCs from several vascular beds [17–19].

With few exceptions [20], there is abundant evidence that the abnormal depolarization of VSMCs during hypertension relies on the loss of at least several K_v channel gene families. For example, voltage-dependent K^+ current is markedly attenuated in patch-clamped VSMCs from genetic, renal and nitric oxide-deficient hypertensive rats compared to normotensive controls [11, 21, 23]. In these studies, the solution in the patch pipette dialyzing the VSMCs contained low or nominally zero calcium. Thus, the reduced K^+ current in VSMCs of the hypertensive rats was attributed to a loss of K_v channels rather than K_v channel inactivation. Additional studies by our laboratories have implicated a loss of *Shaker*-type K_v1 channels as a cause of K^+ current deficit, depolarization and elevated vascular tone in cerebral arteries of SHR and aortic-banded rats with a renal form of hypertension [11]. As reviewed elsewhere [3], the pore-forming structure of K_v channels is composed of four α-subunits ($K_v\alpha$) that may represent different isoforms. For example, the $K_v\alpha1.2$ and $K_v\alpha1.5$ proteins appear to co-assemble to form the pore structure of *Shaker*-type K_v1 channels in VSMCs [24–27]. Based on this knowledge, we confirmed a loss of these two specific α-subunits in arteries from genetic and aortic-banded hypertensive rats [11]. The corresponding transcripts coding for the two pore-forming proteins, $K_v\alpha 1.2$ and $K_v\alpha 1.5$, also were deficient in the arteries of aortic-banded hypertensive rats, whereas only the $K_v\alpha 1.2$ transcript was reduced in arteries of SHR. Mesenteric arteries of *N*ω-nitro-L-arginine-induced hypertensive rats also exhibited decreased expression of $K_v\alpha 1.5$ concomitant with K_v channel-mediated K^+ current [22, 23]. Collectively, these findings implicate reduced gene expression of K_v1 channels as a mechanism of K^+ channel loss and VSMC depolarization during hypertension. They also emphasize that the specific α-subunits targeted for loss in VSMCs appear to vary between different experimental models of hypertension.

Interestingly, although the pulmonary circulation is characterized by a radically lower arterial pressure compared to the systemic circulation, the abnormality of K_v1 channel loss also is a component of pulmonary hypertension. Potassium current attributed to K_v1 channels is attenuated in VSMCs from rats with hypoxia-induced pulmonary hypertension, and this abnormality is associated with a loss of $K_v\alpha 1.2$ and $K_v\alpha 1.5$ transcript and protein [28]. A similar loss of K_v1 channel proteins and transcripts was reported in pulmonary arteries of humans with primary pulmonary arterial hypertension [29]. Notably, the loss of K_v1 channels associated with vasoconstriction may serve unique physiological functions in different vascular beds. During pulmonary hypertension, the vasoconstriction of the pulmonary circulation mediated by loss of K_v1 channels may initially avert ventilation-perfusion mismatching prior to the pathogenic rise in pulmonary arterial pressure. In systemic hypertension, vasoconstriction of the cerebral, coronary and renal circulations mediated by K_v1 channel loss may help to buffer the transmission of the high systemic pressure to the microcirculation of the brain, heart and kidneys, respectively, to prevent pressure-induced injury to the fragile capillary beds. However, ultimately, the loss of K_v channels in VSMCs of systemic resistance arteries and arterioles will elevate peripheral vascular resistance and drive increases in systemic blood pressure, thereby contributing to the development of hypertension and its lethal pattern of end organ damage that includes stroke, renal failure and cardiac disease.

In addition to the loss of vascular K_v1 channels during hypertension, a reduced abundance of other K_v channel types has been implicated in the abnormal vasoconstriction that underlies hypertension. Initial findings infer a loss of *Shab*-type K_v2 channels in VSMCs of angiotensin II (Ang II)-infused hypertensive rats [30]. The isolated cerebral arteries of these animals reveal a diminished vasodilator contribution of K_v2 channels to basal diameter and isolated VSMCs exhibit a lower density of K_v2 channel-mediated K^+ current [30]. The loss of arterial K_v2 channels during hypertension was attributed to activation of the Ca^{2+}-dependent transcription factor NFATc3 enabled by enhanced Ca^{2+} influx through L-type Ca^{2+} channels [30]. More recent studies suggest that the abundance of K_V7 (KCNQ) channels also is reduced in systemic arteries of SHR and in Ang II-infused hypertensive mice as an additional mechanism of VSMC excitation and enhanced arterial tone dependent on K^+ channel loss [31, 32]. Collectively, the loss of K_v channels from different gene families results in reduced voltage-gated K^+ efflux, which promotes VSMC depolarization and heightened vasoconstriction of the affected resistance arteries (Fig. 8.2).

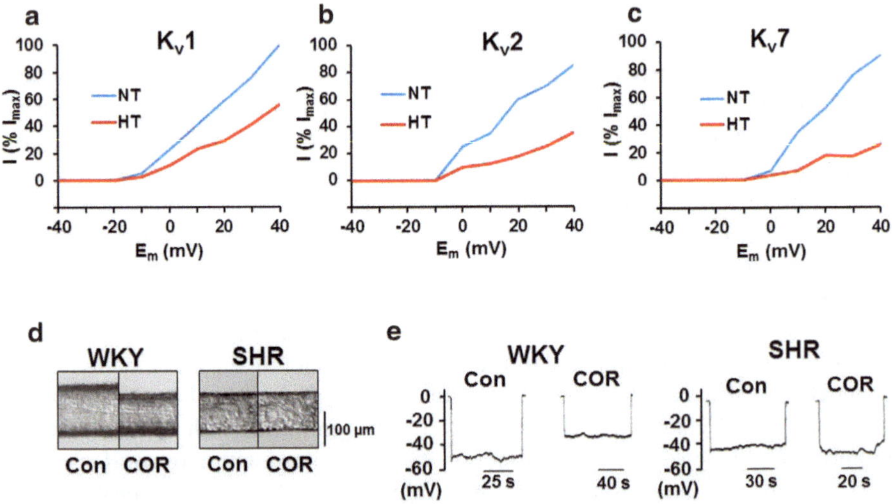

Fig. 8.2 Loss of voltage-gated (K_v) channels in VSMCs is implicated as a mechanism of membrane depolarization and increased vascular tone in hypertension. Current-voltage relationships of (**a**) *Shaker*-type K_v1 channels, (**b**) *Shab*-type K_v2 channels and (**c**) *KCNQ*-encoded K_v7 channels in VSMCs of normotensive (NT) and hypertensive (HT) animals modeled on data presented in [11, 30, 31], respectively. A reduced current density was evident in VSMCs of HT animals. (**d**) Effect of correolide (COR), a K_v1 channel inhibitor, on the resting diameter of isolated, pressurized (80 mm Hg) WKY and SHR cerebral arteries. COR (1 μmol/l) elicited constriction of the WKY artery, suggesting a dilator influence of K_v1 channels. In contrast, COR failed to constrict the SHR artery, suggesting a minimal contribution of K_v1 channels to resting diameter. (**e**) Effect of COR on resting membrane potential (E_m) of WKY and SHR cerebral arteries. Microelectrode recordings of E_m were obtained under control (Con) conditions and after COR (1 μmol/l) was added to the superfusate. The WKY artery exhibited a more negative E_m value compared to the SHR artery, suggesting higher membrane K^+ permeability in cerebral VSMCs of WKY. Only arteries of WKY rats depolarized in response to K_v1 channel inhibition by COR, suggesting a hyperpolarizing influence of K_v1 channels in WKY but not SHR arteries. Recordings of E_m and diameter modified from Figs. 2 and 4 of [11]

In summary, not all studies have confirmed a loss of vascular K_v channels during hypertension [33], but most reports concur that K_v channel abundance and function are reduced in VSMCs of hypertensive animal and human subjects, resulting in depolarization and activation of VSMCs. The underlying mechanism appears to involve a down-regulation of one or more α-subunits composing the pore-forming structures of the K_v1, K_v2 and K_v7 channel families, which are encoded by the *Shaker*, *Shab* and *KCNQ* genes, respectively. However, even within the same K_v channel family (i.e., K_v1), the precise α-subunit transcript in VSMCs that is attenuated during hypertension may differ between experimental models of hypertension, suggesting that the complex pathways that mediate K_v channel loss during hypertension have not been fully identified. Future research into facets of K_v channel regulation for which we have limited knowledge may pinpoint additional events that compromise K_v channel-mediated vasodilation during hypertension. For example, the precise cellular stimuli that trigger the concomitant remodeling of several vascular K_v channel families during the development of hypertension are unknown, although pharmacological inhibition of this event could potentially prevent K_v channel remodeling and its apparent contribution to the development of hypertension. The role and fate of ancillary $K_v\beta$ subunits during hypertension also are unclear, although these small subunits are important regulators of K_v channel function. The $K_v\alpha$ subunits co-assemble in 1:1 stoichiometry with $K_v\beta$ subunits, with the latter assuming responsibility for modulating the kinetics of the fully functional K_v channel and its responsiveness to signaling molecules [34–37]. To date, only a single report has suggested that the expression levels of $K_v\beta$ transcripts are altered in mesenteric arteries of hypertensive rats [33]. Finally, the cellular processes that determine the biogenesis, trafficking and half-life of K_v channels in VSMCs have received little attention. A recent study suggests that the vasodilator influence of K_v1 channels in VSMCs relies on their localization at the plasma membrane by the molecular scaffolding protein, post-synaptic density 95 (PSD95) [38, 39], thereby revealing another new checkpoint at which K_v channel expression and function may be modulated in the vasculature during cardiovascular disease states.

Abnormalities of High-Conductance Ca^{2+}-Sensitive K^+ (BK) Channels

In contrast to the growing body of evidence implicating loss of K_v channels as a contributor to the enhanced vascular tone of hypertension, the fate of high-conductance, Ca^{2+}-sensitive K^+ (BK) channels during hypertension remains a matter of debate. BK channels encoded by the Slo1 (*KCNMA1*) gene are densely expressed in the plasma membrane of VSMCs in 1:1 stoichiometry with regulatory BKβ1 subunits. The latter ancillary subunits increase the Ca^{2+}-sensitivity of the pore-forming BKα subunit. Membrane depolarization and increases in $[Ca^{2+}{}_i]$ synergistically activate BK channels to promote K^+ efflux as a homeostatic mechanism to counteract arterial contraction and restore the resting E_m in VSMCs [1, 3, 12–14]. Thus, BK channels provide a molecular "brake" to limit the development of excessive vascular tone. At face value, it would seem that the fate of the vascular BK channel during cardiovascular diseases including hypertension would be simple to evaluate, since the four α-subunits (BKα) that assemble to form its pore-forming structure are encoded by the same *Slo* gene, inferring a homogenous population of BK channels in the plasma membrane [40]. However, a high level of phenotypic diversity is generated by alternative splicing to establish novel populations of BK channels in different cell types [41]. For example, smooth muscle expresses BK channels that exhibit unique Ca^{2+}-sensitive, mechano-sensitive and oxygen-sensitive properties [42–46]. Although the precise distribution of different BKα isoforms in the vasculature has not been defined, it is appreciated that the properties of BK channels including its hallmark feature of Ca^{2+}-sensitivity, differ between VSMCs of different vascular beds [47–49], prohibiting a simplistic correlation between BK channel expression and functional impact on vascular tone. Additionally, as discussed later, there is growing evidence that BK channels are expressed in intracellular organelles in addition to the plasma membrane.

A number of laboratories have explored the fate of BK channels in arteries from different rat models of hypertension. Most authors [1, 51–54] report that the expression level of BK channel α-subunits (BKα) in small arteries and arterioles increases during hypertension in the systemic, cerebral and coronary circulations. A large number of studies have demonstrated a higher membrane density of BK channel-mediated K^+ current in freshly isolated VSMCs from many vascular beds of genetic, renal and salt-sensitive rat models of hypertension [21, 50–58]. Similarly, single-channel studies in freshly isolated VSMCs of SHR reveal a higher number of BK channel openings compared to VSMCs of WKY rats [57–59]. Finally, the higher membrane density of BK channel-mediated K^+ current in VSMCs of SHR is restored to normal after blood pressure is lowered by an antihypertensive drug, implying that BK channel expression positively correlates with blood pressure level [54]. Collectively, these reports imply that an over-abundance of BK channels in VSMCs results in accentuated Ca^{2+}-dependent K^+ efflux during hypertension as a mechanism to counteract further increases in vascular tone and blood pressure. Indeed, isolated segments of cerebral, femoral, renal and mesenteric arteries from genetic, renal and salt-sensitive rats models of hypertension show accentuated contractions to BK channel antagonists, implying that BK channels exert a stronger dilator influence in arteries from hypertensive animals [54, 55, 60–62]. *In-situ* pial arterioles of SHR also profoundly constrict in response to pharmacological inhibition of BK channels, whereas arterioles of WKY rats show only mild diameter reductions [51]. These findings argue that over-expression of vascular BK channels during hypertension represents a protective mechanism to restore K^+ efflux to the VSMCs, and thereby try to negotiate vasodilation and counteract the abnormal vascular tone that develops during hypertension.

It should be acknowledged, however, that several reports suggest an attenuation of vascular BK channel function during hypertension. A loss of BK channel current was reported in VSMCs of SHR and Ang II-infused hypertensive mice compared to normotensive mice [63–65], an abnormality attributed to the reduced expression of ancillary $BK\beta_1$ subunits caused by activation of the calcineurin-NFATc3 signaling pathway [64]. The loss of $BK\beta_1$ subunits was purported to result in low Ca^{2+}-sensitivity of the BK channel and loss of channel function. A reduced expression level of BK channel proteins also was observed in arteries of Nw-nitro-L-arginine hypertensive rats [23]. Finally, BK channel function was reported recently to be attenuated in human VSMCs from hypertensive patients [66]. In the latter study, arteries of hypertensive patients expressed normal levels of BKα pore-forming proteins but revealed a loss of BKβ1 [66]. However, a detailed analysis of the fidelity of anti-BKβ1 commercial antibodies calls into question their specificity and ability to detect the genuine BKβ1 subunit versus other proteins of similar molecular weight in small arteries [67]. Regardless, a role for BKβ1 loss as a cause of BK channel dysfunction and increased vascular tone received credence from initial studies in

BKβ1 knockout mice. Measurements of systolic blood pressure using tail-cuff plethysmography indicated that these mice exhibited a ≥10 mmHg higher mean arterial pressure compared to wild-type mice [68]. However, direct 24-h monitoring of blood pressure by radio-biotelemetry did not detect a difference in any blood pressure parameters between wild-type and $BK\beta_1$ knockout mice [69]. Still other authors report that gene deletion of BKβ1 in mice results in decreased K^+ excretion by the renal collecting tubules, mild aldosteronism and hypertension of renal origin, inferring that the small rise in blood pressure in BKβ1 knockout mice does not primarily rely on compromised vasodilation but rather relates to renal dysfunction [70]. These discrepant findings emphasize the difficulty of trying to compare ion channel abnormalities between different vascular beds of multiple animal models of hypertension, and also the hazards of relying on commercial antibodies and gene knockout mice to detect changes in channel expression and physiological functions.

An additional challenge is that our knowledge of the mechanisms that regulate BK channel expression in VSMCs is sparse. Investigations of molecular mechanisms that regulate BK channel expression are stymied by the fact that functional BK channels are sparse or absent in VSMC lines (i.e., A7r5) and in primary cultures of VSMCs (personal observations). Some benefit has been provided by studies in heterologous expression systems, which suggest that diverse post-transcriptional mechanisms can influence the expression and function of BK channels. For example, S-acylation (palmitoylation) of the BKα subunit at two distinct sites in the N-and C-termini influences channel trafficking and regulation by protein kinases [71–73]. S-acylation of the N-terminus regulates channel trafficking and surface expression, whereas S-acylation of the C-terminus confers channel responsiveness to protein kinases. Additionally, at least some isoforms of BKβ subunits (β4) also undergo S-acylation, which promotes the exit from the endoplasmic reticulum of BKα pore proteins and their subsequent expression in the plasma membrane [71]. These findings suggest that the traditional approach of trying to correlate the expression levels of BK channel transcript and protein may underestimate the complex and dynamic pathways available to modulate the number of functional BK channels in the plasma membrane of VSMCs.

Another confounding factor, which calls into question the assumption that BK channel function should correlate to the number of channel proteins in the plasma membrane, is the discovery of a mitochondrial BK channel (mitoBK) in a glioma cell line in 1999 [74]. Since its discovery, this mitoBK channel has been extensively studied in brain and heart cells [75, 76]. It is well documented that adult cardiomyocytes are devoid of BK channels in their plasma membrane, yet they have abundant mitoBK channels, which apparently exhibit similar biophysical and functional characteristics as their sister channels on the cell surface. Early studies describing the function of the mitochondrial isoform of BK channels were not without controversy since much of the characterization was based on the use of non-selective inhibitors. However, a recent study established that mitoBK channels are encoded by the *Kcnma1* gene, which confers a C-terminal splice insert required for mitochondrial targeting [76]. Although a mitoBK channel has not been identified in VSMCs, the cardioprotective role of this channel highlights its potential importance in cardiovascular disease. Activation of mitoBK channels protects the heart from ischemia/reperfusion by a pre-conditioning influence that may involve suppression of reactive oxygen species (ROS) [77, 78]. Conversely, loss of BK channel function in BKβ1 knockout mice is linked to ROS generation, increased vascular tone and hypertension [78, 79]. Collectively, these findings raise the possibility that vascular BK channels, either plasmalemmal or mitochondrial, may play a role in suppressing ROS as a novel mechanism of blood pressure control. It is widely accepted that ROS levels are increased during chronic hypertension, and ROS (hydrogen peroxide) can oxidize a critical cysteine (Cys 911) in the BKα subunit, leading to its inactivation [80]. Clearly more work is needed to clarify the presence and function of mitoBK channels in VSMCs as potential participants in the myriad of cellular events that regulate vascular tone and blood pressure.

Ultimately, despite contradictory findings related to the abundance of BK channels in VSMCs exposed to high blood pressure, there is overwhelming evidence that membrane depolarization is a feature of VSMCs from hypertensive animals, suggesting a reduced basal level of membrane K^+ conductance. Clearly, the collective result of K^+ channel remodeling during hypertension appears to be a loss of K^+ efflux near the resting membrane potential of VSMCs. This conclusion is not entirely surprising when one considers that K_v1 channels, and particularly K_v7 channels, exhibit higher voltage-sensitivity than BK channels at physiological $[Ca^{2+}{}_i]$ and thereby are presumed to primarily mediate the basal K^+ efflux that establishes the negative membrane potential in VSMCs [3]. In this regard, restoration of expression of the voltage-sensitive K_v1 and K_v7 channels may be a particularly sound strategy for reversing the anomalous depolarization and vasoconstriction of the arterial circulation that contributes to hypertension. On the other hand, enhancing the expression or function of high-conductance BK channels should elicit a strong K^+ current at more positive membrane potentials to deter additional VSMC excitation super-imposed by cell stretch, vasoconstrictor agonists and other excitatory events associated with hypertension.

Upregulation of Vascular Ca^{2+} Channels During Hypertension

There is strong evidence for a contribution of $Ca_v1.2$ channels in VSMCs to the pathogenesis of experimental and human hypertension. As discussed earlier, the abnormal depolarization of VSMCs during hypertension caused by loss of plasmalemmal K_v channels promotes anomalous vasoconstriction by opening voltage-sensitive "L-type" Ca^{2+} ($Ca_v1.2$) channels. In addition to this contribution of $Ca_v1.2$ channels to vascular activation, the expression level of $Ca_v1.2$ channels increases in VSMCs during hypertension as a further mechanism to drive voltage-dependent Ca^{2+} influx and elevated vascular tone. The $Ca_v1.2$ channels in VSMCs are tri-multimeric protein complexes comprised of an α_{1C} pore-forming subunit ($Ca_v\alpha$) that exhibits voltage-sensitivity, and a cytosolic β3 subunit ($Ca_v\beta3$) and surface-delineated $\alpha_2\delta$ protein dimer that regulate the biophysical properties and trafficking of Ca^{2+} channels

[81–84]. The $Ca_v\alpha$ subunit also is the binding site for clinically-used calcium channel antagonists (i.e., calcium channel blockers or CCBs), which represent drugs of choice for blood pressure control in hypertensive patients [85]. The ensuing paragraphs describe four categories of observations implicating vascular $Ca_v1.2$ channels in the pathogenesis of hypertension, which include: (1) observations of abnormal Ca^{2+}-dependent tone in arteries of hypertensive animals, (2) accentuated antihypertensive effects of CCBs in hypertensive compared to normotensive animal and human subjects, (3) increased expression levels of $Ca_v1.2$ subunit proteins in arteries of hypertensive rats and mice, and (4) higher Ca^{2+} current densities mediated by $Ca_v1.2$ channels in VSMCs of hypertensive animals. Figure 8.3 depicts findings of

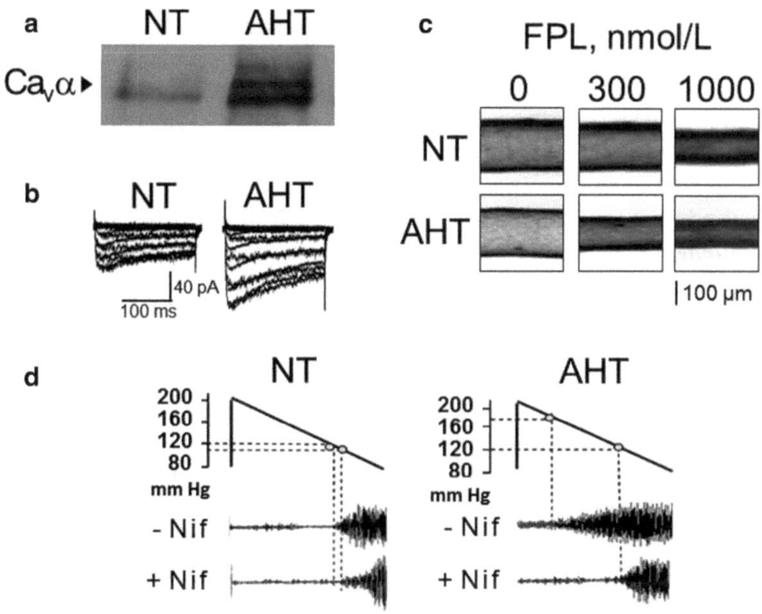

Fig. 8.3 Findings showing enhanced $Ca_v1.2$ channel expression and function in mesenteric VSMCs and arteries from Ang II-infused hypertensive (AHT) mice. (**a**) Western blot reveals that expression of the $Ca_v\alpha$ pore-forming subunit is increased in mesenteric arteries of AHT mice compared to similar arteries from normotensive (NT) mice. (**b**) $Ca_v1.2$ channel–mediated Ca^{2+} currents elicited by progressive 8-mV depolarizing steps from a holding potential of −70 mV are increased in freshly isolated VSMCs of AHT rats compared to NT rats. (**c**) Video-imaging reveals that the vasoconstrictor response to the $Ca_v1.2$ channel agonist, FPL64176 (FPL) is accentuated in an isolated pressurized (80 mmHg) mesenteric artery from an AHT mouse, whereas the mesenteric artery from a NT mouse is less reactive, suggesting fewer functional $Ca_v1.2$ channels in the artery of the normotensive animal. (**d**) Recordings using tail cuff plethysmography to measure systolic blood pressure (SBP) changes in NT and AHT mice in response to acute i.p. administration of the CCB, nifedipine. Top pressures indicate cuff pressure. Lower recordings show return of pulse pressure in the mouse tail artery during cuff deflation; values approximate levels of SBP. Nifedipine failed to significantly lower the control SBP value of ~120 mmHg in the NT mouse, but reduced SBP from ~180 mmHg to ~120 mmHg in the AHT mouse, revealing the dependence of elevated blood pressure on Ca^{2+} influx through $Ca_v1.2$ channels

$Ca_v1.2$ channel over-abundance and evidence of enhanced $Ca_v1.2$ channel function in mesenteric arteries and VSMCs of Ang II-infused hypertensive (AHT) mice compared to normotensive (NT) controls.

First, an accentuated Ca^{2+}-dependent vascular tone reversed by CCBs is observed in isolated and *in situ* arteries of hypertensive rats and mice compared to normotensive animals. Arteries from hypertensive animals also show enhanced contractile responses to calcium channel agonists, including Bay K8644 and FPL64176. Thus, isolated mesenteric and skeletal muscle arteries from SHR develop more Ca^{2+}-dependent spontaneous tone compared to arteries from WKY rats, which can be reversed by application of nifedipine, a specific antagonist of $Ca_v1.2$ channels [86]. Similarly, Ca^{2+} channel agonists elicit strong concentration-dependent contractions in renal arteries from SHR and from aortic-banded hypertensive rats, whereas control arteries exposed to normal blood pressure *in vivo* only weakly respond to the same drugs [10]. The same pattern of over-abundant $Ca_v1.2$ channels and enhanced contractile sensitivity to Ca^{2+} channel agonists is exhibited by isolated, perfused mesenteric arteries from Ang II-infused hypertensive mice [87]. The finding of enhanced Ca^{2+}-dependent vascular tone mediated by $Ca_v1.2$ channels in isolated arteries also extends to arteries *in situ*. Thus, diameters of renal arterioles imaged in kidneys of anesthetized SHR are 18–35 % smaller than in WKY rats, and this difference is abolished by removal of calcium [88]. Similarly, the dose-dependent dilator response of the forearm circulation to the CCB, verapamil, is significantly greater in hypertensive than in normotensive human subjects [89]. In the same sets of subjects, the dilator response to the direct nitric oxide donor, sodium nitroprusside, is reduced in patients with hypertension [89], possibly a reflection of increased arterial stiffness.

The second category of evidence suggesting that $Ca_v1.2$ channels contribute to increased vascular tone during hypertension is the observation that CCBs, including diltiazem, verapamil and dihydropyridine antagonists of $Ca_v1.2$ channels, profoundly lower the elevated blood pressure in hypertensive animal and human subjects, but exert milder antihypertensive effects in subjects with normal blood pressure [89–99]. For example, nifedipine (5 mg/kg p.o.) reduced blood pressure by 22 % of the initial value in normotensive rats, but reduced blood pressure by 49, 44 and 53 % in genetic, renal and salt-sensitive rat models of hypertension [90]. In the same animals, hydralazine, a direct vasodilator of VSMCs, decreased blood pressure similarly between all rat groups [90]. Similarly the hypotensive actions of single and repeated administrations of nifedipine evaluated in normotensive, SHR and renal hypertensive rats revealed a substantial and prolonged reduction of blood pressure after nifedipine (3 mg/kg orally) in the hypertensive animals, which was greater than the blood pressure response of WKY rats [91]. These and other studies [92–96] infer that $Ca_v1.2$ channels partially mediate the increase in vascular tone during experimental hypertension, a finding that extends to humans with essential hypertension [97–99]. For example, proof-of-principle studies supporting the important contribution of vascular $Ca_v1.2$ channels to essential human hypertension were reported by Aoki and colleagues [97], who observed that elevated blood pressure and increased total peripheral resistance in hypertensive patients are reduced to normal levels by CCBs, and this response is attributed to arterial vasodilation.

The third series of observations to implicate vascular $Ca_v1.2$ channels in the pathogenesis of hypertension centers on evidence that $Ca_V1.2$ channel α-subunits ($Ca_v\alpha$) are markedly upregulated in arteries and arterioles during the development of hypertension, suggesting a molecular basis for the increased $Ca_v1.2$ channel-mediated vascular tone observed in hypertensive subjects. Immunoblot analyses reveal that the expression level of the $Ca_v\alpha$ protein is higher in mesenteric, renal and skeletal muscle arteries of SHR compared to WKY rats [10, 86], and $Ca_v\alpha$ expression also is higher in mesenteric arteries of DOCA-salt hypertensive rats compared to control animals [100]. Interestingly, the over-abundant $Ca_v\alpha$ proteins in arteries of SHR represent exon 1-encoded α_{1C} protein, which is a $Ca_v\alpha$ isoform specific to the vasculature and absent in heart, brain and gastrointestinal smooth muscle [101]. This mechanism of alternative splicing activated during hypertension drives the expression of $Ca_v1.2$ channels selectively in the vasculature, whereas $Ca_v1.2$ channel expression in other tissues is unaffected during hypertension. More recent studies have extended findings of $Ca_v1.2$ channel over-abundance to arteries of Ang II-infused hypertensive mice [87]. Ultimately, a definitive, causative link between elevated blood pressure and vascular $Ca_v1.2$ channel expression was established in studies in which the aorta of control rats was banded between the right (upper) and left (lower) renal arteries to immediately establish high and low levels of arterial pressure in the right and left renal circulations, respectively, of the same animal [10]. By 48 h after aortic banding, arteries dissected from the right renal circulation exposed to high pressure exhibited an increased expression of the $Ca_v\alpha$ pore-forming subunit compared to arteries of the left renal circulation exposed to lower pressure [10], implying that even a short-term rise in blood pressure can trigger the expression of vascular $Ca_V1.2$ channels as an early event in the development of hypertension.

A final category of evidence implicating increased $Ca_v1.2$ channel function in the pathogenesis of hypertension consists of observations that $Ca_v1.2$ channel-mediated Ca^{2+} current is higher in freshly isolated VSMCs from arteries and arterioles of hypertensive rats and mice. Patch-clamp studies have demonstrated elevated Ca^{2+} currents in freshly isolated VSMCs of mesenteric, renal, femoral, cerebral and pulmonary arteries from several different rat and mouse models of hypertension [10, 102–106]. Single-channel recordings of unitary Ca^{2+} channel currents indicate that the biophysical properties of $Ca_v1.2$ channels are unchanged during hypertension, suggesting that the enhanced Ca^{2+} current in VSMCs of hypertensive animals primarily relates to a denser membrane expression of apparently normal $Ca_v1.2$ channels [107]. Finally, the membrane density of $Ca_v1.2$ channel-mediated Ca^{2+} current is positively correlated to blood pressure elevation in VSMCs of SHR [105], providing further evidence for a close association of these variables.

Although over-abundant $Ca_v1.2$ channels are heavily implicated in the pathogenesis of hypertension, the mechanisms that drive the expression of vascular $Ca_V1.2$ channels during the development of hypertension are largely unknown. Efforts to resolve these mechanisms continue unabated, due to the realization that preventing the pathogenic increase of vascular $Ca_v1.2$ channels during hypertension may represent an effective therapeutic approach to lower blood pressure and prevent its lethal side effects. Findings from several studies concur that increases in the expression level of $Ca_v\alpha$ transcript cannot account for the marked over-expression of vascular $Ca_v1.2$ channel proteins observed in hypertension. For example, $Ca_v\alpha$ protein increases 3- to 4-fold in mesenteric arteries of SHR compared to WKY rats, whereas only a small (1.53-fold) increase in the expression level of $Ca_v\alpha$ transcript is detected in SHR vessels [10]. Similarly, mesenteric arteries of Ang II-infused mice exhibit a profound increase in expression of the $Ca_v\alpha$ proteins, whereas levels of $Ca_v\alpha$ transcript are unchanged compared to arteries of normotensive mice [87]. The pronounced increase in expression of $Ca_v\alpha$ protein in pulmonary arteries from a porcine model of neonatal pulmonary hypertension also is not associated with an increase in the corresponding $Ca_v\alpha$ transcript [106].

Based on these findings, efforts to identify the molecular mediators that drive the over-expression of vascular $Ca_v1.2$ channels during hypertension have focused on post-transcriptional mechanisms. MicroRNAs (miRNAs) have drawn attention as possible modulators of $Ca_v1.2$ channel abundance during different forms of hypertension. In studies to date, a loss of miR-328-mediated post-transcriptional repression of $Ca_v\alpha$ expression in the pulmonary vasculature of hypoxic mice with pulmonary hypertension was associated with increased numbers of functional $Ca_v1.2$ channels concomitant with enhanced pulmonary vascular reactivity [108]. In contrast, mice overexpressing miR-328 had markedly lower right ventricular systolic pressures and pulmonary vascular remodeling in response to a hypoxic environment [108]. New evidence also implicates the regulatory $Ca_v\beta3$ subunit in VSMCs as a protein required for the over-expression of $Ca_v1.2$ channels during the development of hypertension. As depicted in Fig. 8.4, the $Ca_v\beta$ subunits bind to $Ca_v\alpha$ pore-forming proteins during Ca^{2+} channel biogenesis, and enable trafficking of the multi-protein channel complex from the endoplasmic reticulum (ER) to the plasma membrane [81–84]. Perhaps related to this chaperone function of $Ca_v\beta$ subunits, arteries of $Ca_v\beta3$ knockout mice fail to show increased $Ca_v1.2$ channel expression in response to Ang II infusion to induce hypertension, and blood pressure elevation is markedly less in these mice compared to wild-type mice [87].

Studies in heterologous expression systems are shedding light on the important role of $Ca_v\beta$ subunits in promoting the expression of vascular $Ca_v1.2$ channels, which is accomplished by preventing ubiquitination and degradation of the $Ca_v\alpha$ subunit during channel biogenesis [109]. Ubiquitination targets specific proteins for degradation with specificity conferred by E3 ubiquitin ligases; this activity is antagonized by deubiquitinating enzymes (DUBs), providing another checkpoint for regulating ion channel expression [110–120]. As discussed in the preceding paragraph, and depicted in Fig. 8.4, the expression of functional $Ca_v1.2$ channels requires the trafficking of $Ca_v\alpha$ pore-forming subunits to the plasma membrane by $Ca_v\beta$ subunits, with complementary domains on each subunit providing sites of interaction [81–84]. However, new evidence suggests that the E3 ubiquitin ligase, Nedd4-1, can disrupt the trafficking of $Ca_v\alpha$ subunits to the plasma membrane by interrupting interactions with chaperone β subunits at the ER/Golgi level [109, 121]. Thus, $Ca_v\beta$ subunits bind to $Ca_v\alpha$ subunits to prevent them from entering the ubiquitination and degradation pathway, and an increased expression of $Ca_v\beta$ subunits

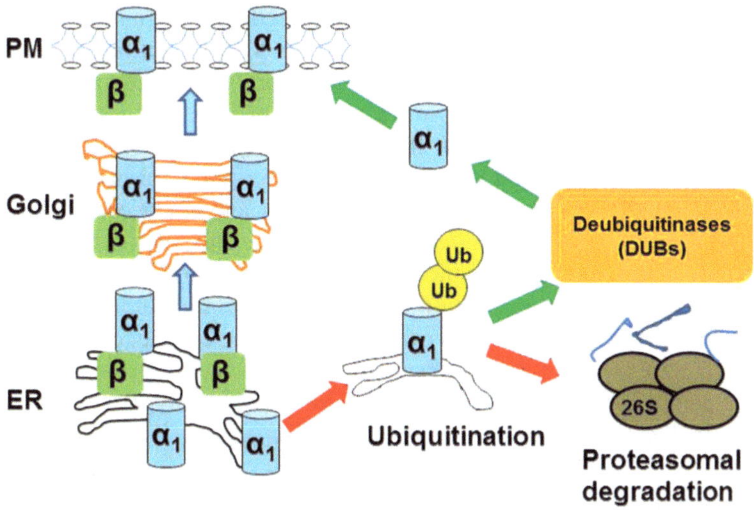

Fig. 8.4 Schematic representation of $Ca_v\beta$ subunit-mediated regulation of $Ca_v1.2$ channel surface expression. After synthesis in the endoplasmic reticulum (ER), $Ca_v\alpha$ subunits associate with accessory $Ca_v\beta$ subunits, an interaction that confers a stable configuration to the channel complex to prevent ubiquitination and permit channel trafficking to the plasma membrane (PM) via the Golgi complex. However, in the absence of $Ca_v\beta$ subunits, the $Ca_v\alpha$ subunits are recognized as unassembled proteins and targeted for ubiquitination and proteasomal degradation. There may be a role for DUBs in removing the ubiquitin (Ub) moieties from $Ca_v\alpha$ subunits to reroute $Ca_v\beta$ subunit-free channels to the PM

or a compromised ubiquitination enzymatic cascade may potentially result in an over-abundance of $Ca_v1.2$ channels at the plasma membrane. Ultimately, identification of the complex processes in VSMCs responsible for the post-transcriptional regulation of $Ca_v1.2$ channels will provide a basis for preventing the development of abnormal vascular tone during hypertension.

Conclusions

The VSMCs of small arteries and arterioles are central players in the regulation of vascular tone and blood pressure. During the development of hypertension, the delicate balance between relaxation and contraction of VSMCs is disrupted, and ion channel remodeling contributes to Ca^{2+}-dependent activation of the VSMCs. Thus, a Ca^{2+}-dependent elevation of vascular tone is a hallmark finding shared by most models of experimental hypertension and also is a feature of essential hypertension in humans. This chapter has reviewed the scientific evidence suggesting that a loss of voltage-gated K^+ (K_v) channels encoded by the *Shaker*, *Shab* and *KCNQ* genes in VSMCs contributes to vascular activation during hypertension. The reduced abundance of K_v channels in VSMCs appears to be at least partly mediated by transcriptional regulation involving NFATc3 signaling. In contrast, the increased abundance of $Ca_v1.2$ channels in VSMCs during hypertension, which is an event that accentuates voltage-dependent Ca^{2+} influx and anomalous vasoconstriction, occurs primarily in the absence of increased transcript expression. Emerging evidence suggests that the over-expression of vascular $Ca_v1.2$ channels during hypertension relates to post-transcriptional processes including regulation of the $Ca_v1.2$ transcript by microRNAs and/or by cellular processes including regulatory $Ca_v\beta3$ subunits involved in the biogenesis, trafficking and membrane stability of $Ca_v1.2$ channels. New studies by many laboratories designed to define mechanisms governing ion channel expression, including the introduction of intracellular ion channels as modulators of vascular tone, continue to inject excitement into the hunt for mechanisms of vascular ion channel remodeling during hypertension and ultimately, the promise of new antihypertensive therapeutics to normalize blood pressure in hypertensive patients.

Acknowledgements This work was supported in part by grants HL064806 (N.J.R.), DK089659 (L.M.) and GM106419 (L.M.) from the National Institutes of Health. Further support was provided by a pilot grant (UL1TR000039) from the UAMS Translational Research Institute (L.M.).

References

1. Cox RH, Rusch NJ. New expression profiles of voltage-gated ion channels in arteries exposed to high blood pressure. Microcirculation. 2002;9:243–57.
2. Sonkusare S, Palade PT, Marsh JD, Telemaque S, Pesic A, Rusch NJ. Vascular calcium channels and high blood pressure: pathophysiology and therapeutic implications. Vascul Pharmacol. 2006;44:131–42.
3. Joseph BK, Thakali KM, Moore CL, Rhee SW. Ion channel remodeling in vascular smooth muscle during hypertension: implications for novel therapeutic approaches. Vasc Res. 2013;70:126–38.
4. Harder DR, Brann L, Halpern W. Altered membrane electrical properties of smooth muscle cells from small cerebral arteries of hypertensive rats. Blood Vessels. 1983;20:154–60.
5. Harder DR, Smeda J, Lombard J. Enhanced myogenic depolarization in hypertensive cerebral arterial muscle. Circ Res. 1985;57:319–22.
6. Stekiel WJ, Contney SJ, Lombard JH. Small vessel membrane potential, sympathetic input, and electrogenic pump rate in SHR. Am J Physiol. 1986;250:C547–56.
7. Stekiel WJ, Contney SJ, Lombard JH. Sympathetic neural control of vascular muscle in reduced renal mass hypertension. Hypertension. 1991;17:1185–91.
8. Stekiel WJ, Contney SJ, Rusch NJ. Altered β-receptor control of in situ membrane potential in hypertensive rats. Hypertension. 1993;21:1005–9.
9. Fujii K, Ohmori S, Onaka U, Abe I, Fujishima M. Effects of salt-loading on membrane potentials in mesenteric arteries of spontaneously hypertensive rats. Hypertens Res. 1999;22:181–6.
10. Pesic A, Madden JA, Pesic M, Rusch NJ. High blood pressure upregulates arterial L-type Ca^{2+} channels: Is membrane depolarization the signal? Circ Res. 2004;94:e97–104.
11. Tobin AA, Joseph BK, Al-Kindi HN, Albarwani S, Madden JA, Nemetz LT, Rusch NJ, Rhee SW. Loss of cerebrovascular *Shaker*-type K^+ channels: a shared vasodilator defect of genetic and renal hypertensive rats. Am J Physiol. 2009;297:H293–303.

12. Nelson MT, Patlak JB, Worley JF, Standen NB. Calcium channels, potassium channels, and voltage dependence of arterial smooth muscle tone. Am J Physiol. 1990;259:C3–18.
13. Brayden JE. Potassium channels in vascular smooth muscle. Clin Exp Pharmacol Physiol. 1996;23:1069–76.
14. Jackson WF. Ion channels and vascular tone. Hypertension. 2000;35:173–8.
15. Post JM, Gelband CH, Hume JR. $[Ca^{2+}]_i$ inhibition of K^+ channels in canine pulmonary artery. Novel mechanism for hypoxia-induced membrane depolarization. Circ Res. 1995;77:131–9.
16. Cox RH, Petrou S. Ca^{2+} influx inhibits voltage-dependent and augments Ca^{2+}-dependent K^+ currents in arterial myocytes. Am J Physiol. 1999;277:C51–63.
17. Coussin F, Scott RH, Nixon GF. Sphingosine 1-phosphate induces CREB activation in rat cerebral artery via a protein kinase C-mediated inhibition of voltage-gated K^+ channels. Biochem Pharmacol. 2003;66:1861–70.
18. Park WS, Son YK, Han J, Kim N, Ko JH, Bae YM, Earm YE. Staurosporine inhibits voltage-dependent K^+ current through a PKC-independent mechanism in isolated coronary arterial smooth muscle cells. J Cardiovasc Pharmacol. 2005;45:260–9.
19. Ko EA, Park WS, Firth AL, da Hong H, Choi SW, Heo HJ, Kim MH, Noh JH, Ko JH, Kim N, Earm YE, Song DK, Han J. Increased sensitivity of serotonin on the voltage-dependent K^+ channels in mesenteric arterial smooth muscle cells of OLETF rats. Prog Biophys Mol Biol. 2010;103:88–94.
20. Cox RH, Fromme SJ, Folander KL, Swanson RJ. Voltage gated K^+ channel expression in arteries of Wistar-Kyoto and spontaneously hypertensive rats. Am J Hypertens. 2008;21:213–8.
21. Martens JR, Gelband CH. Alterations in rat interlobar artery membrane potential and K^+ channels in genetic and nongenetic hypertension. Circ Res. 1996;79:295–301.
22. Bratz IN, Swafford Jr AN, Kanagy NL, Dick GM. Reduced functional expression of K^+ channels in vascular smooth muscle cells from rats made hypertensive with Nw-nitro-L-arginine. Am J Physiol. 2005;289:H1284–90.
23. Bratz IN, Dick GM, Partridge LD, Kanagy NL. Reduced molecular expression of K^+ channel proteins in vascular smooth muscle from rats made hypertensive with Nw-nitro-L-arginine. Am J Physiol. 2005;289:H1277–83.
24. Albarwani S, Nemetz LT, Madden JA, Tobin AA, England SK, Pratt PF, Rusch NJ. Voltage-gated K^+ channels in rat small cerebral arteries: molecular identity of the functional channels. J Physiol. 2003;551:751–63.
25. Thorneloe KS, Chen TT, Kerr PM, Grier EF, Horowitz B, Cole WC, Walsh MP. Molecular composition of 4-aminopyridine-sensitive voltage-gated K^+ channels of vascular smooth muscle. Circ Res. 2001;89:1030–7.
26. Plane F, Johnson R, Kerr P, Wiehler W, Thorneloe K, Ishii K, Chen T, Cole W. Heteromultimeric K_v1 channels contribute to myogenic control of arterial diameter. Circ Res. 2005;96:216–24.
27. Chen TT, Luykenaar KD, Walsh EJ, Walsh MP, Cole WC. Key role of K_v1 channels in vasoregulation. Circ Res. 2006;99:53–60.
28. Wang J, Juhaszova M, Rubin LJ, Yuan XJ. Hypoxia inhibits gene expression of voltage-gated K^+ channel α subunits in pulmonary artery smooth muscle cells. J Clin Invest. 1997;100:2347–53.
29. Yuan JX, Aldinger AM, Juhaszova M, Wang J, Conte Jr JV, Gaine SP, Orens JB, Rubin LJ. Dysfunctional voltage-gated K^+ channels in pulmonary artery smooth muscle cells of patients with primary pulmonary hypertension. Circulation. 1998;98:1400–6.
30. Amberg GC, Rossow CF, Navedo MF, Santana LF. NFATc3 regulates $K_V2.1$ expression in arterial smooth muscle. J Biol Chem. 2004;279:47326–34.
31. Jepps TA, Chadha PS, Davis AJ, Harhun MI, Cockerill GW, Olesen SP, Hansen RS, Greenwood IA. Downregulation of $K_V7.4$ channel activity in primary and secondary hypertension. Circulation. 2011;124:602–11.
32. Chadha PS, Zunke F, Zhu HL, Davis AJ, Jepps TA, Olesen SP, Cole WC, Moffatt JD, Greenwood IA. Reduced KCNQ4-encoded voltage-dependent potassium channel activity

underlies impaired β-adrenoceptor-mediated relaxation of renal arteries in hypertension. Hypertension. 2012;59:877–84.
33. Cox RH, Folander K, Swanson R. Differential expression of voltage-gated K^+ channel genes in arteries from spontaneously hypertensive and Wistar-Kyoto rats. Hypertension. 2001;37:1315–22.
34. Accili EA, Kuryshev YA, Wible BA, Brown AM. Separable effects of human Kvβ1.2 N- and C-termini on inactivation and expression of human Kv1.4. J Physiol. 1998;512:325–36.
35. Bhatnagar A, Kumar R, Tipparaju SM, Liu SQ. Differential pyridine nucleotide coenzyme binding to the β-subunit of the voltage-sensitive K^+ channel: a mechanism for redox regulation of excitability? Chem Biol Interact. 2003;143:613–20.
36. Pan Y, Weng J, Levin EJ, Zhou M. Oxidation of NADPH on Kvβ1 inhibits ball-and-chain type inactivation by restraining the chain. Proc Natl Acad Sci U S A. 2011;108:5885–90.
37. Kilfoil PJ, Tipparaju SM, Barski OA, Bhatnagar A. Regulation of ion channels by pyridine nucleotides. Circ Res. 2013;112:721–41.
38. Joseph BK, Thakali KM, Pathan AR, Kang E, Rusch NJ, Rhee SW. Postsynaptic density-95 scaffolding of Shaker-type K+ channels in smooth muscle cells regulates the diameter of cerebral arteries. J Physiol. 2011;589:5143–52.
39. Moore CL, Nelson PL, Parelkar NK, Rusch NJ, Rhee SW. Protein kinase A-phosphorylated Kv1 channels in PSD95 signaling complex contribute to the resting membrane potential and diameter of cerebral arteries. Circ Res. 2014;114:1258–67.
40. Meera P, Wallner M, Song M, Toro L. Large conductance voltage- and calcium-dependent K^+ channel, a distinct member of voltage-dependent ion channels with seven N-terminal transmembrane segments (S0-S6), an extracellular N terminus, and an intracellular (S9-S10) C terminus. Proc Natl Acad Sci U S A. 1997;94:14066–71.
41. Shipston MJ. Alternative splicing of potassium channels: a dynamic switch of cellular excitability. Trends Cell Biol. 2001;11:353–8.
42. Lorca RA, Prabagaran M, England SK. Functional insights into modulation of BK_{Ca} channel activity to alter myometrial contractility. Front Physiol. 2014;5:289.
43. Lorca RA, Stamnes SJ, Pillai MK, Hsiao JJ, Wright ME, England SK. N-terminal isoforms of the large-conductance Ca^{2+}-activated K^+ channel are differentially modulated by the auxiliary β1-subunit. J Biol Chem. 2014;289:10095–103.
44. Korovkina VP, Brainard AM, England SK. Translocation of an endoproteolytically cleaved maxi-K channel isoform: mechanisms to induce human myometrial cell repolarization. J Physiol. 2006;573:329–41.
45. Wang W, Huang H, Hou D, Liu P, Wei H, Fu X, Niu W. Mechanosensitivity of STREX-lacking BKCa channels in the colonic smooth muscle of the mouse. Am J Physiol. 2010;299:G1231–40.
46. Poulsen AN, Wulf H, Hay-Schmidt A, Jansen-Olesen I, Olesen J, Klaerke DA. Differential expression of BK channel isoforms and β-subunits in rat neurovascular tissues. Biochim Biophys Acta. 2009;1788:380–9.
47. Jackson WF, Blair KL. Characterization and function of Ca^{2+}-activated K^+ channels in arteriolar muscle cells. Am J Physiol. 1998;274:H27–34.
48. Yang Y, Sohma Y, Nourian Z, Ella SR, Li M, Stupica A, Korthuis RJ, Davis MJ, Braun AP, Hill MA. Mechanisms underlying regional differences in the Ca^{2+} sensitivity of BK_{Ca} current in arteriolar smooth muscle. J Physiol. 2013;591:1277–93.
49. Yang Y, Murphy TV, Ella SR, Grayson TH, Haddock R, Hwang YT, Braun AP, Peichun G, Korthuis RJ, Davis MJ, Hill MA. Heterogeneity in function of small artery smooth muscle BK_{Ca}: involvement of the β1-subunit. J Physiol. 2009;587:3025–44.
50. Liu Y, Pleyte K, Knaus HG, Rusch NJ. Increased expression of Ca^{2+}-sensitive K^+ channels in aorta of hypertensive rats. Hypertension. 1997;30:1403–9.
51. Liu Y, Hudetz AG, Knaus HG, Rusch NJ. Increased expression of Ca^{2+}-sensitive K^+ channels in the cerebral microcirculation of genetically hypertensive rats: evidence for their protection against cerebral vasospasm. Circ Res. 1998;82:729–37.

52. Xu H, Bian X, Watts SW, Hlavacova A. Activation of vascular BK channel by tempol in DOCA-salt hypertensive rats. Hypertension. 2005;46:1154–62.
53. Xu H, Jackson WF, Fink GD, Galligan JJ. Activation of potassium channels by tempol in arterial smooth muscle cells from normotensive and deoxycorticosterone acetate-salt hypertensive rats. Hypertension. 2006;48:1080–7.
54. Rusch NJ, Runnells AM. Remission of high blood pressure reverses arterial potassium channel alterations. Hypertension. 1994;23:941–5.
55. Rusch NJ, De Lucena RG, Wooldridge TA, England SK, Cowley Jr AW. A Ca^{2+}-dependent K^+ current is enhanced in arterial membranes of hypertensive rats. Hypertension. 1992;19:301–7.
56. Liu Y, Jones AW, Sturek M. Increased barium influx and potassium current in stroke-prone spontaneously hypertensive rats. Hypertension. 1994;23:1091–5.
57. Liu Y, Jones AW, Sturek M. Ca^{2+}-dependent K^+ current in arterial smooth muscle cells from aldosterone-salt hypertensive rats. Am J Physiol. 1995;269:H1246–57.
58. Pabbidi MR, Mazur O, Fan F, Farley JM, Gebremedhin D, Harder DR, Roman RJ. Enhanced large conductance K^+ channel activity contributes to the impaired myogenic response in the cerebral vasculature of Fawn Hooded Hypertensive rats. Am J Physiol. 2014;306:H989–1000.
59. England SK, Wooldridge TA, Stekiel WJ, Rusch NJ. Enhanced single-channel K^+ current in arterial membranes from genetically hypertensive rats. Am J Physiol. 1993;264:H1337–45.
60. Asano M, Masuzawa-Ito K, Matsuda T, Imaizumi Y, Watanabe M, Ito K. Functional role of Ca^{2+}-activated K^+ channels in resting state of carotid arteries from SHR. Am J Physiol. 1993;265:H843–51.
61. Asano M, Masuzawa-Ito K, Matsuda T. Charybdotoxin-sensitive K^+ channels regulate the myogenic tone in the resting state of arteries from spontaneously hypertensive rats. Br J Pharmacol. 1993;108:214–22.
62. Kolias TJ, Chai S, Webb RC. Potassium channel antagonists and vascular reactivity in stroke-prone spontaneously hypertensive rats. Am J Hypertens. 1993;6:528–33.
63. Amberg GC, Santana LF. Downregulation of the BK channel β1 subunit in genetic hypertension. Circ Res. 2003;93:965–71.
64. Nieves-Cintron M, Amberg GC, Nichols CB, Molkentin JD, Santana LF. Activation of NFATc3 down-regulates the β1 subunit of large conductance, calcium-activated K^+ channels in arterial smooth muscle and contributes to hypertension. J Biol Chem. 2007;282:3231–40.
65. Amberg GC, Bonev AD, Rossow CF, Nelson MT, Santana LF. Modulation of the molecular composition of large conductance, Ca^{2+} activated K^+ channels in vascular smooth muscle during hypertension. J Clin Invest. 2003;112:717–24.
66. Yang Y, Li PY, Cheng J, Mao L, Wen J, Tan XQ, Liu ZF, Zeng XR. Function of BK_{Ca} channels is reduced in human vascular smooth muscle cells from Han Chinese patients with hypertension. Hypertension. 2013;61:519–25.
67. Bhattarai Y, Fernandes Rm Kadrofske MM, Lockwood LR, Galligan JJ, Hui X. Western blot analysis of BK channel β1-subunit expression should be interpreted cautiously when using commercially available antibodies. Physiol Rep. 2014;2:e12189.
68. Brenner R, Perez GJ, Bonev AD, Eckman DM, Kosek JC, Wiler SW, Patterson AJ, Nelson MT, Jiang Z, Wallner M, Meera P, Toro L. Human and rodent MaxiK channel beta-subunit genes: cloning and characterization. Genomics. 1999;55:57–67.
69. Xu H, Garver H, Galligan JJ, Fink GD. Large-conductance Ca^{2+}-activated K^+ channel β1-subunit knockout mice are not hypertensive. Am J Physiol. 2011;300:H476–85.
70. Holtzclaw JD, Grimm PR, Sansom SC. Role of BK channels in hypertension and potassium secretion. Curr Opin Nephrol Hypertens. 2011;20:512–7.
71. Chen L, Bi D, Tian L, McClafferty H, Steeb F, Ruth P, Knaus HG, Shipston MJ. Palmitoylation of the β4-subunit regulates surface expression of large conductance calcium-activated potassium channel splice variants. J Biol Chem. 2013;288:13136–44.

72. Shipston MJ. Regulation of large conductance calcium- and voltage-activated potassium (BK) channels by S-palmitoylation. Biochem Soc Trans. 2013;41:67–71.
73. Shipston MJ. S-acylation dependent post-translational cross-talk regulates large conductance calcium- and voltage-activated potassium (BK) channels. Front Physiol. 2014;5:281.
74. Siemen D, Loupatatzis C, Borecky J, Gulbins E, Lang F. Ca^{2+}-activated K channel of the BK-type in the inner mitochondrial membrane of a human glioma cell line. Biochem Biophys Res Commun. 1999;257:549–54.
75. Balderas E, Zhang J, Stefani E, Toro L. Mitochondrial BK_{Ca} channel. Front Physiol. 2015;6:104.
76. Singh H, Lu R, Bopassa JC, Meredith AL, Stefani E, Toro L. $MitoBK_{Ca}$ is encoded by the Kcnma1 gene, and a splicing sequence defines its mitochondrial location. Proc Natl Acad Sci U S A. 2013;110:10836–41.
77. Xu W, Liu Y, Wang S, McDonald T, Van Eyk JE, Sidor A, O'Rourke B. Cytoprotective role of Ca^{2+}-activated K^+ channels in the cardiac inner mitochondrial membrane. Science. 2002;298:1029–33.
78. Soltysinska E, Bentzen BH, Barthmes M, Hattel H, Thrush AB, Harper ME, Qvortrup K, Larsen FJ, Schiffer TA, Losa-Reyna J, Straubinger J, Kniess A, Thomsen MB, Bruggemann A, Fenske S, Biel M, Ruth P, Wahl-Schott C, Boushel RC, Olesen SP, Lukowski R. KCNMA1 encoded cardiac BK channels afford protection against ischemia-reperfusion injury. PLoS One. 2014;9:e103402.
79. Pluger S, Faulhaber J, Furstenau M, Lohn M, Waldschutz R, Gollasch M, Haller H, Luft FC, Ehmke H, Pongs O. Mice with disrupted BK channel β1 subunit gene feature abnormal Ca^{2+} spark/STOC coupling and elevated blood pressure. Circ Res. 2000;87:E53–60.
80. Tang XD, Garcia ML, Heinemann SH, Hoshi T. Reactive oxygen species impair *Slo1* BK channel function by altering cysteine-mediated calcium sensing. Nat Struct Mol Biol. 2004;11:171–8.
81. Birnbaumer L, Qin N, Olcese R, Tareilus E, Platano D, Costantin J, Stefani E. Structures and functions of calcium channel β subunits. J Bioenerg Biomembr. 1998;30:357–75.
82. Chen YH, Li MH, Zhang Y, He LL, Yamada Y, Fitzmaurice A, Shen Y, Zhang H, Tong L, Yang J. Structural basis of the α1-β subunit interaction of voltage-gated Ca^{2+} channels. Nature. 2004;429(6992):675–80.
83. Catterall WA. Structure and regulation of voltage-gated Ca^{2+} channels. Annu Rev Cell Dev Biol. 2000;16:521–55.
84. Dolphin AC. β subunits of voltage-gated calcium channels. J Bioenerg Biomembr. 2003;35:599–620.
85. James PA, Oparil S, Carter BL, et al. Evidence-based guideline for the management of high blood pressure in adults: report by the panel appointed to the Eighth Joint National Committee (JNC 8). JAMA. 2014;311:507–20. doi:10.1001/JAMA.2013.284427.
86. Pratt PF, Bonnet S, Ludwig LM, Bonnet P, Rusch NJ. Upregulation of L-type Ca^{2+} channels in mesenteric and skeletal arteries of SHR. Hypertension. 2002;40:214–9.
87. Kharade SV, Sonkusare SK, Srivastava AK, Thakali KM, Fletcher TW, Rhee SW, Rusch NJ. The β3 subunit contributes to vascular calcium channel upregulation and hypertension in angiotensin II-infused C57BL/6 mice. Hypertension. 2013;61:137–42.
88. Gebremedhin D, Fenoy FJ, Harder DR, Roman RJ. Enhanced vascular tone in the renal vasculature of spontaneously hypertensive rats. Hypertension. 1990;16:648–54.
89. Robinson BF, Dobbs RJ, Bayley S. Responses of forearm resistance vessels to verapamil and sodium nitroprusside in normotensive and hypertensive men: Evidence for a functional abnormality of vascular smooth muscle in primary hypertension. Clin Sci. 1982;63:33–42.
90. Ishii H, Itoh K, Nose T. Different antihypertensive effects of nifedipine in conscious experimental hypertensive and normotensive rats. Eur J Pharmacol. 1980;64:21–9.
91. Kubo T, Fujie K, Yamashita M, Misu Y. Antihypertensive effects of nifedipine on conscious normotensive and hypertensive rats. J Pharmacobiodyn. 1981;4:294–300.

92. Garthoff B, Kazda S. Calcium antagonist nifedipine normalizes high blood pressure and prevents mortality in salt-loaded DS substrain of DAHL rats. Eur J Pharmacol. 1981;74:111–2.
93. Thievant P, Baranes J, Le Hegarat M, Clostre F, DeFeudis FV. Effects of diltiazem on renovascular-hypertensive and on normotensive rats. Gen Pharmacol. 1982;13:165–7.
94. Lederballe Pedersen C, Mikkelsen E, Jespersen LT. Treatment with verapamil reduces blood pressure and tends to normalize vascular responsiveness and ion transport in the spontaneously hypertensive rat. J Cardiovasc Pharmacol. 1982;4:S294–7.
95. Narita H, Nagao T, Yabana H, Yamaguchi I. Hypotensive and diuretic actions of diltiazem in spontaneously hypertensive and Wistar Kyoto rats. J Pharmacol Exp Ther. 1983;227:472–7.
96. Takata Y, Hutchinson JB. Exaggerated hypotensive responses to calcium antagonists in spontaneously hypertensive rats. Clin Exp Hypertens. 1983;A5:827–47.
97. Aoki K, Kawaguchi Y, Sato K, Kondo S, Yamamoto M. Clinical and pharmacological properties of calcium antagonists in essential hypertension in humans and spontaneously hypertensive rats. J Cardiovasc Pharmacol. 1982;4:S298–302.
98. Lederballe Pedersen O, Christiansen NJ, Ramsch KD. Comparison of acute effects of nifedipine in normotensive and hypertensive man. J Cardiovasc Pharmacol. 1980;2:357–66.
99. Hulthen UL, Bolli F, Amann FW, Kiowski W, Buhler FR. Enhanced vasodilatation in essential hypertension by calcium channel blockade with verapamil. Hypertension. 1982;4:S26–31.
100. Molero MM, Giulumian AD, Reddy VB, Ludwig LM, Pollock JS, Pollock DM, Rusch NJ, Fuchs LC. Decreased endothelin binding and [Ca^{2+}]i signaling in microvessels of DOCA-salt hypertensive rats. J Hypertens. 2002;20:1799–805.
101. Wang WZ, Saada N, Dai B, Pang L, Palade P. Vascular-specific increase in exon 1B-encoded Cav1.2 channels in spontaneously hypertensive rats. Am J Hypertens. 2006;19:823–31.
102. Ohya Y, Abe I, Fujii K, Takata Y, Fujishima M. Voltage-dependent Ca^{2+} channels in resistance arteries from spontaneously hypertensive rats. Circ Res. 1993;73:1090–9.
103. Cox RH, Lozinskaya IM. Augmented calcium currents in mesenteric artery branches of the spontaneously hypertensive rat. Hypertension. 1995;26:1060–4.
104. Simard JM, Li X, Tewari K. Increase in functional Ca^{2+} channels in cerebral smooth muscle with renal hypertension. Circ Res. 1998;82:1330–7.
105. Cox RH, Lozinskaya I, Matsuda K, Dietz NJ. Ramipril treatment alters Ca^{2+} and K^+ channels in small mesenteric arteries from Wistar-Kyoto and spontaneously hypertensive rats. Am J Hypertens. 2002;15:879–90.
106. Hirenallur SD, Haworth ST, Leming JT, Chang J, Hernandez G, Gordon JB, Rusch NJ. Upregulation of vascular calcium channels in neonatal piglets with hypoxia-induced pulmonary hypertension. Am J Physiol. 2008;295:L915–24.
107. Ohya Y, Tsuchihashi T, Kagiyama S, Abe I, Fujishima M. Single L-type calcium channels in smooth muscle cells from resistance arteries of spontaneously hypertensive rats. Hypertension. 1998;31:1125–9.
108. Guo L, Qui Z, Wei L, Yu X, Gao X, Jiang S, Tian H, Jiang C, Zhu D. The microRNA-328 regulates hypoxic pulmonary hypertension by targeting at insulin growth factor 1 receptor and L-type calcium channel α1C. Hypertension. 2012;59:1006–13.
109. Altier C, Garcia-Caballero A, Simms B, You H, Chen L, Walcher J, Tedford HW, Hermosilla T, Zamponi GW. The Cavβ subunit prevents RFP2-mediated ubiquitination and proteasomal degradation of L-type channels. Nat Neurosci. 2011;14:173–80.
110. Hochstrasser M. Evolution and function of ubiquitin-like protein-conjugation systems. Nat Cell Biol. 2000;2:E153–7.
111. Pickart CM. Mechanisms underlying ubiquitination. Annu Rev Biochem. 2001;70:503–33.
112. Komander D. The emerging complexity of protein ubiquitination. Biochem Soc Trans. 2009;37:937–53.

113. Ye Y, Rape M. Building ubiquitin chains: E2 enzymes at work. Nat Rev Mol Cell Biol. 2009;10:755–64.
114. Deshaies RJ, Joazeiro CA. RING domain E3 ubiquitin ligases. Annu Rev Biochem. 2009;78:399–434.
115. Nagy V, Dikic I. Ubiquitin ligase complexes: from substrate selectivity to conjugational specificity. Biol Chem. 2010;391:163–9.
116. Wenzel DM, Stoll KE, Klevit RE. E2s: structurally economical and functionally replete. Biochem J. 2011;433:31–42.
117. Behrends C, Harper JW. Constructing and decoding unconventional ubiquitin chains. Nat Struct Mol Biol. 2011;18:520–8.
118. Komander D, Clague MJ, Urbe S. Breaking the chains: structure and function of the deubiquitinases. Nat Rev Mol Cell Biol. 2009;10:550–63.
119. Durcan TM, Kontogiannea M, Thorarinsdottir T, Fallon L, Williams AJ, Djarmati A, Fantaneanu T, Paulson HL, Fon EA. The Machado-Joseph disease-associated mutant form of ataxin-3 regulates parkin ubiquitination and stability. Hum Mol Genet. 2011;20:141–54.
120. Fraile JM, Quesada V, Rodriguez D, Freije JM, Lopez-Otin C. Deubiquitinases in cancer: new functions and therapeutic options. Oncogene. 2011;31:2373–88.
121. Rougier JS, Albesa M, Abriel H, Viard P. Neuronal precursor cell-expressed developmentally down-regulated 4-1 (NEDD4-1) controls the sorting of newly synthesized $Ca_v1.2$ calcium channels. J Biol Chem. 2011;286:8829–38.

Chapter 9
Kv7 Potassium Channels as Therapeutic Targets in Cerebral Vasospasm

Bharath K. Mani, Lyubov I. Brueggemann, Sarkis Morales-Vidal, Christopher M. Loftus, and Kenneth L. Byron

Abstract Cerebral vasospasm is a devastating medical consequence of aneurysmal subarachnoid hemorrhage (aSAH), which is associated with a high level of morbidity and mortality and for which current therapies are inadequate. Ion channels in the smooth muscle cells of the cerebral vasculature have been considered as important molecular targets for anti-vasospastic therapies, but, other than the calcium channel blocker nimodipine, no therapies directed toward these targets have been demonstrated to improve patient outcomes. Recent research has revealed that a family of previously identified neuronal voltage-activated potassium channels, Kv7 (KCNQ) channels, is expressed in arterial smooth muscle cells, including cerebral arterial myocytes. This finding, along with the availability of clinically used drugs that were developed to target the neuronal Kv7 channels, provides an opportunity to develop novel therapeutic strategies to relieve or prevent cerebral vasospasm. Recent studies using both *in vitro* and *in vivo* approaches provide evidence to support the utility of Kv7 channel activators as anti-vasospastic agents, while revealing a number of additional potentially beneficial effects that could attenuate other neurological or inflammatory co-morbidities associated with aSAH.

Keywords Celecoxib • Retigabine • Stroke • Subarachnoid hemorrhage • Vascular smooth muscle • Voltage-activated potassium channel

---

B.K. Mani • L.I. Brueggemann
Department of Molecular Pharmacology & Therapeutics, Loyola University Chicago, Maywood, IL, USA

S. Morales-Vidal
Department of Neurology, Loyola University Chicago, Maywood, IL, USA

C.M. Loftus
Department of Neurosurgery, Loyola University Chicago, Maywood, IL, USA

K.L. Byron, Ph.D. (✉)
Health Sciences Division, Loyola University Chicago,
2160S. First Avenue, Maywood, IL 60153, USA

Department of Molecular Pharmacology & Therapeutics, Loyola University Chicago, Maywood, IL, USA
e-mail: kbyron@luc.edu

I. Levitan, A.M. Dopico (eds.), *Vascular Ion Channels in Physiology and Disease*, DOI 10.1007/978-3-319-29635-7_9

Introduction

Saccular cerebral aneurysms (focal rounded dilations or balloonings of a cerebral artery) are present in 1–5 % of adults in the United States [15]. It is unclear whether the natural history of small unruptured aneurysms is benign; therefore, treatment to secure a detected aneurysm is often recommended on a prophylactic basis. Of more obvious medical significance, approximately 1 out of every 10,000 persons in the United States experiences a rupture of a cerebral aneurysm. The result, called aneurysmal subarachnoid hemorrhage (aSAH), more often than not has catastrophic medical consequences. Approximately 45 % of aSAH patients die within 45 days and among those who survive, approximately 30 % experience moderate to severe disability [15].

The most common presentation of ruptured aneurysmal subarachnoid hemorrhage is severe thunderclap headache (frequently described as the "worst ever"); associated symptoms include nausea, vomiting, neck pain, photophobia and loss of consciousness [116]. The earliest consequences of a ruptured cerebral aneurysm include: (1) the presence of blood in the subarachnoid space; (2) a transient decrease in perfusion of the brain; (3) hydrocephalus, leading to an increase in intracranial pressure. Cardiac, pulmonary, and electrolyte abnormalities often occur; fever, anemia, and hyperglycemia are also common [96]. Depending on the severity of these events, death or coma may rapidly ensue.

Patients diagnosed with aSAH are typically maintained in a critical care setting, with the main goals of treatment being the prevention of re-bleeding, the prevention and management of vasospasm, and the treatment of other medical and neurological complications [116]. Although interventional procedures (e.g. surgical clipping or insertion of an endovascular coil) are usually employed to stabilize the aneurysm and prevent further bleeding into the subarachnoid space [15, 116], a number of additional pathologies often develop days to weeks after the initial event. Elevated local concentrations of vasoconstrictor substances in the vicinity of the blood clot, combined with impaired vasodilatory responses, can induce cerebral vasospasm [48, 54, 94], a sustained constriction of the cerebral vasculature that may persist for up to 16 days [94]. Vascular remodeling may also occur [90], along with impaired cerebral autoregulation [93], inflammation [78] and edema [48, 54]. This constellation of additional ensuing pathologies, in combination with a prolonged decrease in perfusion of the brain, is often associated with cortical spreading depression and delayed ischemic neurological deficits [61, 106]. Seizures also occur in approximately 25 % of aSAH patients [77].

Cerebral Vasospasm

Cerebral vasospasm is a central event that often determines the severity of illness following aSAH [54, 68]. Vasospasm can occur in the large arteries comprising the Circle of Willis (intracranial internal carotid artery, anterior cerebral artery, anterior communicating artery, middle cerebral artery, posterior communicating artery,

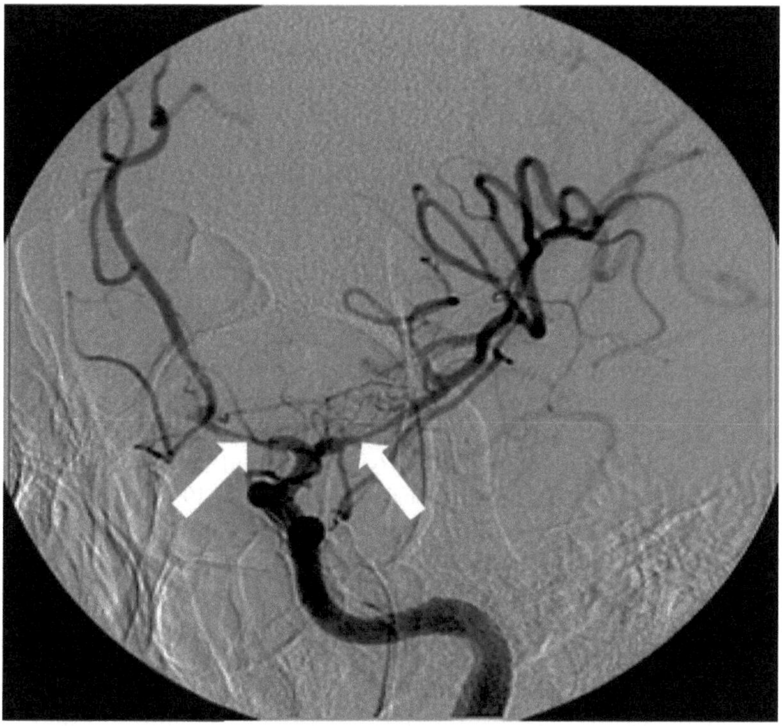

Fig. 9.1 Angiogram showing cerebral vasospasm (*arrows*) in an aSAH patient

intracranial vertebral artery, basilar artery, posterior inferior cerebellar artery, anterior inferior cerebellar artery, superior cerebellar artery, and posterior cerebral artery) and the main branches arising from this vascular ring. Vasospasm is most accurately detected by high-quality catheter angiography, which typically reveals a focal constriction in one of the large conduit arteries within the subarachnoid space (Fig. 9.1). Cerebral vasospasm most commonly begins 3–4 days after the rupture of an aneurysm and persists for about 2 weeks before gradually dissipating [87]. Although the large conduit arteries are the sites in which cerebral vasospasm is usually detected angiographically, it is likely that the more distal cerebral microvasculature is also hyperconstricted to some extent [48, 54, 87].

The only pharmacological treatment for cerebral vasospasm currently approved by the United States Food and Drug Administration is the calcium channel blocker nimodipine [29, 87]. In the United States and Europe, nimodipine is a standard component of treatment regimens for patients with aSAH. It is typically administered orally at a dose of 60 mg every 4 h for 21 days [48, 94, 116]. The rationale for use of nimodipine is to reduce influx of calcium ions (Ca^{2+}) into the vascular smooth muscle cells by blocking L-type voltage-dependent Ca^{2+} channels (VDCCs). Influx of Ca^{2+} via L-type VDCCs is recognized as one of the primary determinants of the vascular smooth muscle contraction that contributes to cerebral vasospasm.

Additional anti-platelet and neuroprotective effects of nimodipine may account for some of its therapeutic benefits in aSAH [87]. Although a modest but statistically significant therapeutic benefit of nimodipine in aSAH patients has been documented [35], its efficacy in relaxing spastic arteries is not clearly established [29]. A limitation of nimodipine treatment is its tendency to lower systemic blood pressure, which can in turn decrease cerebral perfusion. Moreover, the cerebral vasculature expresses other classes of Ca^{2+} channels (e.g. R-type VDCCs) that are insensitive to nimodipine; these may continue to conduct Ca^{2+} into vascular smooth muscle cells and thereby maintain some level of vasospasm despite nimodipine therapy [53].

Considering the modest benefits of nimodipine for alleviation of cerebral vasospasm it is not surprising that investigators have been actively pursuing more effective treatments. One controversial strategy that has been used extensively, called Triple H therapy (hypervolemia, hypertension, hemodilution), seeks to restore cerebral perfusion by inducing hypervolemia and reducing hematocrit to the range of 0.33–0.38 [94], while inducing hypertension (systolic blood pressure as high as 200 mmHg) with phenylephrine, norepinephrine, or dopamine [94, 116]. The therapeutic benefit of Triple H therapy, which targets the symptoms rather than the underlying cause of cerebral vasospasm, has not been established by randomized clinical trials. Endovascular treatments, such as transluminal angioplasty, usually in conjunction with direct vasodilators, are sometimes employed, particularly when other strategies have failed [94, 116]. New experimental therapies include endothelin receptor antagonists (e.g. clazosentan [122]), magnesium sulfate [126], and the rho kinase inhibitor fasudil [137], all of which have had some promising results in clinical trials. None of these experimental treatments have yet been approved for routine use, leaving the medical community still in need of more effective treatments for cerebral vasospasm.

Potassium Channels as Regulators of Vascular Tone and Potential Mediators of Cerebral Vasospasm

Influx of Ca^{2+} into smooth muscle cells through VDCCs activates smooth muscle contraction and vasoconstriction. In particular, L-type VDCCs open in a steeply voltage-dependent manner when membrane voltages exceed a threshold between −40 mV and −20 mV. The opening of these channels is the primary mechanism for increasing Ca^{2+} influx to activate smooth muscle contraction. This process is normally opposed by efflux of potassium ions (K^+), which maintains a hyperpolarized (negative) resting membrane voltage, below the threshold for L-type VDCC activation [83] (Fig. 9.2a). Membrane voltage is highly sensitive to physiological or pharmacological manipulation of K^+ channel activity; inhibition of K^+ channels in vascular smooth muscle cells causes membrane depolarization [82, 83]. During experimental SAH-induced vasospasm, resting membrane voltage in basilar artery myocytes becomes depolarized to the range that facilitates opening of L-type VDCCs, consistent with compromised function of K^+ channels [43, 123] (Fig. 9.2b).

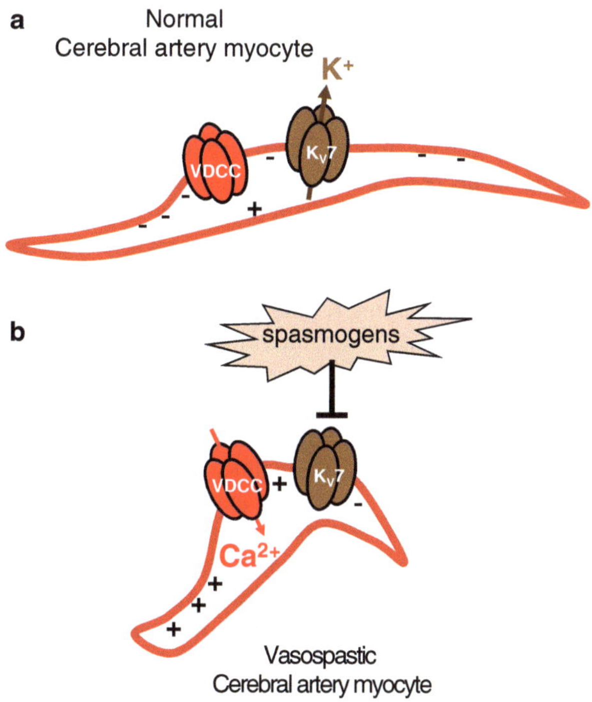

Fig. 9.2 Schematic illustration of ionic movements regulating smooth muscle contraction in the cerebral vasculature. (**a**) In a normal relaxed cerebral artery myocyte, negative membrane voltage is maintained by efflux of K^+. Voltage-dependent Ca^{2+} channels (VDCCs) remain predominantly closed at these negative voltages. (**b**) Spasmogens present at elevated concentrations following aSAH lead to suppression of K^+ efflux via Kv7 channels, resulting in membrane depolarization, activation of VDCCs, Ca^{2+} influx, and activation of smooth muscle contraction

Pharmacological K^+ channel openers have been considered as potential therapeutics for the treatment of cerebral vasospasm [53, 124], though so far none have been approved for this application in aSAH patients. K^+ channel openers are drugs that stabilize the open state of K^+ channels and increase the efflux of K^+ at physiological voltages. They typically act on specific subtypes of K^+ channels. Functional expression of four different classes of K^+ channels has been identified in cerebral artery myocytes: ATP-sensitive (K_{ATP}), inwardly rectifying (K_{IR}), large conductance Ca^{2+}-activated (BK_{Ca}), and voltage-activated (delayed rectifier) K^+ channels [83, 124]. The rationale and limitations for therapeutic targeting of each of these classes of K^+ channels is considered below.

K_{ATP} channels couple electrical activity of the cell to its metabolic status by activating in response to a decrease in cytosolic adenosine triphosphate (ATP) concentration [101]. In vascular smooth muscle cells, K_{ATP} channels open in response to hypoxic conditions, limiting the contractile force to conserve ATP [30]. K_{ATP} channels consist of the pore-forming inwardly-rectifying K^+ channel, Kir6, in complex with sulfonylurea receptor (SUR) subunits, with predominant expression of Kir6.1 and SUR2B in rat basilar and middle cerebral arteries [99]. SAH-induced changes in K_{ATP} channel expression in cerebral arteries have not been examined directly, but enhanced responses of cerebral arteries to pharmacological K_{ATP} channel activators were observed, indicating an increase in functional K_{ATP} channels in cerebral artery myocytes after SAH [117]. Treatment with pharmacological K_{ATP} channel openers has been demonstrated to reduce cerebral vasospasm in several animal studies of SAH [53, 124], but to date, clinical trials failed to show an improvement over standard care with nimodipine [1]. SUR1, a component of an ATP-sensitive non-selective cation channel, has been found to be upregulated following experimentally induced SAH in rats [112]. In that study, the K_{ATP} channel inhibitor glibenclamide, which also inhibits SUR1-containing cation channels, was found to decrease cerebral microvascular leakiness. Glibenclamide also reduced inflammation associated with the upregulation of SUR1 in the rat SAH study [112]. Thus, it is not clear whether K_{ATP} channel activators or inhibitors would have greater therapeutic benefit following SAH.

Inwardly rectifying K^+ (K_{IR}) channels in cerebral arteries are tetramers comprised of $K_{IR}2.1$ and $K_{IR}2.2$ subunits, with $K_{IR}2.1$ being essential [14, 114, 134]. K_{IR} channels conduct inward K^+ current at membrane voltages negative to the reversal potential for K^+ (E_K). At normal K^+ concentrations, small amplitude outward currents are conducted in the physiological voltage range up to 20 mV positive to E_K. The outward current amplitude through K_{IR}, and the voltage range over which outward currents are observed, depend on extracellular K^+ concentration ($[K^+]_o$) [65, 66, 75]. Enhanced neuronal activity can induce a modest increase in $[K^+]_o$ (up to 16 mM), which can in turn increase outward K_{IR} currents in the physiological voltage range; the resulting increase in K^+ efflux hyperpolarizes the cerebral artery myocytes and leads to K_{IR}-dependent vasodilation of small cerebral arteries, increasing blood flow to the active brain region [37, 56]. Apart from the regulation of vascular tone in small cerebral arteries and parenchymal arterioles to meet the metabolic needs of neurons, the physiological significance of K_{IR} channels in larger cerebral arteries remains poorly characterized. Nevertheless, after experimentally-induced SAH, expression of $K_{IR}2.1$ and K_{IR} channel-mediated conductance were found to be robustly increased in dog basilar artery myocytes, indicating a compensatory upregulation of these channels [3, 125]. There are no clinically useful pharmacological activators of K_{IR} channels currently available.

Large conductance Ca^{2+}-activated K^+ channels (BK_{Ca}) form as tetrameric assemblies of pore forming α-subunits in complex with regulatory β-subunits. Expression of BK_{Ca} α-subunit and β1-subunit was detected at the messenger RNA (mRNA) and protein levels in basilar and middle cerebral arteries [128]. The open probability of BK_{Ca} channels depends on both membrane depolarization and an increase in cytosolic free Ca^{2+} concentration ($[Ca^{2+}]_{cyt}$). At resting $[Ca^{2+}]_{cyt}$ (~100 nM), the voltage-dependence of the channel opening is such that channel activity is detectable only at very positive voltages (~+40 mV). In contrast, at elevated $[Ca^{2+}]_{cyt}$ (~10 μM), the voltage-dependence is shifted such that BK_{Ca} channels become active at negative membrane voltages (~−50 mV) [9]. The presence of β-subunits further increases Ca^{2+}- and voltage-sensitivity of BK_{Ca} channels [120]. The function of BK_{Ca} channels was preserved in vasospastic basilar artery myocytes from a dog model of SAH [45–47]. Although pharmacological enhancement of BK_{Ca} channel activity would be expected to reduce cerebral vasospasm, there are no clinically useful activators of BK_{Ca} channels currently available.

Voltage-activated (delayed rectifier) K^+ (Kv) channels mediate efflux of K^+ in response to a positive change (depolarization) of the resting membrane voltage.

Their activity provides negative feedback that promotes membrane hyperpolarization and thereby limits the activation of VDCCs. The voltage range over which Kv channels activate and the extent to which they undergo voltage- or time-dependent inactivation depends upon their subunit composition. Kv currents with a half-maximal activation voltage ($V_{0.5}$) in the range of −1.3 mV to −10 mV, and half-maximal inactivation voltage in the range of −15 mV to −44 mV, were detected in basilar artery myocytes from dog and cerebral artery myocytes from rabbits [45, 103]. A small steady-state Kv current, sensitive to the Kv channel blocker 4-aminopyridine (4-AP), was detected in depolarized rabbit cerebral artery myocytes [10]. In canine basilar artery myocytes, mRNA for Kv1.2, Kv1.5, Kv2.2, Kv3.1, Kv3.4 and Kv4.3 were detected with expression of Kv1.5 and Kv2.2 being predominant [3]. After experimentally-induced SAH, the mRNA, protein level, and current amplitudes of several 4-AP-sensitive Kv channels were reduced in canine basilar artery myocytes [3, 45, 47]. This reduction in Kv currents would be expected to exacerbate cerebral vasospasm, and hence pharmacological activators of these channels might be therapeutically useful to restore normal channel activity. Unfortunately, there are no activators of 4-AP-sensitive Kv channels currently available for clinical use.

Kv7 channels are a distinct sub-family of voltage-activated K^+ channels that are insensitive to 4-AP. There are five mammalian Kv7 channel subtypes (Kv7.1–Kv7.5) encoded by five KCNQ genes (KCNQ1–5) [8, 12, 27, 59, 64, 110, 113, 121]. Kv7 channels play major roles in regulation of membrane voltage and cellular excitability [102]. Their unique biophysical properties include a very negative threshold of activation, near the resting membrane voltage (usually around −60 mV), and the absence of time-dependent inactivation [102]. The channels form as tetrameric assemblies of the individual KCNQ gene products. Four Kv7.1 subunits combine with ancillary KCNE1 subunits to form channels which conduct a slowly activating delayed rectifier K^+ current (I_{Ks}) in cardiac myocytes; I_{Ks} is an important determinant of the electrocardiographic QT-interval [8, 107]. In neurons, Kv7.2/Kv7.3 and Kv7.3/Kv7.5 heteromeric channels conduct "M-currents", named for their inhibition in response to muscarinic acetylcholine receptor stimulation [64, 110, 121]. M-currents function as a "brake", limiting repetitive neuronal action potential discharges and thus regulating neuronal excitability [31]. Upon stimulation of G protein-coupled receptors (GPCRs), such as the M1 muscarinic acetylcholine receptor, neuronal M-currents are inhibited, resulting in slow membrane depolarization and increased action potential firing [31].

A number of pharmacological agents modulate Kv7 channels. Inhibitors of all known Kv7 channels include linopirdine (1-phenyl-3,3-bis(pyridin-4-ylmethyl)-1,3-dihydro-2H-indol-2-one) and its more potent analog XE991 (10,10-bis(4-Pyridinylmethyl)-9(10H)-anthracenone) [4, 121]. Several other compounds were identified as selective inhibitors of cardiac I_{Ks} current, including chromanol 293B (trans-6-cyano-4-(N-ethylsulfonyl-N-methylamino)-3-hydroxy-2,2-dimethyl-chroman), its more potent analogue HMR 1556 ((3R,4S)-(+)-N-[3-hydroxy-2,2-dimethyl-6-(4,4,4-trifluorobutoxy) chroman-4-yl]-N-methylmethanesulfonamide), BMS-208782 ((S)-4-(3-Butyl-1,2,4-oxadiazol-5-yl)-N-(2,2-dimethylcyclopentyl methyl)benzamide), L-768,673 (2-(2,4-trifluoromethyl)-N-(2-oxo-5-phenyl-1-(2,2,2-trifluoroethyl)-2,3-dihydro-1H-benzo(e)(1,4)diazepin-3-yl)acetamide) and JNJ 303 (2-(4-Chlorophenoxy)-2-methyl-N-[5-[(methylsulfonyl)amino]tricyclo[3.3.1.13,7]dec-2-yl]-propanamide) [118].

There are also several clinically used pharmacological Kv7 channel openers. Development of these drugs has been directed primarily toward activators of neuronal M-channels, in hopes of finding new drugs to treat epilepsy and neuropathic pain. Flupirtine (ethyl [2-amino-6-[(4-fluorobenzyl)amino]pyridin-3-yl]carbamate) has been used clinically in Europe as a non-opioid analgesic drug for more than 30 years. Its more potent structural analog, retigabine (ethyl N-[2-amino-4-[(4-fluorophenyl)methylamino]phenyl]carbamate), is an anticonvulsant drug recently approved by the United States Food and Drug Administration (FDA) and European Medical Agency for treatment of partial epilepsies [41]. Retigabine and flupirtine display some selectivity among Kv7 isoforms, activating only channels formed by Kv7.2-7.5 subunits (not Kv7.1).

Expression of Kv7 channels was recently detected in smooth muscle cells from multiple arterial beds [26, 52, 84, 132], including basilar and cerebral arteries [72, 138]. Flupirtine, retigabine, and other Kv7 channel activators were found to be effective vasodilators ([24, 25, 51, 52, 55, 72, 74, 84, 132]; Table 9.1).

Another clinically used drug, the nonsteroidal anti-inflammatory drug (NSAID), celecoxib, was also found to be a Kv7 channel activator and a vasodilator [18, 21, 72, 111]. The effect of celecoxib on native Kv7 channels could be reproduced in expression systems. Celecoxib activates homomeric channels composed of Kv7.2, Kv7.3, Kv7.4, or Kv7.5 subunits, as well as heteromeric Kv7.2/Kv7.3, or Kv7.3/Kv7.5 channels, but inhibits Kv7.1 and Kv7.1/KCNE1 channels [18, 33]. The mechanism by which celecoxib produces these ion channel effects is not clear; homology modeling suggests the possibility of direct binding of celecoxib to the Kv7 channel subunits, similar to retigabine binding [33]. Functionally, however, differences between celecoxib and retigabine actions on Kv7 channels exist. Celecoxib induced a leftward shift in the voltage-dependence of activation of channels constituted by either Kv7.2 or Kv7.3, but it did not induce a shift in channels constituted by Kv7.4 or Kv7.5 or in native myocytes [18, 33, 72]. The possibility that signal transduction-mediated effects contribute to the actions of celecoxib on Kv7 channels has yet to be ruled out, though there is compelling evidence that inhibition of cyclooxygenase-2 (COX-2, the target of celecoxib for its anti-inflammatory effects) is not required (see below).

Recent evidence also suggests that other COX inhibitors, including diclofenac and meclofenamate, also act as potent Kv7 channel modulators [19, 98]. Although an appreciable commonality in COX inhibition and Kv7 current activation with these drugs is evident, the effect on Kv7 channels is independent of COX inhibition. For example, 2,5-dimethyl-celecoxib, a structural analog of celecoxib that lacks its ability to inhibit COX-2, was as effective as celecoxib as an activator of Kv7 channels, whereas rofecoxib, a highly efficacious COX-2 inhibitor, was inactive as a Kv7 channel opener [18, 33]. Therefore the spectrum of ion channel modulatory effects of these molecules is drug-specific and not a class-effect [18, 21]. The actions of 2,5-dimethyl-celecoxib on vascular ion channels may present an opportunity for

Table 9.1 Relaxation of arteries by different Kv7 channel activators

Vascular bed	Vasoconstrictor	Kv7 channel activator	Clinical use	Kv7 channel specificity
Rat basilar artery [72, 74]	Serotonin (5HT), Vasopressin (AVP), Endothelin (ET-1)	Flupirtine	Analgesic	Kv7.2–7.5 activators
		Celecoxib	NSAID	
		Retigabine	Anticonvulsant	
Rat carotid artery, femoral artery, mesenteric artery [132]	Phenylephrine	Retigabine	Anticonvulsant	Kv7.2–7.5 activator
Human visceral adipose tissue artery, mesenteric artery [84]	Phenylephrine	Retigabine	Anticonvulsant	Kv7.2–7.5 activators
		S-1	N/A	
Rat pulmonary artery [52]	Phenylephrine	Retigabine	Anticonvulsant	Kv7.2–7.5 activators
		Flupirtine	Analgesic	
Rat coronary artery [55]	U46619	Retigabine	Anticonvulsant	Kv7.2–7.5 activators
		S-1	N/A	
		BMS-204352	N/A	
Rat middle cerebral artery [24]	U46619	Retigabine	Anticonvulsant	Kv7.2–7.5 activators
		S-1	N/A	
Rat thoracic aorta, mouse mesenteric and tibial arteries [109]	Serotonin (5HT), U46619	Retigabine	Anticonvulsant	Kv7.2–7.5 activators
		VRX0530727	N/A	
		VRX0621238	N/A	
		VRX0621688	N/A	
Rat mesenteric artery, thoracic aorta and intrapulmonary artery [25, 51]	Methoxamine	L-364 373	N/A	Kv7.1 activators
		(R-L3)	N/A	
		Mefenamic acid	Anticonvulsant	Kv7.2–7.5 activators
		Retigabine	N/A	
		S-1	N/A	
		BMS-204352		
Rat mesenteric artery [71]	Vasopressin (AVP)	Flupirtine	Analgesic	Kv7.2–Kv7.5 activator
Rat mesenteric artery [20]	Vasopressin (AVP)	Zinc pyrithione	Anti-fungal agent used in dandruff shampoo	Kv7.1, Kv7.2, Kv7.4, Kv7.5 activator [130]
Rat gracillis artery [135]	Serotonin (5HT)	VRX0530727	N/A	Kv7.2–7.5 activators
		VRX0621238	N/A	
		VRX0621688	N/A	
Rat aorta, renal, and mesenteric arteries [50]	Methoxamine	ML213, NS15370	N/A	Kv7.2–7.5 activators
			N/A	
Murine basilar and coronary arteries [62]	U46619	Retigabine	Anticonvulsant	Kv7.2–7.5 activator

development of chemical derivatives that would have vasodilatory effects without inhibiting COX-2 activity. Such a class of compounds might be useful for treating vasospastic conditions in patients who cannot tolerate NSAID therapy.

The remainder of this chapter will review the evidence that Kv7 channels play a prominent role in maintenance of basilar and cerebral artery tone and will also consider whether pharmacological Kv7 channel activators might be re-purposed as novel therapeutics for the relief of cerebral vasospasm.

Kv7 Channels in the Vasculature

The Kv7 channel isoforms found to be universally expressed in arterial myocytes, including rat basilar artery myocytes [72, 138], are Kv7.1, Kv7.4 and Kv7.5 [22, 52, 71, 84, 91, 132]. All available evidence suggests that functional channels are predominantly composed of Kv7.4 and Kv7.5 homomers or Kv7.4/7.5 heteromers [17, 19, 24]. Recent studies using dominant negative Kv7.4 or Kv7.5 subunits and proximal ligation assays support Kv7.4/Kv7.5 heterotetramers as functional channels in mesenteric artery myocytes [17]. The threshold for voltage-dependent activation of vascular smooth muscle Kv7 currents closely matches the resting membrane voltage (around −60 mV) and pharmacological activation or inhibition of Kv7 currents is associated with corresponding changes in membrane voltage [52, 71, 72].

Like neuronal M-currents, vascular smooth muscle cell Kv7 currents are suppressed upon activation of certain GPCRs. The first evidence for GPCR-dependent regulation of endogenous vascular smooth muscle Kv7 currents was obtained using A7r5 rat aortic smooth muscle cells natively expressing the Kv7.5 channel isoform. Activation of endogenous V_{1a} vasopressin receptors in A7r5 cells by 100 pM arginine vasopressin (AVP) suppressed Kv7.5 currents in a protein kinase C- (PKC-) dependent manner [22]. Suppression of Kv7.5 currents by AVP, or knockdown of Kv7.5 expression by shRNA, led to membrane depolarization and action potential firing in A7r5 cells [22, 73]. AVP was also found to induce PKC-dependent suppression of native Kv7 currents in mesenteric artery smooth muscle cells, an effect which appears to be essential for constriction of pressurized mesenteric arteries by physiological concentrations of AVP [71]. Activators of Kv7 channels were able to relax mesenteric arteries pre-constricted with AVP *in vitro* or *in vivo* [16, 71]. Work from a number of laboratories revealed similar vasorelaxant effects of Kv7 channel activators on arteries from different vascular beds pre-constricted with a variety of GPCR agonists (Table 9.1) [24, 25, 51, 52, 55, 72, 74, 84, 132].

Kv7 Channels in Basilar Artery Myocytes

Kv7 channels are expressed in rat basilar and cerebral artery smooth muscle cells [72, 138]. Proximity ligation assays support formation of Kv7.4/Kv7.5 heteromeric channels in rat cerebral artery myocytes [24], though Kv7.5 mRNA expression was nearly undetectable in murine basilar artery, while Kv7.4 expression was robustly detected at both mRNA and protein levels in the same tissues [62]. So far, there is no evidence suggestive of Kv7 channel subunits interacting with subunits of other Kv channel families to form K^+ conducting channels. Whole cell Kv7 currents, recorded from rat basilar artery myocytes using patch-clamp electrophysiology, exhibited biophysical properties typical of Kv7 currents (activation threshold around −60 mV, minimal time-dependent inactivation) [72, 74], as previously shown in neurons [2, 121] and vascular smooth muscle cells [22, 71]. The currents also displayed sensitivity to pharmacological Kv7 channel activators (retigabine (Fig. 9.3), flupirtine, and celecoxib) and were blocked by XE991 [72, 74], similar to what was observed for mesenteric artery myocytes [18, 71].

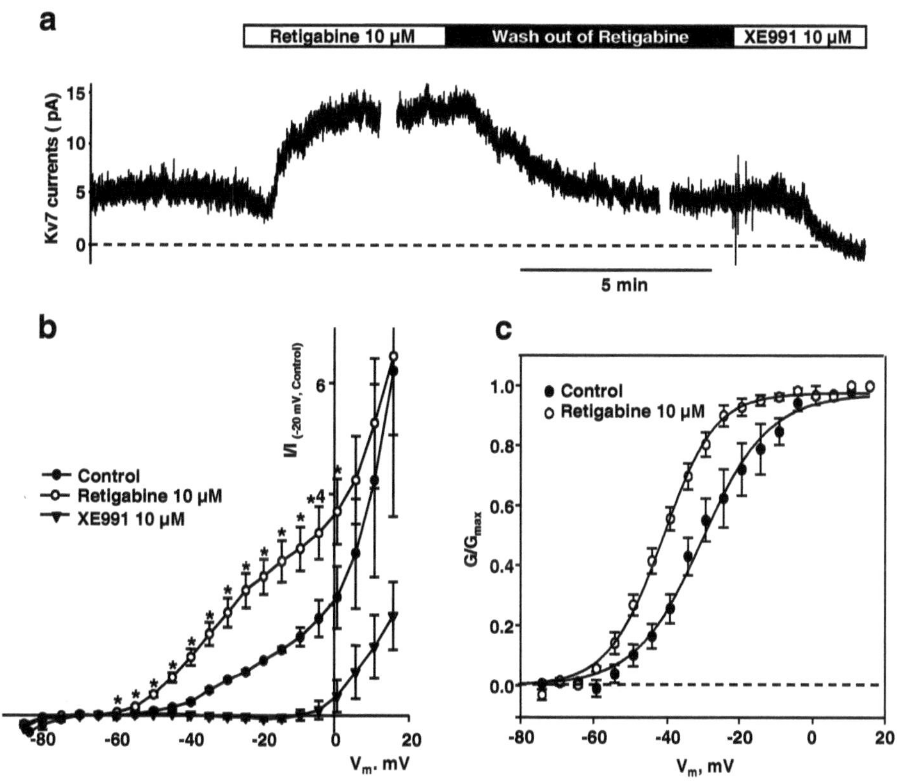

Fig. 9.3 Retigabine enhances Kv7 currents in freshly isolated rat basilar artery myocytes. (**a**) Time-course of XE991-sensitive Kv7 currents with the addition of 10 μM retigabine, recorded from a single basilar artery myocyte by continuous holding at −20 mV. (**b**) Current-voltage curves (I-V) of mean steady-state Kv7 currents before (control; *filled circles*) and during treatment with 10 μM retigabine (*open circles*) and 10 μM XE991 (*filled inverted triangles*), normalized to control current measured at −20 mV before treatment. *Significantly different from control at indicated voltages using paired Student's "t" test, $p<0.05$, $n=5$. (**c**) Normalized conductance plotted against membrane voltage was fitted by a single Boltzmann function to show the effect of retigabine on voltage dependence of channel activation. Retigabine induced a significant hyperpolarizing shift in the steady-state voltage dependence of activation; $V_{0.5}$ (control) was -30.2 ± 1.4 mV, $V_{0.5}$ (retigabine) was -41.4 ± 0.5 mV, $p<0.05$ based on paired Student's "t" test, $n=5$. Reproduced from [74] with permission

Kv7 Channels Are Common Signal Transduction Intermediates in Spasmogen-Induced Constriction

Suppression of Kv7 channels using the selective Kv7 channel blocker XE991 induced membrane depolarization in basilar artery myocytes and constricted basilar artery segments [72]. Cerebral artery spasmogens (e.g. serotonin, AVP, and endothelin) which act on Gq-coupled GPCRs, are present at elevated concentrations in the subarachnoid space after aSAH and have been implicated in the development of cerebral vasospasm [5, 11, 86, 119, 139]. Serotonin, AVP, and endothelin each significantly suppressed Kv7 currents in freshly dissociated rat basilar artery myocytes, supporting the hypothesis that these spasmogens suppress Kv7 currents to induce cerebral vasospasm [72]. Furthermore, a demonstrated contribution of Kv7 currents to maintenance of resting membrane voltage in basilar artery myocytes [72] indicates that the suppression of Kv7 currents by the spasmogens is likely to be a key contributing factor to the persistent depolarization observed in basilar artery myocytes after SAH [43, 123, 140].

The signal transduction events activated in response to binding of the spasmogens to the GPCRs, which eventually lead to the suppression of Kv7 channels in vascular myocytes, are just beginning to be understood. There is abundant evidence that activation of protein kinase C (PKC) is both necessary and sufficient for the GPCR-mediated suppression of Kv7 currents in vascular smooth muscle cells [17, 22, 71, 74]. This contrasts to the GPCR-mediated suppression of Kv7 currents in neurons, which has been shown to predominantly involve hydrolysis of phosphatidylinositol 4,5-bisphosphate (PIP_2) and/or elevation of $[Ca^{2+}]_{cyt}$ [31]. The sensitivity of vascular Kv7 channels to activated PKC differs depending upon the tetrameric subunit composition—human Kv7.5 (hKv7.5), and hKv7.4/hKv7.5 channels when overexpressed in A7r5 cells, were regulated by the AVP-PKC pathway (the Kv7 currents were suppressed by AVP and by 4β-Phorbol 12-myristate 13-acetate (PMA), a PKC activator), but hKv7.4 channels were not [17]. Similarly, a corresponding increase in PKC-dependent phosphorylation of hKv7.5 was detected, but not for hKv7.4 subunits, except when they were co-expressed with hKv7.5 [17]. The requirement of Kv7.5 channel phosphorylation to mediate the AVP-induced suppression of currents through the channel is still yet to be confirmed. These results suggest that diversity of Kv7 channel subunit composition among different vascular beds could be a determinant for differential sensitivity to vasoconstrictors.

Activation and translocation of PKC from the cytosol to the membrane has been demonstrated in myocytes within vasospastic arteries from SAH models [88, 89]. There are several pieces of evidence implicating PKC in the development and maintenance of cerebral vasospasm [60]. Constriction of basilar artery in response to spasmogens like AVP, serotonin and endothelin was shown to be dependent on PKC activation [79, 80]. Intracisternal injection of a PKC activator mimicked basilar artery vasospasm in a canine model [105]. Finally, PKC inhibitors, H7 and staurosporine, inhibited spasmogen-induced constrictor responses and reversed basilar artery vasospasm in a dog model of SAH [76]. Taken together with our recent results regarding PKC-dependent regulation of Kv7 channels, these findings suggest that PKC is a common signaling intermediate involved in the suppression of Kv7 currents by spasmogens, leading to persistent membrane depolarization of arterial myocytes and chronic arterial constriction.

Kv7 Channel Openers as Novel Therapeutics for Treatment of Cerebral Vasospasm

Considering that cerebral vasospasm following aSAH likely results at least in part from suppression of Kv7 currents, it stands to reason that enhancement of Kv7 currents might have a beneficial opposing effect. In support of this notion, basilar artery segments pre-constricted with a spasmogen (serotonin, endothelin, or vasopressin) were dilated by addition of a Kv7 channel activator (retigabine, flupirtine, or celecoxib (Fig. 9.4)) [72, 74]. The dilation can be attributed to activation of Kv7 channels, which opposes spasmogen-induced membrane depolarization and influx of Ca^{2+} via L-type VDCCs. Addition of the Kv7 channel opener retigabine prevented spasmogen-induced membrane depolarization in basilar artery myocytes [74]. The vasodilatory effects of retigabine and celecoxib in basilar arteries might also be attributable to direct blocking of VDCCs, independent of their actions on Kv7 channels [72, 74], though flupirtine was still an effective vasodilator [72] despite evidence that it does not block VDCCs [18]. Celecoxib was more efficacious in inhibiting voltage-dependent calcium currents compared to the clinically used calcium channel blocker nimodipine in rat basilar artery myocytes (Fig. 9.5; [72]). The higher efficacy of celecoxib could be due to inhibition of all or most types of VDCCs expressed in the basilar artery myocytes compared to more selective inhibition of L-type calcium channels by nimodipine. Celecoxib has been recently shown to inhibit T-type VDCCs in other

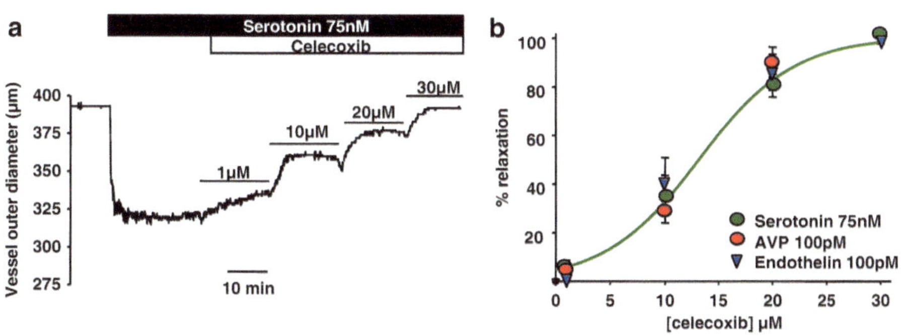

Fig. 9.4 Celecoxib reverses the spasmogen-induced constriction in basilar artery. (**a**) Representative trace shows the concentration-dependent effect of celecoxib on a basilar artery segment pre-constricted with 75 nM serotonin. (**b**) Mean concentration-response curves of the percentage relaxation produced by celecoxib in arteries pre-constricted with the spasmogens, serotonin, AVP and ET-1. Reproduced from [72] with permission

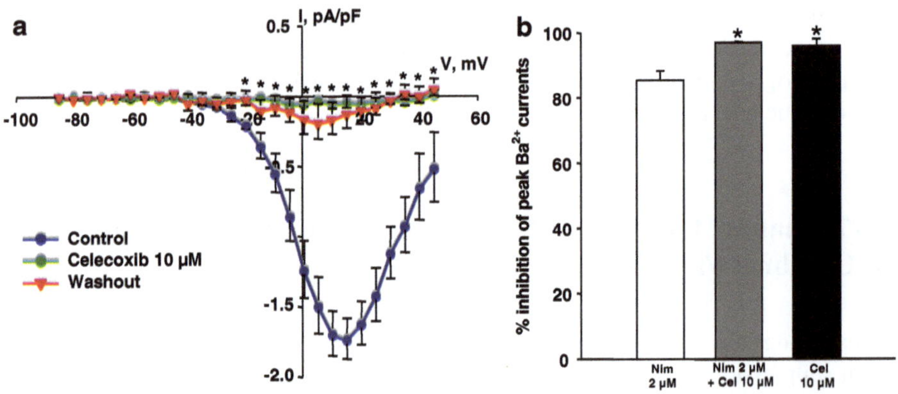

Fig. 9.5 Celecoxib inhibits VDCCs in basilar artery myocytes. (**a**) Summarized I-V curves show the effect of 10 μM celecoxib on L-type voltage-dependent Ba^{2+} currents (I_{Ba}). Celecoxib significantly inhibited I_{Ba} at all tested voltages from −20.2 to +39.8 mV ($n=5$, $*P<0.05$, Student's t-test). (**b**) Percentage inhibition of peak I_{Ba} with the application of 2 μM nimodipine alone, 2 μM nimodipine + 10 μM celecoxib, and 10 μM celecoxib alone. Celecoxib either alone or when added along with nimodipine produced significantly more inhibition of I_{Ba} compared with nimodipine alone ($n=5$–6, $*P<0.05$ using one-way ANOVA followed by *post hoc* Holm-Sidak test). Reproduced from [72] with permission

cell types [7], but its effect on R-, P/Q-, and N-type VDCCs, which are all known to be expressed in basilar artery myocytes [81, 85] is unknown.

The efficacy of retigabine and celecoxib to dilate basilar artery segments led us to investigate whether activation of vascular Kv7 channels could be used as a strategy to treat cerebral vasospasm [72, 74]. We used a rat model of SAH in which the presence of blood after SAH was simulated by injection of fresh autologous arterial blood into the cisterna magna [74]. Basilar artery spasm was significant after 48 h in rats injected with autologous blood compared with rats injected with artificial cerebrospinal fluid or untreated controls [74]. Administration of retigabine or celecoxib by intraperitoneal injection twice daily for 48 h prevented spasm of the basilar artery in animals injected with autologous blood (Fig. 9.6; [74]). Based on these findings, we propose that drugs which act as Kv7 channel openers are promising as novel therapeutic agents to alleviate cerebral vasospasm.

Retigabine also induced significant hypothermia in the rats with experimental SAH [74], confirming an earlier report of the same effect when retigabine was injected in mice [58]. The hypothermia induced by retigabine could be beneficial in the case of cerebral vasospasm, as clinical induction of mild hypothermia in patients has been shown to have neuroprotective effects and to improve the functional outcome in patients with delayed cerebral vasospasm [6, 36, 92]. Kv7 channel openers could also be potentially combined with other classes of drugs that target other cell types or signal transduction pathways to treat the vasospasm and associated pathology after aSAH, which is increasingly recognized as multifactorial [69, 97, 104].

With the approval of retigabine to treat epilepsy, and increasing evidence from clinical trials in the US and EU for beneficial effects of retigabine and flupirtine in

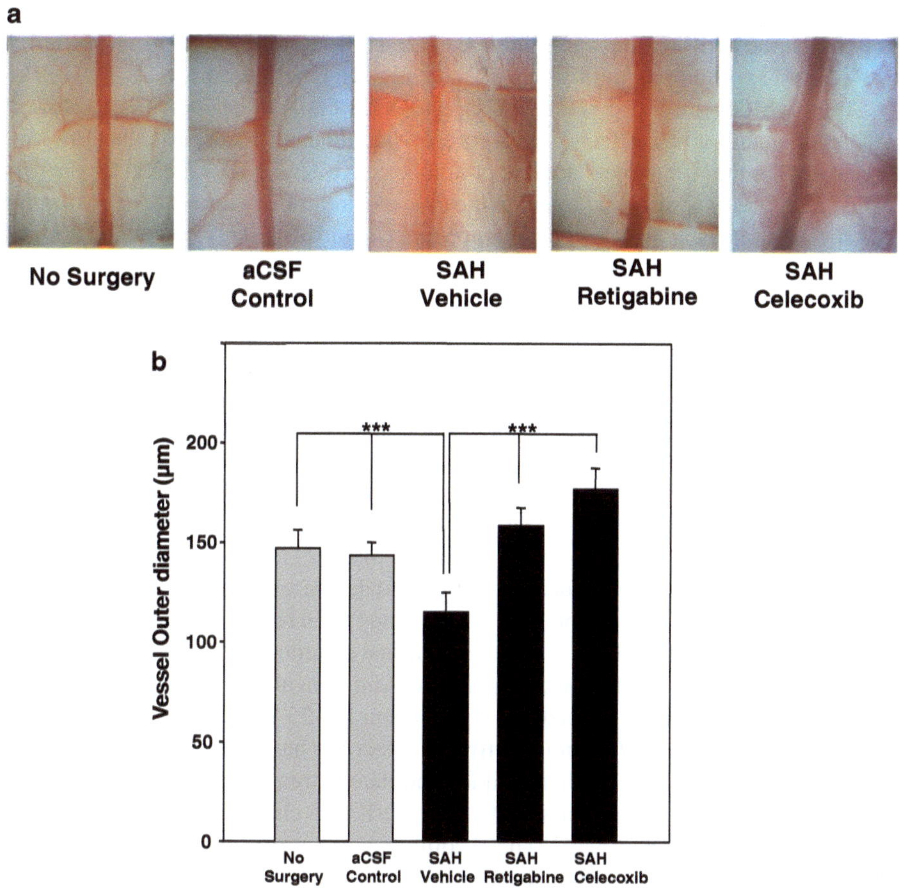

Fig. 9.6 Kv7 channel openers attenuate basilar artery spasm in a rat model of SAH. (**a**) Illustrative photographs show basilar artery in the base of the brain from rats in different treatment groups. Brains were photographed 48 h after injection of blood or aCSF and compared with brains of control rats (no surgery). (**b**) Bar graph shows the mean basilar artery outer diameter measured in each of the treatment groups—*Grey bars*: Control (no surgery), aCSF control (sham surgery), *Black bars*: SAH rats treated with vehicle (DMSO, 1 ml/kg, i.p., b.i.d), SAH rats treated with retigabine (7.5 mg/kg, i.p., b.i.d) and SAH rats treated with celecoxib (20 mg/kg, i.p., b.i.d). ***Significant difference between the treatment groups as indicated using one-way ANOVA followed by post hoc Holm-Sidak test, $p<0.001$, $n=9$. Reproduced from [74] with permission

the treatment of pain associated with fibromyalgia and neuropathic pain, there has been intense interest in developing more potent and selective Kv7 channel activators. Drugs like retigabine and flupirtine act as openers of neuronal Kv7.2/Kv7.3 channels, decreasing neuronal excitability without activating cardio-selective Kv7.1 channels [108, 129]. However, retigabine and flupirtine clearly do not discriminate among Kv7 channels expressed in neurons (predominantly Kv7.2/Kv7.3) and those expressed in vascular myocytes (predominantly Kv7.4/Kv7.5). Presently, we do not know of any compounds that display either neuronal or vascular Kv7 channel selectivity.

Recently Jensen and colleagues were successful in synthesizing an acrylamide compound, named SMB-1, that can inhibit Kv7.2, but activate Kv7.4 [13], indicating a possibility for development of tissue-specific Kv7 channel openers.

Although developing a selective Kv7.4/Kv7.5 channel opener might prevent off-target neurological effects, while preserving the anti-vasospastic actions, we speculate that the "off-target" actions of non-selective Kv7.2-Kv7.5 channel openers would likely be of additional benefit in the case of cerebral vasospasm. The reason for this speculation is that pan-Kv7 channel openers would also likely counteract cortical spreading depression (CSD), another devastating co-morbid consequence after aSAH [61, 106]. This self-propagating slow wave of depolarization in neurons and glial cells is accompanied by depression of electrocorticographic activity [106]. The spreading depolarization is triggered by pathological repartitioning of ions across the cell membrane, with marked elevation of extracellular K^+ and vasoactive neurotransmitter concentrations after SAH [57]. The loss of neuronal function and neurovascular coupling due to CSD, along with the increased energy demand by neurons to restore the ionic balances, contributes to vasospasm and ischemia after SAH [57, 106]. Kv7 channel activators decrease neuronal excitability by increasing neuronal M-currents mediated primarily by Kv7.2/Kv7.3 channels [49, 121], an effect that would be expected to inhibit the wave of depolarization and decrease the energy demands of the gray matter. Pan-Kv7 channel openers have already been reported to counteract cortical spreading events associated with other neuronal disorders [127, 136] and therefore are likely to be beneficial in aSAH patients. Anti-convulsant effects of drugs like retigabine may also alleviate seizures, which occur in ~25 % of aSAH patients [77].

The non-selective Kv7 channel activator celecoxib has several pharmacological properties, which make it a rather unique candidate to treat cerebral vasospasm after aSAH. Celecoxib acts as a non-selective Kv7.2-Kv7.5 channel opener [33] as well as a blocker of VDCCs [72, 74]. These combined effects can account for its ability to dilate the vasospastic basilar artery both *ex vivo* and *in vivo* [72, 74]. If these effects are also exhibited in neurons (likely, but not yet tested), celecoxib would also be expected to inhibit the co-morbid development of cortical spreading depression in aSAH. While the dual ion channel activity noted above would be expected to lower blood pressure, celecoxib, likely owing to pro-hypertensive effects downstream of COX-2 inhibition [44, 133], has not been found to significantly lower blood pressure in animal models or in clinical subjects [28]. In fact, celecoxib tends to modestly elevate blood pressure with chronic use, though to a lesser extent than other COX-2 inhibitors, which lack its vascular ion channel effects [21, 28, 115]. This blood pressure neutrality is favored, because one of the serious limitations to use of K^+ channel openers for cardiovascular diseases, in general, is their propensity to produce hypotension. Specifically, hypotension would be detrimental as a side-effect of drugs to treat cerebral vasospasm, as it would further compromise the already constrained blood flow to the brain (due to vasospasm) and would therefore exacerbate ischemia. Hence, celecoxib, which dilates the cerebral arteries while maintaining sustained mean arterial pressure [74] may possibly be an ideal candidate to treat cerebral vasospasm. And, as an additional benefit, celecoxib may simultaneously treat the inflammatory component of aSAH pathology. The COX-2 inhibitory actions of celecoxib would likely oppose inflammation at the hemorrhagic site, thereby attenuating the contributions of the immune system to vasospasm after aSAH [23, 34, 78, 100]. Celecoxib has been extensively used as an anti-inflammatory agent in the clinical setting, including a recent small scale clinical trial in patients with intracerebral hemorrhage (ICH), in which celecoxib treatment was found to have beneficial effects (reduction of hematoma and edema volumes) without increasing the incidence of adverse events [63, 95].

Conclusions

Cerebral vasospasm is a devastating medical condition for which current therapies are inadequate. Kv7 channel openers, already in clinical use to treat neurological conditions such as epilepsy and pain, may be re-purposed to relieve or even prevent cerebral vasospasm, while simultaneously attenuating other co-morbidities associated with aSAH, such as cortical spreading depression, delayed ischemic neurological deficits, seizures, inflammation, and edema formation. There is increasing recognition among the medical community that development of more effective therapeutic strategies for aSAH will require drugs or combinations of drugs that target multiple pathological pathways rather than simply alleviating cerebral vasospasm [32, 42]. The established safety profiles of Kv7 channel activators, and their broad spectrum of beneficial therapeutic effects, suggest that these drugs may be effectively re-purposed, perhaps in combination with other treatments, to develop a more effective therapeutic strategy for improving outcomes in patients following aSAH.

References

1. Effect of calcitonin-gene-related peptide in patients with delayed postoperative cerebral ischaemia after aneurysmal subarachnoid haemorrhage. European CGRP in Subarachnoid Haemorrhage Study Group. Lancet 1992;339:831–4.
2. Adams PR, Brown DA. Luteinizing hormone-releasing factor and muscarinic agonists act on the same voltage-sensitive K^+-current in bullfrog sympathetic neurones. Br J Pharmacol. 1980;68:353–5.
3. Aihara Y, Jahromi BS, Yassari R, Nikitina E, Agbaje-Williams M, Macdonald RL. Molecular profile of vascular ion channels after experimental subarachnoid hemorrhage. J Cereb Blood Flow Metab. 2004;24:75–83.
4. Aiken SP, Lampe BJ, Murphy PA, Brown BS. Reduction of spike frequency adaptation and blockade of M-current in rat CA1 pyramidal neurones by linopirdine (DuP 996), a neurotransmitter release enhancer. Br J Pharmacol. 1995;115:1163–8.
5. Allen GS, Gold LH, Chou SN, French LA. Cerebral arterial spasm. 3. In vivo intracisternal production of spasm by serotonin and blood and its reversal by phenoxybenzamine. J Neurosurg. 1974;40:451–8.
6. Badjatia N, Fernandez L, Schmidt JM, Lee K, Claassen J, Connolly ES, Mayer SA. Impact of induced normothermia on outcome after subarachnoid hemorrhage: a case-control study. Neurosurgery. 2010;66:696–700. discussion 700–1.

7. Balderas E, Sanchez-Cardenas C, Chavez JC, de la Vega Beltran JL, Gomez-Lagunas F, Trevino CL, Darszon A. The anti-inflammatory drug celecoxib inhibits T-type Ca^{2+} currents in spermatogenic cells yet it elicits the acrosome reaction in mature sperm. FEBS Lett. 2013;587:2412–9.
8. Barhanin J, Lesage F, Guillemare E, Fink M, Lazdunski M, Romey G. K(V)LQT1 and lsK (minK) proteins associate to form the I(Ks) cardiac potassium current. Nature. 1996;384:78–80.
9. Barrett JN, Magleby KL, Pallotta BS. Properties of single calcium-activated potassium channels in cultured rat muscle. J Physiol. 1982;331:211–30.
10. Baylis PH, Robertson GL. Osmotic and non-osmotic stimulation of vasopressin release [proceedings]. J Endocrinol. 1979;83:39P–40.
11. Beg SS, Hansen-Schwartz JA, Vikman PJ, Xu CB, Edvinsson LI. Protein kinase C inhibition prevents upregulation of vascular ET(B) and 5-HT(1B) receptors and reverses cerebral blood flow reduction after subarachnoid haemorrhage in rats. J Cereb Blood Flow Metab. 2007;27:21–32.
12. Biervert C, Schroeder BC, Kubisch C, Berkovic SF, Propping P, Jentsch TJ, Steinlein OK. A potassium channel mutation in neonatal human epilepsy. Science. 1998;279:403–6.
13. Blom SM, Rottlander M, Kehler J, Bundgaard C, Schmitt N, Jensen HS. From pan-reactive Kv7 channel opener to subtype selective opener/inhibitor by addition of a methyl group. PLoS One. 2014;9:e100209.
14. Bradley KK, Jaggar JH, Bonev AD, Heppner TJ, Flynn ER, Nelson MT, Horowitz B. Kir2.1 encodes the inward rectifier potassium channel in rat arterial smooth muscle cells. J Physiol. 1999;515:639–51.
15. Brisman JL, Song JK, Newell DW. Cerebral Aneurysms. N Engl J Med. 2006;355:928–39.
16. Brueggemann LI, Byron KL. Use of patch clamp electrophysiology to identify off-target effects of clinically used drugs. In: Kaneez FS, editor. Patch clamp technique. Rijeka, Croatia: Intech; 2012, pp 267–282.
17. Brueggemann LI, Mackie AR, Cribbs LL, Freda J, Tripathi A, Majetschak M, Byron KL. Differential protein kinase C-dependent modulation of Kv7.4 and Kv7.5 subunits of vascular Kv7 channels. J Biol Chem. 2013;289:2099–111.
18. Brueggemann LI, Mackie AR, Mani BK, Cribbs LL, Byron KL. Differential effects of selective cyclooxygenase-2 inhibitors on vascular smooth muscle ion channels may account for differences in cardiovascular risk profiles. Mol Pharmacol. 2009;76:1053–61.
19. Brueggemann LI, Mackie AR, Martin JL, Cribbs LL, Byron KL. Diclofenac distinguishes among homomeric and heteromeric potassium channels composed of KCNQ4 and KCNQ5 subunits. Mol Pharmacol. 2011;79:10–23.
20. Brueggemann LI, Mani BK, Haick J, Byron KL. Exploring arterial smooth muscle Kv7 potassium channel function using patch clamp electrophysiology and pressure myography. J Vis Exp. 2012;14:4263.
21. Brueggemann LI, Mani BK, Mackie AR, Cribbs LL, Byron KL. Novel actions of nonsteroidal anti-inflammatory drugs on vascular ion channels: accounting for cardiovascular side effects and identifying new therapeutic applications. Mol Cell Pharmacol. 2010;2:15–9.
22. Brueggemann LI, Moran CJ, Barakat JA, Yeh JZ, Cribbs LL, Byron KL. Vasopressin stimulates action potential firing by protein kinase C-dependent inhibition of KCNQ5 in A7r5 rat aortic smooth muscle cells. Am J Physiol Heart Circ Physiol. 2007;292:H1352–63.
23. Carr KR, Zuckerman SL, Mocco J. Inflammation, cerebral vasospasm, and evolving theories of delayed cerebral ischemia. Neurol Res Int. 2013;2013:506584.
24. Chadha PS, Jepps TA, Carr G, Stott JB, Zhu H-L, Cole WC, Greenwood IA. Contribution of Kv7.4/Kv7.5 heteromers to intrinsic and calcitonin gene-related peptide—induced cerebral reactivity. Arterioscler Thromb Vasc Biol. 2014;34:887–93.
25. Chadha PS, Zunke F, Davis AJ, Jepps TA, Linders JT, Schwake M, Towart R, Greenwood IA. Pharmacological dissection of K(v)7.1 channels in systemic and pulmonary arteries. Br J Pharmacol. 2012;166:1377–87.
26. Chadha PS, Zunke F, Zhu H-L, Davis AJ, Jepps TA, Olesen SP, Cole WC, Moffatt JD, Greenwood IA. Reduced KCNQ4-encoded voltage-dependent potassium channel activity

underlies impaired β-adrenoceptor-mediated relaxation of renal arteries in hypertension. Hypertension. 2012;59:877–84.
27. Charlier C, Singh NA, Ryan SG, Lewis TB, Reus BE, Leach RJ, Leppert M. A pore mutation in a novel KQT-like potassium channel gene in an idiopathic epilepsy family. Nat Genet. 1998;18:53–5.
28. Cho J, Cooke CE, Proveaux W. A retrospective review of the effect of COX-2 inhibitors on blood pressure change. Am J Ther. 2003;10:311–7.
29. Connolly Jr ES, Rabinstein AA, Carhuapoma JR, Derdeyn CP, Dion J, Higashida RT, Hoh BL, Kirkness CJ, Naidech AM, Ogilvy CS, Patel AB, Thompson BG, Vespa P. Guidelines for the management of aneurysmal subarachnoid hemorrhage: a guideline for healthcare professionals from the American Heart Association/American Stroke Association. Stroke. 2012;43:1711–37.
30. Daut J, Maier-Rudolph W, von Beckerath N, Mehrke G, Gunther K, Goedel-Meinen L. Hypoxic dilation of coronary arteries is mediated by ATP-sensitive potassium channels. Science. 1990;247:1341–4.
31. Delmas P, Brown DA. Pathways modulating neural KCNQ/M (Kv7) potassium channels. Nat Rev Neurosci. 2005;6:850–62.
32. Diringer MN. Controversy: does prevention of vasospasm in subarachnoid hemorrhage improve clinical outcome? Stroke. 2013;44:S29–30.
33. Du XN, Zhang X, Qi JL, An HL, Li JW, Wan YM, Fu Y, Gao HX, Gao ZB, Zhan Y, Zhang HL. Characteristics and molecular basis of celecoxib modulation on K_v7 potassium channels. Br J Pharmacol. 2011;164:1722–37.
34. Dumont AS, Dumont RJ, Chow MM, Lin CL, Calisaneller T, Ley KF, Kassell NF, Lee KS. Cerebral vasospasm after subarachnoid hemorrhage: putative role of inflammation. Neurosurg. 2003;53:123–33. discussion 133–5.
35. Feigin VL, Rinkel GJ, Algra A, Vermeulen M, van Gijn J. Calcium antagonists in patients with aneurysmal subarachnoid hemorrhage: a systematic review. Neurology. 1998;50:876–83.
36. Fernandez A, Schmidt JM, Claassen J, Pavlicova M, Huddleston D, Kreiter KT, Ostapkovich ND, Kowalski RG, Parra A, Connolly ES, Mayer SA. Fever after subarachnoid hemorrhage: risk factors and impact on outcome. Neurology. 2007;68:1013–9.
37. Filosa JA, Bonev AD, Straub SV, Meredith AL, Wilkerson MK, Aldrich RW, Nelson MT. Local potassium signaling couples neuronal activity to vasodilation in the brain. Nat Neurosci. 2006;9:1397–403.
38. Frolov RV, Berim IG, Singh S. Inhibition of delayed rectifier potassium channels and induction of arrhythmia: a novel effect of celecoxib and the mechanism underlying it. J Biol Chem. 2008;283:1518–24.
39. Frolov RV, Bondarenko VE, Singh S. Mechanisms of Kv2.1 channel inhibition by celecoxib—modification of gating and channel block. Br J Pharmacol. 2010;159:405–18.
40. Frolov RV, Singh S. Celecoxib and ion channels: a story of unexpected discoveries. Eur J Pharmacol. 2014;730:61–71.
41. Grunnet M, Strobaek D, Hougaard C, Christophersen P. Kv7 channels as targets for anti-epileptic and psychiatric drug-development. Eur J Pharmacol. 2014;726:133–7.
42. Hansen-Schwartz J, Vajkoczy P, Macdonald RL, Pluta RM, Zhang JH. Cerebral vasospasm: looking beyond vasoconstriction. Trends Pharmacol Sci. 2007;28:252–6.
43. Harder DR, Dernbach P, Waters A. Possible cellular mechanism for cerebral vasospasm after experimental subarachnoid hemorrhage in the dog. J Clin Invest. 1987;80:875–80.
44. Hinz B, Dormann H, Brune K. More pronounced inhibition of cyclooxygenase 2, increase in blood pressure, and reduction of heart rate by treatment with diclofenac compared with celecoxib and rofecoxib. Arthritis Rheum. 2006;54:282–91.
45. Jahromi BS, Aihara Y, Ai J, Zhang ZD, Nikitina E, Macdonald RL. Voltage-gated K+ channel dysfunction in myocytes from a dog model of subarachnoid hemorrhage. J Cereb Blood Flow Metab. 2008;28:797–811.
46. Jahromi BS, Aihara Y, Ai J, Zhang ZD, Weyer G, Nikitina E, Yassari R, Houamed KM, Macdonald RL. Preserved BK channel function in vasospastic myocytes from a dog model of subarachnoid hemorrhage. J Vasc Res. 2008;45:402–15.

47. Jahromi BS, Aihara Y, Ai J, Zhang ZD, Weyer G, Nikitina E, Yassari R, Houamed KM, Macdonald RL. Temporal profile of potassium channel dysfunction in cerebrovascular smooth muscle after experimental subarachnoid haemorrhage. Neurosci Lett. 2008;440:81–6.
48. Janjua N, Mayer SA. Cerebral vasospasm after subarachnoid hemorrhage. Curr Opin Crit Care. 2003;9:113–9.
49. Jentsch TJ. Neuronal KCNQ potassium channels: physiology and role in disease. Nat Rev Neurosci. 2000;1:21–30.
50. Jepps TA, Bentzen BH, Stott JB, Povstyan OV, Sivaloganathan K, Dalby-Brown W, Greenwood IA. Vasorelaxant effects of novel Kv 7.4 channel enhancers ML213 and NS15370. Br J Pharmacol. 2014;171:4413–24.
51. Jepps TA, Chadha PS, Davis AJ, Harhun MI, Cockerill GW, Olesen SRP, Hansen RS, Greenwood IA. Downregulation of Kv7.4 channel activity in primary and secondary hypertension. Circulation. 2011;124:602–11.
52. Joshi S, Sedivy V, Hodyc D, Herget J, Gurney AM. KCNQ modulators reveal a key role for KCNQ potassium channels in regulating the tone of rat pulmonary artery smooth muscle. J Pharmacol Exp Ther. 2009;329:368–76.
53. Kamp MA, Dibue M, Schneider T, Steiger HJ, Hanggi D. Calcium and potassium channels in experimental subarachnoid hemorrhage and transient global ischemia. Stroke Res Treat. 2012;2012:382146.
54. Kassell NF, Sasaki T, Colohan AR, Nazar G. Cerebral vasospasm following aneurysmal subarachnoid hemorrhage. Stroke. 1985;16:562–72.
55. Khanamiri S, Soltysinska E, Jepps TA, Bentzen BH, Chadha PS, Schmitt N, Greenwood IA, Olesen SP. Contribution of Kv7 channels to basal coronary flow and active response to ischemia. Hypertension. 2013;62:1090–7.
56. Knot HJ, Zimmermann PA, Nelson MT. Extracellular K^+-induced hyperpolarizations and dilatations of rat coronary and cerebral arteries involve inward rectifier K^+ channels. J Physiol. 1996;492:419–30.
57. Koide M, Sukhotinsky I, Ayata C, Wellman GC. Subarachnoid hemorrhage, spreading depolarizations and impaired neurovascular coupling. Stroke Res Treat. 2013;2013:819340.
58. Kristensen LV, Sandager-Nielsen K, Hansen HH. Kv7 (KCNQ) channel openers induce hypothermia in the mouse. Neurosci Lett. 2011;488:178–82.
59. Kubisch C, Schroeder BC, Friedrich T, Lutjohann B, El-Amraoui A, Marlin S, Petit C, Jentsch TJ. KCNQ4, a novel potassium channel expressed in sensory outer hair cells, is mutated in dominant deafness. Cell. 1999;96:437–46.
60. Laher I, Zhang JH. Protein kinase C and cerebral vasospasm. J Cereb Blood Flow Metab. 2001;21:887–906.
61. Lauritzen M, Dreier JP, Fabricius M, Hartings JA, Graf R, Strong AJ. Clinical relevance of cortical spreading depression in neurological disorders: migraine, malignant stroke, subarachnoid and intracranial hemorrhage, and traumatic brain injury. J Cereb Blood Flow Metab. 2011;31:17–35.
62. Lee S, Yang Y, Tanner MA, Li M, Hill MA. Heterogeneity in Kv7 channel function in the cerebral and coronary circulation. Microcirculation. 2014;2014:12183.
63. Lee SH, Park HK, Ryu WS, Lee JS, Bae HJ, Han MK, Lee YS, Kwon HM, Kim CK, Park ES, Chung JW, Jung KH, Roh JK. Effects of celecoxib on hematoma and edema volumes in primary intracerebral hemorrhage: a multicenter randomized controlled trial. Eur J Neurol. 2013;20:1161–9.
64. Lerche C, Scherer CR, Seebohm G, Derst C, Wei AD, Busch AE, Steinmeyer K. Molecular cloning and functional expression of KCNQ5, a potassium channel subunit that may contribute to neuronal M-current diversity. J Biol Chem. 2000;275:22395–400.
65. Liu TA, Chang HK, Shieh RC. Extracellular K^+ elevates outward currents through Kir2.1 channels by increasing single-channel conductance. Biochim Biophys Acta. 2011;1808:1772–8.
66. Lopatin AN, Makhina EN, Nichols CG. Potassium channel block by cytoplasmic polyamines as the mechanism of intrinsic rectification. Nature. 1994;372:366–9.

67. Lu Y, Hanna ST, Tang G, Wang R. Contributions of Kv1.2, Kv1.5 and Kv2.1 subunits to the native delayed rectifier K^+ current in rat mesenteric artery smooth muscle cells. Life Sci. 2002;71:1465–73.
68. Macdonald RL, Hunsche E, Schüler R, Wlodarczyk J, Mayer SA. Quality of life and health-care resource use associated with angiographic vasospasm after aneurysmal subarachnoid hemorrhage. Stroke. 2012;43:1082–8.
69. Macdonald RL, Pluta RM, Zhang JH. Cerebral vasospasm after subarachnoid hemorrhage: the emerging revolution. Nat Clin Pract Neurol. 2007;3:256–63.
70. Macías A, Moreno C, Moral-Sanz J, Cogolludo A, David M, Alemanni M, Pérez-Vizcaíno F, Zaza A, Valenzuela C, González T. Celecoxib blocks cardiac Kv1.5, Kv4.3 and Kv7.1 (KCNQ1) channels. Effects on cardiac action potentials. J Mol Cell Cardiol. 2010;49:984–92.
71. Mackie AR, Brueggemann LI, Henderson KK, Shiels AJ, Cribbs LL, Scrogin KE, Byron KL. Vascular KCNQ potassium channels as novel targets for the control of mesenteric artery constriction by vasopressin, based on studies in single cells, pressurized arteries, and in vivo measurements of mesenteric vascular resistance. J Pharmacol Exp Ther. 2008;325:475–83.
72. Mani BK, Brueggemann LI, Cribbs LL, Byron KL. Activation of vascular KCNQ (Kv7) potassium channels reverses spasmogen-induced constrictor responses in rat basilar artery. Br J Pharmacol. 2011;164:237–49.
73. Mani BK, Brueggemann LI, Cribbs LL, Byron KL. Opposite regulation of KCNQ5 and TRPC6 channels contributes to vasopressin-stimulated calcium spiking responses in A7r5 vascular smooth muscle cells. Cell Calcium. 2009;45:400–11.
74. Mani BK, O'Dowd J, Kumar L, Brueggemann LI, Ross M, Byron KL. Vascular KCNQ (Kv7) potassium channels as common signaling intermediates and therapeutic targets in cerebral vasospasm. J Cardiovasc Pharmacol. 2013;61:51–62.
75. Matsuda H, Hayashi M, Okada M. Voltage-dependent block by internal spermine of the murine inwardly rectifying K^+ channel, Kir2.1, with asymmetrical K+ concentrations. J Physiol. 2010;588:4673–81.
76. Matsui T, Takuwa Y, Johshita H, Yamashita K, Asano T. Possible role of protein kinase C-dependent smooth muscle contraction in the pathogenesis of chronic cerebral vasospasm. J Cereb Blood Flow Metab. 1991;11:143–9.
77. Mayberg MR, Batjer HH, Dacey R, Diringer M, Haley EC, Heros RC, Sternau LL, Torner J, Adams Jr HP, Feinberg W, et al. Guidelines for the management of aneurysmal subarachnoid hemorrhage. A statement for healthcare professionals from a special writing group of the Stroke Council, American Heart Association. Stroke. 1994;25:2315–28.
78. Miller BA, Turan N, Chau M, Pradilla G. Inflammation, vasospasm, and brain injury after subarachnoid hemorrhage. Biomed Res Int. 2014;2014:384342.
79. Murray MA, Faraci FM, Heistad DD. Effect of protein kinase C inhibitors on endothelin- and vasopressin-induced constriction of the rat basilar artery. Am J Physiol Heart Circ Physiol. 1992;263:H1643–9.
80. Murray MA, Faraci FM, Heistad DD. Role of protein kinase C in constrictor responses of the rat basilar artery in vivo. J Physiol. 1992;445:169–79.
81. Navarro-Gonzalez MF, Grayson TH, Meaney KR, Cribbs LL, Hill CE. Non-L-type voltage-dependent calcium channels control vascular tone of the rat basilar artery. Clin Exp Pharmacol Physiol. 2009;36:55–66.
82. Nelson MT, Patlak JB, Worley JF, Standen NB. Calcium channels, potassium channels, and voltage dependence of arterial smooth muscle tone. Am J Physiol Cell Physiol. 1990;259:C3–18.
83. Nelson MT, Quayle JM. Physiological roles and properties of potassium channels in arterial smooth muscle. Am J Physiol Cell Physiol. 1995;268:C799–822.
84. Ng FL, Davis AJ, Jepps TA, Harhun MI, Yeung SY, Wan A, Reddy M, Melville D, Nardi A, Khong TK, Greenwood IA. Expression and function of the K^+ channel KCNQ genes in human arteries. Br J Pharmacol. 2011;162:42–53.

85. Nikitina E, Zhang ZD, Kawashima A, Jahromi BS, Bouryi VA, Takahashi M, Xie A, Macdonald RL. Voltage-dependent calcium channels of dog basilar artery. J Physiol. 2007;580:523–41.
86. Nishihashi T, Trandafir CC, Wang A, Ji X, Kurahashi K. Enhanced reactivity to vasopressin in rat basilar arteries during vasospasm after subarachnoid hemorrhage. Eur J Pharmacol. 2005;513:93–100.
87. Nishizawa S, Laher I. Signaling mechanisms in cerebral vasospasm. Trends Cardiovasc Med. 2005;15:24–34.
88. Nishizawa S, Obara K, Nakayama K, Koide M, Yokoyama T, Yokota N, Ohta S. Protein kinase cdelta and alpha are involved in the development of vasospasm after subarachnoid hemorrhage. Eur J Pharmacol. 2000;398:113–9.
89. Nishizawa S, Yamamoto S, Yokoyama T, Ryu H, Uemura K. Chronological changes of arterial diameter, cGMP, and protein kinase C in the development of vasospasm. Stroke. 1995;26:1916–20.
90. Ohkuma H, Suzuki S, Ogane K. Phenotypic modulation of smooth muscle cells and vascular remodeling in intraparenchymal small cerebral arteries after canine experimental subarachnoid hemorrhage. Neurosci Lett. 2003;344:193–6.
91. Ohya S, Sergeant GP, Greenwood IA, Horowitz B. Molecular variants of KCNQ channels expressed in murine portal vein myocytes: a role in delayed rectifier current. Circ Res. 2003;92:1016–23.
92. Oliveira-Filho J, Ezzeddine MA, Segal AZ, Buonanno FS, Chang Y, Ogilvy CS, Rordorf G, Schwamm LH, Koroshetz WJ, McDonald CT. Fever in subarachnoid hemorrhage: relationship to vasospasm and outcome. Neurology. 2001;56:1299–304.
93. Otite F, Mink S, Tan CO, Puri A, Zamani AA, Mehregan A, Chou S, Orzell S, Purkayastha S, Du R, Sorond FA. Impaired cerebral autoregulation is associated with vasospasm and delayed cerebral ischemia in subarachnoid hemorrhage. Stroke. 2014;45:677–82.
94. Oyama K, Criddle L. Vasospasm after aneurysmal subarachnoid hemorrhage. Crit Care Nurse. 2004;24:58–60. 62, 64–7.
95. Park H-K, Lee S-H, Chu K, Roh J-K. Effects of celecoxib on volumes of hematoma and edema in patients with primary intracerebral hemorrhage. J Neurol Sci. 2009;279:43–6.
96. Patel VN, Samuels OB. The critical care management of aneurysmal subarachnoid hemorrhage. In: Murai Y, editor. Aneurysm. Croatia: InTech; 2012. p. 383–402.
97. Penn DL, Witte SR, Komotar RJ, Sander Connolly Jr E. Pathological mechanisms underlying aneurysmal subarachnoid haemorrhage and vasospasm. J Clin Neurosci. 2015;22:1–5.
98. Peretz A, Degani N, Nachman R, Uziyel Y, Gibor G, Shabat D, Attali B. Meclofenamic acid and diclofenac, novel templates of KCNQ2/Q3 potassium channel openers. Depress cortical neuron activity and exhibit anticonvulsant properties. Mol Pharmacol. 2005;67:1053–66.
99. Ploug KB, Edvinsson L, Olesen J, Jansen-Olesen I. Pharmacological and molecular comparison of K_{ATP} channels in rat basilar and middle cerebral arteries. Eur J Pharmacol. 2006;553:254–62.
100. Pradilla G, Chaichana KL, Hoang S, Huang J, Tamargo RJ. Inflammation and cerebral vasospasm after subarachnoid hemorrhage. Neurosurg Clin N Am. 2010;21:365–79.
101. Quayle JM, Nelson MT, Standen NB. ATP-sensitive and inwardly rectifying potassium channels in smooth muscle. Physiol Rev. 1997;77:1165–232.
102. Robbins J. KCNQ potassium channels: physiology, pathophysiology, and pharmacology. Pharmacol Ther. 2001;90:1–19.
103. Robertson BE, Nelson MT. Aminopyridine inhibition and voltage dependence of K^+ currents in smooth muscle cells from cerebral arteries. Am J Physiol Cell Physiol. 1994;267:C1589–97.
104. Rowland MJ, Hadjipavlou G, Kelly M, Westbrook J, Pattinson KT. Delayed cerebral ischaemia after subarachnoid haemorrhage: looking beyond vasospasm. Br J Anaesth. 2012;109:315–29.
105. Sako M, Nishihara J, Ohta S, Wang J, Sakaki S. Role of protein kinase C in the pathogenesis of cerebral vasospasm after subarachnoid hemorrhage. J Cereb Blood Flow Metab. 1993;13:247–54.

106. Sanchez-Porras R, Zheng Z, Santos E, Scholl M, Unterberg AW, Sakowitz OW. The role of spreading depolarization in subarachnoid hemorrhage. Eur J Neurol. 2013;20:1121–7.
107. Sanguinetti MC, Curran ME, Zou A, Shen J, Spector PS, Atkinson DL, Keating MT. Coassembly of K(V)LQT1 and minK (IsK) proteins to form cardiac I(Ks) potassium channel. Nature. 1996;384:80–3.
108. Schenzer A, Friedrich T, Pusch M, Saftig P, Jentsch TJ, Grotzinger J, Schwake M. Molecular determinants of KCNQ (Kv7) K^+ channel sensitivity to the anticonvulsant retigabine. J Neurosci. 2005;25:5051–60.
109. Schleifenbaum J, Köhn C, Voblova N, Dubrovska G, Zavarirskaya O, Gloe T, Crean CS, Luft FC, Huang Y, Schubert R, Gollasch M. Systemic peripheral artery relaxation by KCNQ channel openers and hydrogen sulfide. J Hypertens. 2010;28:1875–82.
110. Schroeder BC, Hechenberger M, Weinreich F, Kubisch C, Jentsch TJ. KCNQ5, a novel potassium channel broadly expressed in brain mediates M-type currents. J Biol Chem. 2000;275:24089–95.
111. Shapiro MS. An ion channel hypothesis to explain divergent cardiovascular safety of cyclooxygenase-2 inhibitors: the answer to a hotly debated puzzle? Mol Pharmacol. 2009;76:942–5.
112. Simard JM, Geng Z, Woo SK, Ivanova S, Tosun C, Melnichenko L, Gerzanich V. Glibenclamide reduces inflammation, vasogenic edema, and caspase-3 activation after subarachnoid hemorrhage. J Cereb Blood Flow Metab. 2009;29:317–30.
113. Singh NA, Charlier C, Stauffer D, DuPont BR, Leach RJ, Melis R, Ronen GM, Bjerre I, Quattlebaum T, Murphy JV, McHarg ML, Gagnon D, Rosales TO, Peiffer A, Anderson VE, Leppert M. A novel potassium channel gene, KCNQ2, is mutated in an inherited epilepsy of newborns. Nat Genet. 1998;18:25–9.
114. Smith PD, Brett SE, Luykenaar KD, Sandow SL, Marrelli SP, Vigmond EJ, Welsh DG. KIR channels function as electrical amplifiers in rat vascular smooth muscle. J Physiol. 2008;586:1147–60.
115. Solomon SD, Pfeffer MA, McMurray JJ, Fowler R, Finn P, Levin B, Eagle C, Hawk E, Lechuga M, Zauber AG, Bertagnolli MM, Arber N, Wittes J. Effect of celecoxib on cardiovascular events and blood pressure in two trials for the prevention of colorectal adenomas. Circulation. 2006;114:1028–35.
116. Suarez JI, Tarr RW, Selman WR. Aneurysmal subarachnoid hemorrhage. N Engl J Med. 2006;354:387–96.
117. Sugai K, Yaganisawa T, Motohashi O, Suzuki M, Yoshimoto T. Levcromakalim decreases cytoplasmic Ca^{2+} and vascular tone in basilar artery of SAH model dogs. J Cardiovasc Pharmacol. 1999;33:868–75.
118. Towart R, Linders JT, Hermans AN, Rohrbacher J, van der Linde HJ, Ercken M, Cik M, Roevens P, Teisman A, Gallacher DJ. Blockade of the I(Ks) potassium channel: an overlooked cardiovascular liability in drug safety screening? J Pharmacol Toxicol Methods. 2009;60:1–10.
119. Trandafir CC, Nishihashi T, Wang A, Murakami S, Ji X, Kurahashi K. Participation of vasopressin in the development of cerebral vasospasm in a rat model of subarachnoid haemorrhage. Clin Exp Pharmacol Physiol. 2004;31:261–6.
120. Tseng-Crank J, Godinot N, Johansen TE, Ahring PK, Strobaek D, Mertz R, Foster CD, Olesen SP, Reinhart PH. Cloning, expression, and distribution of a Ca^{2+}-activated K^+ channel β-subunit from human brain. Proc Natl Acad Sci U S A. 1996;93:9200–5.
121. Wang HS, Pan Z, Shi W, Brown BS, Wymore RS, Cohen IS, Dixon JE, McKinnon D. KCNQ2 and KCNQ3 potassium channel subunits: molecular correlates of the M-channel. Science. 1998;282:1890–3.
122. Wang X, Li YM, Li WQ, Huang CG, Lu YC, Hou LJ. Effect of clazosentan in patients with aneurysmal subarachnoid hemorrhage: a meta-analysis of randomized controlled trials. PLoS One. 2012;7:e47778.
123. Waters A, Harder DR. Altered membrane properties of cerebral vascular smooth muscle following subarachnoid hemorrhage: an electrophysiological study. I. Changes in resting

membrane potential (Em) and effect on the electrogenic pump potential contribution to Em. Stroke. 1985;16:990–7.

124. Wellman GC. Ion channels and calcium signaling in cerebral arteries following subarachnoid hemorrhage. Neurol Res. 2006;28:690–702.
125. Weyer GW, Jahromi BS, Aihara Y, Agbaje-Williams M, Nikitina E, Zhang ZD, Macdonald RL. Expression and function of inwardly rectifying potassium channels after experimental subarachnoid hemorrhage. J Cereb Blood Flow Metab. 2006;26:382–91.
126. Wong GK, Boet R, Poon WS, Chan MT, Gin T, Ng SC, Zee BC. Intravenous magnesium sulphate for aneurysmal subarachnoid hemorrhage: an updated systemic review and meta-analysis. Crit Care. 2011;15:R52.
127. Wu YJ, Boissard CG, Greco C, Gribkoff VK, Harden DG, He H, L'Heureux A, Kang SH, Kinney GG, Knox RJ, Natale J, Newton AE, Lehtinen-Oboma S, Sinz MW, Sivarao DV, Starrett Jr JE, Sun LQ, Tertyshnikova S, Thompson MW, Weaver D, Wong HS, Zhang L, Dworetzky SI. (S)-N-[1-(3-morpholin-4-ylphenyl)ethyl]- 3-phenylacrylamide: an orally bioavailable KCNQ2 opener with significant activity in a cortical spreading depression model of migraine. J Med Chem. 2003;46:3197–200.
128. Wulf H, Hay-Schmidt A, Poulsen AN, Klaerke DA, Olesen J, Jansen-Olesen I. Molecular investigations of BK(Ca) channels and the modulatory beta-subunits in porcine basilar and middle cerebral arteries. J Mol Histol. 2009;40:87–97.
129. Wuttke TV, Seebohm G, Bail S, Maljevic S, Lerche H. The new anticonvulsant retigabine favors voltage-dependent opening of the Kv7.2 (KCNQ2) channel by binding to its activation gate. Mol Pharmacol. 2005;67:1009–17.
130. Xiong Q, Sun H, Li M. Zinc pyrithione-mediated activation of voltage-gated KCNQ potassium channels rescues epileptogenic mutants. Nat Chem Biol. 2007;3:287–96.
131. Xu C, Lu Y, Tang G, Wang R. Expression of voltage-dependent K^+ channel genes in mesenteric artery smooth muscle cells. Am J Physiol. 1999;277:G1055–63.
132. Yeung SY, Pucovsky V, Moffatt JD, Saldanha L, Schwake M, Ohya S, Greenwood IA. Molecular expression and pharmacological identification of a role for K(v)7 channels in murine vascular reactivity. Br J Pharmacol. 2007;151:758–70.
133. Yu Y, Ricciotti E, Scalia R, Tang SY, Grant G, Yu Z, Landesberg G, Crichton I, Wu W, Pure E, Funk CD, FitzGerald GA. Vascular COX-2 modulates blood pressure and thrombosis in mice. Sci Transl Med. 2012;4:132ra154.
134. Zaritsky JJ, Eckman DM, Wellman GC, Nelson MT, Schwarz TL. Targeted disruption of Kir2.1 and Kir2.2 genes reveals the essential role of the inwardly rectifying K^+ current in K^+-mediated vasodilation. Circ Res. 2000;87:160–6.
135. Zavaritskaya O, Zhuravleva N, Schleifenbaum J, Gloe T, Devermann L, Kluge R, Mladenov M, Frey M, Gagov H, Fésüs G, Gollasch M, Schubert R. Role of KCNQ channels in skeletal muscle arteries and periadventitial vascular dysfunction. Hypertension. 2013;61:151–9.
136. Zhang X, Jakubowski M, Buettner C, Kainz V, Gold M, Burstein R. Ezogabine (KCNQ2/3 channel opener) prevents delayed activation of meningeal nociceptors if given before but not after the occurrence of cortical spreading depression. Epilepsy Behav. 2013;28:243–8.
137. Zhao J, Zhou D, Guo J, Ren Z, Zhou L, Wang S, Zhang Y, Xu B, Zhao K, Wang R, Mao Y, Xu B, Zhang X. Efficacy and safety of fasudil in patients with subarachnoid hemorrhage: final results of a randomized trial of fasudil versus nimodipine. Neurol Med Chir (Tokyo). 2011;51:679–83.
138. Zhong XZ, Harhun MI, Olesen SP, Ohya S, Moffatt JD, Cole WC, Greenwood IA. Participation of KCNQ (Kv7) potassium channels in myogenic control of cerebral arterial diameter. J Physiol. 2010;588:3277–93.
139. Zimmermann M, Seifert V. Endothelin and subarachnoid hemorrhage: an overview. Neurosurgery. 1998;43:863–75. discussion 875–6.
140. Zuccarello M, Bonasso CL, Lewis AI, Sperelakis N, Rapoport RM. Relaxation of subarachnoid hemorrhage-induced spasm of rabbit basilar artery by the K+ channel activator cromakalim. Stroke. 1996;27:311–6.

Chapter 10
Lysosomal Transient Receptor Potential Mucolipin (TRPML) Channels in Vascular Regulation and Diseases

Fan Zhang and Pin-Lan Li

Abstract Transient Receptor Potential Mucolipin 1 (TRPML1) channel protein, one of the TRPML family members, is a Ca^{2+}-permeable membrane channel with a predominant location on the late endo/lysosomes. In arterial myocytes, it has been found that this ligand-gated channel is sensitive to NAADP (Nicotinic Acid Adenine Dinucleotide Phosphate), a secondary Ca^{2+} messenger, to generate Ca^{2+} release from lysosomes followed by a global Ca^{2+} release response from the sarcoplasmic reticulum through Ca^{2+}-induced Ca^{2+} release mechanism (CICR), which is referred to as an intracellular two-phase Ca^{2+} release. In coronary artery, recent studies demonstrated that NAADP-sensitive lysosomal TRPML1 channels can be activated by a variety of pro-pathogenic stimuli such as endothelin 1 (ET-1), reactive oxidative species (ROS), and death receptor ligand (FasL). The defects or derangement of the lysosomal TRPMLl channel activity have been implicated in vascular injury or diseases such as endothelial dysfunction, apoptosis and even atherosclerosis.

Keywords TRPML channels • NAADP • Lysosomes • Ca^{2+} • Vascular tone and vasomotor • Atherosclerosis

Introduction

Transient Receptor Potential Mucolipin (TRPML) channel proteins have been reported to have three family members including TRPML1, TRPML2 and TRPML3. These TRPMLs are exclusively resided on the late endo/lysosomal organelles with cationic permeability to Ca^{2+}. Functionally, TRPMLs have been implicated in the regulation of lysosomes, for example, facilitating the processes in both endocytosis and autophagy in a variety of cells. The genetic defects of TRPML1 have been

F. Zhang, M.D., Ph.D. • P.-L. Li, M.D., Ph.D. (✉)
Department of Pharmacology and Toxicology, Virginia Commonwealth University, School of Medicine, 410 N. 12th Street, Richmond, VA 23298, USA
e-mail: pli@vcu.edu

I. Levitan, A.M. Dopico (eds.), *Vascular Ion Channels in Physiology and Disease*, DOI 10.1007/978-3-319-29635-7_10

demonstrated to be a major cause of the mucolipidosis type IV disease, one of the lysosomal storage disorders. Recently, we found that the TRPML1 channels are present in coronary arterial myocytes (CAMs), which are sensitive to nicotinic acid adenine dinucleotide phosphate (NAADP), a potent Ca^{2+}-mobilizing second messenger, producing Ca^{2+} release from lysosomes. It has been also confirmed that this TRPML1/NAADP Ca^{2+} signaling pathway responds to a variety of pro-pathogenic stimuli, such as vasoconstrictor, ET-1, reactive oxidative species (ROS) and apoptotic reagent, FasL. In this chapter, we will briefly summarize some important current findings regarding the role of lysosomal TRPML1 in vascular regulation and diseases, providing some novel insights into the physiological and pathological relevance of this lysosomal channel.

Lysosomal TRPML Channels in Arterial Myocytes

Lysosomal TRPML1 Channel

The transient receptor potential (TRP) channels are a group of non-selective ion channels with variable degrees in Ca^{2+} permeability. It has been reported that a total number of 28 TRP channel proteins are expressed in different mammalian cells, and they are grouped into six subfamilies: TRPC (Canonical), TRPV (Vanilloid), TRPM (Melastatin), TRPA (Ankyrin), TRPP (Polycystin) and TRPML (Mucolipin). TRP channel proteins contain six transmembrane domains, with a pore loop region between domain 5 and 6. Among these TRP channels, TRPMLs are primarily localized at membrane-bounded vesicles, which mediate the endocytic and exocytic processes. TRPMLs are relatively small proteins consisting of <600 amino acids with expected molecular size of approximate 56–65 kDa. TRPML1 is encoded by the TRPML1 gene, *MCOLN1*. The product of this gene may be detected in different sizes from 36 to 75 kDa in various native cells due to cleavage and other modifications [1, 2]. TRPML1 is a channel protein characterized by a TRP channel-homologous region located within amino acids 331–521, an internal Ca^{2+} and Na^+ channel pore region between amino acids 496 and 521, and cytosolic location of both NH_2 and COOH terminal chains. TRPML1 is widely expressed and mainly resided on late endo/lysosomes [3–5]. Mutations of *MCOLN1* are implicated in the pathogenesis of a neurological disease, namely, mucolipidosis Type IV (MLIV). This disease is a lysosomal storage disorder with severe neurologic and ophthalmologic abnormalities [1, 5–8]. Since the mutation of *MCOLN1* was identified as the genetic cause of MLIV disease [1, 9, 10] and *MCOLN1* products were shown to belong to TRP superfamily [1, 8–10], extensive studies have been conducted to understand their biophysical properties and related functions. Compared with TRPML1, TRPML2 and TRPML3 are not yet extensively characterized until recently [11]. Given the predominant endo/lysosomal locations, it is logical to assume that the physiological or pathological relevance of TRPML channels are largely associated with the lysosome-related signaling mechanisms and functions such as lysosome Ca^{2+} release, molecular trafficking, endocytic and phagocytic processes.

TRPML1-Mediated Ca^{2+} Release from Lysosomes

TRPMLs are not voltage sensitive channels but gated by ligands. It has been reported that phosphatidylinositol 3,5-bisphosphate (PI(3,5)P2) activates lysosomal TRPML1 channels to produce Ca^{2+}, Fe^{2+}, Mn^{2+} and H^+ currents [12, 13]. The ion selectivity of human TRPML1 has been demonstrated to be $Ba^{2+} > Mn^{2+} > Fe^{2+} = Ca^{2+} = Mg^{2+} > Ni^{2+} = Co^{2+} = Cd^{2+} > Zn^{2+} \gg Cu^{2+}$ at pH 4.6 and its pS is at 32–40 in the presence of 30–105 mM Fe^{2+}[12, 13]. Many studies have done to elucidate the physiological relevance of the Ca^{2+} conductivity of TRPML1, which relates to the regulation of endosome maturation, retrograde trafficking of lysosomal membrane from late endosomes or lysosomes (LEL) to trans-Golgi-network (TGN), autophagosome-lysosome fusion, lysosome exocytosis, and Ca^{2+} signaling-related cellular activities [14]. With respect to TRPML1 conductance of other ions, there were very few studies that addressed their functional significance. Some assumptions were proposed based on the ion conductance of TRPML1. For example, due to its Fe^{2+} conductance TRPML1 has been assumed to regulate the cellular homeostasis of Fe^{2+} and then contribute to the hematological and degenerative abnormality in mucolipidosis type IV disease [12, 15]. In addition, some studies reported that TRPML1 channels were impermeable to protons and thereby played a role in the regulation of lysosomal pH. This H^+ conductance of TRPML1 is critical in the regulation of lysosomal pH homeostasis, which may serve as important regulator of lysosome function [16, 17]. However, the functional relevance of TRPML1 conductance of ions besides Ca^{2+} is far from clarification.

In searching for the molecular target of nicotinic acid adenine dinucleotide phosphate (NAADP), a potent Ca^{2+} releasing second messenger, we demonstrated that NAADP-sensitive Ca^{2+} channels in lysosomes of hepatocytes had characteristics of that lysosomal TRPML1 channels both biochemically and pharmacologically [18]. This NAADP-sensitive TRPML1 channels were also found in the lysosomal compartments in coronary arterial myocytes, which play an important in mediating two-phase Ca^{2+} release, a novel Ca^{2+} regulatory pattern in smooth muscle cells that may contribute to the regulation of vascular tone and cell surviving under physiological and pathological conditions [19].

Compared with other two well-established intracellular Ca^{2+} mobilizing messengers, namely, inositol trisphosphate (IP_3) and cyclic ADP-ribose (cADPR), NAADP identified in 1995 is more potent in mobilization of intracellular Ca^{2+} and able to mediate Ca^{2+} release from the sarcoplasmic reticulum (SR) by its direct action on ryanodine receptors (RyRs) or via lysosome-mediated two-phase Ca^{2+} release, depending upon cell types [20, 21]. Similar to cADPR production, soluble protein Aplysia ADP-ribosylcyclase and its membrane-bound homologs, CD38 and CD157, are also involved in the production of NAADP. These enzymes were shown to catalyze the exchange of the terminal nicotinamide group in $NADP^+$ with nicotinic acid to produce NAADP in a variety of cells and tissues such as sea urchin eggs, pancreatic acinar cells, human T lymphocytes, rat brain, and smooth muscle cells [22–25]. However, in myometrial cells (uterine smooth muscle cells), NAADP was found to be produced by an unidentified pathway independent of CD38 or base-exchange reaction [26]. In our recent studies, CD38 and its cytosolic isoforms are found to be abundant in mouse coronary arterial myocytes (CAMs), where NAADP is produced in responses to the treatments of death receptor ligands, endothelin-1 or oxidants [27, 28]. The capability of CAMs to produce NAADP in response to various stimuli suggests that Ca^{2+} release through lysosomal TRPML1 by NAADP may be involved in different processes of physiological regulations or pathological changes in these cells.

Lysosomal TRPML1 in the Regulation of Vascular Tone and Vasomotor Response

TRPML1 Activated by NAADP and Two-Phase Ca^{2+} Release in CAMs

In the resistance arteries, endothelial cells (ECs) and smooth muscle cells (SMCs) have opposite responses to the mechanic forces exerted by the circulating blood. The shear stress on ECs from blood flow promotes vasodilation, while the direct impact of blood pressure on the smooth muscle itself tends to cause vasoconstriction. Recent studies have indicated that both types of responses in determining vascular tension are associated with different TRP channels. In myogenic arteries, TRPV4 has been found expressed in the vascular ECs and activation of this channel promotes vasodilation. TRPC6, TRPM4 and TRPV4 are mainly expressed on SMCs, which can sense the mechanical changes from blood flow to produce vasoconstriction, transducing mechanical stimuli into vascular tone [29]. Over the last 5 years, we have demonstrated that TRPML1 is mainly expressed in lysosome membrane of both bovine and mouse CAMs. We have also shown that an endogenous activator of this lysosomal TRPML1 in CAMs is NAADP, which induces lysosomal Ca^{2+} bursts and consequently produce large Ca^{2+} release through Ca^{2+}-induced Ca^{2+} release (CICR). However, there is no evidence that this TRPML1 channel is not sensitive to mechanical forces to regulate vascular tone and mediate corresponding vasomotor response.

NAADP as a Ca^{2+} messenger is significantly different from the well-defined other two messengers, IP_3 and cADPR in terms of targeting Ca^{2+} stores and the Ca^{2+} release patterns. Both IP_3 and cADPR act on SR/ER Ca^{2+} store, and they directly bind to inositol trisphosphate receptor (IP_3R) and ryanodine receptors (RyRs), respectively, to generate a robust Ca^{2+} response. In contrast, NAADP-mediated Ca^{2+} discharges have been shown to be able to produce a two-phase Ca^{2+} response with a delayed second phase Ca^{2+} release (within seconds) in addition to its direct effect on

RyRs in some cells such as lymphocytes. This two-phase Ca^{2+} release induced by NAADP is demonstrated to first activate Ca^{2+} bursts from lysosomes as a trigger, which subsequently leads to global Ca^{2+} mobilization from SR/ER through CICR, a so-called two-pool mechanism. NAADP-sensitive Ca^{2+} store or acidic Ca^{2+} store, which is lately shown to be lysosomes, is responsible for a localized signal, which consequently triggers CICR to cause global Ca^{2+} increases through RyRs or IP_3Rs on the SR [28, 30].

In vascular SMCs, this two-pool mechanism has been demonstrated to function upon stimulations by ET-1, FasL or direct delivery of NAADP into the cells [28, 31, 32]. By Ca^{2+} fluorescence imaging analysis, it has been shown that ET-1 induced a sequential two-phase increase in $[Ca^{2+}]_i$ in bovine CAMs [31], where the first phase Ca^{2+} spike occurred 1 min after ET-1 was added into the bath solution whereas the second phase Ca^{2+} transient occurred 3 min later, representing a global increase in $[Ca^{2+}]_i$. In the presence of bafilomycin A1(Baf), both phases of Ca^{2+} response were markedly attenuated, suggesting the involvement of lysosome organelles because Baf is a lysosomal vacuolar H^+-ATPase antagonist and inhibition of vacuolar H^+-ATPase de-energizes H^+/Ca^{2+} exchanger and prevents Ca^{2+} from sequestering into lysosomes [31], which leads to lysosomal Ca^{2+} depletion. However, a relatively high concentration of Rya as a classic inhibitor of RyR/Ca^{2+} channels on the SR only abolished the late Ca^{2+} increase, which was functionally related to Ca^{2+} release from SR Ca^{2+} store. These results suggest that ET-1 may first stimulate Ca^{2+} release from lysosomes through NAADP action and then activate Ca^{2+} release from the SR via RyRs-sensitive CICR mechanism.

In mouse CAMs, we also demonstrated Ca^{2+} release in response to FasL, which was characterized by a typical two-phase Ca^{2+} response pattern, namely, a local Ca^{2+} burst from lysosomes followed by a global Ca^{2+} response. In isolated perfused mouse septal coronary arteries, FasL was found to significantly increase U-46619-induced vasoconstriction, which was abolished by lysosomal function inhibitor, Baf. It has been indicated that the early response of CAMs to FasL is to increase intracellular Ca^{2+} levels from lysosomes. This lysosome-mediated Ca^{2+} response to FasL was also blocked by silencing TRPML1 gene, but markedly enhanced with overexpression of this gene [32] (Fig. 10.1). It is clear that lysosomal TRPML1 channels are critical to Ca^{2+} release responses in CAMs. This TRPML1 channel and its role in mediating a two-phase Ca^{2+} release are briefly summarized in Fig. 10.2.

Lysosome (Lyso)-SR Junction

The two-pool mechanism in mediating intracellular Ca^{2+} movements is interesting in regard to a two phase Ca^{2+} release induced by NAADP. However, the enigma remains why NAADP-induced CICR has a long-delayed second Ca^{2+} release phase (in seconds to minutes) [28], which is very different from the classical CICR (in milliseconds). It also remains unknown what physiological role this delayed second phase Ca^{2+} in CICR may play in the regulation of cell function. Although some reports have indicated that there may be lysosome (lyso)-SR junction, which

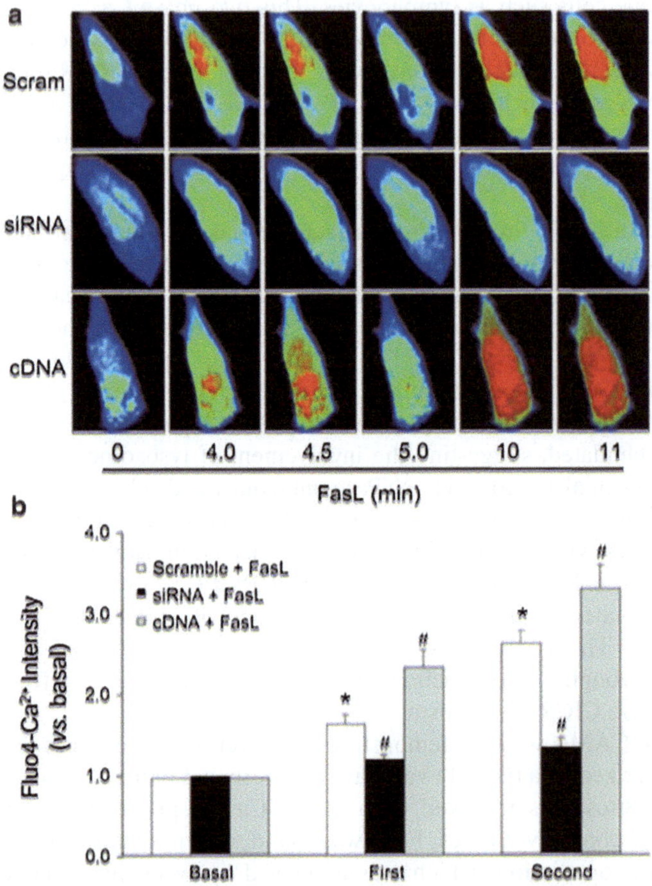

Fig. 10.1 Effects of TRPML1 channel gene silencing or overexpression on Fas ligand (FasL)-induced lysosomal Ca^{2+} bursts and Ca^{2+} release from the sarcoplasmic reticulum (SR) of mouse CAMs. (**a**) Representative Fluo4-Ca^{2+} imaging record showed an early Ca^{2+} release (first phase, at 4.5 min; *red spots* in the image), followed by a global increase in intracellular Ca^{2+} (*red* cytosol; second phase, at 10 min) upon FasL (10 ng/ml) stimulation in CAMs before and after transfection of TRPML1 siRNA or cDNA. (**b**) Summarized data showing the effects of TRPML1 siRNA or its overexpression on FasL-induced 2-phase Ca^{2+} release in CAMs (n=6; *P<0.05 vs. basal condition; #P<0.05 vs. scramble sRNA-treated CAMs with FasL stimulation)

contributes to CICR to lead to a two-phase Ca^{2+} release [30, 33, 34], however, the anatomical and molecular characteristics of this lyso-SR junction remains poorly understood. Given the great mobility of lysosomes within cells, recent studies have demonstrated that there may not be any sustained lyso-SR junctions in SMCs, but a dynamic lyso-SR junction is possibly functioning to mediate CICR depending upon lysosome trafficking and aggregation toward the SR [19]. Based on this model of so called dynamic functional lyso-SR junction, the global Ca^{2+} release following small Ca^{2+} bursts from lysosomes may be associated with lysosomal trafficking to the

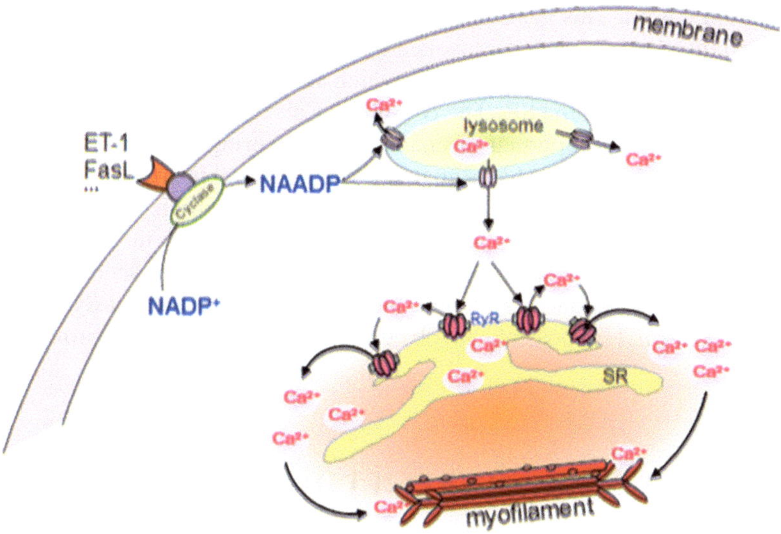

Fig. 10.2 NAADP-mediated Ca^{2+} release from lysosomal TRPML1 channels produces a local Ca^{2+} burst, followed by global Ca^{2+} release responses from the sarcoplasmic reticulum through a Ca^{2+}-induced-Ca^{2+} mechanism in coronary arterial smooth muscle cells (CASMCs). NAADP is produced in the cytosol of CASMCs from $NADP^+$ by ADP-ribosylcyclase in response to different agonists, such as ET-1 and FasL. NAADP serves as a second messenger to activate TRPML1 channels on lysosomes to produce small Ca^{2+} bursts, which then trigger a robust global Ca^{2+} release via CICR through RyRs or IP_3Rs, thereby, resulting in activation of these vascular cells and producing vasoconstriction

SR. Small lysosomal Ca^{2+} bursts may not be enough to activate global Ca^{2+} release from the SR; however, this Ca^{2+} bursts may be sufficient to drive lysosome movement or aggregation. These clustered or aggregated lysosomes work together to activate global Ca^{2+} release from the SR [19]. This interaction of lysosome and SR was also proposed later in pulmonary VSMCs [35]. This coordination of lysosomes and SR was confirmed in CAMs [32] by detected colocalization of lysosomal marker (GFP-Lamp1) with RyR3 or SR tracker (red fluorescent labeled) during stimulation. By more dynamic and direct measurements, NAADP was indeed shown to stimulate lysosome trafficking through its Ca^{2+} mobilizing action [36].

Role of TRPML1 in the Regulation of Vascular Tone and Vasomotor Responses

We have demonstrated that ET-1-induced-vasoconstriction of coronary arteries was dependent on an intact NAADP-lysosome signaling pathway. In addition, FasL was shown to enhance U-46619-induced vasoconstriction, which was consistent with a time frame of the FasL-induced two-phase Ca^{2+} response as observed in the Ca^{2+}

release from CAMs. This FasL-induced slow vasomotor response is different from that of instant vasoconstriction provoked by some classic vasoactive agonists (e.g., norepinephrine, angiotensin II, or ATP), which are dependent on the PLC-mediated direct Ca^{2+} release from the SR. The two-phase Ca^{2+} release and slow vasoconstriction may represent an important mechanism underscoring the maintenance of vascular tone. It is assumed that lysosomal TRPML1 channel in arterial SMs is critical for the development of sustained vascular tone. This hypothesis needs to be further tested. In addition, a chronic elevation of intracellular Ca^{2+} in SMCs has been reported to produce histopathological alterations leading to vascular remodeling and arteriosclerotic lesions [37], which implies the derangement or dysfunction of lysosomal TRPML1 is possibly involved in the pathogenesis of coronary artery diseases.

Implication of Lysosomal TRPML1 Dysfunction in Atherosclerosis

TRPML1 and Lysosomal Cholesterol Transportation

Endocytic and phagocytic pathways serve as the basic mechanisms in degradation of extracellular particles and dysfunctional cellular components. As a primary digestive organelle, lysosome is destined to receive all the endocytosed membranes, complex lipids, membrane proteins and polysaccharides, as well as the autophagocytosed organelles for degradation. However, lysosomes are not only dead-end metabolizing digestive organelles but also play essential roles in signal transduction by dynamically engaging in the membrane fusion, lipid trafficking and Ca^{2+}-involved sequential events. There is evidence that the Ca^{2+} release from lysosomes, which TRPML Ca^{2+} channels are mainly located on, is an important Ca^{2+} signaling or regulatory mechanism. It has shown that during endocytosis lipid trafficking from endosomes to lysosomes is Ca^{2+}-dependent process and that intracellular Ca^{2+} elevation is required for the formation of hybrid compartments between endosome and lysosome [38, 39]. The insufficiency of Ca^{2+} release via TRPML1 due to TRPML1 mutation and dysfunction may cause the delay in delivery of endocytosed material to lysosomes. In mammalian cells, the mutation of TMPML1 leads to the delayed fusion of autophagic vacuoles with lysosomes [40]. The loss of TRPML1 was indeed found to obstruct the delivery of endocytosed lipids to lysosomes in MLIV cells, while TRPML3 was shown to be a regulator of membrane trafficking along both endocytic and phagocytic pathways mainly through their lysosomal Ca^{2+} release and consequent activation of trafficking molecules such as motor protein dynein [41, 42]. With respect to the sources of Ca^{2+} for TRPMLs to conduct during endocytosis and autophagy, it has been proposed that lysosomal lumenal Ca^{2+} rather than cytosolic Ca^{2+} is the candidate, which is now referred to as lysosomal Ca^{2+} store [43, 44].

Despite extensive investigations conducted in elaborating the correlation between lysosomal TRPMLs Ca^{2+} release channels and the associated heredity diseases, the implication of lysosomal TRPMLs channels in vascular pathology remains poorly understood. In atherosclerosis, LDL in its oxidized form is taken up by macrophages through scavenger receptors and delivered to lysosomes for degradation by an endocytic process. Under normal conditions, the cholesteryl esters are hydrolyzed by lysosomal acid lipase into free cholesterol, which is actively transported out of lysosomes to the cytosol, presumably through lysosomal NPC1 protein [45–47]. It has been found that the pathological hallmark of atherosclerosis is the excessive accumulation of lipids in macrophages, which is featured by the lipid-swollen lysosomes [48, 49] dominated by cholesterol in both the free and esterified form [50, 51].

There are reports that the foam cell formation during atherosclerosis has features of inherited lysosomal storage disorders [52], such as MLIV [6, 8] and Niemann-Pick type C1(NPC1) diseases [53], both of which are characterized by insufficient lysosomal Ca^{2+} release. TRPML1, as a Ca^{2+}-permeable channel [4] and NPC1 as a cholesterol transporter, both are exclusively expressed on the late endosomes and lysosomes, which may regulate cholesterol transport within cells. Indeed, the lack of lysosomal Ca^{2+} release due to TRPML1 mutation and the lysosomal Ca^{2+} storage depletion due to the defects of NPC1 have been identified as the underlying pathological mechanisms in lipid trafficking deterrence and lysosome lipid deposition during MLIV and NPC1 diseases [54, 55], respectively.

Given the similarity of the endocytic trafficking derangement in lysosomes during atherosclerosis to that occurred in MLIV or NPC1 diseases, we recently tested the hypothesis that the CD38-NAADP-mediated Ca^{2+} signaling pathway plays a critical role in the removal of free cholesterol from lysosomes in macrophages and that the abnormalities in such CD38-associated lysosome regulation may contribute to the lysosomal cholesterol accumulation, macrophage foam cell formation and atherogenesis. It has been demonstrated that the free cholesterol egression from lysosomes was profoundly attenuated in the macrophages with disruption of CD38/NAADP Ca^{2+} pathway, which resulted in the foam cell formation and mouse coronary artery atherosclerosis. Rescuing of CD38 gene in $CD38^{-/-}$ macrophages or direct supplementation of NAADP to these cells prevented the deposition of free cholesterol in lysosomes and the formation of foam cells. We also demonstrated that the lysosomal TRPML1-released Ca^{2+} plays an essential role in the facilitation of lipid endocytic trafficking in human fibroblast cells, and that supplement of the Ca^{2+} messenger of NAADP profoundly promoted this process in prevention of lipid accumulation in lysosomes [36]. These results suggest that dysfunction of lysosomal TRPML1 channels due to the lack of CD38 regulation indeed results in derangement of cholesterol trafficking out of lysosomes in macrophages, which may be a major mechanism leading to the formation of foam cells during atherosclerosis. Figure 10.3 summarizes this lysosomal TRPML1-mediated cholesterol clearance in macrophages.

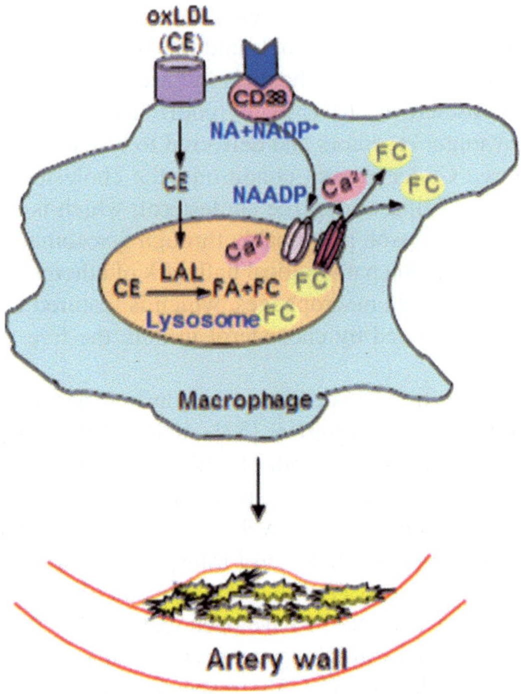

Fig. 10.3 Ca^{2+} release from NAADP-activated lysosomal TRPML1 channel promotes oxLDL derived-free cholesterol transportation out of lysosomes in macrophage cells. The endocytosed oxLDL is trafficked into lysosomes where the cholesterol ester is hydrolyzed to free cholesterol. The CD38 enzymatic product, NAADP, serves as a Ca^{2+} messenger to release Ca^{2+} from TRPML1 channels. This local Ca^{2+} increase activates free cholesterol transporters (such as Niemann-Pick type C1) and facilitates the egression of cholesterol from lysosomes. A deficiency in NAADP-mediated Ca^{2+} release from TRPML1 will lead to free cholesterol accumulation in lysosomes. This free cholesterol buildup will compromise lysosomal lumen acidity, Ca^{2+} storage and lysosomal acid lipase (LAL) activity and exacerbates cholesterol segregation in lysosomes, which results in foam cell formation and atherosclerosis

TRPML1 and Autophagic Flux

In autophagy, lysosomal TRPMLs was found to facilitate autophagosome biogenesis. The defects of TRPMLs lead to marked accumulation of late endosomes, lysosomes and autophagosomes (APs), which are associated with reduced autophagy flux [38, 40]. Although multiple studies documented the occurrence of autophagy in cardiovascular tissues [56–62], much less has been reported regarding autophagy occurring in vascular smooth muscle cells [63], of which only a few of them in coronary artery myocytes [64, 65]. It has been shown that in coronary artery NAADP generated Ca^{2+} release response from lysosomes, an essential component in the facilitation of completing autophagy process. Our laboratory has recently been working on the potential contribution of NAADP-mediated regulation of lysosome function in the development of atherosclerosis. Our preliminary studies show that NAADP regulates lysosome function through its action on TRPML1, which controls lysosome trafficking or fusion to autophagosomes (APs) and thereby regulates autophagic flux. The failure of autophagic flux induced by defect of TRPML1 or its abnormal regulation in arterial SMCs has been implicated in the atherogenesis. Under normal condition, lysosome trafficking and fusion to APs are controlled by NAADP-regulated TRPML1 channel activity, which leads to the formation of autophagolysosomes (APLs) and subsequent breakdown of autophagic vesicles within the cells. This regulated autophagic process via NAADP signaling pathway protects SMCs from atherosclerotic injury upon atherogenic stimulations. When the controlling mechanism of lysosome function is insufficient, such as impaired CD38-ADP-ribosylcyclase activity, reduced NAADP production, or defect of TRPML1 channels, the formation of APLs and breakdown of autophagic vesicles becomes impaired. The cell dedifferentiation and proliferation are thus activated, which lead to the enhancement of extracellular matrix production and result in the atherogenesis or acceleration of atherosclerosis [66].

Lysosomal TRPML1 in Apoptosis

Fas ligand (FasL) is a type II transmembrane protein that belongs to the tumor necrosis factor (TNF) superfamily. It is ubiquitously expressed in the vasculature and exists as a membrane molecule (mFasL) as well as a soluble form (sFasL), which is proteolytically generated by metalloproteinase (MMP7) cleavage. Functionally, activation of FasL has been demonstrated to implicate in the pathophysiological processes of various cardiovascular diseases, such as coronary heart disease, arteriosclerosis, and ischemia-reperfusion injury [67–69]. These vascular pathogenic roles of FasL are mainly attributed to its effect in inducing apoptosis.

Recent studies have demonstrated that activation of the death receptor Fas by FasL produced a two-phase intracellular Ca^{2+} release response in CAMs, which consists of local Ca^{2+} bursts via lysosomes and consequent Ca^{2+} release from the SR [28, 32]. In this FasL-induced two-phase Ca^{2+} response, the initial pilot phase is mediated by lysosomal TRPML1 channel, which functionally relates to apoptosis of CAMs, a common effect with the stimulation of FasL. By confocal microscopy, silencing CAMs with TRPML1 small interfering siRNA was found to substantially inhibit FasL-induced lysosome Ca^{2+} burst and consequent SR Ca^{2+} release. In contrast, transfection of CAMs with plasmids containing a full-length TRPML1 gene enhanced FasL-induced two-phase Ca^{2+} release. Correspondingly, FasL-induced apoptosis and activation of calpain and calcineurin, two Ca^{2+}-sensitive apoptotic proteins, were significantly attenuated by silencing TRPML1 gene, but enhanced by overexpression of TRPML1 channel protein. These results suggest that TRPML1 channel-mediated lysosomal Ca^{2+} bursts upon FasL stimulation promote apoptosis of CAMs via a Ca^{2+}-dependent mechanism.

Conclusion

There is increasing evidence that lysosomal TRPMLs are implicated in the regulation of vascular function and may be involved in the development of vascular injury and even atherosclerosis. It has been shown that these lysosomal TRPML channels, in particular, TRPML1 participates in the regulation of intracellular Ca^{2+} levels in response to different agonists or stimuli through a two-phase Ca^{2+} release. TRPML1-mediated lysosome Ca^{2+} release not only determines basal vascular tone and some vasomotor response, but also regulates cell surviving mechanisms in SMCs such as autophagy, apoptosis and lipid transportation. Future studies are needed to elucidate the molecular mechanisms regulating lysosomal TRPML channel activity and clarify their physiological and pathological relevance in response to different pro-pathologic factors. It is anticipated that TRPML channels may be a potential therapeutic target for prevention and treatment of vascular injury and disease such as atherosclerosis.

Acknowledgment The works cited in the authors' laboratory were supported by grants from the National Institutes of Health (HL075316, HL057244 and HL115068).

Author Disclosure Statement: none.

References

1. Sun M, Goldin E, Stahl S, Falardeau JL, Kennedy JC, Acierno Jr JS, Bove C, Kaneski CR, Nagle J, Bromley MC, Colman M, Schiffmann R, Slaugenhaupt SA. Mucolipidosis type IV is caused by mutations in a gene encoding a novel transient receptor potential channel. Hum Mol Genet. 2000;9:2471–8.
2. Yamaguchi S, Jha A, Li Q, Soyombo AA, Dickinson GD, Churamani D, Brailoiu E, Patel S, Muallem S. Transient receptor potential mucolipin 1 (TRPML1) and two-pore channels are functionally independent organellar ion channels. J Biol Chem. 2011;286:22934–42.
3. Bach G. Mucolipin 1: endocytosis and cation channel—a review. Pflugers Arch. 2005;451:313–7.
4. LaPlante JM, Falardeau J, Sun M, Kanazirska M, Brown EM, Slaugenhaupt SA, Vassilev PM. Identification and characterization of the single channel function of human mucolipin-1 implicated in mucolipidosis type IV, a disorder affecting the lysosomal pathway. FEBS Lett. 2002;532:183–7.
5. LaPlante JM, Ye CP, Quinn SJ, Goldin E, Brown EM, Slaugenhaupt SA, Vassilev PM. Functional links between mucolipin-1 and Ca2+-dependent membrane trafficking in mucolipidosis IV. Biochem Biophys Res Commun. 2004;322:1384–91.
6. Bach G. Mucolipidosis type IV. Mol Genet Metab. 2001;73:197–203.
7. Kiselyov K, Chen J, Rbaibi Y, Oberdick D, Tjon-Kon-Sang S, Shcheynikov N, Muallem S, Soyombo A. TRP-ML1 is a lysosomal monovalent cation channel that undergoes proteolytic cleavage. J Biol Chem. 2005;280:43218–23.
8. Slaugenhaupt SA. The molecular basis of mucolipidosis type IV. Curr Mol Med. 2002;2:445–50.
9. Bargal R, Avidan N, Ben-Asher E, Olender Z, Zeigler M, Frumkin A, Raas-Rothschild A, Glusman G, Lancet D, Bach G. Identification of the gene causing mucolipidosis type IV. Nat Genet. 2000;26:118–23.

10. Bassi MT, Manzoni M, Monti E, Pizzo MT, Ballabio A, Borsani G. Cloning of the gene encoding a novel integral membrane protein, mucolipidin-and identification of the two major founder mutations causing mucolipidosis type IV. Am J Hum Genet. 2000;67:1110–20.
11. Nilius B, Owsianik G, Voets T, Peters JA. Transient receptor potential cation channels in disease. Physiol Rev. 2007;87:165–217.
12. Dong XP, Cheng X, Mills E, Delling M, Wang F, Kurz T, Xu H. The type IV mucolipidosis-associated protein TRPML1 is an endolysosomal iron release channel. Nature. 2008;455:992–6.
13. Feng X, Huang Y, Lu Y, Xiong J, Wong CO, Yang P, Xia J, Chen D, Du G, Venkatachalam K, Xia X, Zhu MX. Drosophila TRPML forms PI(3,5)P2-activated cation channels in both endolysosomes and plasma membrane. J Biol Chem. 2014;289:4262–72.
14. Wang WY, Zhang XL, Guo Q, Xu HX. TRPML1: an ion channel in the lysosome. In: Nilius B, Flockerzi V, editors. Mammalian Transient Receptor Potential (TRP) cation channels, handbook of experimental pharmacology 222. Berlin Heidelberg: Springer-Verlag; 2014.
15. Dong XP, Wang X, Shen D, Chen S, Liu M, Wang Y, Mills E, Cheng X, Delling M, Xu H. Activating mutations of the TRPML1 channel revealed by proline-scanning mutagenesis. J Biol Chem. 2009;284:32040–52.
16. Cheng X, Shen D, Samie M, Xu H. Mucolipins: intracellular TRPML1-3 channels. FEBS Lett. 2010;584:2013–21.
17. Puertollano R, Kiselyov K. TRPMLs: in sickness and in health. Am J Physiol Renal Physiol. 2009;296:F1245–54.
18. Zhang F, Li PL. Reconstitution and characterization of a nicotinic acid adenine dinucleotide phosphate (NAADP)-sensitive Ca2+ release channel from liver lysosomes of rats. J Biol Chem. 2007;282:25259–69.
19. Zhang F, Jin S, Yi F, Li PL. TRP-ML1 functions as a lysosomal NAADP-sensitive Ca2+ release channel in coronary arterial myocytes. J Cell Mol Med. 2009;13:3174–85.
20. Chini EN, Beers KW, Dousa TP. Nicotinate adenine dinucleotide phosphate (NAADP) triggers a specific calcium release system in sea urchin eggs. J Biol Chem. 1995;270:3216–23.
21. Lee HC, Aarhus R. A derivative of NADP mobilizes calcium stores insensitive to inositol trisphosphate and cyclic ADP-ribose. J Biol Chem. 1995;270:2152–7.
22. Ge ZD, Li PL, Chen YF, Gross GJ, Zou AP. Myocardial ischemia and reperfusion reduce the levels of cyclic ADP-ribose in rat myocardium. Basic Res Cardiol. 2002;97:312–9.
23. Ge ZD, Zhang DX, Chen YF, Yi FX, Zou AP, Campbell WB, Li PL. Cyclic ADP-ribose contributes to contraction and Ca2+ release by M1 muscarinic receptor activation in coronary arterial smooth muscle. J Vasc Res. 2003;40:28–36.
24. Lee HC, Aarhus R. Functional visualization of the separate but interacting calcium stores sensitive to NAADP and cyclic ADP-ribose. J Cell Sci. 2000;113(Pt 24):4413–20.
25. Li PL, Tang WX, Valdivia HH, Zou AP, Campbell WB. cADP-ribose activates reconstituted ryanodine receptors from coronary arterial smooth muscle. Am J Physiol Heart Circ Physiol. 2001;280:H208–15.
26. Soares S, Thompson M, White T, Isbell A, Yamasaki M, Prakash Y, Lund FE, Galione A, Chini EN. NAADP as a second messenger: neither CD38 nor base-exchange reaction are necessary for in vivo generation of NAADP in myometrial cells. Am J Physiol Cell Physiol. 2007;292:C227–39.
27. Xu M, Zhang Y, Xia M, Li XX, Ritter JK, Zhang F, Li PL. NAD(P)H oxidase-dependent intracellular and extracellular O2*- production in coronary arterial myocytes from CD38 knockout mice. Free Radic Biol Med. 2012;52:357–65.
28. Zhang F, Xia M, Li PL. Lysosome-dependent Ca(2+) release response to Fas activation in coronary arterial myocytes through NAADP: evidence from CD38 gene knockouts. Am J Physiol Cell Physiol. 2010;298:C1209–16.
29. Hill-Eubanks DC, Gonzales AL, Sonkusare SK, Nelson MT. Vascular TRP channels: performing under pressure and going with the flow. Physiology. 2014;29:343–60.
30. Kinnear NP, Wyatt CN, Clark JH, Calcraft PJ, Fleischer S, Jeyakumar LH, Nixon GF, Evans AM. Lysosomes co-localize with ryanodine receptor subtype 3 to form a trigger zone for calcium signalling by NAADP in rat pulmonary arterial smooth muscle. Cell Calcium. 2008;44:190–201.

31. Zhang F, Zhang G, Zhang AY, Koeberl MJ, Wallander E, Li PL. Production of NAADP and its role in Ca2+ mobilization associated with lysosomes in coronary arterial myocytes. Am J Physiol Heart Circ Physiol. 2006;291:H274–82.
32. Xu M, Li X, Walsh SW, Zhang Y, Abais JM, Boini KM, Li PL. Intracellular two-phase Ca2+ release and apoptosis controlled by TRP-ML1 channel activity in coronary arterial myocytes. Am J Physiol Cell Physiol. 2013;304:C458–66.
33. Fameli N, Ogunbayo OA, van Breemen C, Evans AM. Cytoplasmic nanojunctions between lysosomes and sarcoplasmic reticulum are required for specific calcium signaling. F1000Res. 2014;3:93.
34. Kinnear NP, Boittin FX, Thomas JM, Galione A, Evans AM. Lysosome-sarcoplasmic reticulum junctions. A trigger zone for calcium signaling by nicotinic acid adenine dinucleotide phosphate and endothelin-1. J Biol Chem. 2004;279:54319–26.
35. Zhu MX, Ma J, Parrington J, Calcraft PJ, Galione A, Evans AM. Calcium signaling via two-pore channels: local or global, that is the question. Am J Physiol Cell Physiol. 2010;298: C430–41.
36. Zhang F, Xu M, Han WQ, Li PL. Reconstitution of lysosomal NAADP-TRP-ML1 signaling pathway and its function in TRP-ML1(−/−) cells. Am J Physiol Cell Physiol. 2011;301:C421–30.
37. Fleckenstein-Grun G, Frey M, Thimm F, Hofgartner W, Fleckenstein A. Calcium overload—an important cellular mechanism in hypertension and arteriosclerosis. Drugs. 1992;44 Suppl 1:23–30.
38. Wong CO, Li R, Montell C, Venkatachalam K. Drosophila TRPML is required for TORC1 activation. Curr Biol. 2012;22:1616–21.
39. Venkatachalam K, Wong CO, Montell C. Feast or famine: role of TRPML in preventing cellular amino acid starvation. Autophagy. 2013;9:98–100.
40. Vergarajauregui S, Connelly PS, Daniels MP, Puertollano R. Autophagic dysfunction in mucolipidosis type IV patients. Hum Mol Genet. 2008;17:2723–37.
41. Martina JA, Lelouvier B, Puertollano R. The calcium channel mucolipin-3 is a novel regulator of trafficking along the endosomal pathway. Traffic. 2009;10:1143–56.
42. Kim HJ, Soyombo AA, Tjon-Kon-Sang S, So I, Muallem S. The Ca(2+) channel TRPML3 regulates membrane trafficking and autophagy. Traffic. 2009;10:1157–67.
43. Pryor PR, Mullock BM, Bright NA, Gray SR, Luzio JP. The role of intraorganellar Ca(2+) in late endosome-lysosome heterotypic fusion and in the reformation of lysosomes from hybrid organelles. J Cell Biol. 2000;149:1053–62.
44. Luzio JP, Bright NA, Pryor PR. The role of calcium and other ions in sorting and delivery in the late endocytic pathway. Biochem Soc Trans. 2007;35:1088–91.
45. Carstea ED, Morris JA, Coleman KG, Loftus SK, Zhang D, Cummings C, Gu J, Rosenfeld MA, Pavan WJ, Krizman DB, Nagle J, Polymeropoulos MH, Sturley SL, Ioannou YA, Higgins ME, Comly M, Cooney A, Brown A, Kaneski CR, Blanchette-Mackie EJ, Dwyer NK, Neufeld EB, Chang TY, Liscum L, Strauss III JF, Ohno K, Zeigler M, Carmi R, Sokol J, Markie D, O'Neill RR, van Diggelen OP, Elleder M, Patterson MC, Brady RO, Vanier MT, Pentchev PG, Tagle DA. Niemann-Pick C1 disease gene: homology to mediators of cholesterol homeostasis. Science. 1997;277:228–31.
46. Carstea ED, Polymeropoulos MH, Parker CC, Detera-Wadleigh SD, O'Neill RR, Patterson MC, Goldin E, Xiao H, Straub RE, Vanier MT, et al. Linkage of Niemann-Pick disease type C to human chromosome 18. Proc Natl Acad Sci U S A. 1993;90:2002–4.
47. Liscum L, Sturley SL. Intracellular trafficking of Niemann-Pick C proteins 1 and 2: obligate components of subcellular lipid transport. Biochim Biophys Acta. 2004;1685:22–7.
48. Haley NJ, Shio H, Fowler S. Characterization of lipid-laden aortic cells from cholesterol-fed rabbits. I. Resolution of aortic cell populations by metrizamide density gradient centrifugation. Lab Invest. 1977;37:287–96.
49. Jerome WG, Cash C, Webber R, Horton R, Yancey PG. Lysosomal lipid accumulation from oxidized low density lipoprotein is correlated with hypertrophy of the Golgi apparatus and trans-Golgi network. J Lipid Res. 1998;39:1362–71.

50. Yancey PG, Jerome WG. Lysosomal sequestration of free and esterified cholesterol from oxidized low density lipoprotein in macrophages of different species. J Lipid Res. 1998;39: 1349–61.
51. Yancey PG, Miles S, Schwegel J, Jerome WG. Uptake and trafficking of mildly oxidized LDL and acetylated LDL in THP-1 cells does not explain the differences in lysosomal metabolism of these two lipoproteins. Microsc Microanal. 2002;8:81–93.
52. Jerome WG. Advanced atherosclerotic foam cell formation has features of an acquired lysosomal storage disorder. Rejuvenation Res. 2006;9:245–55.
53. Loftus SK, Morris JA, Carstea ED, Gu JZ, Cummings C, Brown A, Ellison J, Ohno K, Rosenfeld MA, Tagle DA, Pentchev PG, Pavan WJ. Murine model of Niemann-Pick C disease: mutation in a cholesterol homeostasis gene. Science. 1997;277:232–5.
54. Butler JD, Vanier MT, Pentchev PG. Niemann-Pick C disease: cystine and lipids accumulate in the murine model of this lysosomal cholesterol lipidosis. Biochem Biophys Res Commun. 1993;196:154–9.
55. te Vruchte D, Lloyd-Evans E, Veldman RJ, Neville DC, Dwek RA, Platt FM, van Blitterswijk WJ, Sillence DJ. Accumulation of glycosphingolipids in Niemann-Pick C disease disrupts endosomal transport. J Biol Chem. 2004;279:26167–75.
56. Choi AM, Ryter SW, Levine B. Autophagy in human health and disease. N Engl J Med. 2013;368:1845–6.
57. Martinet W, Knaapen MW, Kockx MM, De Meyer GR. Autophagy in cardiovascular disease. Trends Mol Med. 2007;13:482–91.
58. Martinet W, De Meyer GR. Autophagy in atherosclerosis. Curr Atheroscler Rep. 2008;10:216–23.
59. Gustafsson AB, Gottlieb RA. Autophagy in ischemic heart disease. Circ Res. 2009;104: 150–8.
60. Cao DJ, Gillette TG, Hill JA. Cardiomyocyte autophagy: remodeling, repairing, and reconstructing the heart. Curr Hypertens Rep. 2009;11:406–11.
61. Gottlieb RA, Mentzer RM. Autophagy during cardiac stress: joys and frustrations of autophagy. Annu Rev Physiol. 2010;72:45–59.
62. Nemchenko A, Chiong M, Turer A, Lavandero S, Hill JA. Autophagy as a therapeutic target in cardiovascular disease. J Mol Cell Cardiol. 2011;51:584–93.
63. Salabei JK, Hill BG. Implications of autophagy for vascular smooth muscle cell function and plasticity. Free Radic Biol Med. 2013;65:693–703.
64. Geer JC, Mc Jr GH, Strong JP. The fine structure of human atherosclerotic lesions. Am J Pathol. 1961;38:263–87.
65. Parker F, Odland GF. A light microscopic, histochemical and electron microscopic study of experimental atherosclerosis in rabbit coronary artery and a comparison with rabbit aorta atherosclerosis. Am J Pathol. 1966;48:451–81.
66. Zhang Y, Xu M, Xia M, Li X, Boini KM, Wang M, Gulbins E, Ratz PH, Li PL. Defective autophagosome trafficking contributes to impaired autophagic flux in coronary arterial myocytes lacking CD38 gene. Cardiovasc Res. 2014;102:68–78.
67. Feng QZ, Zhao YS, Abdelwahid E. The role of Fas in the progression of ischemic heart failure: prohypertrophy or proapoptosis. Coron Artery Dis. 2008;19:527–34.
68. Henriques-Pons A, de Oliveira GM. Is the Fas/Fas-L pathway a promising target for treating inflammatory heart disease? J Cardiovasc Pharmacol. 2009;53:94–9.
69. Sata M, Suhara T, Walsh K. Vascular endothelial cells and smooth muscle cells differ in expression of Fas and Fas ligand and in sensitivity to Fas ligand-induced cell death: implications for vascular disease and therapy. Arterioscler Thromb Vasc Biol. 2000;20:309–16.

Part III
Ion Channels in the Regulation of Cell Proliferation, Remodeling, Hypertrophy and Angiogenesis

Chapter 11
Calcium Mobilization via Intracellular Ion Channels, Store Organization and Mitochondria in Smooth Muscle

John G. McCarron, Susan Chalmers, Calum Wilson, and Mairi E. Sandison

Abstract In smooth muscle, Ca^{2+} release from the internal store into the cytoplasm occurs via inositol trisphosphate (IP_3R) and ryanodine receptors (RyR). The internal Ca^{2+} stores containing IP_3R and RyR may be arranged as multiple separate compartments with various IP_3R and RyR arrangements, or there may be a single structure containing both receptors. The existence of multiple stores is proposed to explain several physiological responses which include the progression of Ca^{2+} waves, graded Ca^{2+} release from the store and various local responses and sensitivities. We suggest that, rather than multiple stores, a single luminally-continuous store exists in which Ca^{2+} is in free diffusional equilibrium throughout. Regulation of Ca^{2+} release via IP_3R and RyR by the local Ca^{2+} concentration within the stores explains the apparent existence of multiple stores and physiological processes such as graded Ca^{2+} release and Ca^{2+} waves. Close positioning of IP_3R on the store with mitochondria or with receptors on the plasma membrane creates 'IP_3 junctions' to generate local responses on the luminally-continuous store.

Keywords Smooth muscle • Calcium signalling • Calcium stores • IP3 receptors • Ryanodine receptors • Quantal calcium release • Mitochondria

Introduction

Ca^{2+} regulates several smooth muscle functions including contraction, proliferation and the changes in muscle performance that accompanies disease [1]. The characteristics of the Ca^{2+} signal (e.g. the amplitude, duration, frequency and location) determine the nature of the biological response. A major Ca^{2+} source in smooth muscle is

The original version of this chapter was revised. An erratum to this chapter can be found at DOI 10.1007/978-3-319-29635-7_19

J.G. McCarron (✉) • S. Chalmers • C. Wilson • M.E. Sandison
Strathclyde Institute of Pharmacy and Biomedical Sciences, University of Strathclyde, SIPBS Building, 161 Cathedral Street, Glasgow G4 0RE, UK
e-mail: john.mccarron@strath.ac.uk

I. Levitan, A.M. Dopico (eds.), *Vascular Ion Channels in Physiology and Disease*, DOI 10.1007/978-3-319-29635-7_11

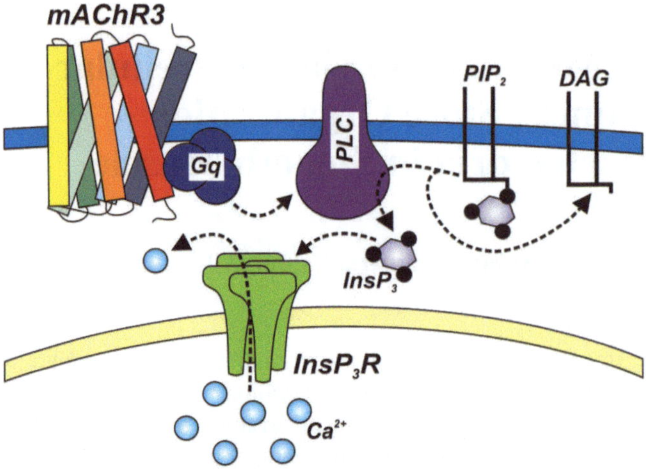

Fig. 11.1 Receptor activation and generation of IP_3 and Ca^{2+} release. Muscarinic receptors (mAChR3), phospholipase C (PLC) and IP_3R may be co-localized to create junctions in which IP_3 acts as a highly localized signal by being rapidly delivered to IP_3R. PIP2, phosphatidylinositol 4,5-bisphosphate; DAG diacylglycerol

an internal storage compartment which accumulates Ca^{2+} via sarco/endoplasmic reticulum Ca^{2+}-ATPases (SERCA). Ca^{2+} is released from the store into the cytoplasm via the ligand-gated channel/receptor complexes, the inositol trisphosphate (IP_3R) and ryanodine receptors. Release of Ca^{2+} via IP_3R is activated by IP_3 generated in response to many G-protein or tyrosine kinase-linked receptor activators including drugs (Fig. 11.1). RyR may be activated pharmacologically (e.g. caffeine), by Ca^{2+} influx from outside the cell in the process of Ca^{2+}-induced Ca^{2+} release (CICR), or when the stores's Ca^{2+} content exceeds normal physiological values, i.e. in 'store overload' [2–6]. Activation of either receptor allows diffusion of Ca^{2+} from the store to increase the cytoplasmic Ca^{2+} concentration ($[Ca^{2+}]_c$) from the resting value of ~100 nM to ~1 μM for many seconds throughout the cell and briefly (e.g. 100 ms) to much higher values (e.g. 50 μM) in small parts of the cytoplasm.

Physiological Functions Proposed to Be Explained by the Structure of the Store

The amplitude and duration of the Ca^{2+} signal depends on the quantity of Ca^{2+} available for release, which is determined in large part by the structural arrangement of the store. The store appears as an interconnected network of tubules [7] with a single lumen in which Ca^{2+} is in free diffusional equilibrium throughout (Fig. 11.2) [e.g. 8, 9]. However, considerable controversy persists about the stores structural and functional continuity or discontinuity. Rather than a store with a single lumen, multiple separate smaller Ca^{2+} storage units may exist (Fig. 11.2) [e.g. 7, 10–12]. Although the structure is unresolved, the arrangement of the store is proposed to account for several characteristics of Ca^{2+}

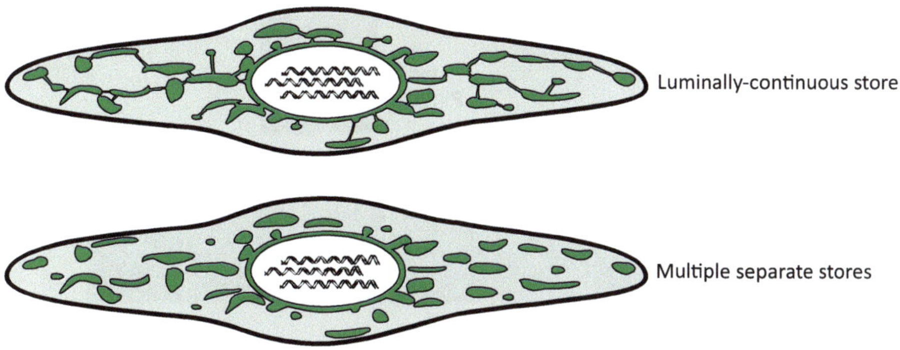

Fig. 11.2 Arrangement of the store. The store may be a single luminally-continuous structure with Ca^{2+}in free diffusional equilibrium throughout (*top*) or a series of multiple separate elements (*bottom*)

signals, such as the graded concentration-dependence of IP_3-mediated Ca^{2+} release, the variation in sensitivity in different parts of the cell to generate local responses and the progression of Ca^{2+} signals through the cell. For example, while Ca^{2+} entry via voltage-dependent Ca^{2+} channels generates quite uniform rises in Ca^{2+} (Fig. 11.3; [13, 14]), Ca^{2+} release from internal stores may generate complex patterns, such as travelling spatial gradients of Ca^{2+} ('Ca^{2+} waves'; Fig. 11.3). For Ca^{2+} waves to progress through the cell, sequential activation of IP_3R [13], by Ca^{2+} itself, occur in a repeating positive feedback CICR-like process [15, 16], i.e. Ca^{2+} release from one IP_3R activates neighbouring receptors to progress the wave. An explanation put forward to explain wave movement, rather than there being a persistent Ca^{2+} release at one site on the cell, is that store is arranged as several stores along the length of the cell, each with a limited amount of Ca^{2+}. Each store is activated and depleted in turn (Fig. 11.4a).

A discontinuous structure of the store has also been proposed to explain the graded IP_3 concentration-dependent Ca^{2+} release process [17, 18]. Low concentrations of IP_3 release only part of the overall available Ca^{2+} content of the store [17, 19–22]. As the IP_3 concentration increases, a further release of Ca^{2+} occurs [reviewed 23]. Such a graded release seems incompatible with the positive feedback CICR-like facility at IP_3R [24], which would be anticipated to fully deplete the store when activated. To explain graded Ca^{2+} release, the store has been proposed to assemble in multiple separate units, each endowed with a finite Ca^{2+} storage capacity and sensitivity to IP_3 (Fig. 11.4b). At any given concentration of IP_3 only some stores will be activated to release Ca^{2+} [17, 18, 25] (Fig. 11.4b). This same feature of the store may also explain the reported variations in sensitivity different parts of the cell to IP_3 [19, 26, 27].

Structure of the Ca^{2+} Stores

There are several different RyR and IP_3R arrangements which may exist on each of the proposed separate stores to explain the various experimental observations. Indeed, the Ca^{2+} stores have been classified on the arrangement of IP_3R and RyR and proposals for one, two, or more, stores with a variety of complex receptor

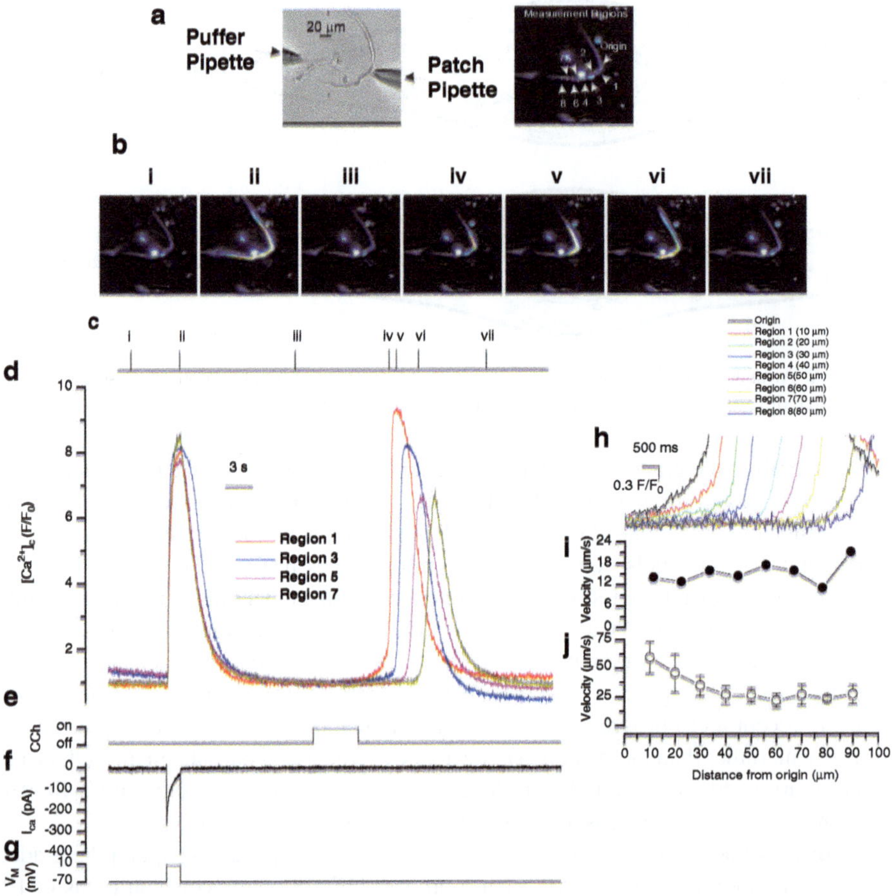

Fig. 11.3 Depolarization and IP_3-evoked increases in $[Ca^{2+}]_c$. Depolarization (−70 mV to +10 mV; **g**), activated a voltage-dependent Ca^{2+} current (I_{Ca}; **f**) to evoke a relatively uniform rise in $[Ca^{2+}]_c$ (**b**, **d**). In contrast, $[Ca^{2+}]_c$ increases in response to and IP_3-generating agonist began in one part of the cell and progressed from that site (**b**, **d** and expanded time base **h**). The $[Ca^{2+}]_c$ images (**b**) are derived from the time points indicated by the corresponding *numerals* in **c**. $[Ca^{2+}]_c$ changes in **b** are represented by *colour*; *blue* low and *red*/*white* high $[Ca^{2+}]_c$. Changes in the fluorescence ratio with time (**d**, **h**) are derived from 1 pixel lines ('origin' and regions 1–8 in **a**, *right* panel; drawn at a 3 pixel width to facilitate visualization). (**a**) *Left* panel shows a bright field image of the cell; see also whole cell electrode (*right* side) and puffer pipette containing agonist (*left* side). The velocity of wave progression is shown in **i** for the data presented in (**d**, **h**). Summarized velocity data is presented (**j** n = 5). From McCarron et al. 2010 [13] with permission

arrangements have been made (Fig. 11.5). There may be multiple stores each containing both IP_3R and RyR [28–32], or there may be stores which contain only RyR and separate stores only IP_3R [12, 28, 32–34] (e.g. basilar mesenteric or pulmonary arteries; Fig. 11.5i, ii). In other studies, there may be Ca^{2+} stores containing IP_3R and RyR together on some stores along with other separate stores in the same cell with either IP_3R alone (e.g. pulmonary artery and aorta [29, 35]; Fig. 11.5iii) or RyR

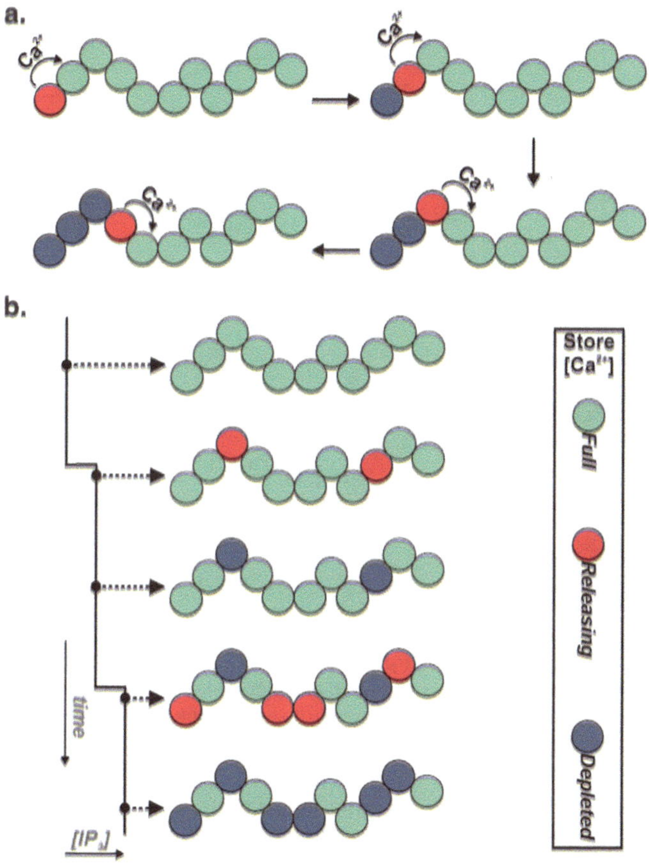

Fig. 11.4 Wave progression and store arrangement. (**a**) The store may function as a series of discontinuous compartments that are activated and depleted in turn to explain wave progression. (**b**) Separate stores with various sensitivities to IP_3 are activated and depleted as the IP_3 concentration (*left*-side) increases

alone (e.g. mesenteric artery [30]; Fig. 11.5iv). Stores have also been differentiated by their sensitivity to the SERCA pump inhibitors cyclopiazonic acid (CPA) and thapsigargin. In A7r5 cells (a cell line derived from thoracic aorta tissue) there are stores containing RyR that are insensitive to thapsigargin and separate stores in the same cells (also with RyR) that are sensitive to thapsigargin [12]. In an alternative proposal for store arrangement in A7r5 cells, a thapsigargin-insensitive store with IP_3R but not RyR may exist [36]. In murine bladder smooth muscle, three types of Ca^{2+} store are proposed: two sensitive to thapsigargin, one with IP_3R and one without, and a third store insensitive to IP_3 and thapsigargin [37]. In tracheal myocytes three types of Ca^{2+} stores are proposed which were refilled by different pathways. Ca^{2+} influx through voltage-dependent Ca^{2+} channels and CPA sensitive pumps refilled 80 % of the IP_3R-containing stores. The remaining 20 % were not refilled by CPA-sensitive pumps or Ca^{2+} influx through voltage-dependent Ca^{2+} channels and

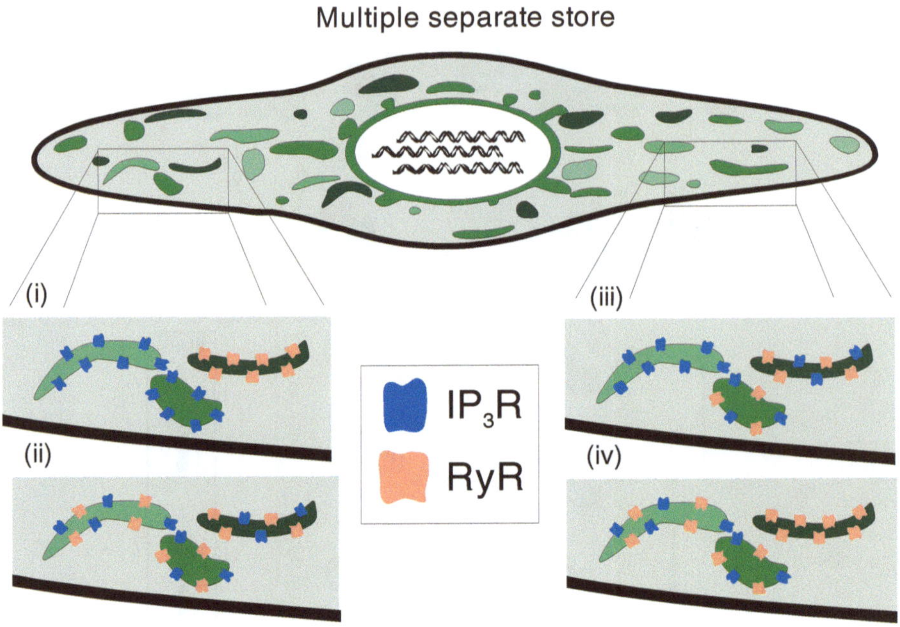

Fig. 11.5 Arrangement of RyR and IP_3R on the store(s). There may be store with RyR (*blue*) alone or IP_3R (*red*) alone (i), or stores with both receptors (ii) or a combination of the two (iii, iv). Although the cartoon shows the different proposed store receptor arrangements in the same cell, the proposed stores have been described for different cell types

neither was the RyR-containing store. Instead, thapsigargin depleted the CPA/voltage-dependent Ca^{2+} channels insensitive IP_3R store fully and the RyR store by more than 50 % [38]. These differences in refilling mechanisms of the stores are proposed to demonstrate pharmacologically distinct Ca^{2+} stores which play an important role in the generation of Ca^{2+}signals in airway smooth muscle cells [38].

Thus, data from various functional studies suggest there may be structural discontinuities in the store and that different types of receptor arrangements on those stores exist. Proposals for stores which contain only IP_3R or RyR exist as do proposals for stores with RyR and IP_3R together and in combination with additional separate stores in the same cells containing only either IP_3R or RyR. The questions arise, why is such a diversity of stores and receptor arrangement required and do functional experiments unambiguously reveal structural discontinuities in the store?

Methods Used to Investigate Stores May Create the Appearance of Multiple Stores

It could be the case that the experimental conditions used to investigate the stores may contribute to the diversity of proposals on arrangement. In native cells, methods for studying Ca^{2+} store subcompartments are limited. The main experimental approach is to define the structural organisation of the Ca^{2+} stores from functional (Ca^{2+} response) data. To do this, the store is depleted typically via one receptor (RyR or IP_3R) by repeated activation with a *single* concentration of either IP_3 or caffeine under conditions which prevent store refilling with Ca^{2+}. After depletion via one receptor (e.g. RyR), whether or not Ca^{2+} is available to be released via the other receptor (e.g. IP_3R) is then determined. If depletion via one receptor abolishes Ca^{2+} release from the other, the receptors are suggested to be co-localized on a single store and access a common Ca^{2+} source. However, if depletion of the stores from one receptor leaves the other receptor's response largely unaffected, the two channels are suggested to be localized on different stores. With this approach, some investigations (e.g. on portal vein and pulmonary artery) have shown a single store containing both RyR and IP_3R, since depletion of the Ca^{2+} store by caffeine (which activates RyR) prevented IP_3-mediated Ca^{2+} release [31, 32, 39, 40]. On the other hand, other studies on pulmonary artery have suggested there may be separate stores for each receptor since depletion of the RyR-containing store did not abolish agonist-evoked IP_3-mediated Ca^{2+} release and *vice versa* [41]. In yet other studies (e.g. portal vein, pulmonary artery and taenia caeci), one store may express RyR and IP_3R and other stores, in the same cell, only IP_3R [11, 35, 42]. This conclusion came from the finding that depletion of the IP_3R-containing store abolished Ca^{2+} release via RyR, while depletion of the RyR-containing store did not abolish Ca^{2+} release via IP_3R. In further studies in other cell types (mesenteric artery) and in our own investigations in colonic smooth muscle [43], some stores may express both RyR and IP_3R while others only RyR [30, 43]. In this case, depletion of the RyR-containing store abolished Ca^{2+} release via IP_3R, while depletion of the IP_3R-containing store did not abolish Ca^{2+} release via RyR—a result apparently consistent with there being a store which contained RyR alone.

However, in our own later experiments examining the structure of the store [44–46] we found unexpectedly that the entire store appeared to be a single luminally-continuous entity rather than a series of separate stores. In these later experiments, to examine luminal continuity, the store was depleted at *one small site* in the cell by repetitively applying IP_3 to a small (10 μm) region under conditions preventing store refilling. Even though only a small site in the cell was activated, the store depleted throughout the cell [44]. This result suggested that Ca^{2+} was in free diffusional equilibrium in the store (Fig. 11.6) i.e. a luminally-continuous store. In keeping with these findings, the IP_3-sensitive store also could be refilled from one small site on the cell (Fig. 11.7); a result suggesting there was a single store in which Ca^{2+} was able to diffuse freely throughout.

Depletion of the RyR-sensitive store at one site also depleted the entire store [44, 46]. In this case the RyR-containing store was depleted by attaching a pipette containing ryanodine to one small site of the cell to deplete the store there. Caffeine was applied to the entire cell. If the RyR containing store comprised separate elements, depletion of one aspect of the store should not affect the Ca^{2+} available to be released in another area of the store. However, caffeine-evoked Ca^{2+} transients decreased uniformly throughout the cell [44, 46] suggesting that ryanodine, acting at one part of the cell, had depleted the entire store i.e. a single luminally-continuous store exists.

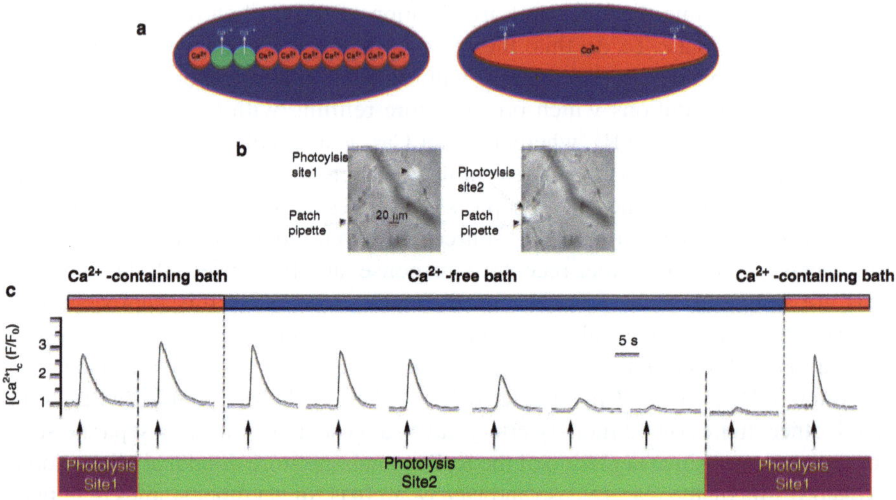

Fig. 11.6 Store luminal continuity:depletion of the IP_3-sensitive Ca^{2+} store in a localized area depletes the entire store of Ca^{2+}. (**a**) If the store was a series of luminal discontinuous elements (*left*) then Ca^{2+} release at one site would not alter the Ca^{2+} available for release from another. However if the store was luminally continuous, then Ca^{2+} release from one site would decrease the Ca^{2+} available for release from another site. To test this, at −70 mV, locally-photolyzed IP_3 (↑, **c**) in a 10 µm diameter region, (photolysis site 1; *bright spot* in **b** *left-hand* panel; see also patch electrode, *left* side) evoked Ca^{2+} transients (**c**). Results from photolysis site 1 are indicated by the *magenta bar* below the $[Ca^{2+}]_c$ trace in **c**. When repositioned to photolysis site 2 (**b**; *right hand* panel) subsequent photolysis ~90 s later produced a $[Ca^{2+}]_c$ increase (**c**). Photolysis site 2 is indicated by the *green line* below the $[Ca^{2+}]_c$ trace (**c**). In a Ca^{2+} free solution (containing EGTA (1 mM) and $MgCl_2$ (3 mM); *blue bar* above the $[Ca^{2+}]_c$ trace) the $[Ca^{2+}]_c$ increase evoked by IP_3 at photolysis site 2 declined in amplitude as the store was depleted of Ca^{2+} (**c**). When the store content had been substantially reduced at photolysis site 2 (**b**) (as revealed by the smaller Ca^{2+} transients **c**) IP_3 was liberated by photolysis at site 1 (**b**). Again as at photolysis site 2 the response was now almost abolished compared to control. On restoring external Ca^{2+} (**c**, *right hand* side) the Ca^{2+} increase evoked by IP_3 at photolysis site 1 was restored towards control values. These results suggest that the SR is luminally-continuous and within it Ca^{2+} is freely diffusible. $[Ca^{2+}]_c$ measurements were made from a 5 µm diameter circle at the photolysis site. Thus when photolysis occurred at photolysis site 1 $[Ca^{2+}]_{c,}$ measurements were made from a 5 µm diameter circle at the photolysis site 1. When photolysis occurred at photolysis site 2, $[Ca^{2+}]_c$ measurements were made from a 5 µm diameter circle at the photolysis site 2. (**b**, **c**) These results were original published in McCarron & Olson 2008 [44]

The question of whether there is a single store with luminal continuity or multiple stores has also been addressed in other cell types (HeLa, RBL, CHO) using a Ca^{2+} store-located green fluorescent protein (GFP) [47, 48]. Prolonged GFP photobleaching in a small restricted region of the cell resulted in the disappearance of fluorescence throughout store, suggesting GFP could move freely around the store to be eventually photobleached. Short periods of photobleaching were followed by a rapid restoration of fluorescence by the diffusion of GFP from sites neighbouring the photobleached region [47, 49]. A single store with luminal continuity throughout was also suggested by the diffusion of Ca^{2+} in pancreatic acinar cells [8].

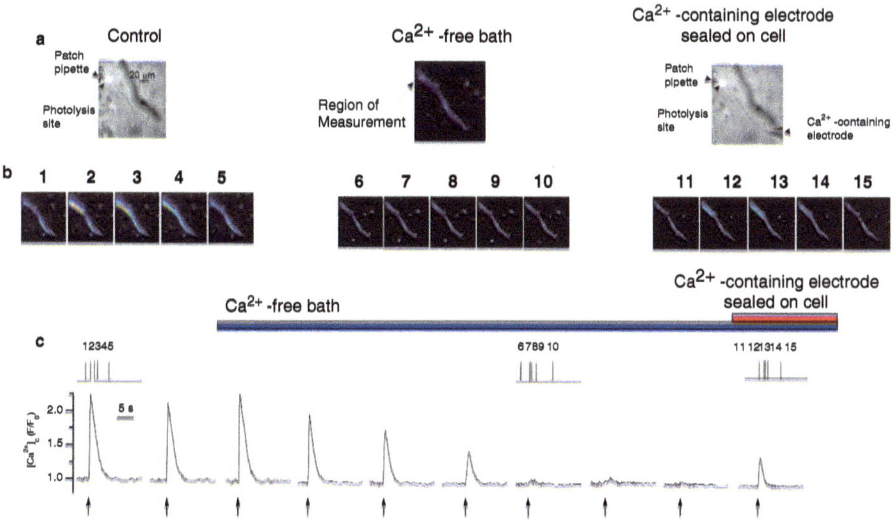

Fig. 11.7 Ca^{2+} can move through the SR to replenish a site previously depleted of the ion. At −70 mV, locally-photolyzed IP_3 (↑, **c**) in a 10 μm diameter region (*bright spot* in **a** *left-hand* panel; see also whole cell patch electrode (*left* side)) increased $[Ca^{2+}]_c$ (**b** and **c**). The $[Ca^{2+}]_c$ images (**b**) are derived from the time points indicated by the corresponding numbers in C. $[Ca^{2+}]_c$ changes in **b** are represented by colour; blue low and red high $[Ca^{2+}]_c$ A second photolysis of IP_3 ~ 60 s later at the same site (**c**) generated an approximately comparable $[Ca^{2+}]_c$ increase. In a Ca^{2+} free solution (containing 1 mM EGTA and 3 mM $MgCl_2$; *blue bar above* the trace) the $[Ca^{2+}]_c$ increase evoked by IP_3 declined and was abolished as the store became depleted of Ca^{2+}. When the Ca^{2+} containing patch electrode was subsequently sealed onto the cell in 'cell-attached' mode (**a** *right hand* panel; **c** *red bar*) there was no measurable increase in $[Ca^{2+}]_c$ yet the Ca^{2+} increase to IP_3 at the photolysis region (**a**) was subsequently restored partially (**c**). This result suggests that Ca^{2+} had diffused through the store lumen to replenish the store. The position of the region of $[Ca^{2+}]_c$ measurement is shown as a *white circle* in **a**, *center* panel. These results were original published in McCarron & Olson 2008 [44]

The Ca^{2+} store in the apical region was refilled with Ca^{2+} originating from a pipette attached to the opposite side of the cell on the basolateral membrane [see also 9]. Together, these experiments suggest the store is a luminally-continuous entity in which Ca^{2+} can diffuse freely throughout. How then does the appearance of multiple stores [43] occur on a single luminally-continuous store structure?

Complex RyR and IP_3R Regulation Characteristics and Apparent Store Configuration

IP_3R and RyR are each regulated by the Ca^{2+}concentration within the lumen of the store ('luminal Ca^{2+} regulation') [4, 45]. As the luminal Ca^{2+} concentration increases so does the activity of the store release channels [3–6]. Conversely, the activity of

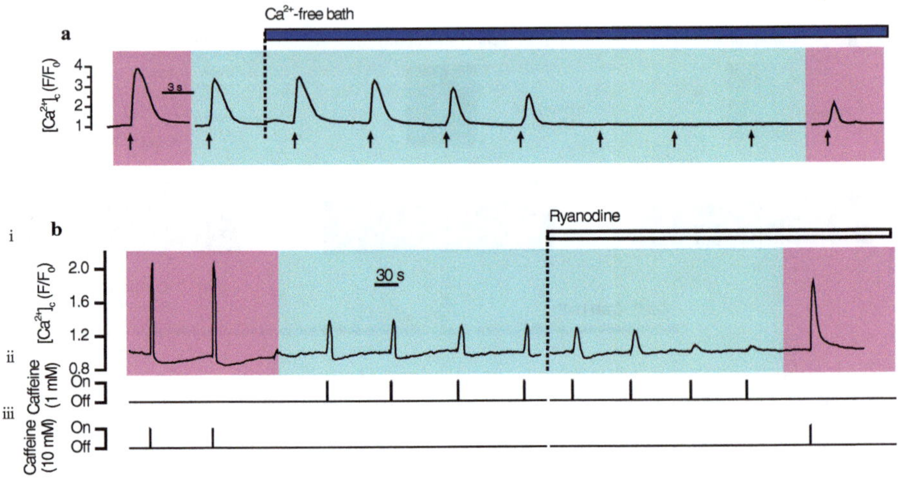

Fig. 11.8 The store may contain substantial residual Ca^{2+}after apparently being depleted. (**a**) At −70 mV high [IP_3] (*pink*; photolysed using a high lamp intensity; ↑) increased $[Ca^{2+}]_c$. A lower [IP_3] (*light blue*; ↑) evoked a submaximal $[Ca^{2+}]_c$ increase. In a Ca^{2+} free bath solution (containing 1 mM EGTA and 3 mM $MgCl_2$; *dark blue bar*) these increases declined then disappeared. The absence of a response to [IP_3] was not due to depletion of the store. Increasing [IP_3] (*pink*; *right side*; ↑) evoked further Ca^{2+} release. A mechanism, other than depletion of the store of Ca^{2+}, e.g. 'luminal' regulation of IP_3R, may have accounted for the loss of response to IP_3. The time between each IP_3 challenge was approximately 1 min. (**b**) Caffeine (10 mM; iii) indicated by *pink* (i) evoked approximately reproducible increases in $[Ca^{2+}]_{c.}$ (i). Caffeine (1 mM; ii) indicated by *light blue* (i) evoked submaximal $[Ca^{2+}]_c$ increases (i). In ryanodine (50 μM; for the duration of the unfilled bar) these increases declined to 12 % of their control value (i). However, after the substantial reduction in response to submaximal caffeine (1 mM; ii), caffeine (10 mM; iii) evoked a $[Ca^{2+}]_c$ rise of 77 % of its control value. The break in the record is ~90 s in which a new data recording file was established. These results were original published in McCarron & Olson 2008 [44]

RyR and IP_3R each decrease as the store Ca^{2+} content declines. Ca^{2+} release evoked by IP_3 or caffeine may substantially decline or stop as the store content falls, even when this store retains a significant residual quantity of Ca^{2+}. To examine this possibility, a series of experiments were carried out in which the store was depleted of Ca^{2+} (Fig. 11.8). When the store had been 'depleted', as revealed by the inhibition of response to IP_3 or caffeine, the concentration of each activator was increased and a substantial Ca^{2+} release occurred [44]. These experiments suggest that after apparent depletion the store retained significant quantities of Ca^{2+} and that residual Ca^{2+} is available for release with increased concentrations of IP_3 or caffeine.

Interpreting the amplitude of a Ca^{2+} response to a *single* repeatedly applied concentration of either IP_3 or caffeine as the store content declines is problematic as the amplitude of the response depends (1) on the position of the activator concentration on the concentration-response relationship curve and (2) the store luminal Ca^{2+} concentration. The absence of a response to a single concentration of IP_3 or caffeine, therefore, may not reflect an absence of available Ca^{2+} within the store but rather termination of channel activity by luminal regulation of the store release channels as the store Ca^{2+} content declines.

Luminal regulation may explain the appearance of multiple stores when pharmacological agents and functional data are used to define store subcompartments. Indeed, we reproduced data previously interpreted as various different store arrangements in a single smooth muscle cell type. For example, after depletion of the Ca^{2+} stores with caffeine and ryanodine, the response to IP_3 disappeared (Fig. 11.9a). This result suggest RyR and IP_3R access a single Ca^{2+} pool. However, in the same cell type, after depletion of the Ca^{2+} stores with caffeine and ryanodine, when a higher concentration of IP_3 (125 μM vs. 250 μM) was subsequently applied, a substantial Ca^{2+} increase occurred (Fig. 11.9b). This result suggests IP_3R accesses a different Ca^{2+} pool from RyR. On the other hand, after the store had been apparently depleted of Ca^{2+} by IP_3 (at a concentration which produced a maximal response) a substantial response to caffeine persisted (Fig. 11.9c), suggesting there was a store which only contains RyR [30, 43]. In yet other experiments, in the same cell type, when the concentration of IP_3 used to deplete the store of Ca^{2+} was increased, no Ca^{2+} response to caffeine occurred i.e. the apparently separate stores for RyR disappeared (Fig. 11.9d).

Rather than there being various separate stores with different receptor arrangements, these results suggests that partial depletion of the store terminates activity of the channels by luminal channel regulation by $[Ca^{2+}]$ within the store.

These results (Figs. 11.6, 11.8, 11.9) do not dispute the existence of multiple stores but suggest that care is required when interpreting results from functional data in terms of store structure. In some cells, multiple stores do exist unequivocally. Different Ca^{2+}concentrations have been measured in various regions of the store using recombinant aequorin [47], electron microscopic determination of Ca^{2+} content [50] or fluorescent indicators loaded into the cell [34], suggesting that discontinuities exist within the structures surrounding the lumen itself. The store [34] may adopt different configurations within the cell and components may even detach and reattach, so influencing the pattern and distribution of Ca^{2+} release channel [51]. In Purkinje neurons, for example, IP_3R-expressing regions may separate off from other internal store elements [52]. Store compartments exist which accumulate and release Ca^{2+} but are luminally-discontinuous from the bulk of the store have been observed in cultured hippocampal dendrites [53]. Life cycle stage or prior experimental conditions of the cell may influence the appearance of subcompartments. $[Ca^{2+}]_c$ increases which persisted for at least 10 min, led to the breakdown of the Ca^{2+} store into subcompartments in rat basophilic leukaemia cells [49]. Store structural changes are also associated with fertilization and mitosis [54]. Fertilization leads to a reorganization of the store, measured as a slowing of the diffusion of membrane probes and luminal proteins, in sea urchin eggs [55, 56]. In mitosis, significant Ca^{2+} store changes also occur, which include the structure itself fragmenting into subcompartments [57, 58].

Other structures within the cell such as Golgi, mitochondria, granules and the nucleus may also contribute to Ca^{2+} storage [59–63] and generate subregions which appear to have various Ca^{2+} concentrations, especially when lipophilic Ca^{2+} indicators are used to image the distribution of $[Ca^{2+}]$ through the cell.

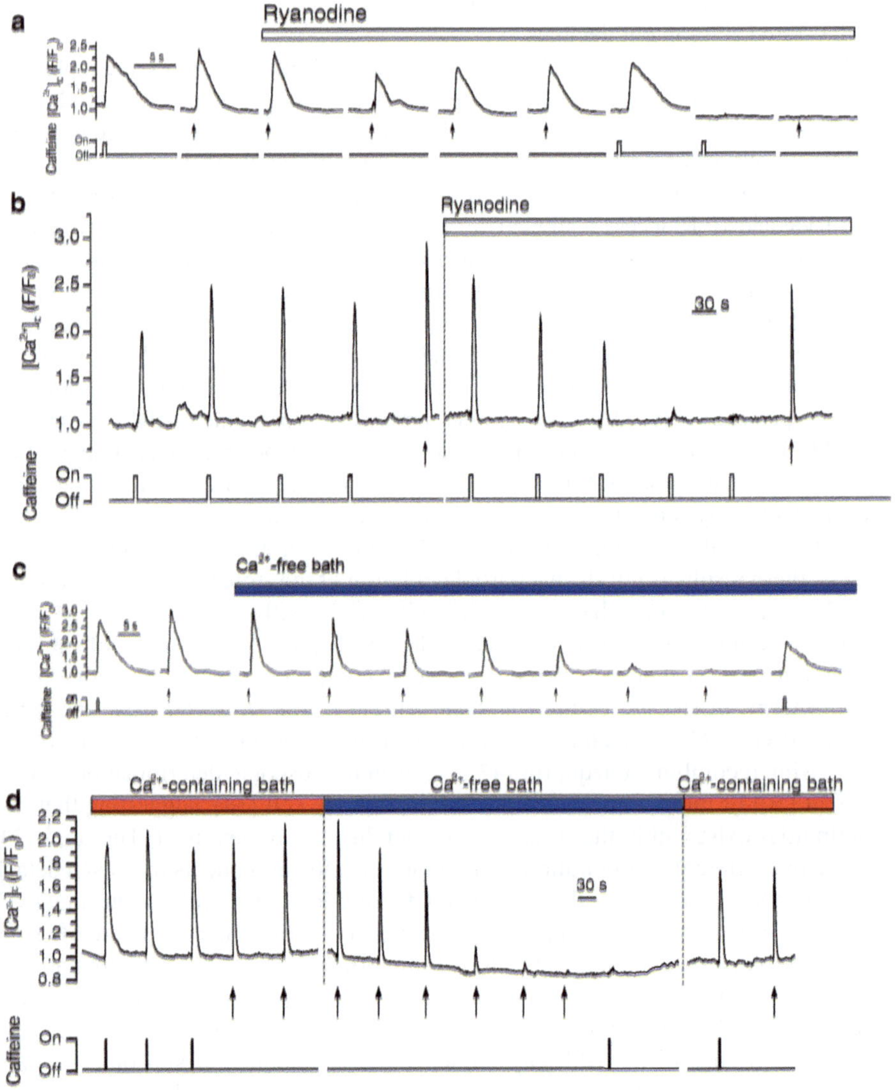

Fig. 11.9 Various apparent SR receptor arrangements. All the following experiments were performed on the same cell type (colonic smooth muscle) (**a**) *IP_3R and RyR access a single Ca^{2+}pool.* Caffeine (10 mM by pressure ejection lower trace) evoked a rise in Ca^{2+}. IP_3-evoked Ca^{2+} increases (125 μM; ↑) were not significantly reduced by ryanodine (50 μM; *open bar* above the trace). Activation of RyR by caffeine (10 mM), in the continued presence of ryanodine, initially increased $[Ca^{2+}]_c$. A second application of caffeine to the same cell however some 90 s later, generated little increase in $[Ca^{2+}]_c$ presumably because of SR store depletion; ryanodine's effects on RyR require prior channel activation. The IP_3 response was also subsequently inhibited (↑). Because the IP_3-evoked Ca^{2+} transient was not blocked by ryanodine alone (only after RyR activation with caffeine), IP_3-mediated Ca^{2+} release did not activate RyR. IP_3R and RyR may share a common Ca^{2+} store; this is depleted of Ca^{2+} by ryanodine, after activation of RyR by caffeine, to reduce the Ca^{2+} available for IP_3-mediated Ca^{2+} release to occur. (**b**) *IP_3R accesses a separate Ca^{2+}pool from RyR.*

Graded Ca^{2+} Release, Ca^{2+} Waves and Local Ca^{2+} Events from a Luminally-Continuous Store

If the Ca^{2+} store in smooth muscle is indeed a single, luminally-continuous entity, how do the various physiological events (waves, graded release, local responses) previously explained with multiple separate stores occur?

Ca2+waves: Ca^{2+}waves are the progressive movement of Ca^{2+} through the cell following Ca^{2+} release from the internal store. Using localized activation of IP_3R, the forward movement of the Ca^{2+} wave was shown to arise from CICR at the IP_3R [13, 16]. The decline in $[Ca^{2+}]_c$—the back of the wave—occurred not because of depletion of separate stores but from a functional compartmentalization of the store which rendered the site of IP_3-mediated Ca^{2+} release—and only this site—refractory to IP_3 after Ca^{2+} release . A localized feedback deactivation of IP_3R produced by an increased $[Ca^{2+}]_c$ caused the functional compartmentalization [16]. The deactivation of the IP_3R was delayed in onset, compared with the time of the rise in $[Ca^{2+}]_c$ and persisted (>30 s) even when $[Ca^{2+}]_c$ had been restored to resting levels [13, 16]. This feedback deactivation ensures the wave's progressive movement in a single direction [16].

Graded Ca2+release: There are several proposals for graded IP_3-mediated Ca^{2+} release that do not require the presence of numerous stores with various sensitivities to IP_3. Rather, at any given $[IP_3]$ the entire Ca^{2+} store is activated and releases a fraction of its content, becoming partially depleted. Partial depletion may deactivate Ca^{2+} release [64, 65]. Raising the $[IP_3]$ reactivates IP_3R to renew the Ca^{2+} release process. This proposal does not require multiple stores but a complex adaptive change in IP_3R activity. Negative feedback processes operating either at the cytoplasmic or the luminal aspects of IP_3R may explain the adaptive behaviour. In one proposal the binding of IP_3 to IP_3R may initially activate, then partially inactivate IP_3R in a concentration-dependent way to produce graded Ca^{2+} release [66–68]. To test this proposal we examined the time course of IP_3R activation at a constant $[IP_3]$ but under conditions in which there was varying amplitude of Ca^{2+} release [45]. The latter was achieved by buffering the cytoplasmic Ca^{2+} concentration (BAPTA) or partial depletion of the store (Ca^{2+} free bath solution). If IP_3 inactivated IP_3R to prevent release, then at

Fig. 11.9 (continued) Caffeine (1 mM; by pressure ejection, lower trace) evoked approximately reproducible increases in $[Ca^{2+}]_c$. IP_3 (250 μM; ↑) also increased $[Ca^{2+}]_c$. Ryanodine (50 μM; *open bar*) inhibited caffeine-evoked $[Ca^{2+}]_c$ increases by depletion of the SR. After the apparent depletion of caffeine-sensitive Ca^{2+} store, IP_3-evoked a substantial $[Ca^{2+}]_c$ increase (in contrast to the results in **a**).(**c**) *RyR accesses a different Ca^{2+}pool from IP_3R.* Caffeine (10 mM) and photolyzed IP_3 (↑) increased $[Ca^{2+}]_c$. In a Ca^{2+} free solution (containing 1 mM EGTA and 3 mM $MgCl_2$; *blue bar* above the trace) the IP_3-evoked Ca^{2+} transient decrease as the store was depleted of Ca^{2+}. Following depletion of the IP_3-sensitive store, caffeine evoked a substantial Ca^{2+} transient. (**d**) *RyR and IP_3R access a single Ca^{2+}pool.* Caffeine (2 mM) and IP_3 (125 μM) each evoked approximately reproducible increases in $[Ca^{2+}]_c$. Removal of external Ca^{2+} (and addition of 1 mM EGTA and 3 mM $MgCl_2$; *blue bar*) reduced the IP_3-evoked Ca^{2+} transient. Following depletion of the IP_3-sensitive store, the caffeine-evoked $[Ca^{2+}]_c$ transient was inhibited (in contrast to the results in **c**). Reintroduction of Ca^{2+} (*red bar*) restored the IP_3- and caffeine-evoked Ca^{2+} transients towards control values. These results were original published in McCarron & Olson 2008 [44]

constant [IP_3], release should stop at approximately the same time regardless of the amplitude of the [Ca^{2+}]$_c$ rise. However, as the amplitude of the [Ca^{2+}]$_c$ rise declined (in either BAPTA or in Ca^{2+}-free solution) the time course of release became more prolonged [45]. This result suggests that mechanisms other than IP_3 inactivation of IP_3R would appear responsible for terminating IP_3-mediated Ca^{2+} release.

In another proposal, the sensitivity of IP_3R to IP_3 is controlled by the luminal [Ca^{2+}] so that as the concentration of the ion within the store lumen falls so does IP_3R activity [e.g. 65, 69]. For example, decreasing the store [Ca^{2+}] to below 80 % of the steady-state level abolished IP_3-mediated Ca^{2+}release in rat uterine myoctes [70] [see also 65, 69]. However, it is unclear whether or not the control of IP_3R activity by luminal Ca^{2+} operates over the store's physiological Ca^{2+} concentration range. The threshold for luminal regulation to begin altering the activity of IP_3R is depletion of the store by >70 % of the steady-state luminal Ca^{2+} concentration (500–600 μM; [71]) in HeLa cells. The store [Ca^{2+}] must also be substantially depleted in hepatocytes (>45 or 95 %) [72, 73] and in A7r5 cells by >70 % [74] before IP_3R sensitivity changes are detected. In each case, control of IP_3R activity by Ca^{2+} binding to the luminal aspect of the receptor, is unlikely to explain 'quantal' Ca^{2+} release when store [Ca^{2+}] exceeds 55, 5, or 30 % of the normal steady-state value respectively in these cells [72–74].

On the other hand, IP_3R might not be controlled by luminal Ca^{2+} at all. Single channel IP_3R activity, measured in planar lipid bilayers, *increased* when the [Ca^{2+}] at the luminal aspect of the channel declined [75]. In the latter study a luminal [Ca^{2+}] exceeding 1 mM inhibited IP_3R activity [75] (see also [76]). In other studies in permeabilized cells (e.g. portal vein; [18] or hepatocytes; [77]), decreases in store [Ca^{2+}] failed to reduce the sensitivity of IP_3-mediated Ca^{2+} release or alter Ca^{2+} leak when pumps were blocked in permeabilized avian supraorbital nasal gland cells [78]. Together, these results suggest that regulation of IP_3R by Ca^{2+} at the luminal aspect of the channel may, at best, operate over a limited range of store [Ca^{2+}].

Our results (Fig. 11.8) [44–46] suggest that as the store content falls IP_3R become less responsive to IP_3. However, rather than luminal regulation being expressed from within the store at the luminal aspect of IP_3R, detection of [Ca^{2+}] within the store may lie at the *cytoplasmic aspect of IP_3R* [45]. The Ca^{2+} current flowing through IP_3R evokes further release by a positive feedback effect of the ion at the cytoplasmic aspect of the channel, i.e. a Ca^{2+}-dependent positive feedback loop. Reduction of the store Ca^{2+} content reduces the Ca^{2+} current flowing through IP_3R and will result in a falling positive feedback at the cytoplasmic aspect of IP_3R until release eventually stops. Ca^{2+} release is renewed by an increased [IP_3]. In this case, the co-incidental activation of several neighboring IP_3Rs within a cluster offsets the declining IP_3R Ca^{2+} current to renew positive feedback and Ca^{2+} release and accounts for graded IP_3-mediated Ca^{2+} release.

Alternatively, the rise in cytoplasmic [Ca^{2+}]$_c$, which derives from the activity of IP_3R, may itself inactivate the receptor [79–81]. However, if Ca^{2+}-dependent inactivation terminated release [16, 79] to explain the graded IP_3-mediated Ca^{2+} release, the Ca^{2+} chelator BAPTA, would have been expected to have potentiated IP_3-evoked [Ca^{2+}]$_c$ increase; BAPTA decreased IP_3-mediated Ca^{2+} release [45].

Localized Ca2+responses IP_3 is a rapidly diffusing messenger and IP_3R are subject to positive feedback CICR on a single luminally-continuous entity, so how do highly-localized Ca^{2+} changes occur? In heart cells, the store is also a continuous network [82] in which Ca^{2+} can rapidly redistribute [83, 84] and positive feedback CICR occurs at RyR, yet highly localized Ca^{2+} release events occur. The highly localized responses arise in specialized domains formed by a junction of the store with the plasmalemma ('peripheral couplings') or the store and transverse (T)-tubules ('Dyads'). A number of proteins accrue at these specialized store domains: the L-type channel dihydropyridine receptors of the plasmalemma and T-tubules; the RyRs of store; triadin and junctin, of the store membrane; and calsequestrin (CSQ), the internal calcium binding protein [82]. The close coupling of dihydropyridine receptors and RyR provides control of Ca^{2+} release by Ca^{2+} influx. The quaternary complexes between triadin, junctin, RyR, and CSQ provides the luminal Ca^{2+} sensing capabilities that regulates RyR activity[85].

IP_3-mediated Ca^{2+} signaling may also generate highly localized responses even though IP_3 is a messenger that can diffuse quickly to evoke activity throughout the cell. To do this, certain receptors co-localize with IP_3R to form a local signalling complex [86–89]. In cultured sympathetic neurons, although muscarinic and bradykinin receptors each stimulate phospholipase C, only bradykinin receptors co-immunoprecipitate with, and activate, IP_3R to evoke Ca^{2+} release [86]. The arrangement enables PLC activation by muscarinic and bradykinin receptors to evoke different cellular responses. In SH-SY5Y cells the positioning of IP_3R near the plasma membrane provides a mechanism which may enable agonist activation, acting via IP_3, to target specific types of cellular response i.e. by generating Ca^{2+} rises in specific regions of the cell [90]. The clustering of agonist-activated surface receptors in certain regions on the plasma membrane (e.g. the *Escherichia coli* chemotaxis receptor) may contribute further, by providing areas with increased sensitivity to extracellular stimuli [91].

Smooth muscle also assembles IP_3 Ca^{2+} release components into specialized Ca^{2+} domains [92] (Fig. 11.1). This conclusion came initially from the observation that Ca^{2+} waves, triggered by agonists applied to the entire cell, began consistently at the same site on successive activations in smooth muscle i.e. there appeared to be regions with preferential IP_3-mediated Ca^{2+} release. Using centre of mass co-localization analysis of the distribution of the surface membrane receptors (for ACh) and IP_3R, a small percentage (~10 %) of sites showed co-localization. Significantly, the extent of co-localization was greatest at the Ca^{2+} wave initiation site. At these sites of co-localization, wave initiation may arise from a preferential delivery of IP_3 from mAChR3 activity to particular IP_3R clusters to generate faster local $[Ca^{2+}]_c$ increases. When the Ca^{2+} rise at the initiation site was rapidly and selectively attenuated (using photolysis of the caged Ca^{2+} buffer diazo-2) the Ca^{2+} wave shifted and initiated at a new site. Conversely, when a localized subthreshold 'priming' IP_3 concentration was applied rapidly to regions distant from the initiation site, the wave initiation site shifted to the site of priming IP_3 release. These results indicate that Ca^{2+} waves initiate where the most rapid Ca^{2+} change occurs at sites in which there is a structural and functional coupling of ACh receptors and IP_3R (Fig. 11.1). The coupling generates junctions in which IP_3 acts as a highly localized signal by being rapidly and selectively delivered to IP_3R.

Role of Mitochondria in Modulating Ca^{2+}Signals

Away from the plasma membrane, IP_3R activity in smooth muscle is also tightly regulated by mitochondria. Mitochondria have a well-developed Ca^{2+} uptake facility and may modulate bulk cytoplasmic Ca^{2+} signals [93–96] derived from Ca^{2+} entry and release [97]. Mitochondria also provide tight local control of Ca^{2+} release via IP_3R [93, 94, 98] but Ca^{2+} influx via voltage-dependent Ca^{2+} channels or release via RyR appears to be less tightly controlled at a local level by mitochondria [93, 94].

Mitochondrial control of IP_3R arises at IP_3-mediated release sites. IP_3-sensitive Ca^{2+} release initiates at discrete sites on the store that contain a few tens of IP_3R from which the local increase in $[Ca^{2+}]$ is called a 'puff'. Ca^{2+} puffs are spatially restricted events and of short duration but may interact and coalesce to generate a global release in Ca^{2+}. Mitochondria are positioned close to IP_3R and regulate activity of the channels [99]; inhibition of mitochondrial Ca^{2+} uptake attenuated the magnitude of Ca^{2+} puffs [100]. Indeed mitochondrial Ca^{2+} uptake was rapid enough to influence Ca^{2+} communication within an IP_3R cluster. Mitochondrial Ca^{2+} uptake appears to prevent the negative feedback effect of high $[Ca^{2+}]_c$ on IP_3R activity within a cluster to prolong Ca^{2+} release from the store [100]. As a consequence of the control at IP_3R, mitochondrial Ca^{2+} uptake exerts a pronounced effect on IP_3-mediated Ca^{2+} release throughout the cell [93, 94, 98, 101].

Mitochondria and IP_3R appear to be close, and perhaps tethered, to allow mitochondrial Ca^{2+} uptake, ATP supply, ROS production and or redox/antioxidant control to influence IP_3R activity. Conversely, mitochondrial division (required to maintain mitochondrial population health and allow cell proliferation) involves encircling of the dividing mitochondria by a store membrane tubule at the point of mitochondrial constriction [102]. During smooth muscle proliferation IP_3R expression and activity are increased [103–105] and there is a marked switch in mitochondrial phenotype from stationary to highly motile [106]. Inhibiting either IP_3R activity [104, 107] or mitochondrial motility and division [106, 108] inhibits smooth muscle proliferation. The interplay between mitochondria and IP_3R in smooth muscle thus presents an interesting potential therapeutic avenue by which pathological smooth muscle proliferation in vascular disease may be targeted.

Acknowledgements This work was funded by the Wellcome Trust (092292/Z/10/Z) and British Heart Foundation (PG/11/70/29086)

References

1. Westcott EB, Jackson WF. Heterogeneous function of ryanodine receptors, but not ip3 receptors, in hamster cremaster muscle feed arteries and arterioles. Am J Physiol Heart Circ Physiol. 2011;300:H1616–30.
2. McCarron JG, Chalmers S, Bradley KN, Macmillan D, Muir TC. Ca^{2+} microdomains in smooth muscle. Cell Calcium. 2006;40:461–93.

3. McCarron JG, Craig JW, Bradley KN, Muir TC. Agonist-induced phasic and tonic responses in smooth muscle are mediated by $InsP_3$. J Cell Sci. 2002;115:2207–18.
4. ZhuGe R, Tuft RA, Fogarty KE, Bellve K, Fay FS, Walsh Jr JV. The influence of sarcoplasmic reticulum ca^{2+} concentration on Ca^{2+} sparks and spontaneous transient outward currents in single smooth muscle cells. J Gen Physiol. 1999;113:215–28.
5. Nelson MT, Cheng H, Rubart M, Santana LF, Bonev AD, Knot HJ, Lederer WJ. Relaxation of arterial smooth muscle by calcium sparks. Science. 1995;270:633–7.
6. Burdyga T, Wray S. Action potential refractory period in ureter smooth muscle is set by Ca sparks and BK channels. Nature. 2005;436:559–62.
7. Somlyo AP. Excitation-contraction coupling and the ultrastructure of smooth muscle. Circ Res. 1985;57:497–507.
8. Mogami H, Nakano K, Tepikin AV, Petersen OH. Ca^{2+} flow via tunnels in polarized cells: Recharging of apical Ca^{2+} stores by focal Ca^{2+} entry through basal membrane patch. Cell. 1997;88:49–55.
9. Park MK, Petersen OH, Tepikin AV. The endoplasmic reticulum as one continuous Ca^{2+} pool: visualization of rapid Ca^{2+} movements and equilibration. EMBO J. 2000;19:5729–39.
10. Blaustein MP, Golovina VA, Song H, Choate J, Lencesova L, Robinson SW, Wier WG. Organization of Ca^{2+} stores in vascular smooth muscle: functional implications. Novartis Found Symp. 2002;246:125–37. discussion 137–41, 221–7.
11. Yamazawa T, Iino M, Endo M. Presence of functionally different compartments of the Ca^{2+} store in single intestinal smooth muscle cells. FEBS Lett. 1992;301:181–4.
12. Tribe RM, Borin ML, Blaustein MP. Functionally and spatially distinct Ca^{2+} stores are revealed in cultured vascular smooth muscle cells. Proc Natl Acad Sci U S A. 1994;91:5908–12.
13. McCarron JG, Chalmers S, MacMillan D, Olson ML. Agonist-evoked Ca^{2+} wave progression requires Ca^{2+} and IP_3. J Cell Physiol. 2010;224:334–44.
14. McCarron JG, Olson ML, Currie S, Wright AJ, Anderson KI, Girkin JM. Elevations of intracellular calcium reflect normal voltage-dependent behavior, and not constitutive activity, of voltage-dependent calcium channels in gastrointestinal and vascular smooth muscle. J Gen Physiol. 2009;133:439–57.
15. Bootman M, Niggli E, Berridge M, Lipp P. Imaging the hierarchical Ca^{2+} signalling system in hela cells. J Physiol. 1997;499:307–14.
16. McCarron JG, MacMillan D, Bradley KN, Chalmers S, Muir TC. Origin and mechanisms of Ca^{2+} waves in smooth muscle as revealed by localized photolysis of caged inositol 1,4,5-trisphosphate. J Biol Chem. 2004;279:8417–27.
17. Muallem S, Pandol SJ, Beeker TG. Hormone-evoked calcium release from intracellular stores is a quantal process. J Biol Chem. 1989;264:205–12.
18. Hirose K, Iino M. Heterogeneity of channel density in inositol-1,4,5-trisphosphate-sensitive Ca^{2+} stores. Nature. 1994;372:791–4.
19. Shin DM, Luo X, Wilkie TM, Miller LJ, Peck AB, Humphreys-Beher MG, Muallem S. Polarized expression of g protein-coupled receptors and an all-or-none discharge of Ca^{2+} pools at initiation sites of $[Ca^{2+}]_i$ waves in polarized exocrine cells. J Biol Chem. 2001;276:44146–56.
20. Meyer T, Stryer L. Transient calcium release induced by successive increments of inositol 1,4,5-trisphosphate. Proc Natl Acad Sci U S A. 1990;87:3841–5.
21. Bootman MD, Cheek TR, Moreton RB, Bennett DL, Berridge MJ. Smoothly graded ca^{2+} release from inositol 1,4,5-trisphosphate-sensitive Ca^{2+} stores. J Biol Chem. 1994;269:24783–91.
22. Oldershaw KA, Nunn DL, Taylor CW. Quantal Ca^{2+} mobilization stimulated by inositol 1,4,5-trisphosphate in permeabilized hepatocytes. Biochem J. 1991;278:705–8.
23. Parys JB, Missiaen L, Smedt HD, Sienaert I, Casteels R. Mechanisms responsible for quantal Ca^{2+} release from inositol trisphosphate-sensitive calcium stores. Pflugers Arch. 1996;432:359–67.

24. Iino M. Biphasic Ca^{2+} dependence of inositol 1,4,5-trisphosphate-induced Ca^{2+} release in smooth muscle cells of the guinea pig taenia caeci. J Gen Physiol. 1990;95:1103–22.
25. Parker I, Ivorra I. Localized all-or-none calcium liberation by inositol trisphosphate. Science. 1990;250:977–9.
26. Kasai H, Li YX, Miyashita Y. Subcellular distribution of Ca^{2+} release channels underlying Ca^{2+} waves and oscillations in exocrine pancreas. Cell. 1993;74:669–77.
27. Fogarty KE, Kidd JF, Tuft DA, Thorn P. Mechanisms underlying InsP3-evoked global Ca^{2+} signals in mouse pancreatic acinar cells. J Physiol. 2000;526(Pt 3):515–26.
28. Urena J, Smani T, Lopez-Barneo J. Differential functional properties of Ca^{2+} stores in pulmonary arterial conduit and resistance myocytes. Cell Calcium. 2004;36:525–34.
29. Noguera MA, D'Ocon MP. Different and common intracellular calcium-stores mobilized by noradrenaline and caffeine in vascular smooth muscle. Naunyn Schmiedebergs Arch Pharmacol. 1992;345:333–41.
30. Baro I, Eisner DA. Factors controlling changes in intracellular Ca^{2+} concentration produced by noradrenaline in rat mesenteric artery smooth muscle cells. J Physiol. 1995;482:247–58.
31. Pacaud P, Loirand G. Release of Ca^{2+} by noradrenaline and ATP from the same Ca^{2+} store sensitive to both $InsP_3$ and Ca^{2+} in rat portal vein myocytes. J Physiol. 1995;484:549–55.
32. Janiak R, Wilson SM, Montague S, Hume JR. Heterogeneity of calcium stores and elementary release events in canine pulmonary arterial smooth muscle cells. Am J Physiol Cell Physiol. 2001;280:C22–33.
33. Haddock RE, Hill CE. Differential activation of ion channels by inositol 1,4,5-trisphosphate (IP_3)- and ryanodine-sensitive calcium stores in rat basilar artery vasomotion. J Physiol. 2002;545:615–27.
34. Golovina VA, Blaustein MP. Spatially and functionally distinct Ca^{2+} stores in sarcoplasmic and endoplasmic reticulum. Science. 1997;275:1643–8.
35. Iino M, Kobayashi T, Endo M. Use of ryanodine for functional removal of the calcium store in smooth muscle cells of the guinea-pig. Biochem Biophys Res Commun. 1988; 152:417–22.
36. Missiaen L, Vanoevelen J, Parys JB, Raeymaekers L, De Smedt H, Callewaert G, Erneux C, Wuytack F. Ca^{2+} uptake and release properties of a thapsigargin-insensitive nonmitochondrial Ca^{2+} store in A7r5 and 16HBe14o- cells. J Biol Chem. 2002;277:6898–902.
37. Bian JH, Ghosh TK, Wang JC, Gill DL. Identification of intracellular calcium pools. Selective modification by thapsigargin. J Biol Chem. 1991;266:8801–6.
38. Liu X, Farley JM. Depletion and refilling of acetylcholine- and caffeine-sensitive Ca^{++} stores in tracheal myocytes. J Pharmacol Exp Ther. 1996;277:789–95.
39. Boittin FX, Macrez N, Halet G, Mironneau J. Norepinephrine-induced Ca^{2+} waves depend on $InsP_3$ and ryanodine receptor activation in vascular myocytes. Am J Physiol. 1999;277:C139–51.
40. Bolton TB, Lim SP. Properties of calcium stores and transient outward currents in single smooth muscle cells of rabbit intestine. J Physiol. 1989;409:385–401.
41. Jabr RI, Toland H, Gelband CH, Wang XX, Hume JR. Prominent role of intracellular ca^{2+} release in hypoxic vasoconstriction of canine pulmonary artery. Br J Pharmacol. 1997;122:21–30.
42. Iino M. Calcium dependent inositol trisphosphate-induced calcium release in the guinea-pig taenia caeci. Biochem Biophys Res Commun. 1987;142:47–52.
43. Flynn ER, Bradley KN, Muir TC, McCarron JG. Functionally separate intracellular Ca^{2+} stores in smooth muscle. J Biol Chem. 2001;276:36411–8.
44. McCarron JG, Olson ML. A single luminally continuous sarcoplasmic reticulum with apparently separate Ca^{2+} stores in smooth muscle. J Biol Chem. 2008;283:7206–18.
45. McCarron JG, Chalmers S, Muir TC. "Quantal" Ca^{2+} release at the cytoplasmic aspect of the ins(1,4,5)p_3r channel in smooth muscle. J Cell Sci. 2008;121:86–98.

46. Rainbow RD, Macmillan D, McCarron JG. The sarcoplasmic reticulum Ca^{2+} store arrangement in vascular smooth muscle. Cell Calcium. 2009;46:313–22.
47. Montero M, Alvarez J, Scheenen WJ, Rizzuto R, Meldolesi J, Pozzan T. Ca^{2+} homeostasis in the endoplasmic reticulum: Coexistence of high and low [Ca^{2+}] subcompartments in intact hela cells. J Cell Biol. 1997;139:601–11.
48. Dayel MJ, Hom EF, Verkman AS. Diffusion of green fluorescent protein in the aqueous-phase lumen of endoplasmic reticulum. Biophys J. 1999;76:2843–51.
49. Subramanian K, Meyer T. Calcium-induced restructuring of nuclear envelope and endoplasmic reticulum calcium stores. Cell. 1997;89:963–71.
50. Pezzati R, Bossi M, Podini P, Meldolesi J, Grohovaz F. High-resolution calcium mapping of the endoplasmic reticulum-golgi-exocytic membrane system. Electron energy loss imaging analysis of quick frozen-freeze dried PC12 cells. Mol Biol Cell. 1997;8:1501–12.
51. Lee C, Chen LB. Dynamic behavior of endoplasmic reticulum in living cells. Cell. 1988;54:37–46.
52. Takei K, Mignery GA, Mugnaini E, Sudhof TC, De Camilli P. Inositol 1,4,5-trisphosphate receptor causes formation of ER cisternal stacks in transfected fibroblasts and in cerebellar purkinje cells. Neuron. 1994;12:327–42.
53. Bannai H, Inoue T, Nakayama T, Hattori M, Mikoshiba K. Kinesin dependent, rapid, bi-directional transport of ER sub-compartment in dendrites of hippocampal neurons. J Cell Sci. 2004;117:163–75.
54. Boulware MJ, Marchant JS. IP_3 receptor activity is differentially regulated in endoplasmic reticulum subdomains during oocyte maturation. Curr Biol. 2005;15:765–70.
55. Terasaki M, Jaffe LA. Organization of the sea urchin egg endoplasmic reticulum and its reorganization at fertilization. J Cell Biol. 1991;114:929–40.
56. Terasaki M, Jaffe LA, Hunnicutt GR, Hammer 3rd JA. Structural change of the endoplasmic reticulum during fertilization: Evidence for loss of membrane continuity using the green fluorescent protein. Dev Biol. 1996;179:320–8.
57. Koch GL, Booth C, Wooding FB. Dissociation and re-assembly of the endoplasmic reticulum in live cells. J Cell Sci. 1988;91:511–22.
58. Henson JH, Begg DA, Beaulieu SM, Fishkind DJ, Bonder EM, Terasaki M, Lebeche D, Kaminer B. A calsequestrin-like protein in the endoplasmic reticulum of the sea urchin: Localization and dynamics in the egg and first cell cycle embryo. J Cell Biol. 1989;109:149–61.
59. Missiaen L, Dode L, Vanoevelen J, Raeymaekers L, Wuytack F. Calcium in the golgi apparatus. Cell Calcium. 2007;41:405–16.
60. Dolman NJ, Tepikin AV. Calcium gradients and the golgi. Cell Calcium. 2006;40:505–12.
61. Rizzuto R, Pozzan T. Microdomains of intracellular Ca^{2+}: molecular determinants and functional consequences. Physiol Rev. 2006;86:369–408.
62. Pinton P, Pozzan T, Rizzuto R. The golgi apparatus is an inositol 1,4,5-trisphosphate-sensitive Ca^{2+} store, with functional properties distinct from those of the endoplasmic reticulum. EMBO J. 1998;17:5298–308.
63. Brini M, Murgia M, Pasti L, Picard D, Pozzan T, Rizzuto R. Nuclear Ca^{2+} concentration measured with specifically targeted recombinant aequorin. EMBO J. 1993;12:4813–9.
64. Missiaen L, De Smedt H, Droogmans G, Casteels R. Ca^{2+} release induced by inositol 1,4,5-trisphosphate is a steady-state phenomenon controlled by luminal Ca^{2+} in permeabilized cells. Nature. 1992;357:599–602.
65. Tanimura A, Turner RJ. Calcium release in hsy cells conforms to a steady-state mechanism involving regulation of the inositol 1,4,5-trisphosphate receptor Ca^{2+} channel by luminal [ca^{2+}]. J Cell Biol. 1996;132:607–16.
66. Hajnoczky G, Thomas AP. The inositol trisphosphate calcium channel is inactivated by inositol trisphosphate. Nature. 1994;370:474–7.

67. Marchant JS, Taylor CW. Rapid activation and partial inactivation of inositol trisphosphate receptors by inositol trisphosphate. Biochemistry. 1998;37:11524–33.
68. Wilcox RA, Strupish J, Nahorski SR. Quantal calcium release in electropermeabilized sh-sy5y neuroblastoma cells perfused with myo-inositol 1,4,5-trisphosphate. Cell Calcium. 1996;20:243–55.
69. Irvine RF. 'Quantal' Ca^{2+} release and the control of Ca^{2+} entry by inositol phosphates--a possible mechanism. FEBS Lett. 1990;263:5–9.
70. Shmygol A, Wray S. Modulation of agonist-induced Ca^{2+} release by SR Ca^{2+} load: Direct sr and cytosolic Ca^{2+} measurements in rat uterine myocytes. Cell Calcium. 2005;37:215–23.
71. Barrero MJ, Montero M, Alvarez J. Dynamics of [Ca^{2+}] in the endoplasmic reticulum and cytoplasm of intact hela cells. A comparative study. J Biol Chem. 1997;272:27694–9.
72. Beecroft MD, Taylor CW. Incremental Ca^{2+} mobilization by inositol trisphosphate receptors is unlikely to be mediated by their desensitization or regulation by luminal or cytosolic ca^{2+}. Biochem J. 1997;326:215–20.
73. Combettes L, Cheek TR, Taylor CW. Regulation of inositol trisphosphate receptors by luminal Ca^{2+} contributes to quantal Ca^{2+} mobilization. EMBO J. 1996;15:2086–93.
74. Parys JB, Missiaen L, De Smedt H, Casteels R. Loading dependence of inositol 1,4,5-trisphosphate-induced Ca^{2+} release in the clonal cell line a7r5. Implications for the mechanism of quantal Ca^{2+} release. J Biol Chem. 1993;268:25206–12.
75. Bezprozvanny I, Ehrlich BE. Inositol (1,4,5)-trisphosphate $InsP_3$-gated Ca channels from cerebellum: conduction properties for divalent cations and regulation by intraluminal calcium. J Gen Physiol. 1994;104:821–56.
76. Thrower EC, Mobasheri H, Dargan S, Marius P, Lea EJ, Dawson AP. Interaction of luminal calcium and cytosolic atp in the control of type 1 inositol (1,4,5)-trisphosphate receptor channels. J Biol Chem. 2000;275:36049–55.
77. Combettes L, Claret M, Champeil P. Do submaximal $InsP_3$ concentrations only induce the partial discharge of permeabilized hepatocyte calcium pools because of the concomitant reduction of intraluminal ca^{2+} concentration? FEBS Lett. 1992;301:287–90.
78. Shuttleworth TJ. Ca^{2+} release from inositol trisphosphate-sensitive stores is not modulated by intraluminal [Ca^{2+}]. J Biol Chem. 1992;267:3573–6.
79. Oancea E, Meyer T. Reversible desensitization of inositol trisphosphate-induced calcium release provides a mechanism for repetitive calcium spikes. J Biol Chem. 1996; 271:17253–60.
80. Iino M, Tsukioka M. Feedback control of inositol trisphosphate signalling by calcium. Mol Cell Endocrinol. 1994;98:141–6.
81. Adkins CE, Taylor CW. Lateral inhibition of inositol 1,4,5-trisphosphate receptors by cytosolic Ca^{2+}. Curr Biol. 1999;9:1115–8.
82. Franzini-Armstrong C, Protasi F, Tijskens P. The assembly of calcium release units in cardiac muscle. Ann N Y Acad Sci. 2005;1047:76–85.
83. Wu X, Bers DM. Sarcoplasmic reticulum and nuclear envelope are one highly interconnected Ca^{2+} store throughout cardiac myocyte. Circ Res. 2006;99:283–91.
84. Swietach P, Spitzer KW, Vaughan-Jones RD. Ca^{2+}-mobility in the sarcoplasmic reticulum of ventricular myocytes is low. Biophys J. 2008;95:1412–27.
85. Gyorke I, Hester N, Jones LR, Gyorke S. The role of calsequestrin, triadin, and junctin in conferring cardiac ryanodine receptor responsiveness to luminal calcium. Biophys J. 2004;86:2121–8.
86. Delmas P, Wanaverbecq N, Abogadie FC, Mistry M, Brown DA. Signaling microdomains define the specificity of receptor-mediated $InsP_3$ pathways in neurons. Neuron. 2002;34:209–20.
87. Hur EM, Park YS, Huh YH, Yoo SH, Woo KC, Choi BH, Kim KT. Junctional membrane inositol 1,4,5-trisphosphate receptor complex coordinates sensitization of the silent EGF-induced Ca^{2+} signaling. J Cell Biol. 2005;169:657–67.

88. Yuan Z, Cai T, Tian J, Ivanov AV, Giovannucci DR, Xie Z. Na^+/K^+-ATPase tethers phospholipase C and IP_3 receptor into a calcium-regulatory complex. Mol Biol Cell. 2005;16:4034–45.
89. Lur G, Sherwood MW, Ebisui E, Haynes L, Feske S, Sutton R, Burgoyne RD, Mikoshiba K, Petersen OH, Tepikin AV. Inspreceptors and orai channels in pancreatic acinar cells: Co-localization and its consequences. Biochem J. 2011;436:231–9.
90. Smith IF, Wiltgen SM, Parker I. Localization of puff sites adjacent to the plasma membrane: Functional and spatial characterization of Ca^{2+} signaling in SH-SY5Y cells utilizing membrane-permeant caged IP_3. Cell Calcium. 2009;45:65–76.
91. Thomason PA, Wolanin PM, Stock JB. Signal transduction: receptor clusters as information processing arrays. Curr Biol. 2002;12:R399–401.
92. Olson ML, Sandison ME, Chalmers S, McCarron JG. Microdomains of muscarinic acetylcholine and Ins(1,4,5)P_3 receptors create 'Ins(1,4,5)P_3 junctions' and sites of Ca^{2+} wave initiation in smooth muscle. J Cell Sci. 2012;125:5315–28.
93. McCarron JG, Muir TC. Mitochondrial regulation of the cytosolic Ca^{2+} concentration and the $InsP_3$-sensitive Ca^{2+} store in guinea-pig colonic smooth muscle. J Physiol. 1999; 516:149–61.
94. Chalmers S, McCarron JG. The mitochondrial membrane potential and Ca^{2+} oscillations in smooth muscle. J Cell Sci. 2008;121:75–85.
95. Kamishima T, McCarron JG. Ca^{2+} removal mechanisms in rat cerebral resistance size arteries. Biophys J. 1998;75:1767–73.
96. McGeown JG, McCarron JG, Drummond RM, Fay FS. Calcium-calmodulin-dependent mechanisms accelerate calcium decay in gastric myocytes from bufo marinus. J Physiol. 1998;506(Pt 1):95–107.
97. Drummond RM, Fay FS. Mitochondria contribute to Ca^{2+} removal in smooth muscle cells. Pflugers Arch. 1996;431:473–82.
98. Chalmers S, McCarron JG. Inhibition of mitochondrial calcium uptake rather than efflux impedes calcium release by inositol-1,4,5-trisphosphate-sensitive receptors. Cell Calcium. 2009;46:107–13.
99. Rizzuto R, Brini M, Murgia M, Pozzan T. Microdomains with high Ca^{2+} close to IP_3-sensitive channels that are sensed by neighboring mitochondria. Science. 1993;262:744–7.
100. Olson ML, Chalmers S, McCarron JG. Mitochondrial Ca^{2+} uptake increases Ca^{2+} release from inositol 1,4,5-trisphosphate receptor clusters in smooth muscle cells. J Biol Chem. 2010;285:2040–50.
101. Chalmers S, Caldwell ST, Quin C, Prime TA, James AM, Cairns AG, Murphy MP, McCarron JG, Hartley RC. Selective uncoupling of individual mitochondria within a cell using a mitochondria-targeted photoactivated protonophore. J Am Chem Soc. 2012;134:758–61.
102. Friedman JR, Lackner LL, West M, DiBenedetto JR, Nunnari J, Voeltz GK. Er tubules mark sites of mitochondrial division. Science. 2011;334:358–62.
103. Berra-Romani R, Mazzocco-Spezzia A, Pulina MV, Golovina VA. Ca^{2+} handling is altered when arterial myocytes progress from a contractile to a proliferative phenotype in culture. Am J Physiol Cell Physiol. 2008;295:C779–90.
104. Wilkerson MK, Heppner TJ, Bonev AD, Nelson MT. Inositol trisphosphate receptor calcium release is required for cerebral artery smooth muscle cell proliferation. Am J Physiol Heart Circ Physiol. 2006;290:H240–7.
105. Moses S, Dreja K, Lindqvist A, Lovdahl C, Hellstrand P, Hultgardh-Nilsson A. Smooth muscle cell response to mechanical injury involves intracellular calcium release and ERK1/ERK2 phosphorylation. Exp Cell Res. 2001;269:88–96.
106. Chalmers S, Saunter C, Wilson C, Coats P, Girkin JM, McCarron JG. Mitochondrial motility and vascular smooth muscle proliferation. Arterioscler Thromb Vasc Biol. 2012; 32:3000–11.

107. Wang Y, Chen J, Taylor CW, Hirata Y, Hagiwara H, Mikoshiba K, Toyo-oka T, Omata M, Sakaki Y. Crucial role of type 1, but not type 3, inositol 1,4,5-trisphosphate ip_3 receptors in IP_3-induced Ca^{2+} release, capacitative Ca^{2+} entry, and proliferation of A7r5 vascular smooth muscle cells. Circ Res. 2001;88:202–9.
108. Marsboom G, Toth PT, Ryan JJ, Hong Z, Wu X, Fang YH, Thenappan T, Piao L, Zhang HJ, Pogoriler J, Chen Y, Morrow E, Weir EK, Rehman J, Archer SL. Dynamin-related protein 1-mediated mitochondrial mitotic fission permits hyperproliferation of vascular smooth muscle cells and offers a novel therapeutic target in pulmonary hypertension. Circ Res. 2012;110:1484–97.

Chapter 12
Role of Mechanosensitive TRP Channels in Abnormal Vasculature of Tumors

Holly C. Cappelli, Roslin J. Thoppil, Ravi K. Adapala, J. Gary Meszaros, Sailaja Paruchuri, and Charles K. Thodeti

Abstract Solid tumors necessitate vascularization for metabolic support and metastasis, relying on the process of angiogenesis to form new blood vessels. However, the constant stimulation of endothelial cells from pro-angiogenic soluble factors and mechanical forces creates a tumor vasculature that is structurally and functionally abnormal. Most anti-angiogenic therapies have focused on targeting VEGF signaling to pursue the tumor vasculature. However, these anti-VEGF therapies have been met with limited success in clinical trials. Hence, recent studies have started to investigate the role of mechanical signaling in tumor angiogenesis as it occurs in a mechanically dynamic environment. This chapter focuses on mechanosensitive ion channels that belong to the transient receptor potential (TRP) superfamily, with special emphasis on the role of TRPV4 in the endothelium, as well as deregulation of TRPV4 signaling within the tumor endothelium, and its potential as a target for normalization of tumor vasculature to improve cancer therapy.

Keywords Angiogenesis • Calcium • Endothelial cells • Extracellular matrix • Mechanotransduction • TRPV4 • Tumor

Introduction

The oxygen and nutrient demands of solid tumors make them dependent on angiogenesis for new blood vessels to grow and metastasize. However, the continuous needs of the tumor create vessels that are structurally and functionally abnormal

H.C. Cappelli, B.S. • R.J. Thoppil, Ph.D. • R.K. Adapala, M.S. • J.G. Meszaros, Ph.D.
C.K. Thodeti, Ph.D. (✉)
Department of Integrative Medical Sciences, Northeast Ohio Medical University, Rootstown, OH 44272, USA

School of Biomedical Sciences, Kent State University, Kent, OH 44240, USA
e-mail: cthodeti@neomed.edu

S. Paruchuri, Ph.D.
Department of Chemistry, University of Akron, Akron, OH 44325, USA

I. Levitan, A.M. Dopico (eds.), *Vascular Ion Channels in Physiology and Disease*, DOI 10.1007/978-3-319-29635-7_12

[1, 2] and result in a hypoxic environment due to insufficient delivery of nutrients and reduced waste clearance [3]. Tumor angiogenesis is driven by numerous factors in attempts of sprouting new vessels [4] which include acidosis, inflammatory cytokines, as well as the activation of oncogenes that occur at the cellular level, upstream of VEGF and other tumor promoting hormones [5, 6]. Importantly, the tumor extracellular matrix (ECM) becomes stiffer due to the release of plasma components through the leaky tumor vessels as well as dynamic remodeling of these components by tumor and stromal cells. Further, an increase in interstitial fluid pressure in tumors creates turbid blood flow [7] impeding the delivery of cancer drugs. Together, these mechanical forces create an abnormal micromechanical environment to which the tumor vasculature does not respond properly. Therefore, recent studies have started to target the mechanical aspects of the tumor microenvironment for cancer therapies, with specific interest on ion channels. The transient receptor potential (TRP) family of ion channels has already gained much attention in cancer, especially those that exhibit mechanosensitive properties. This chapter will cover the abnormal tumor vasculature, and aims to examine the role of mechanosensitive ion channels involved in the aberrant environment, with a special focus on TRPV4.

Tumor Vasculature

As with any organ, the endothelium can be subjected to dysfunction and failure. Such is the case of the tumor vasculature, the structure of which is controlled by irregular intrinsic and extrinsic factors. These abnormalities generate microvasculature that contain regions of heterogeneity, with areas within the tumor being "hyper" or "hypo" vascular in both function and density [8, 9]. At the cellular level, pro-angiogenic factors generate the weakening of VE-cadherin-mediated endothelial cell (EC) junctions, which in turn distort the vessel wall structure and promote EC migration [10]. Pericytes that cover the endothelium, generally create a stable and mature vascular network, but in tumors have been found reduced in number, and loosely attached to the EC, causing vessels to become immature and leaky. This effect is the result of increased growth factor signaling, such as vascular endothelial growth factor (VEGF), which has been found to impede the adherence of pericytes to the surrounding endothelium [11–13]. Overall, these new vessels are not only tortuous and dilated, but lack structural support [8], endorsing dysfunctional blood vessels that promote erratic blood flow [8, 14] and hyper-permeability that allow protein and fluid extravasation into the extracellular space, all of which increase matrix deposition [8]. Additionally, the growing tumor mass often times leads to the compression and eventual collapsing of the existing vessels, further diminishing blood supply and increasing hypoxic and acidic conditions within the tumor [8, 15]. Altogether, the structural and functional aberrations of the tumor vasculature are prompted by the persistent needs of the tumor cells to vascularize and the constant stimulation of angiogenesis.

Tumor Angiogenesis

The progression of solid tumors relies heavily on angiogenesis and requires the endothelium to switch from a quiescent, or resting, phenotype to one that is more invasive, termed the "angiogenic switch"[16]. To begin this process, the basement membrane must be injured or wounded, triggering destruction and hypoxia of the surrounding tissue. Pro-angiogenic factors are released to stimulate EC to migrate, proliferate, and stabilize. However, in tumor angiogenesis, not only are pro-angiogenic factors up-regulated, but angiogenic inhibitors must be down-regulated to maintain adequate stimulation of angiogenesis [17]. Overall, the local equilibrium of the tumor microenvironment is unbalanced, and considering the abnormalities of the tumor vasculature, makes it pertinent to address the complexities associated with tumor EC (TEC) themselves.

Tumor Endothelial Cells (TEC)

Altogether, the structural and functional aberrations of the tumor vasculature promote drastic genetic and morphological changes within the cells [18]. These factors can then generate tumor cells that demonstrate a more hostile phenotype with potential for metastasis [19]. It is now known to a large extent that the tumor endothelium is defective, with TEC showing distinct irregularities with respect to shape and size when compared to normal endothelial cells (NEC). These cells tend to have long cytoplasmic projections that can extend across the lumen, with the tips of some TEC protruding into the lumen, creating intercellular gaps within the vessel wall that most often result in the leakage or pooling of blood [20].

At the molecular level, the tumor endothelium, unlike the normal endothelium, express a host of genes, recently identified as transmembrane proteins called tumor endothelial markers (TEM) [21, 22]. In-depth studies on TEC isolated from different carcinomas, show that these cells do not undergo senescence, are resistant to serum starvation and apoptosis, and are structurally abnormal compared to NEC. TEC isolated from mouse xenograft tumors were found to have variable DNA content, not only between its normal counterpart cell types, but between individual cells as well, indicating the existence of heterogeneity within TEC. These cells have larger nuclei and exhibit characteristic cytogenic and structural abnormalities, such as aneuploidy and chromosomal aberrations including deletions, non-reciprocal translocations, and abnormal centrosomes [23–26]. Furthermore, TEC isolated from mouse prostate tumors were found to express both hematopoietic and mesenchymal stem cell markers, verifying the heterogeneity of these cells. Additionally, it was found that these cells were also able to undergo unusual mesenchymal differentiation into cartilage and bone-like tissues in conditioned medium, confirming the ability of TEC to adapt to their surrounding environment [27]. Based on these studies, it can be interpreted that the cellular and molecular aberrations of TEC can invariably contribute to the abnormal angiogenic process and ultimately tumor growth and metastasis.

However, when grown in defined endothelial medium, TEC (derived from mouse prostate tumors) express endothelial markers and show morphological features similar to NEC [27]. Interestingly, we have demonstrated that TEC show aberrant mechanosensing and abnormal angiogenesis *in vitro*, suggesting deregulation of mechanosensing mechanisms in these cells [28]. A fascinating question, that has yet to be considerably studied, pertains to the origin of these TEC. A variety of studies [29, 30] have highlighted that the source is from stem-cell tumor cells, to "circulating $CD34^+$/$VEGFR\text{-}2^+$ endothelial progenitor cells"; however, there has been limited progress in conclusively identifying the source of these cells, making it difficult to develop any specific targets as part of an anti-angiogenic strategy.

Current Approaches to Target Tumor Angiogenesis

Since first described by Dr. Folkman [3], the concept of angiogenesis and anti-angiogenic therapies has revolutionized the way cancer has been studied and clinically treated. Inhibition of angiogenic activators such as VEGF, placental growth factor (PIGF), fibroblast growth factor (FGF) [31] and associated growth factor signaling mechanisms, have provided new avenues in successfully treating as well as studying tumor angiogenesis. Anti-VEGF therapies, including anti-VEGF neutralizing monoclonal antibodies and receptor tyrosine kinase inhibitors (RTKIs), dominate current approaches in treating malignant tumors [32, 33], initially achieving great success. One such study found that treating patients of metastatic colorectal cancer with anti-VEGF antibody, Bevacizumab, in combination with systemic chemotherapy, increased patient survival [34], which was later attributed to tumor 'vascular normalization.' The principle of vascular normalization seeks to restore the balance of angiogenic factors within the tumor microenvironment to regulate vessel growth and maturity [8, 35]. This allows for improved delivery of chemotherapeutic agents by reestablishing a more normal vascular network. While anti-VEGF-mediated approaches hold promise, it has a narrow window in terms of the transient nature of the resulting 'normal' vessel. Furthermore, these growth factor mediated therapies have become redundant, ineffective, and in some instances, detrimental to the treatment of cancer, owing to inherent or acquired drug resistance, the potential for metastatic capability, and the general absence of predictive markers to monitor tumor responses in select patient populations [36]. These findings were further substantiated in a recent finding that demonstrated a rapid decrease in delivery of chemotherapy to the tumor after anti-VEGF therapy in non-small cell lung cancer (NSCLC) patients using PET imaging [37].

Limitations in targeting soluble factors have led to the idea that other factors, such as mechanical forces, may be contributing to the abnormal tumor vasculature. In fact, defects in mechanotransduction, the conversion of mechanical forces into biochemical signals, have been reported as the basis of diverse pathological conditions [38, 39]. A number of studies have also described a balance of underlying mechanical forces as driving factors in sensitizing capillary EC to angiogenic growth factors, to form functional networks of blood vessels. Furthermore, EC are exposed and respond to mechanical forces such as shear stress and cyclic strain imposed by blood flow. In fact, it was postulated that local micromechanical forces modulate the endothelial response to growth factors [40]. One of the fastest responses of EC to mechanical forces is Ca^{2+} influx through mechanosensitive ion channels. Ion channels have already been the subject of various reviews, as they have recently been found up-regulated/down-regulated or simply dysfunctional in different pathological diseases. However, the role of mechanosensitive ion channels in physiological or pathological angiogenesis is largely unknown. Thus, targeting ion channels in the mechanically dynamic tumor microenvironment may lead to novel therapies among the abnormal tumor vasculature.

TRP Channels

The onset of neovascularization and subsequent tumor progression is associated with the generation of cell populations that differ phenotypically, which often arise from a deregulation of key signaling pathways and mutations or deletion of several proteins. Transient receptor potential (TRP) channels represent a superfamily of proteins that have, over the years, been understood to affect or be affected by a variety of pathological conditions, including cancer. In addition to the transcriptional regulation of TRPs by hormones and growth factors produced by the tumor microenvironment, alternative splicing of genes leads to the generation of protein isoforms with altered functions and variations in subcellular localization. Such effects have been documented in diverse tumor types where a decrease in Ca^{2+} influx may be attributed to decreased expression of TRP channels [41].

A growing body of evidence has identified members of the TRP family, mainly members of TRPC (canonical), TRPM (melastatin) and TRPV (vanilloid), as key regulators of mechanotransduction [42–45]. TRP channels have a profound effect on EC function, and any dysregulation of these channels can result in EC dysfunction [46]. Because TRP channels are found in the plasma membrane, they are easily and directly accessible to the blood stream, which make these channels potential molecular targets when vascular diseases arise. Below, we specifically describe mechanosensitive TRP channels expressed in the endothelium, most of which are involved in a variety of cancers. Although we have tried to incorporate all relevant studies of these channels in tumor angiogenesis, much of this field remains vastly unexplored, with potential to investigate new targets.

TRPC

TRP canonical channels (TRPC), the founding member of the TRPs, are made up of seven family members (1–7) that are expressed in the endothelium [47], as well as the surrounding smooth muscle cells [48]. TRPC channels support endothelial function such as vascular regulation, permeability, and the endothelial-derived nitric oxide (NO) mediated vasorelaxation of smooth muscle cells. While all of these essential functions may be necessary for the angiogenic process, TRPC1 and TRPC6 have been implicated in mediating mechanotransduction. Studies propose TRPC1 is a stretch-activated channel while others have confirmed TRPC6 is activated by pressure as well as osmotically or mechanically induced plasma membrane stretch [47]. While the role of TRPC1 in angiogenesis has been confirmed, most of the evidence suggests that the mechanosensitive properties may not be obligatory, but that TRPC1 may act alongside other TRPC channels to carry out mechanotransduction. Additionally, TRPC1 and TRPC6 have been found to respond to inflammatory agonists resulting in EC cell shape changes [49, 50]. TRPC6 has also been found in human pulmonary arterial endothelial cells (HPAEC) to control endothelial contraction, cell shape, and permeability [49, 51]. In regards to the cancer environment, many TRPC channels have been found to exist among several different types of cancer tissue, including breast cancer [52], ovarian cancer [53, 54], hepatoma [55], prostate cancer [56], basal cell carcinoma [57], renal cell carcinoma [58], malignant gliomas [59], glioblastoma [60], gastric tumors [61], and lung cancer [62]. However, there have not been any studies to date that suggest TRPC channels are directly involved in the tumor vasculature.

TRPM

TRP melastatin channels (TRPM) are the largest and most diverse among the TRP superfamily. Made up of eight family members (1–8), these channels are widely expressed in endothelial cells and vascular smooth muscle [47], making them important for normal vascular function. When it comes to vascular mechanosensing, both TRPM4 and TRPM7 have been found to contribute to mechanotransduction. In cerebral artery myocytes, TRPM4 is activated by membrane stretch [63] and in cerebral artery smooth muscle cells, take part in mediating membrane depolarization and myogenic vascular tone [64]. Furthermore, some studies have suggested that TRPM4 activation by stretch may be secondary to Ca^{2+} responses [63–65]. Additionally, TRPM7, which acts as an ion channel and a functional kinase, can be activated by cell stretch and swelling to carry out mechanotransduction [66]. TRPM7 is highly permeable to Mg^{2+}, which contributes to the role TRPM7 plays in angiogenesis and vascular remodeling due to the diverse effect of Mg^{2+} on EC function [67]. Pathologically, some studies suggest that TRPM4 may be involved in decreased NO production [68], while TRPM7 plays a role in angiogenesis and oxidative stress induced cell death [47, 69]. Recent TRPM7 studies in ovarian cancer have found that TRPM7 is needed for cancer cell growth, migration, and invasion [70], but TRPM4 nor TRPM7 channels have been found to affect the tumor endothelium.

TRPP and TRPA

The TRP polycystin (TRPP) ion channel family is well known in polycystic kidney disease (PKD) [71]. Expressed ubiquitously among vertebrates, these channels are involved in mechanotransduction due to TRPP1 and TRPP2 channel activation by shear stress [47]. These channels are important for flow-induced vascular response and NO production [47, 72].

TRPA1 is the only mammalian member to belong to the TRP ankyrin (TRPA) family. Although predominately expressed in nociceptive neurons of the peripheral ganglia and the mechanosensory epithelium of the inner ear [73], this mechanosensitive ion channel has recently been found expressed in the endothelium as well as the surrounding perivascular cells in cerebral circulation [74]. Overall, TRPA1 mainly plays a role in nociception [75], mechanotransduction [76], and thermal [77], and oxygen sensing [78]. The only studies of TRPA1 in the endothelium were performed by Earley and colleagues [79], which found activated TRPA1 stimulated endothelium-dependent smooth muscle cell hyperpolarization and vasodilation. Pathologically, TRPA1 is involved in acute and chronic pain and possibly chronic inflammatory diseases, and only few studies have been performed in regards to TRPA1 in cancer [80, 81].

TRPV1

The TRP vanilloid (TRPV) family of channels has received the most attention in regards to mechanotransduction in the vasculature, specifically TRPV1 and TRPV4, both of which are expressed in the endothelium [47]. The first family member, TRPV1, is activated by capsaicin [82], anandamide, arachidonic acid (AA) [83], PIP2 hydrolysis [84], acidity, and noxious heat ($T>43$ °C) [82], and is important for NO production and endothelium-dependent vascular relaxation [85, 86]. TRPV1 has been implicated to mediate the later stages of angiogenesis from various studies. In human umbilical vascular EC (HUVEC), TRPV1 activation inhibits VEGF-induced proliferation, DNA synthesis, and capillary-like tube formation. Additionally, activation of TRPV1 via capsaicin also inhibited VEGF-induced vessel formation and vessel sprouting in mouse Matrigel plug and rat aortic ring assays, respectively [87]. When it comes to mechanosensation, TRPV1 was originally discovered to take part in inflammatory thermal hyperalgesia, nociception, and pain sensation [88, 89]. TRPV1 expression has been found altered in cancers of the prostate [90, 91], colon, pancreas [92, 93], and bladder. Expression is increased in all of these except bladder cancer, in which the cancerous urothelium causes a decrease in TRPV1 due to cells becoming more de-differentiated as the carcinoma cells progress to a more aggressive state [92]. Moreover, TRPV1 in cancer has been often associated with pain, especially due to TRPV1 expression in neurons [94]. These findings make TRPV1 a good target for pharmacological inhibition of cancer pain in several types of cancers.

TRPV4: A Novel Target for Cancer Therapy

In recent years, TRPV4 has emerged as a widely accepted mechanosensor, contributing significantly to the process of mechanotransduction. TRPV4 is a non-specific Ca^{2+} permeable channel activated by a variety of physical and chemical stimuli such as temperature, hypotonicity, phorbol esters, endocannabinoids, arachidonic acid (AA), and epoxy eicosatrienoic acids (EETs) [95–97]. The TRPV4 protein is composed of a cytosolic N-terminal region, six transmembrane domains including the pore region, and an intracellular C-terminal tail (Fig. 12.1). The N-terminal region contains the ankyrin repeat domains (ARD), which consists of six ankyrin repeats (ANK) (ANK1–6) [98, 99]. A proline-rich domain (PRD) has been implicated in the mechanosensitivity of the TRPV4 channel, preceding the first ankyrin repeat. Within this PRD, proline residues at positions 142, 143, and 144, interact with pacsin 3, a protein implicated in vesicular membrane transport, endocytosis, and cytoskeleton reorganization [100, 101]. The TRPV4 C-terminal tail contains additional functional domains such as a TRP box, a calmodulin-binding site, and a binding site for cytoskeletal proteins such as MAP7, actin, and tubulin [102, 103].

TRPV4 is ubiquitously expressed among various tissues including lung, liver, heart, trachea, and the vascular endothelium [104–107]. Structurally similar to the *osm-9* gene found in *Caenorhabditis elegans*, mammalian TRPV4 is thought to

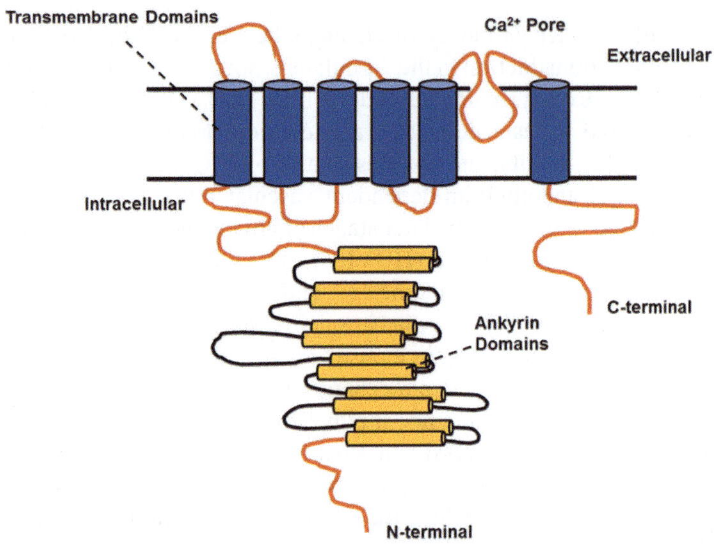

Fig. 12.1 Schematic of the structure of the mechanosensitive ion channel TRPV4. The TRPV4 channel is a tetramer and each subunit is composed of a cytosolic N-terminal region and six transmembrane domains, including the pore region, and an intracellular C-terminal tail. The N-terminal region contains the ankyrin repeat domain (ARD), which consists of six ankyrin repeats

share functional properties in mechanosensation. Previous studies found that when human embryonic kidney (HEK) 293 cells transfected with TRPV4 were exposed to shear stress, the increase in intracellular Ca^{2+} was due to TRPV4 activation, which was inhibited by TRPV antagonist, ruthenium red [108]. Liedtke et al. [109] found that sensory neurons in *C. elegans* responded to osmotic and mechanical stimuli via TRPV4 channels, providing *in vivo* support of the mechanosensitive properties of TRPV4. The first evidence indicating TRPV4 as a mechanosensor in the endothelium was reported by Kohler and coworkers [110], where they showed that shear stress-induced Ca^{2+} influx and vasodilation were mediated by TRPV4, which was confirmed by their later study using TRPV4-null mice [111]. A separate study has also revealed that TRPV4-mediated Ca^{2+} influx is critical for flow-induced release of mitochondrial reactive oxygen species (ROS) and vasodilation in human coronary arteries [112]. We and others have reported that TRPV4-dependent Ca^{2+} influx plays critical role in agonist (acetylcholine)-induced vasodilation [74, 113, 114]. Together, these studies demonstrate that TRPV4 plays an important role in the mechanical force and agonist-induced regulation of vascular tone. TRPV4 also plays a role in lung microvascular EC in which elevated lung hydrostatic pressures caused EC Ca^{2+} influx, an increase in vascular permeability and lung edema; these effects were abolished upon treatment with TRPV4 inhibitors and in TRPV4KO mice [106, 115].

Although the role of TRPV4 as a mechanosensor has been confirmed in the endothelium and other tissue and organ systems, the molecular mechanisms by which TRPV4 transduces mechanical signals is not well known. Both cell and ECM-generated mechanical forces have been critical signals that dictate normal vessel growth and patterning [40, 116, 117]. A critical event that takes place during neovascularization involves the directional migration of EC towards angiogenic stimuli, characterized by the realignment of the actin cytoskeleton and EC reorientation. Endothelial cells have been shown to reorient perpendicular to the direction of cyclic strain; but align parallel to flow in response to shear stress [118]. This reorientation response of EC to cyclic strain was found to be regulated by stretch-activated (SA) channels [119], however the specific identity of the SA channel or molecular mechanism(s) underlying this reorientation response is not known.

We set out to investigate the precise stretch-activated Ca^{2+} channel involved in regulating EC reorientation and our studies revealed that mechanical strain applied to integrins resulted in a rapid influx of Ca^{2+} via TRPV4 in EC. We further identified TRPV4 channels as the specific SA channel that mediates integrin-to-integrin signaling required for the cyclic strain-induced reorientation of EC [120] (Fig. 12.2). Specifically, we demonstrated that cyclic strain-induced TRPV4-mediated Ca^{2+} influx activates PI3K which in turn stimulates the activation of additional β1 integrins that may regulate Rho/Rac signaling required for the reorganization of the actin cytoskeleton and EC reorientation (Fig. 12.3). We further showed that mechanical strain application on β1 integrins activates TRPV4 through its interaction with a transmembrane CD98 protein, located in focal adhesions [121]. Based on these findings, we postulated that the transfer of mechanical force between cell surface molecules (β1 integrins, CD98 and TRPV4) localizes mechanotransduction almost

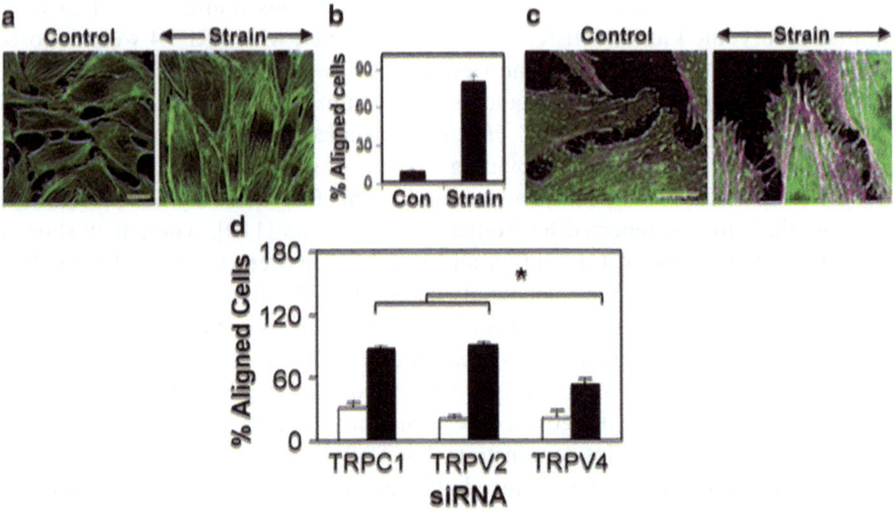

Fig. 12.2 TRPV4 channels mediate cyclic strain-induced endothelial cell reorientation. (**a**) Fluorescence images of endothelial cells (EC) subjected to 0 or 10 % uniaxial cyclic strain (2 h, 1 Hz). Cells were cultured on fibronectin-coated flexible silicone membranes, fixed, and stained with Alexa488-phalloidin to visualize actin stress fibers. *Arrow* indicates the direction of applied strain. Scale bar: 25 μm. (**b**) Quantification of cell alignment in control and strain exposed cells as the percentage of cells oriented 90 ± 30° (aligned) relative to the direction of applied strain ($p < 0.0006$); error bars indicated S.E.M. (**c**) Immunofluorescence images of EC subjected to 0 or 10 % uniaxial cyclic strain. Cells were stained for vinculin (*green*) and actin stress fibers (*magenta*) to show that application of strain causes enhanced recruitment of vinculin to large focal adhesions, that co-localize with the ends of reinforced stress fibers (shown in *white*). Scale bar: 25 μm. (**d**) siRNA knockdown of TRPV4 significantly inhibited cyclic strain-induced EC reorientation, compared to siRNA knockdown of TRPV2 or TRPC1. The quantification of cell alignment was measured as the percentage of cells oriented 90 ± 30° (aligned) relative to the direction of applied strain in control (*white bars*) and strain exposed (*black bars*) in human EC. (*$p < 0.0025$) (reprinted from Thodeti et al., *Circ. Res.* 104:1123–1130, 2009; Fig. 1 and Fig. 6D [120])

instantaneously within focal adhesions [121] and may regulate many complex cell and tissue behaviors. The importance of TRPV4 as a mechanosensor in EC was further validated when we demonstrated that a siRNA-mediated knockdown of TRPV4 in EC resulted in the failure of these cells to reorient perpendicular to the direction of applied cyclic strain [120] (Fig. 12.2d). To our surprise, when tumor-derived endothelial cells (TEC) were exposed to cyclic strain, these cells failed to reorient [28]. This aberrant mechanosensitive response exhibited by TEC was also found manifested in cell-spreading experiments. These cells exhibited uncontrolled spreading on ECM substrates of increasing rigidity (such as within a tumor), as opposed to NEC that stopped spreading at the highest stiffness. Further, *in vitro* angiogenesis assays revealed that TEC exhibited abnormal angiogenic behavior which was demonstrated to be mediated by high basal active-Rho and Rho-kinase (ROCK) [28].

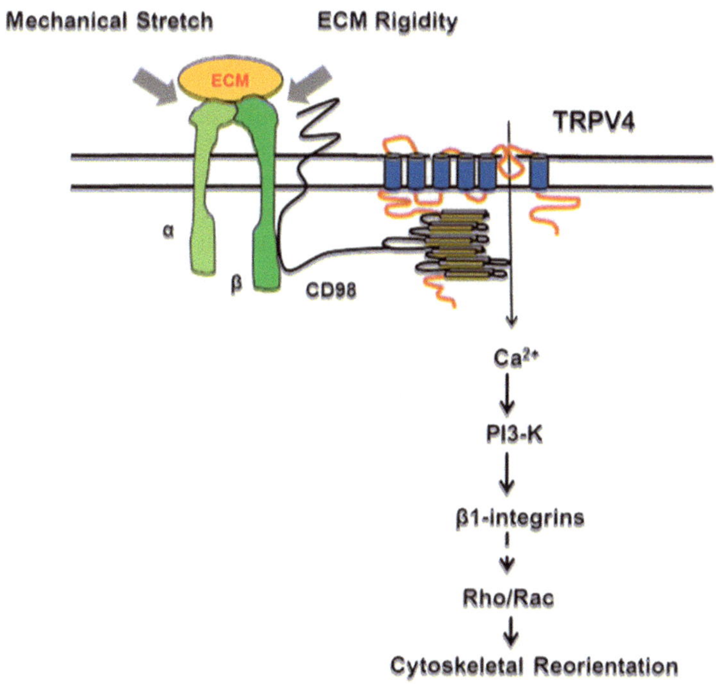

Fig. 12.3 TRPV4 mediated mechanotransduction in endothelial cells. A schematic model showing TRPV4-dependent mechanical signaling in normal endothelial cells. Application of mechanical force (cyclic stretch or ECM stiffness) to integrins activates ultra-rapid calcium influx through TRPV4 via interaction with a transmembrane protein CD98. The released Ca^{2+} activates additional integrins via PI3K. This integrin-to-integrin signaling may further regulate downstream Rho/Rac pathways necessary for reorganization of the actin cytoskeleton and reorientation of EC [120]

Since TEC failed to reorient to applied cyclic strain, a phenomenon reminiscent to NEC subjected to siRNA-mediated TRPV4 silencing, we postulated that aberrant mechanosensitivity in TEC may be caused by altered TRPV4-dependent signaling. Indeed, our recent work has demonstrated that TRPV4 expression and function are impaired in TEC [122]. Specifically, we found a significant reduction in TRPV4 expression (~40 %), while functional assays revealed a 40–50 % decrease in Ca^{2+} influx in TEC when stimulated with specific TRPV4 agonists, GSK1016790A (GSK) and 4α-PDD (Fig. 12.4). An important downstream consequence of tumor growth is an increase in matrix rigidity, owing to the leakage of plasma components from highly permeable blood vessels, and increased synthesis/degradation of matrix components [123]. This increase in ECM stiffness has been shown to influence TEC spreading, migration, and tube formation. Because we have previously shown that TEC display abnormal mechanosensitivity to substrate stiffness [28], and TRPV4 was found functionally impaired, we overexpressed TRPV4 to determine whether we could restore TEC mechanosensitivity towards varying ECM rigidity. We found that overexpression of TRPV4 reduced the abnormal spreading exhibited by TEC on the highest stiffness gelatin gels (2 kPa; comparable to the stiffness of tumors). The ability of cells to migrate is reliant on the mechanosensing efficiency of the cell in response to ECM rigidity, which is an important step in the angiogenic process [124]. We therefore investigated the migratory ability of TEC and found that these cells exhibit abnormal migration (40 μm/h); and restoring TRPV4-dependent mechanosensitivity normalized TEC migration consistent with the migration of NEC (10 μm/h). Additionally, overexpression of TRPV4 in TEC was able to decrease the high basal Rho activity previously observed [28], as well as normalize the abnormal angiogenesis exhibited in both 2D and 3D *in vitro* angiogenesis assays. We further demonstrated that these effects were similarly achieved by modulating TRPV4 activity pharmacologically using a specific TRPV4 activator, GSK1016790A. These findings suggest that TRPV4 signaling regulates tumor angiogenesis via Rho-dependent mechanosensing mechanisms. Finally, to determine the functional role of TRPV4 in tumor angiogenesis *in vivo*, we induced syngeneic tumors in TRPV4KO mice by subcutaneously injecting LLC (mouse Lewis Lung Carcinoma) cells. Tumor growth and angiogenesis was significantly enhanced in TRPV4KO mice compared to WT mice. Immunohistochemical analysis of tumor sections obtained from TRPV4KO tumors revealed the vessels were immature, i.e. large vessels with decreased pericyte coverage. Further, TRITC-dextran perfusion experiments confirmed that these vessels were leaky, as the dextran fluorescence was enhanced in the tumor tissue in TRPV4KO mice compared to WT mice. Notably, these findings suggest that TRPV4 is critical for maintaining vessel structure and integrity and that absence of TRPV4 not only increases tumor angiogenesis but also inhibits vessel maturity.

Thus, our study has not only provided a unique role for the mechanosensitive TRPV4 channel in the tumor vasculature, but opens up an uncharted therapeutic target in vascular normalization strategies. Specifically we have demonstrated TRPV4 activation with GSK1016790A normalized tumor vasculature (increased pericyte coverage) and reduced tumor growth when given in combination with the anti-cancer drug Cisplatin, but not alone. These findings suggest that normalization of tumor vasculature by TRPV4 activation may have improved the delivery of Cisplatin and reduced tumor growth (Fig. 12.5). Thus, our work is the first to study the functional significance of TRPV4 in tumor angiogenesis *in vitro* and *in vivo* and identify a novel role for TRPV4 in the regulation of tumor vessel growth and maturity.

Conclusion and Perspectives

The neovascularization of solid tumors create dysfunctional endothelium, generating a microenvironment in which the vasculature becomes heterogeneous and erratic blood flow ensues. This ultimately affects the delivery and efficacy of systemically administered agents. Although endothelial cells are generally heterogeneous, they are commonly regulated by many soluble, membrane-bound, and mechanical factors, including ion channels. Because most ion channels are found in

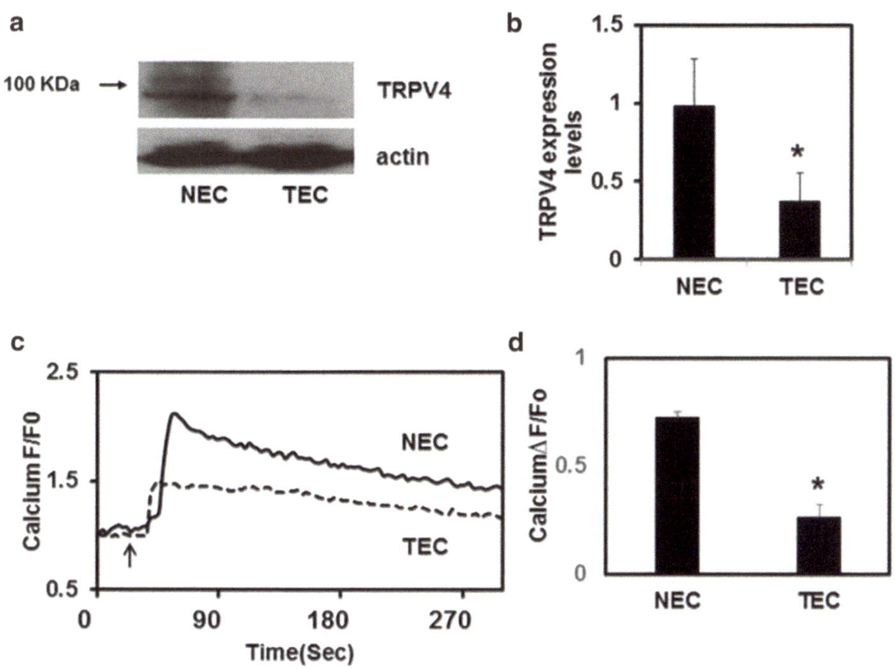

Fig. 12.4 Expression and function of TRPV4 channels in normal and tumor endothelial cells. (**a**) Western blot showing TRPV4 protein expression in normal (NEC) and tumor-derived endothelial cells (TEC). (**b**) Densitometric analysis of the Western blots showing significant ($p \leq 0.05$) reduction in TRPV4 expression in TEC compared to NEC. (**c**) Relative changes in cytosolic calcium in response to a selective TRPV4 agonist, GSK1016790A (100 nM) in Fluo-4 loaded NEC and TEC. *Arrow* denotes the time when the cells were stimulated with the TRPV4 agonist. (**d**) Quantitative analysis of cytosolic calcium influx induced by GSK1016790A in NEC and TEC. (F/F0 = ratio of normalized Fluo-4 fluorescence intensity relative to time 0). The results shown are mean ± SEM from three independent experiments. The significance was set at $p \leq 0.05$ (reprinted from Adapala et al., *Oncogene* 2016; 35:314–22 [122])

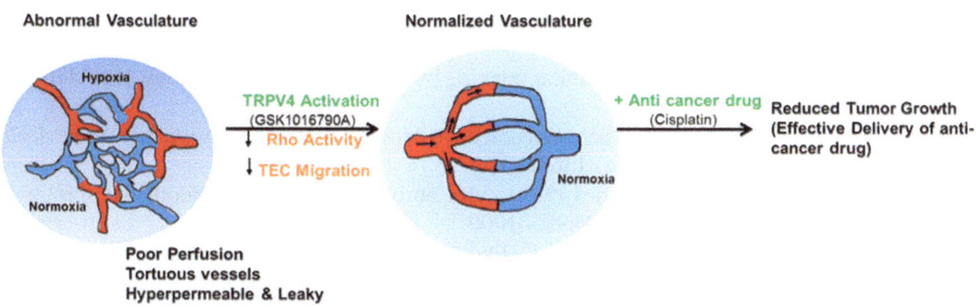

Fig. 12.5 TRPV4 activation normalizes tumor vasculature enhancing chemotherapeutic drug delivery. The tumor vasculature is characterized by tortuous, hyper-permeable, and poorly perfused vessels surrounded by varying regions of hypoxia and normoxia. The treatment with a specific small molecular activator of TRPV4, GSK1016790A, normalizes the tumor vasculature and improves the efficient delivery of chemotherapeutic drug (Cisplatin) leading to reduced tumor growth in WT mice injected subcutaneously with LLC (Lewis Lung Carcinoma) cells (Adapala et al., *Oncogene* 2016; 35:314–22 [122])

the plasma membrane, they are accessible to the blood stream, making them favorable targets when vascular diseases arise. Most current therapies concentrate on targeting the cytokines involved in tumor angiogenesis. Recent pre-clinical and clinical evidence has demonstrated that anti-VEGF drugs, as well as many other direct or indirect angiogenesis inhibitors, can transiently promote the normalization of the tumor vasculature. However, growth factor mediated normalization therapies have largely been unsuccessful due to several challenges, such as drug resistance and redundancy among others. Considering these findings, targeting mechanotransduction may offer a more effective means to treating or reducing solid tumor growth. We have shown mechanosensitive ion channel TRPV4 modulates tumor endothelial function and angiogenesis, and have unraveled a TRPV4-dependent mechanotransduction mechanism in angiogenesis. These findings could lead to the development of novel growth factor-independent therapeutic targets, not only to induce vascular normalization and improve cancer therapy, but also for other angiogenic disorders such as diabetic retinopathy and age-related macular degeneration.

References

1. Nagy JA, et al. Why are tumour blood vessels abnormal and why is it important to know? Br J Cancer. 2009;100(6):865–9.
2. Fukumura D, et al. Tumor microvasculature and microenvironment: novel insights through intravital imaging in pre-clinical models. Microcirculation. 2010;17(3):206–25.
3. Folkman J. Tumor angiogenesis: therapeutic implications. N Engl J Med. 1971;285(21):1182–6.
4. Nagy JA, et al. Vascular permeability factor/vascular endothelial growth factor induces lymphangiogenesis as well as angiogenesis. J Exp Med. 2002;196(11):1497–506.
5. Ferrara N. VEGF as a therapeutic target in cancer. Oncology. 2005;69 Suppl 3:11–6.
6. Carmeliet P, Jain RK. Molecular mechanisms and clinical applications of angiogenesis. Nature. 2011;473(7347):298–307.
7. Boucher Y, Baxter LT, Jain RK. Interstitial pressure gradients in tissue-isolated and subcutaneous tumors: implications for therapy. Cancer Res. 1990;50(15):4478–84.
8. Jain RK. Normalization of tumor vasculature: an emerging concept in antiangiogenic therapy. Science. 2005;307(5706):58–62.
9. Kim P, et al. In vivo wide-area cellular imaging by side-view endomicroscopy. Nat Methods. 2010;7(4):303–5.
10. Hashizume H, et al. Openings between defective endothelial cells explain tumor vessel leakiness. Am J Pathol. 2000;156(4):1363–80.
11. Morikawa S, et al. Abnormalities in pericytes on blood vessels and endothelial sprouts in tumors. Am J Pathol. 2002;160(3):985–1000.
12. Abramsson A, Lindblom P, Betsholtz C. Endothelial and nonendothelial sources of PDGF-B regulate pericyte recruitment and influence vascular pattern formation in tumors. J Clin Invest. 2003;112(8):1142–51.
13. Greenberg JI, et al. A role for VEGF as a negative regulator of pericyte function and vessel maturation. Nature. 2008;456(7223):809–13.
14. Jain RK. Determinants of tumor blood flow: a review. Cancer Res. 1988;48(10):2641–58.
15. Padera TP, et al. Pathology: cancer cells compress intratumour vessels. Nature. 2004; 427(6976):695.

16. Folkman J. Role of angiogenesis in tumor growth and metastasis. Semin Oncol. 2002;29(6 Suppl 16):15–8.
17. Folkman J. The role of angiogenesis in tumor growth. Semin Cancer Biol. 1992;3(2):65–71.
18. Bottaro DP, Liotta LA. Cancer: out of air is not out of action. Nature. 2003;423(6940): 593–5.
19. DeClerck K, Elble RC. The role of hypoxia and acidosis in promoting metastasis and resistance to chemotherapy. Front Biosci (Landmark Ed). 2010;15:213–25.
20. Dudley AC. Tumor endothelial cells. Cold Spring Harb Perspect Med. 2012;2(3):a006536.
21. St Croix B, et al. Genes expressed in human tumor endothelium. Science. 2000;289(5482):1197–202.
22. Nanda A, et al. Identification of a binding partner for the endothelial cell surface proteins TEM7 and TEM7R. Cancer Res. 2004;64(23):8507–11.
23. Bussolati B, et al. Bifunctional role for VEGF-induced heme oxygenase-1 in vivo: induction of angiogenesis and inhibition of leukocytic infiltration. Blood. 2004;103(3):761–6.
24. Allport JR, Weissleder R. Murine Lewis lung carcinoma-derived endothelium expresses markers of endothelial activation and requires tumor-specific extracellular matrix in vitro. Neoplasia. 2003;5(3):205–17.
25. Hida K, et al. Tumor-associated endothelial cells with cytogenetic abnormalities. Cancer Res. 2004;64(22):8249–55.
26. Hida K, Klagsbrun M. A new perspective on tumor endothelial cells: unexpected chromosome and centrosome abnormalities. Cancer Res. 2005;65(7):2507–10.
27. Dudley AC, et al. Calcification of multipotent prostate tumor endothelium. Cancer Cell. 2008;14(3):201–11.
28. Ghosh K, et al. Tumor-derived endothelial cells exhibit aberrant Rho-mediated mechanosensing and abnormal angiogenesis in vitro. Proc Natl Acad Sci U S A. 2008;105(32): 11305–10.
29. Asahara T, et al. Isolation of putative progenitor endothelial cells for angiogenesis. Science. 1997;275(5302):964–7.
30. Hendrix MJ, et al. Vasculogenic mimicry and tumour-cell plasticity: lessons from melanoma. Nat Rev Cancer. 2003;3(6):411–21.
31. Khoury CC, Ziyadeh FN. Angiogenic factors. Contrib Nephrol. 2011;170:83–92.
32. Kim KJ, et al. The vascular endothelial growth factor proteins: identification of biologically relevant regions by neutralizing monoclonal antibodies. Growth Factors. 1992;7(1):53–64.
33. Kim KJ, et al. Inhibition of vascular endothelial growth factor-induced angiogenesis suppresses tumour growth in vivo. Nature. 1993;362(6423):841–4.
34. Saltz LB, et al. Bevacizumab in combination with oxaliplatin-based chemotherapy as first-line therapy in metastatic colorectal cancer: a randomized phase III study. J Clin Oncol. 2008;26(12):2013–9.
35. Jain RK. Normalizing tumor vasculature with anti-angiogenic therapy: a new paradigm for combination therapy. Nat Med. 2001;7(9):987–9.
36. Shojaei F, Ferrara N. Role of the microenvironment in tumor growth and in refractoriness/resistance to anti-angiogenic therapies. Drug Resist Updat. 2008;11(6):219–30.
37. Van der Veldt AA, et al. Rapid decrease in delivery of chemotherapy to tumors after anti-VEGF therapy: implications for scheduling of anti-angiogenic drugs. Cancer Cell. 2012;21(1):82–91.
38. Huang S, Ingber DE. Cell tension, matrix mechanics, and cancer development. Cancer Cell. 2005;8(3):175–6.
39. Suresh S. Biomechanics and biophysics of cancer cells. Acta Biomater. 2007;3(4):413–38.
40. Ingber DE, Folkman J. Mechanochemical switching between growth and differentiation during fibroblast growth factor-stimulated angiogenesis in vitro: role of extracellular matrix. J Cell Biol. 1989;109(1):317–30.
41. Gkika D, Prevarskaya N. Molecular mechanisms of TRP regulation in tumor growth and metastasis. Biochim Biophys Acta. 2009;1793(6):953–8.

42. O'Neil RG, Heller S. The mechanosensitive nature of TRPV channels. Pflugers Arch. 2005;451(1):193–203.
43. Liedtke W, Kim C. Functionality of the TRPV subfamily of TRP ion channels: add mechano-TRP and osmo-TRP to the lexicon! Cell Mol Life Sci. 2005;62(24):2985–3001.
44. Christensen AP, Corey DP. TRP channels in mechanosensation: direct or indirect activation? Nat Rev Neurosci. 2007;8(7):510–21.
45. Pedersen SF, Nilius B. Transient receptor potential channels in mechanosensing and cell volume regulation. Methods Enzymol. 2007;428:183–207.
46. Kwan HY, Huang Y, Yao X. TRP channels in endothelial function and dysfunction. Biochim Biophys Acta. 2007;1772(8):907–14.
47. Yao X, Garland CJ. Recent developments in vascular endothelial cell transient receptor potential channels. Circ Res. 2005;97(9):853–63.
48. Venkatachalam K, Zheng F, Gill DL. Regulation of canonical transient receptor potential (TRPC) channel function by diacylglycerol and protein kinase C. J Biol Chem. 2003;278(31):29031–40.
49. Singh I, et al. Galphaq-TRPC6-mediated Ca2+ entry induces RhoA activation and resultant endothelial cell shape change in response to thrombin. J Biol Chem. 2007;282(11):7833–43.
50. Cioffi DL, Stevens T. Regulation of endothelial cell barrier function by store-operated calcium entry. Microcirculation. 2006;13(8):709–23.
51. Mehta D, Malik AB. Signaling mechanisms regulating endothelial permeability. Physiol Rev. 2006;86(1):279–367.
52. Dhennin-Duthille I, et al. High expression of transient receptor potential channels in human breast cancer epithelial cells and tissues: correlation with pathological parameters. Cell Physiol Biochem. 2011;28(5):813–22.
53. Zeng B, et al. TRPC channels and their splice variants are essential for promoting human ovarian cancer cell proliferation and tumorigenesis. Curr Cancer Drug Targets. 2013;13(1):103–16.
54. Yang SL, et al. Transient receptor potential channel C3 contributes to the progression of human ovarian cancer. Oncogene. 2009;28(10):1320–8.
55. El Boustany C, et al. Capacitative calcium entry and transient receptor potential canonical 6 expression control human hepatoma cell proliferation. Hepatology. 2008;47(6):2068–77.
56. Thebault S, et al. Differential role of transient receptor potential channels in Ca2+ entry and proliferation of prostate cancer epithelial cells. Cancer Res. 2006;66(4):2038–47.
57. Beck B, et al. TRPC channels determine human keratinocyte differentiation: new insight into basal cell carcinoma. Cell Calcium. 2008;43(5):492–505.
58. Veliceasa D, et al. Transient potential receptor channel 4 controls thrombospondin-1 secretion and angiogenesis in renal cell carcinoma. FEBS J. 2007;274(24):6365–77.
59. Bomben VC, Sontheimer HW. Inhibition of transient receptor potential canonical channels impairs cytokinesis in human malignant gliomas. Cell Prolif. 2008;41(1):98–121.
60. Chigurupati S, et al. Receptor channel TRPC6 is a key mediator of Notch-driven glioblastoma growth and invasiveness. Cancer Res. 2010;70(1):418–27.
61. Cai R, et al. Blockade of TRPC6 channels induced G2/M phase arrest and suppressed growth in human gastric cancer cells. Int J Cancer. 2009;125(10):2281–7.
62. Jiang HN, et al. Involvement of TRPC channels in lung cancer cell differentiation and the correlation analysis in human non-small cell lung cancer. PLoS One. 2013;8(6):e67637.
63. Morita H, et al. Membrane stretch-induced activation of a TRPM4-like nonselective cation channel in cerebral artery myocytes. J Pharmacol Sci. 2007;103(4):417–26.
64. Earley S, Straub SV, Brayden JE. Protein kinase C regulates vascular myogenic tone through activation of TRPM4. Am J Physiol Heart Circ Physiol. 2007;292(6):H2613–22.
65. Earley S, Waldron BJ, Brayden JE. Critical role for transient receptor potential channel TRPM4 in myogenic constriction of cerebral arteries. Circ Res. 2004;95(9):922–9.
66. Numata T, Shimizu T, Okada Y. TRPM7 is a stretch- and swelling-activated cation channel involved in volume regulation in human epithelial cells. Am J Physiol Cell Physiol. 2007;292(1):C460–7.

67. Owsianik G, et al. Permeation and selectivity of TRP channels. Annu Rev Physiol. 2006;68:685–717.
68. Suh SH, et al. ATP and nitric oxide modulate a Ca(2+)-activated non-selective cation current in macrovascular endothelial cells. Pflugers Arch. 2002;444(3):438–45.
69. Aarts M, et al. A key role for TRPM7 channels in anoxic neuronal death. Cell. 2003;115(7):863–77.
70. Wang J, et al. TRPM7 is required for ovarian cancer cell growth, migration and invasion. Biochem Biophys Res Commun. 2014;454(4):547–53.
71. Mochizuki T, et al. PKD2, a gene for polycystic kidney disease that encodes an integral membrane protein. Science. 1996;272(5266):1339–42.
72. Kim K, et al. Polycystin 1 is required for the structural integrity of blood vessels. Proc Natl Acad Sci U S A. 2000;97(4):1731–6.
73. Nagata K, et al. Nociceptor and hair cell transducer properties of TRPA1, a channel for pain and hearing. J Neurosci. 2005;25(16):4052–61.
74. Earley S, Gonzales AL, Crnich R. Endothelium-dependent cerebral artery dilation mediated by TRPA1 and Ca2+-activated K+ channels. Circ Res. 2009;104(8):987–94.
75. Kwan KY, et al. TRPA1 contributes to cold, mechanical, and chemical nociception but is not essential for hair-cell transduction. Neuron. 2006;50(2):277–89.
76. Corey DP, et al. TRPA1 is a candidate for the mechanosensitive transduction channel of vertebrate hair cells. Nature. 2004;432(7018):723–30.
77. Cordero-Morales JF, Gracheva EO, Julius D. Cytoplasmic ankyrin repeats of transient receptor potential A1 (TRPA1) dictate sensitivity to thermal and chemical stimuli. Proc Natl Acad Sci U S A. 2011;108(46):E1184–91.
78. Takahashi N, et al. TRPA1 underlies a sensing mechanism for O2. Nat Chem Biol. 2011;7(10):701–11.
79. Earley S. TRPA1 channels in the vasculature. Br J Pharmacol. 2012;167(1):13–22.
80. Schaefer EA, et al. Stimulation of the chemosensory TRPA1 cation channel by volatile toxic substances promotes cell survival of small cell lung cancer cells. Biochem Pharmacol. 2013;85(3):426–38.
81. Oehler B, et al. TRPA1 is functionally expressed in melanoma cells but is not critical for impaired proliferation caused by allyl isothiocyanate or cinnamaldehyde. Naunyn Schmiedebergs Arch Pharmacol. 2012;385(6):555–63.
82. Caterina MJ, et al. The capsaicin receptor: a heat-activated ion channel in the pain pathway. Nature. 1997;389(6653):816–24.
83. Rami HK, Gunthorpe MJ. The therapeutic potential of TRPV1 (VR1) antagonists: clinical answers await. Drug Discov Today Ther Strateg. 2004;1(1):97–104.
84. Prescott ED, Julius D. A modular PIP2 binding site as a determinant of capsaicin receptor sensitivity. Science. 2003;300(5623):1284–8.
85. Poblete IM, et al. Anandamide elicits an acute release of nitric oxide through endothelial TRPV1 receptor activation in the rat arterial mesenteric bed. J Physiol. 2005;568(Pt 2):539–51.
86. Golech SA, et al. Human brain endothelium: coexpression and function of vanilloid and endocannabinoid receptors. Brain Res Mol Brain Res. 2004;132(1):87–92.
87. Min JK, et al. Capsaicin inhibits in vitro and in vivo angiogenesis. Cancer Res. 2004;64(2):644–51.
88. Caterina MJ, et al. Impaired nociception and pain sensation in mice lacking the capsaicin receptor. Science. 2000;288(5464):306–13.
89. Davis JB, et al. Vanilloid receptor-1 is essential for inflammatory thermal hyperalgesia. Nature. 2000;405(6783):183–7.
90. Domotor A, et al. Immunohistochemical distribution of vanilloid receptor, calcitonin-gene related peptide and substance P in gastrointestinal mucosa of patients with different gastrointestinal disorders. Inflammopharmacology. 2005;13(1-3):161–77.
91. Hartel M, et al. Vanilloids in pancreatic cancer: potential for chemotherapy and pain management. Gut. 2006;55(4):519–28.

92. Lazzeri M, et al. Transient receptor potential vanilloid type 1 (TRPV1) expression changes from normal urothelium to transitional cell carcinoma of human bladder. Eur Urol. 2005;48(4):691–8.
93. Sanchez MG, et al. Expression of the transient receptor potential vanilloid 1 (TRPV1) in LNCaP and PC-3 prostate cancer cells and in human prostate tissue. Eur J Pharmacol. 2005;515(1-3):20–7.
94. Nilius B, et al. Transient receptor potential cation channels in disease. Physiol Rev. 2007;87(1):165–217.
95. Vriens J, et al. Cell swelling, heat, and chemical agonists use distinct pathways for the activation of the cation channel TRPV4. Proc Natl Acad Sci U S A. 2004;101(1):396–401.
96. Liedtke WB. TRPV channels' function in osmo- and mechanotransduction. In: Liedtke WB, Heller S, editors. TRP ion channel function in sensory transduction and cellular signaling cascades. Boca Raton, FL: CRC Press; 2007.
97. Nilius B, et al. TRPV4 calcium entry channel: a paradigm for gating diversity. Am J Physiol Cell Physiol. 2004;286(2):C195–205.
98. Owsianik G, et al. Structure-function relationship of the TRP channel superfamily. Rev Physiol Biochem Pharmacol. 2006;156:61–90.
99. Nilius B, Owsianik G. The transient receptor potential family of ion channels. Genome Biol. 2011;12(3):218.
100. Cuajungco MP, et al. PACSINs bind to the TRPV4 cation channel. PACSIN 3 modulates the subcellular localization of TRPV4. J Biol Chem. 2006;281(27):18753–62.
101. D'Hoedt D, et al. Stimulus-specific modulation of the cation channel TRPV4 by PACSIN 3. J Biol Chem. 2008;283(10):6272–80.
102. Strotmann R, Schultz G, Plant TD. Ca2+-dependent potentiation of the nonselective cation channel TRPV4 is mediated by a C-terminal calmodulin binding site. J Biol Chem. 2003;278(29):26541–9.
103. Suzuki M, Hirao A, Mizuno A. Microtubule-associated [corrected] protein 7 increases the membrane expression of transient receptor potential vanilloid 4 (TRPV4). J Biol Chem. 2003;278(51):51448–53.
104. Liedtke W, et al. Vanilloid receptor-related osmotically activated channel (VR-OAC), a candidate vertebrate osmoreceptor. Cell. 2000;103(3):525–35.
105. Delany NS, et al. Identification and characterization of a novel human vanilloid receptor-like protein, VRL-2. Physiol Genomics. 2001;4(3):165–74.
106. Alvarez DF, et al. Transient receptor potential vanilloid 4-mediated disruption of the alveolar septal barrier: a novel mechanism of acute lung injury. Circ Res. 2006;99(9):988–95.
107. Yin J, et al. Negative-feedback loop attenuates hydrostatic lung edema via a cGMP-dependent regulation of transient receptor potential vanilloid 4. Circ Res. 2008;102(8):966–74.
108. Gao X, Wu L, O'Neil RG. Temperature-modulated diversity of TRPV4 channel gating: activation by physical stresses and phorbol ester derivatives through protein kinase C-dependent and -independent pathways. J Biol Chem. 2003;278(29):27129–37.
109. Liedtke W, et al. Mammalian TRPV4 (VR-OAC) directs behavioral responses to osmotic and mechanical stimuli in Caenorhabditis elegans. Proc Natl Acad Sci U S A. 2003;100 Suppl 2:14531–6.
110. Kohler R, et al. Evidence for a functional role of endothelial transient receptor potential V4 in shear stress-induced vasodilatation. Arterioscler Thromb Vasc Biol. 2006;26(7):1495–502.
111. Hartmannsgruber V, et al. Arterial response to shear stress critically depends on endothelial TRPV4 expression. PLoS One. 2007;2(9):e827.
112. Bubolz AH, et al. Activation of endothelial TRPV4 channels mediates flow-induced dilation in human coronary arterioles: role of Ca2+ entry and mitochondrial ROS signaling. Am J Physiol Heart Circ Physiol. 2012;302(3):H634–42.
113. Zhang DX, et al. Transient receptor potential vanilloid type 4-deficient mice exhibit impaired endothelium-dependent relaxation induced by acetylcholine in vitro and in vivo. Hypertension. 2009;53(3):532–8.

114. Adapala RK, et al. PKCalpha mediates acetylcholine-induced activation of TRPV4-dependent calcium influx in endothelial cells. Am J Physiol Heart Circ Physiol. 2011;301(3):H757–65.
115. Hamanaka K, et al. TRPV4 initiates the acute calcium-dependent permeability increase during ventilator-induced lung injury in isolated mouse lungs. Am J Physiol Lung Cell Mol Physiol. 2007;293(4):L923–32.
116. Davies PF. Flow-mediated endothelial mechanotransduction. Physiol Rev. 1995;75(3): 519–60.
117. Chen CS, et al. Geometric control of cell life and death. Science. 1997;276(5317):1425–8.
118. Kaunas R, et al. Cooperative effects of Rho and mechanical stretch on stress fiber organization. Proc Natl Acad Sci U S A. 2005;102(44):15895–900.
119. Naruse K, Yamada T, Sokabe M. Involvement of SA channels in orienting response of cultured endothelial cells to cyclic stretch. Am J Physiol. 1998;274(5 Pt 2):H1532–8.
120. Thodeti CK, et al. TRPV4 channels mediate cyclic strain-induced endothelial cell reorientation through integrin-to-integrin signaling. Circ Res. 2009;104(9):1123–30.
121. Matthews BD, et al. Ultra-rapid activation of TRPV4 ion channels by mechanical forces applied to cell surface beta1 integrins. Integr Biol (Camb). 2010;2(9):435–42.
122. Adapala RK, et al. Activation of mechanosensitive ion channel TRPV4 normalizes tumor vasculature and improves cancer therapy. Oncogene. 2016;35(3):314–22.
123. Paszek MJ, Weaver VM. The tension mounts: mechanics meets morphogenesis and malignancy. J Mammary Gland Biol Neoplasia. 2004;9(4):325–42.
124. Dvorak HF. Rous-Whipple Award Lecture. How tumors make bad blood vessels and stroma. Am J Pathol. 2003;162(6):1747–57.

Chapter 13
TRPC and Orai Channels in Store-Operated Calcium Entry and Vascular Remodelling

David J. Beech, Jing Li, Lynn McKeown, and Hollie L. Appleby

Abstract In this article we address ion channels formed by Transient Receptor Potential Canonical (TRPC) proteins and Orai proteins and the phenomenon of store-operated Ca^{2+} entry (SOCE) as they relate to vascular remodelling and two dominant cell types in this biology, the vascular smooth muscle cells and the endothelial cells. Emphasis is placed on studies published by our group but there is also general introduction for non-experts, critical evaluation, and debate of current concepts, controversies, and future perspectives. Overall we believe that this is an interesting and important area of biology that still raises unsolved questions and substantial technical challenges if full understanding is to be achieved. It is our opinion that within these mechanisms there are important targets for the discovery and development of novel therapeutic agents to address key unsolved disease challenges in the cardiovascular, cancer and many other fields.

Keywords Calcium channel • Store-operated calcium channel • Transient receptor potential • Orai channel • Stromal interaction molecule 1 • Smooth endoplasmic reticulum • Calcium stores • Artery • Vein • Vascular smooth muscle • Cell proliferation • Cell migration • Neointimal hyperplasia • Coronary artery bypass graft

Introduction

SOCE refers to Ca^{2+} influx through plasma membrane channels activated when intracellular Ca^{2+} stores are depleted. Although Ca^{2+} release is commonly the event leading to store-depletion and channels in this class are sometimes referred to as Ca^{2+}-release-activated Ca^{2+} channels (CRAC channels) there is a view that channels should only be considered as SOC channels if they can be activated simply by

D.J. Beech (✉) • J. Li • L. McKeown • H.L. Appleby
School of Medicine, University of Leeds,
LIGHT Building, Clarendon Way, Leeds LS2 9JT, UK
e-mail: d.j.beech@leeds.ac.uk

I. Levitan, A.M. Dopico (eds.), *Vascular Ion Channels in Physiology and Disease*, DOI 10.1007/978-3-319-29635-7_13

store-depletion alone: i.e. only by lowering of the free Ca^{2+} concentration inside the stores. There is also a view that the purpose of SOC channels is to provide a privileged mechanism for refilling the depleted stores, perhaps with little or no impact on cytosolic Ca^{2+}. However the situation is more complex and such definitions are not always helpful: for example, many observations suggest elevation of cytosolic Ca^{2+} in response to opening of SOC channels and importance of this elevation in its own right, for example as a driver of transcription factor translocation to the nucleus.

SOC channels occur in most if not all mammalian cell types as well as cell types in other animal classes. They have striking functional significance in non-excitable cells but they are also relevant in electrically-excitable cells. The channels are not voltage-gated and thus they differ markedly from voltage-gated Ca^{2+} channels which confer increased Ca^{2+} entry by gating in response to membrane depolarisation. Ca^{2+} entry through store-operated Ca^{2+} channels, by contrast, usually decreases with membrane depolarisation and increases with hyperpolarisation: i.e. the opposite to what is commonly seen in excitable cells such as mature contractile cardiac myocytes and contractile vascular smooth muscle cells. This is not because the channels gate in response to hyperpolarisation but because increased negativity at the internal face of the membrane increases the electric drive for entry of the positively-charged calcium ion. Potassium channel opening should, therefore, enhance SOCE.

It is also important to note that the degree of depletion of stores required for channel opening need only be modest and so the channels may be functional in the absence of obvious or sustained Ca^{2+} release events. Moreover Ca^{2+}-release may occur yet store-depletion avoided because of rapid re-uptake of Ca^{2+} into stores; even in this case there can be activity and functional importance of store-operated Ca^{2+} channels as described below. Our work on SOCE in vascular smooth muscle cells has suggested activation by regulatory factors such as platelet-derived growth factor (PDGF) in the absence of significant store-depletion but also activation by profound store-depletion, a stress to the endoplasmic reticulum (ER stress) and threat to cell survival [1].

Vascular remodelling is an over-arching term referring to changes in the structure of blood vessels or vascular beds in response to physiological events or insults arising from chemical or mechanical factors. The remodelling can be important physiologically, providing new or recovered blood supply. It can also be unwanted for example in neointimal hyperplasia which restricts blood flow and leads to tissue ischaemia; an important example is neointimal hyperplasia in saphenous vein segments used as bypass grafts to ameliorate the impact of coronary artery disease. A key phenomenon in many types of remodelling is the reversible switching of the vascular smooth muscle cell from its contractile to non-contractile phenotype [2]. In simple terms the cell switches from an excitable cell that depends heavily on voltage-gated Ca^{2+} channels to a non-excitable cell that depends on SOCE and other non-voltage-gated Ca^{2+} entry mechanisms [3]. In its non-excitable form the cell is able to proliferate and migrate, enabling changes in vessel structure. After the change in structure the cell normally returns to its non-proliferating phenotype to resume contractile function. Unlike smooth muscle cells, endothelial cells use voltage-gated Ca^{2+} channels very little, if at all, whether they are quiescent or proliferating and mobile. They use SOCE and other non-voltage-gated Ca^{2+} entry mechanisms extensively.

To understand SOCE in the vasculature, and indeed in other systems, an approach has been to identify the molecular basis of the underlying Ca^{2+}-permeable channels and the regulatory mechanisms that control these channels. There has been much effort directed here and substantial progress. We now know some of the key proteins. Yet the field is relatively young—14 years old—and so considerable uncertainties and puzzles remain. A big challenge when studying the vasculature is its heterogeneity: one blood vessel is not necessarily the same as another: as there is heterogeneity across neurones and astrocytes, so there is across vascular smooth muscle cells and endothelial cells. There are also species differences and differences across animal models of human diseases and human diseases themselves. There are substantial technical variations in how experiments are performed by different investigators and challenges relating to how best to design the experiments and interpret the data. There also turn out to be multiple multi-subunit channels that are relevant to SOCE, the relationships between which are dynamic and mostly quite poorly understood. It has often been difficult to distinguish the different components and their functional significance because of inherent weaknesses in the genetic manipulation techniques available, the common absence of specific and potent small-molecule modulators, and the often modest specificity of antibodies targeted to the proteins. A key type of SOCE, some investigators argue the only type, is mediated by channels that are not resolvable at the single channel level and which generate an extremely small whole-cell current beneath technical resolution in physiological conditions. The development of better genetic, small-molecule and antibody tools is important if we are to truly understand SOCE and reveal its role and full potential as a target for novel therapeutics.

Here we focus on two types of channel-forming protein: TRPC and Orai proteins. They are very different types of protein which almost certainly have little structural similarity, if any. TRPC proteins assemble to form channels that are Ca^{2+}-permeable non-selective cationic channels and so they enable substantial Na^{+} entry as well as Ca^{2+} entry [4, 5]. Orai proteins assemble to form channels that are capable of high Ca^{2+} selectivity [6–8], although Na^{+} entry has emerged as possible, or associated, and may occur under certain conditions [9, 10]. The two types of protein are different yet they both contribute to Ca^{2+} entry and agonist responses in a single cell. Both are implicated in endothelial cell and vascular smooth muscle cell proliferation or migration and the remodelling of intact blood vessels. The evidence is strongest for Orai1 channels as channels that open purely as a consequence of store-depletion. Other Orai channels, for example formed by Orai3, are probably not normally store-operated [11]. TRPC proteins were discovered before Orai proteins and were the first proposed mediators of SOCE. The relationship of TRPC channels to SOCE is complex and controversial [12–14].

TRPC Channels

Twenty-eight mammalian genes encode TRP proteins. They are widely expressed with functions across multiple cell types [15]. The seven TRPC proteins are a subset but only TRPC1, TRPC3, TRPC4, TRPC5, TRPC6 and TRPC7 occur in humans [4, 5, 16]. Based on structural predictions and cryo-EM analysis of the distantly-related TRPV1 channels, every TRPC protein is thought to have six membrane-spanning segments and large intracellular N- and C-termini [13, 17]. They are thought to assemble as a group of four proteins around a central ion selectivity filter: i.e. each TRPC channel is a tetramer of TRPC proteins [4]. Such tetrameric assembly allows for homomeric and heteromeric channels from the same or related TRPC proteins [4]. The TRPCs appear to be the TRP subfamily that is most promiscuous in forming heteromers and there is even evidence that the promiscuity extends outside the TRPC sub-family to TRPV4 and TRPP2 [4]. TRPCs are ubiquitously expressed and have diverse functions throughout the body [4, 5]. A TRPC-null mammalian cell-type is unknown and may not exist. Positive modulators of TRPC channels include but are not restricted to lipids and redox factors and agonists at G-protein and tyrosine kinase receptors [4]. TRPC1 has, from amongst the TRPCs, been most commonly associated with SOCE, but not consistently in all studies [13]. Its closest relatives by sequence similarity (TRPC4 and TRPC5) have also been associated with SOCE in some but not all studies [12, 14, 18–20]. Intriguingly, TRPC1 does not form functional channels, or forms them poorly, when expressed alone [4, 13]. It does nevertheless participate readily and importantly in heteromers with TRPC4 and TRPC5 [21–23]. TRPC3, TRPC6 and TRPC7 have been relatively less associated with SOCE. Therefore, TRPC1-containing channels are the strongest candidates from amongst the TRPC family, indeed the TRP super-family, as contributors to SOCE.

Orai Channels

A small but commonly described ionic current occurring in store-depleted immune cells is CRAC channel current (I_{CRAC}) [7]. This current is Ca^{2+} specific and inwardly rectifying. These and other biophysical characteristics are quite different from those of TRPC channel currents. It is now recognised that an assembly of Orai1 proteins can explain CRAC channels mediating I_{CRAC} [6, 7]. Based on structural predictions and crystallographic data from truncated drosophila Orai every Orai protein is thought to have four membrane-spanning segments and relatively short intracellular N- and C-termini [24]. They are about a third the size of TRPC proteins but also assemble as a group around a central ion selectivity filter: i.e. each Orai channel is a multimer of Orai proteins. The exact number of Orai proteins in each channel is debated but may be four, five or six [11, 24, 25]. Homomeric and heteromeric channels are possible but the CRAC channel is usually thought to be a homomeric assembly of Orai1 proteins. Its activation occurs following detection of Ca^{2+} depletion within stores by a Ca^{2+}-sensing domain of a protein called stromal interaction molecule 1 (STIM1) which then binds Orai1 directly to cause channel opening. Orai1 and STIM1 are now strongly linked to SOCE in many cell types. Orai3 channels have been described as arachidonic-acid-activated or leukotriene-activated channels which, although also interacting with STIM1, are not CRAC channels or contributors to SOCE [11]. STIM1 also directly binds to the microtubule-plus-end-tracking protein EB1 where it plays a role in ER remodelling [26]. STIM1 also binds TRPC1, TRPC4 and TRPC5 and this binding may be required for these TRP proteins to contribute to SOCE [27]. Therefore Orai1 channels contribute to SOCE as CRAC channels. STIM1 is an important player in the Orai1 channel mechanism but it is not specific for these channels. The Orai1 channels are undoubtedly central players in the SOCE phenomenon and are not known to have other functions. Although I_{CRAC} was revealed through studies of immune cells and Orai1 was discovered substantially through investigation of immune-deficiency, Orai1 was subsequently found to be widely expressed and functionally relevant to SOCE in many cell types including vascular smooth muscle cells and endothelial cells [14].

SOCE, TRPC and Orai in Vascular Remodelling

Why might SOCE be a player in vascular remodelling? Ca^{2+} is critical in the functions of many cell types, including vascular smooth muscle cells and endothelial cells: there is a dogma that elevation of the intracellular free Ca^{2+} concentration drives cell activity until the Ca^{2+} reaches a certain concentration (~10 μM) above which there is triggering of cell death through apoptosis, necrosis and other mechanisms. The cell activity can take the form of contraction or secretion but Ca^{2+} also drives many other activities such as gene expression, proliferation and migration, key features of vascular remodelling. Without Ca^{2+} these processes are much diminished and many cell types simply wither. Of course there are many types of Ca^{2+}-permeable channel through which Ca^{2+} could enter to drive such activities but they are not all expressed and there has been increasing traction on the idea that SOCE has a prominent role. We have previously discussed the evidence that vascular smooth muscle cells lose their L-type (α_{1C}) voltage-gated Ca^{2+} channels or the activity of such channels as they switch from the contractile phenotype to proliferation [3]. Inhibition of $K_V1.3$ potassium channel leads to a decrease in intracellular Ca^{2+} in the non-contractile cells [28] and there is increased SOCE in vascular smooth muscle cells when they are proliferating [29, 30].

TRPC and Orai proteins are expressed when vascular smooth muscle cells and endothelial cells are proliferating and several studies show striking up-regulation as cells shift from quiescence to proliferation both in vitro and in vivo [14, 31]. The channels arising from these TRPC and Orai proteins may be activated by store-depletion but this type of activation is almost certainly not obligatory; indeed, because store-depletion is an ER stress it seems unlikely that the cells reach such an extreme in vascular remodelling unless the blood vessel or vascular bed as a whole is severely compromised by injury or disease. Constitutive activity or activation by other mechanisms may often be more important.

In our studies on this topic we emphasise human vascular smooth muscle cells and human endothelial cells, including, where possible, studies of intact blood vessel and cells from healthy volunteers and patients with coronary artery disease. In this way we have sought not only better understanding but a perspective on the potential for targeting these mechanisms for the discovery and development of new agents for monitoring and treating human cardiovascular disease and unwanted remodelling evoked by invasive clinical procedure.

TRPC and SOCE in Vascular Smooth Muscle Cells

Shortly after mammalian TRPCs were discovered we developed an approach for generating antibodies that target the third extracellular loop (E3) in TRPCs and inhibit the function of channels containing a particular TRPC, providing the first isoform-specific pharmacology for these channels [32–34] (Fig. 13.1a, b). This was an important step because pharmacology for the channels is generally poor, remaining so to this day [16]. The principles of the inhibitor antibody approach were published in later summary articles [32, 34] but our early success was with T1E3 antibody, which binds TRPC1 and inhibits TRPC1-containing channels [33]. We found a small but statistically significant inhibitory effect of T1E3 on Ca^{2+} entry of store-depleted vascular smooth muscle cells, providing the first evidence that endogenous TRPC1-containing channels contribute to SOCE [33]. The initial studies were in freshly-isolated rabbit cerebral arterioles [33] but we later showed a similar effect in human saphenous vein smooth muscle cells from patients with coronary artery disease [31]. Inhibitor antibody targeted to TRPC5 was also effective, TRPC1 and TRPC5 co-immunoprecipitated, and the current-voltage relationship of the activated channels was like that of TRPC1/TRPC5 heteromers [21, 35]. The data suggested that TRPC1-containing heteromeric channels contribute to, but do not fully or even substantially explain, SOCE in vascular smooth muscle cells of these types.

Although we had success in conferring SOCE on HEK 293 cells by over-expressing TRPC5 the effect was relatively difficult to observe and modest in amplitude [19]. Over-expression is likely to generate TRPC5 protein in high abundance relative to its physiological partner proteins. Because the link to stores is a multi-protein mechanism it is quite possible that over-expression biases the experimenter away from seeing a relationship of TRPC proteins to stores. There are also difficulties with large endogenous SOCE in host cell lines. Some investigators have seen a relationship of TRPCs to SOCE through such experiments while others have not. Failure to see reconstitution has been a reason why some investigators have concluded that TRPCs are not relevant to SOCE. However, negative results of this type are not conclusive; if endogenous SOCE is first inhibited, store-dependent

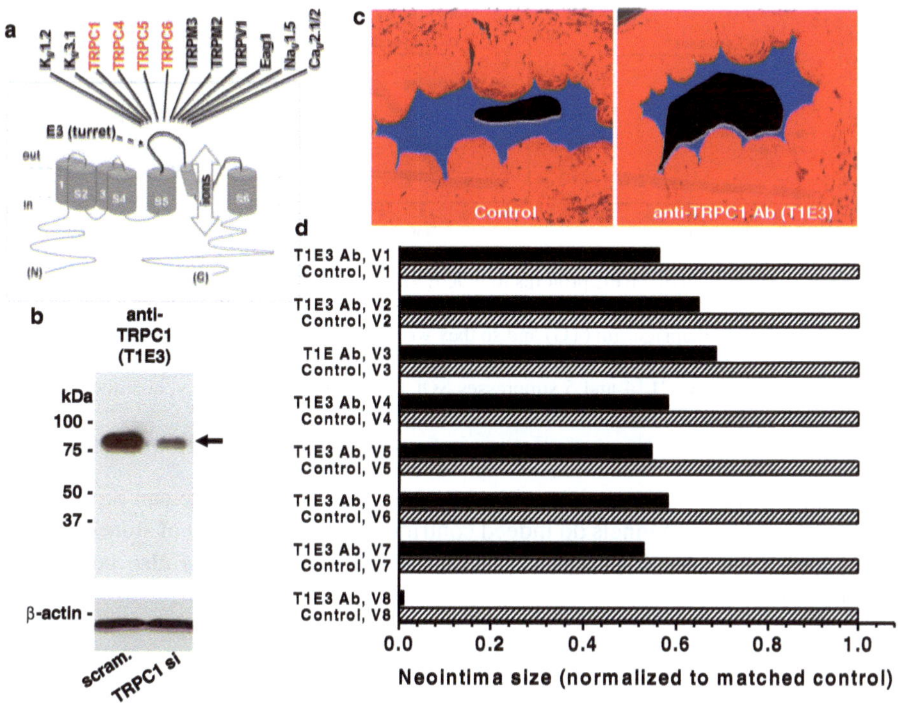

Fig. 13.1 Isoform-selective inhibitor antibodies. (**a**) Summary of inhibitor antibodies generated to the extracellular turret in the indicated channel types. (**b**) Specificity of the anti-TRPC1 inhibitor antibody (T1E3) shown by western blotting. (**c**) Example colourized images for neointimal hyperplasia (*blue*) in human saphenous vein in cross-section. (**d**) As for (**c**) but quantification of neointimal formation in veins (V) from eight patients. Adapted from [31–35]

activity of TRPC5 channels can be observed (Fig. 13.2b, c) [19]. It is nevertheless clear that TRPC5 channels can be activated by other mechanisms that do not involve stores [4, 19] and we can only say that it is possible, albeit not straightforward, to reconstitute a relationship to SOCE. We and others have found that TRPC1 is non-functional when expressed alone [21] while other investigators have suggested that TRPC1 can reconstitute a store-operated signal [36].

Because of the relatively small effects of TRPC inhibitor antibodies on SOCE in vascular smooth muscle cells and the difficulties of the reconstitution approach we sought independent tests of whether TRPCs contribute to endogenous SOCE [37]. Importantly, depletion of TRPC1 or TRPC5 by RNA interference also caused small but significant inhibition of SOCE in human saphenous vein smooth muscle cells and simultaneous depletion of TRPC1, TRPC4 and TRPC5 had a similar inhibitory effect (Fig. 13.2c) [37]. Expression of a dominant negative ion pore mutant of TRPC5 to disrupt permeation in endogenous TRPC channels was also inhibitory

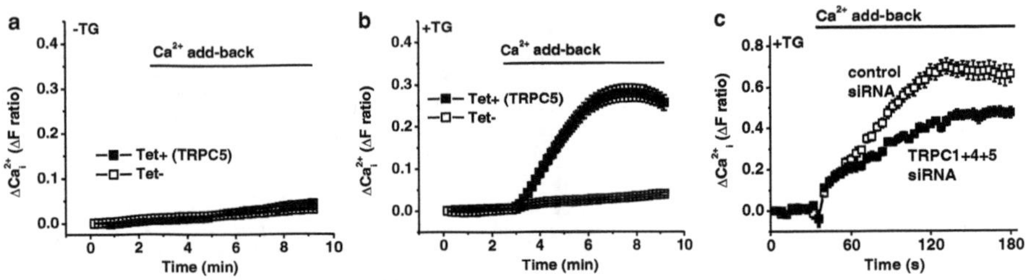

Fig. 13.2 Contribution of TRPC proteins to SOCE. (**a**) Lack of effect of adding extracellular Ca^{2+} back to HEK 293 cells over-expressing TRPC5 (Tet+) or not expressing TRPC5 (Tet–). The cells were not exposed to thapsigargin (TG) and so they were not store-depleted. Extracellular Gd^{3+} was present to inhibit endogenous SOCE. (**b**) As for (**a**) except stores were depleted by TG treatment. (**c**) Knockdown of TRPC1, 4 and 5 suppresses SOCE in human saphenous vein smooth muscle cells. Adapted from [19, 37]

but there was again substantial residual SOCE [37]. Therefore we can be sure that TRPC1-containing channels do indeed contribute to the Ca^{2+} entry of store-depleted saphenous vein smooth muscle cells but that other mechanisms also contribute, substantially.

Other Activators of TRPC in Vascular Smooth Muscle Cells

There are clearly additional activators of TRPC1-containing channels and other TRPC channels. We have contributed to and reviewed this topic extensively [4] and do not revisit it substantially here, but, in short, the TRPC channels can be considered as relatively promiscuous in terms of activators and activation mechanisms. The other activators are likely to be important for the channels in vivo but Ca^{2+} release and store-depletion are also likely to play roles: indeed, a key modulator of the channels is elevation of the intracellular Ca^{2+} concentration. Intriguing activators found by us are the oxidised phospholipids 1-palmitoyl-2-glutaroyl-phosphatidylcholine and 1-palmitoyl-2-oxovaleroyl-phosphatidylcholine [38] (Fig. 13.3). These lipids are described as major drivers of coronary artery disease [39]. It turns out that they activate TRPC1-containing channels of human saphenous vein smooth muscle cells cleanly and reliably [38]. Although they activate the channels indirectly via a G protein mechanisms ($G_{i/o}$ type) they completely fail to cause Ca^{2+} release or store-depletion and so this is a store-independent mode of activation [38]. Strikingly, the T1E3 and T5E3 inhibitor antibodies almost abolished Ca^{2+} entry evoked by these lipids, demonstrating effectiveness of the antibody approach and suggesting TRPC1/TRPC5-containing channels as sole mediators of the Ca^{2+} signal in this case [38] (Fig. 13.3). Therefore TRPC channels are expressed and functional in proliferating human vascular smooth muscle cells that are important contributors to the unsolved clinical problem of coronary artery bypass graft failure due to neointimal hyperplasia.

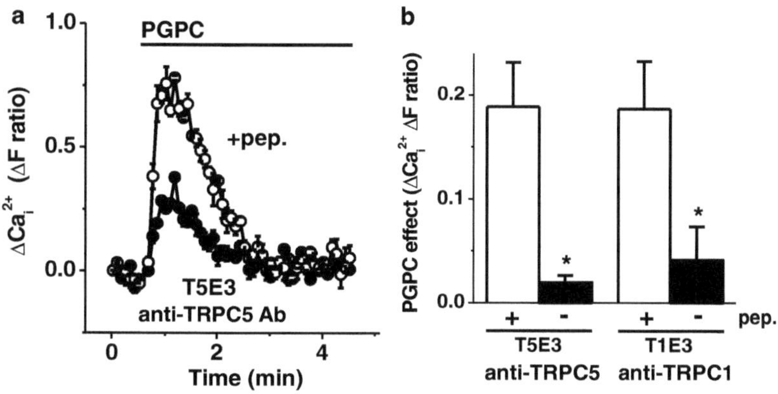

Fig. 13.3 Store-independent activation of TRPC1/5-containing channels by the oxidised phospholipid 1-palmitoyl-2-glutaryl phosphatidylcholine (PGPC) in human saphenous vein smooth muscle cells. (**a**) Ca^{2+} entry evoked by PGPC was blocked by TRPC5 inhibitor antibody (+pep. indicates the control where antibody was pre-adsorbed to its antigenic peptide so that it could not bind TRPC5). (**b**) As for (**a**) but averaged data for TRPC5 and TRPC1 antibodies, suggesting the PGPC response is mediated by heteromeric channels which contain TRPC1 and TRPC5. Adapted from [38]

Role of TRPC1-Containing Channels in Neointimal Hyperplasia of a Coronary Artery Bypass Graft

A reason for developing TRPC inhibitor antibodies was to test the functional relevance of the channels in clinical samples and thus in human disease processes. One study we performed was to test the hypothesis that TRPC1-containing channels contribute positively and importantly to the development of neointimal hyperplasia in saphenous vein segments commonly used as grafts to bypass blocked coronary arteries. There is a window of opportunity in such bypass operations to expose vein to an agent or agents and so an option might be to include a TRPC inhibitor, if the channels are indeed important players. Therefore we embarked on a challenging study to organ-culture matched vein segments with control agent or TRPC inhibitor. As the TRPC1 inhibitor we used T1E3 antibody or the chemical 2-aminophenoxydiphenylborate (2-APB). 2-APB is not specific for TRPC channels but was the best available option. Strikingly, both inhibitors suppressed the neointimal formation in these vein segments, supporting the idea that TRPC1-containing channels are important factors driving the neointimal growth and encouraging efforts to develop TRPC inhibitors that could be used in systematic trials [31] (the T1E3 data are displayed in Fig. 13.1c, d). In support of these findings, short interfering RNA targeted to TRPC1, TRPC4 and TRPC5 or dominant negative mutant TRPC5 suppressed human saphenous vein smooth muscle cell migration and proliferation [37]. Therefore we conclude that TRPC1-containing channels play a positive role in driving neointimal hyperplasia in the vein graft and that inhibitors of the channels might be useful as a way to suppress excessive, unwanted, remodelling.

Orai and SOCE in Vascular Smooth Muscle Cells

As explained above TRPC1-independent SOCE exists in human saphenous vein smooth muscle cells. In May 2006 it was published that Orai1 is a critical player in T cell CRAC channels [7] and by July 2006 we had completed experiments showing clearly that Orai1 is expressed in human saphenous vein smooth muscle cells and that its depletion suppresses SOCE. Publication proved difficult and it was not until 5 years later that the data appeared in the public domain along with data based on an impressive small-molecule CRAC channel inhibitor, as described below [40]. What was evident was that Orai1 was required for part of the SOCE in human saphenous vein smooth muscle cells. We observed about 50 % inhibition of SOCE after knockdown by either of two different short interfering RNAs targeted to Orai1 and this SOCE could be rescued by exogenous Orai1 [40] (Fig. 13.4). Over-expression of exogenous Orai1 on its own had no effect (i.e. without depletion of endogenous Orai1), suggesting critical dependence on a correct balance of Orai1 abundance with an endogenous co-factor or co-factors [40]. A dominant negative mutant of Orai1 was also inhibitory and there was still residual SOCE [40]. Depletion of Orai2 or Orai3 or Orai2 and Orai3 had no effect even though expression of these Orais was readily detectable [40]. Evidence for I_{CRAC} was sought but store-depletion in these cells caused a large non-selective cationic current [37] which could not be blocked to enable measurement of the presumably very small, if detectable, I_{CRAC}. Overall the data suggested that Orai1 explains the TRPC1-independent Ca^{2+} entry of store-depleted human saphenous vein smooth muscle cells.

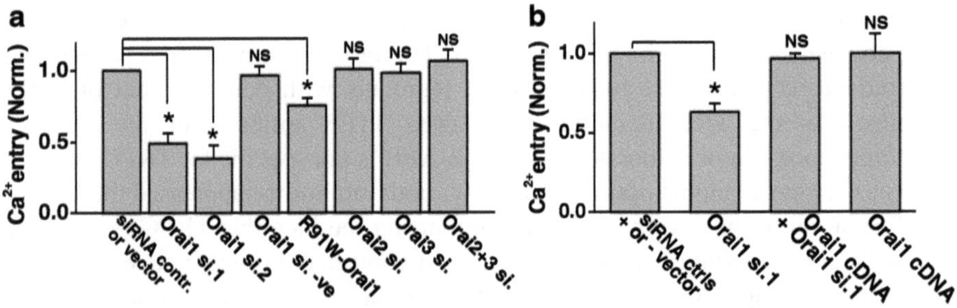

Fig. 13.4 Orai1 dependence of SOCE in human saphenous vein smooth muscle cells. The data are averaged results for paired Ca^{2+} add-back experiments normalised to their own controls. (**a**) Inhibition of SOCE by two different Orai1 siRNAs and the R91W Orai1 dominant-negative mutant. (**b**) As for (**a**) but also showing rescue of the Ca^{2+} entry by exogenous Orai1. Adapted from [40]

Growth Factor Activation of Orai1 in Vascular Smooth Muscle Cells and ER Stress

The CRAC channels generated by Orai1 are usually considered to be highly specialised channels that activate exclusively in response to store-depletion. We therefore sought physiological activators of Orai1 channels in human saphenous vein smooth muscle cells by applying the key growth factor, PDGF, which triggers inositol 1,4,5-triphosphate production and Ca^{2+} release [40]. Ca^{2+} release occurred and was independent of Orai1, as expected [40]. The Ca^{2+} entry that followed was strongly dependent on Orai1, suggesting a major role of the Orai1 channels [40] (Fig. 13.5a).

The mechanism of the PDGF effect is intriguing [1]. The widely accepted activation mechanism for Orai1 channels is that store-depletion causes clustering of STIM1 proteins in the endoplasmic reticulum (ER) membrane at sites immediately juxtapose to the inner surface of the plasma membrane [41]. This STIM1 causes Orai1 channels to cluster at the STIM1 sites and activate by direct conformational coupling with STIM1. In human saphenous vein smooth muscle cells store-depletion evoked by inhibiting Ca^{2+}-ATPase of the stores did indeed lead to clustering of Orai1 yet PDGF did not cause clustering even though it activated Orai1-dependent Ca^{2+} entry [1] (Fig. 13.5a, b). We identified that clustering was avoided by PDGF because the stores were not depleted. They were not depleted because Ca^{2+} entry through the PDGF-activated Orai1 channels maintained the stores replete, the Ca^{2+}-

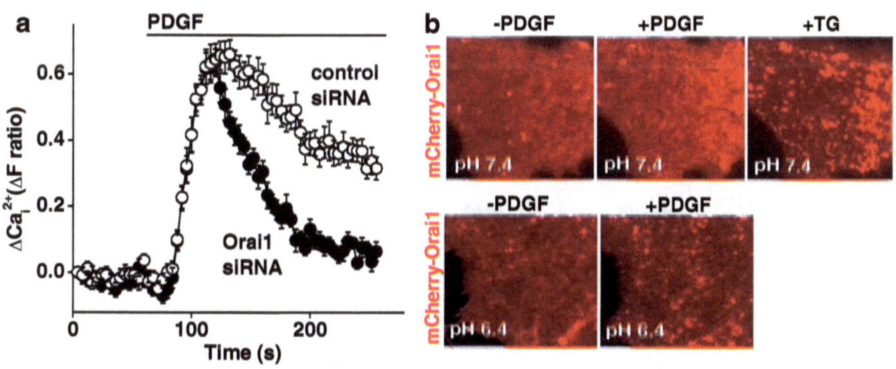

Fig. 13.5 Role of Orai1 in the response to platelet-derived growth factor (PDGF) in human saphenous vein smooth muscle cells. (**a**) PDGF-evoked Ca^{2+} elevation and inhibition of the sustained component of the response by Orai1 knockdown. (**b**) Images of red-fluorescent Orai1 construct (mCherry-Orai1) expressed in the vascular smooth muscle cells, showing that PDGF caused the Orai1 to cluster at pH 6.4 but not 7.4; by contrast thapsigargin (TG) caused clustering at pH 7.4. Adapted from [1]

ATPase remaining active [1]. We reasoned that inhibition of the Orai1 channels would thus trigger the clustering mechanism. Consistent with this hypothesis, acidification to pH 6.4, which inhibited the Orai1-dependent Ca^{2+} entry, conferred on PDGF the ability to evoke Orai1 clustering [1] (Fig. 13.5b). Therefore the widely-accepted clustering mechanism for Orai1 activation was only relevant when the ER stress signal of store-depletion occurred. Physiological activation by a physiological agonist led to Ca^{2+} release but not store-depletion (ER stress) because the stores were kept stress-free by Ca^{2+} entry through Orai1 channels and subsequent Ca^{2+} uptake into the stores by the Ca^{2+}-ATPase. This makes sense for a physiological cell, just like it makes sense for a supermarket to keep its shelves full, balanced against the removal of items by shoppers. The concepts are summarised schematically in Fig. 13.6.

The above studies with PDGF raised important questions about the biological meaning of SOCE. We suggest that SOCE reflects an ER stress response. Many investigators almost certainly deplete stores and cause ER stress in their experiments because they use Ca^{2+}-free extracellular solutions even though they are seeking to understand physiological responses. What we showed was that there can be important activation of Orai1 channels even without ER stress. It remains, however, an unanswered question how the channel activation occurs in the absence of store-depletion while serving, at least in part, to maintain stores replete.

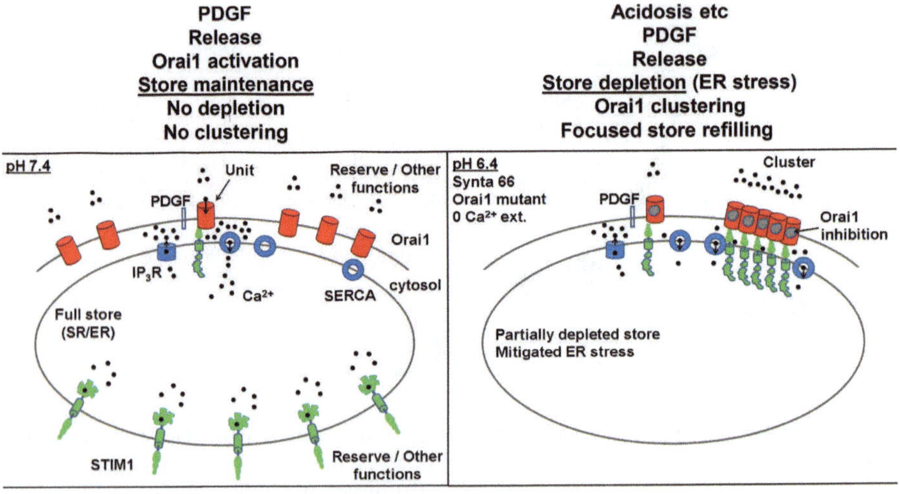

Fig. 13.6 Orai1-dependent Ca^{2+} entry evoked by platelet-derived growth factor (PDGF). Adapted from [1]

Small-Molecule Inhibitor of SOCE in Vascular Smooth Muscle Cells

Synta66 (3-fluoropyridine-4-carboxylic acid (2′,5′-dimethoxybiphenyl-4-yl)amide) is a synthetic molecule with inhibitory effect against SOCE of human saphenous vein smooth muscle cells [40]. It is potent (IC_{50} 26 nM) and selective [40]. There is a solubility limitation above 5 μM but otherwise it is easy to use. It strongly, almost completely, inhibits SOCE of human saphenous vein smooth muscle cells [40]. This is intriguing when we know that SOCE in these cells is partly due to TRPC1-containing channels and partly Orai1 channels. Does it inhibit both types of channel? It might. However, when we tested Synta66 against TRPC5, TRPC6 and TRPV4 channels it had no effect [40, 42]. In these studies the TRP channels were not activated by store-depletion, which might be critical in the effect of Synta66 even though it does not affect clustering [40].

Although Orai1 also mediates SOCE in T cells, Synta66 is almost two orders of magnitude more potent in the saphenous vein smooth muscle cells under identical conditions [40] (Fig. 13.7a). Why there is this greater potency is unclear but it may relate to the exact molecular composition of the channels, their density, or conformational state. Whatever the reason, the result suggests that it is possible to achieve selectivity for SOCE in one tissue over another. It might be possible, for example, to affect vascular remodelling without compromising physiological immunity.

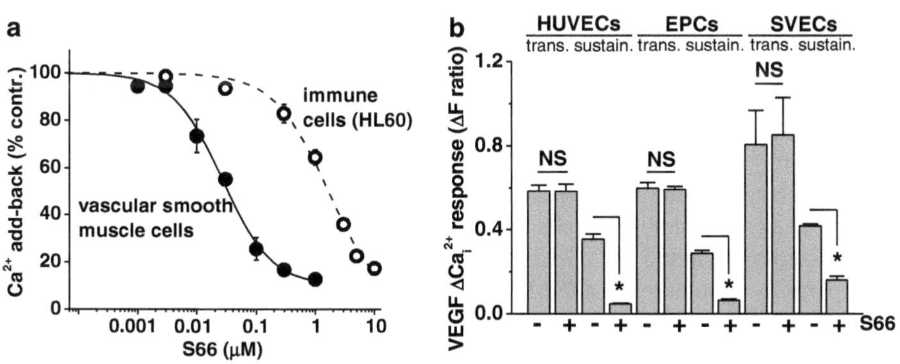

Fig. 13.7 Potent inhibition of SOCE by the Synta66 compound (S66). (**a**) Greater potency of S66 against SOCE in human saphenous vein vascular smooth muscle cells compared with SOCE in the human promyelocytic leukaemia HL 60 cell line. (**b**) Inhibition of sustained Ca^{2+} entry evoked by vascular endothelial growth factor (VEGF) in human umbilical vein endothelial cells (HUVECs), late outgrowth endothelial progenitor cells (EPCs) and human saphenous vein endothelial cells (SVECs). The initial transient (trans.) response to VEGF was unaffected. Adapted from [40, 41]

Role of Orai1 in Neointimal Hyperplasia

We have not yet investigated whether Orai1 has a role in neointimal hyperplasia in intact human saphenous vein bypass grafts because we lack an inhibitor with assured Orai1 specificity. However, Orai1 depletion inhibits migration of smooth muscle cells cultured from these grafts while having little effect on cell proliferation [40]. We would therefore expect Orai1 inhibition to suppress migration of cells into the intima but not suppress the overall hyperplasia of the intact vein. Data from balloon-injured rat carotid artery suggest marked reduction in neointima after depletion of Orai1 [43].

TRPC and SOCE in Endothelial Cells and Angiogenesis

We have not focused our experiments on the role of TRPC proteins in endothelial cells and so do not address this topic here. Numerous reports have indicated roles of TRPC proteins in SOCE of endothelial cells [12, 44–48] but there are also conflicting perspectives on this topic which we have previously described and sought to rationalise [14, 20]. Positive roles of TRPC proteins in the endothelial remodelling process of angiogenesis have been suggested [49–52].

Orai and SOCE in Endothelial Cells

We focussed initially on human umbilical vein endothelial cells (HUVECs) which are a workhorse of the endothelial field and widely used to reveal fundamental properties of endothelial cells which can then be explored further in other endothelial cells and physiological contexts. Unusually, this vein carries oxygen and nutrient-rich blood. We have also investigated late outgrowth endothelial progenitor cells from healthy human volunteers because these cells are candidate contributors to vascular repair and tumour vessel formation. As with human saphenous vein smooth muscle cells the endothelial cells exhibit a clear SOCE that is strongly and potently suppressed by Synta66 (IC_{50} 25.5 nM) [42]. Also they express Orai1 and depletion of it partially suppresses SOCE [42]. Dominant negative mutant Orai1 is also inhibitory [42]. Therefore, Orai1 is a contributor to SOCE in these endothelial cells.

Growth Factor Activation of Orai1 in Endothelial Cells

Our primary purpose in exploring Orai1 in endothelial cells was to explore its relevance to the action of vascular endothelial growth factor (VEGF) which is a major endogenous signal driving physiological and pathological angiogenesis. Although

VEGF evoked Ca^{2+} release and Ca^{2+} entry we had found that T1E3, the TRPC1 inhibitor antibody, had no effect against these events [42]. Therefore we thought there might be a similar situation to that observed for PDGF in vascular smooth muscle cells. There is similarity but also difference. Depletion of Orai1 suppressed the initial transient and the sustained later phases of the Ca^{2+} elevation evoked by VEGF [42]. Synta66 suppressed the sustained phase but not the transient [42] (Fig. 13.7b). Our studies of the sustained Ca^{2+} event showed it to be mediated by Orai1-dependent CRAC channels even though a VEGF-evoked I_{CRAC} was undetectable under physiological conditions [42]. We confirmed that sustained VEGF-evoked Ca^{2+} entry was also inhibited by Synta66 in other endothelial cells, including those from human saphenous veins of patients with coronary artery disease [42]. Therefore Synta66-sensitive and Orai1-dependent channels are part of the VEGF signalling toolbox.

Relevance of SOCE and Orai1 to VEGF-Driven Angiogenesis

Because VEGF triggers Orai1-dependent Ca^{2+} entry in endothelial cells we sought relevance to VEGF effects on endothelial cell function [42]. One of VEGF's key functions in angiogenesis is as a chemotactic agent, putatively causing endothelial cells to move towards a hypoxic core. Importantly, depletion of Orai1 by RNA interference or dominant negative mutant Orai1 suppressed migration of HUVECs towards VEGF [42]. These manipulations also inhibited the distinctive structural formations that HUVECs make on extracellular matrix, often referred to as tube formations and which are an in vitro model of angiogenesis. Exogenous Orai1 rescued these tube formations in HUVECs depleted of Orai1 by RNA interference. Synta66 similarly inhibited the tube formations while having no effect on HUVEC viability [42]. In vivo angiogenesis was evaluated by placing a VEGF-infused gelatin sponge on the chick chorioallantoic membrane. New vessel sprouting was induced by this sponge and suppressed when Synta66 was included in the sponge [42]. Therefore, VEGF can trigger angiogenesis that depends on a Synta66-sensitive mechanism and which probably involves activation of Orai1 channels.

Conclusions and Simplified Model

In summary, there is a compelling case that modestly elevated cytosolic Ca^{2+} concentrations are important and perhaps even required in some circumstances for cell growth, movement, and proliferation, which are in turn necessary for vascular remodelling in its various forms, and that plasma membrane ion channels formed by TRPC and Orai proteins play important roles in enabling the Ca^{2+} entry that elevates this Ca^{2+} in both vascular smooth muscle cells and endothelial cells. Studies of SOCE have been important in enabling identification of these channels yet channel activation by store-depletion is probably not obligatory and store-depletion may not even occur significantly especially in remodelling as part of a physiological process (i.e. without adversity). Store-depletion is undoubtedly an ER stress and the clustering process of Orai1 channels has a key role to play in a cell's efforts to minimise it. The clustering, we hypothesise, evolved as a protective mechanism against this threat to cell viability. Vascular smooth muscle cells and endothelial cells presumably require this protective mechanism (just as many other cell types do) but it seems unlikely that it is used as part of normal physiological processes. It can be generally agreed that TRPC1-containing channels can be activated without store-depletion and we know many mechanisms that can provide this activation. It is certainly not widely conceived that Orai1 CRAC channels open without store-depletion but our data clearly support this idea for responses to the key growth factor PDGF; it is also easy to understand that stores will readily refill via Ca^{2+}-ATPase activity. There is evidence not only from our work but also other work that stores do not deplete significantly when a physiological agonist evokes Ca^{2+} release [53].

A working model is shown in simplified form in Fig. 13.8. Growth factors such as PDGF and VEGF commonly stimulate Orai1 channels, without causing store-depletion or clustering, to maintain stored Ca^{2+} and elevate cytosolic Ca^{2+} which drives cell migration. An increasingly stressful environment triggers the generation of a cocktail which includes factors such as oxidised phospholipids which activate TRPC1-containing channels which in turn further elevates the cytosolic Ca^{2+} concentration and the Na^+ concentration which elevates Ca^{2+} indirectly, for example via Na^+-Ca^{2+} exchange, to drive both cell migration and proliferation. Further strain on the cell may lead to store-depletion and thus ER stress, triggering the clustering phenomenon, which involves not only STIM1 and Orai1 but also TRPC1 [54] and other proteins with the purpose of focussing delivery of Ca^{2+} into the stores at specialised junction sites.

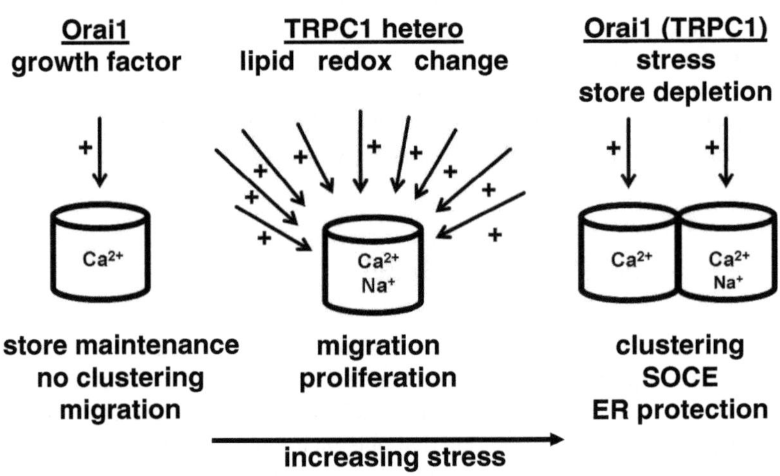

Fig. 13.8 Model for Orai1 channels and TRPC1-containing channels in vascular remodelling. See text for details

Future Perspectives

There remain many aspects of this topic which we do not understand sufficiently well, or even at all. We know little about the mechanisms that control the channel expression, assembly, trafficking and degradation. We do not understand how Synta66 acts and why it is so much more potent against SOCE in vascular smooth muscle cells and endothelial cells compared with immune cells. We cannot properly quantify the relative importance of TRPC/Orai compared with other Ca^{2+} entry mechanisms. We do not know the full composition of the TRPC1-heteromer or any other heteromers of this channel type; indeed we do not currently have the tools or technical capabilities to make such determinations. We have relatively little knowledge of the in vivo significance of TRPC1-containing channels and Orai1 channels in vascular remodelling during development and in its various forms in the adult, especially in disease. A recent study suggested particular importance of TRPC channels in hypercholesterolaemia [55] and this is a promising direction for further studies. There have been no studies in humans and we cannot embark on them until there are better agents with which to modulate the channels.

Small-molecule tool and drug discovery are likely to play crucial roles as we move forward. We already have Ca^{2+} antagonists (a.k.a. Ca^{2+} channel blockers) such as amlodipine which are used to suppress vascular smooth muscle contraction and thus lower peripheral resistance. These agents target the L-type (α_{1C}) voltage-gated Ca^{2+} channel and have specificity for this channel. Therefore the term "Ca^{2+} antagonist" is unfortunate because it implies that the agents block all Ca^{2+} channel types or that there are no other types of Ca^{2+} channel. It is akin to referring to beta blockers as receptor antagonists. An opportunity now is to develop new types of Ca^{2+} channel blockers that target other Ca^{2+} channel types such as TRPC1-containing channels and Orai1 channels. We predict suppression of vascular remodelling by such blockers. This could be useful to suppress clinical problems such as neointimal hyperplasia in vascular grafts and tumour angiogenesis. There would likely need to be caution in the use of such agents because normal physiological vascular remodelling might also be affected. Wound-healing, for example, might be compromised. The channels are broadly expressed and so there might be unwanted effects on other cell types and tissues. However, as our pharmacological studies with Synta66 have shown, there is possibility for a considerable window of selectivity. As greater understanding of TRPC heteromers emerges it might unfold that distinct heteromers mediate distinct effects in different cell types and so it might be possible to design small molecules that are heteromer-specific. Therefore there is interest in developing new types of Ca^{2+} channel antagonists that target other Ca^{2+} channels: we might end up referring to "α1C blockers", "TRPC1 heteromer blockers" and "Orai1 channel blockers". To reach this goal and make good use of it will almost certainly require a great deal of investment and effort. As ever, there could be problems along the way and perhaps even insurmountable obstacles, but we suggest that it will be a path worth taking.

Acknowledgements Our work in this area has been supported by research grants from the Wellcome Trust, Medical Research Council and British Heart Foundation. H.L.A. is supported by a Studentship from the British Heart Foundation.

References

1. McKeown L, et al. Platelet-derived growth factor maintains stored calcium through a nonclustering Orai1 mechanism but evokes clustering if the endoplasmic reticulum is stressed by store depletion. Circ Res. 2012;111:66–76.
2. Owens GK, Kumar MS, Wamhoff BR. Molecular regulation of vascular smooth muscle cell differentiation in development and disease. Physiol Rev. 2004;84:767–801.
3. Beech DJ. Ion channel switching and activation in smooth-muscle cells of occlusive vascular diseases. Biochem Soc Trans. 2007;35:890–4.
4. Beech DJ. Characteristics of transient receptor potential canonical calcium-permeable channels and their relevance to vascular physiology and disease. Circ J. 2013;77:570–9.
5. Abramowitz J, Birnbaumer L. Physiology and pathophysiology of canonical transient receptor potential channels. FASEB J. 2009;23:297–328.
6. Feske S. ORAI1 and STIM1 deficiency in human and mice: roles of store-operated Ca^{2+} entry in the immune system and beyond. Immunol Rev. 2009;231:189–209.
7. Feske S, et al. A mutation in Orai1 causes immune deficiency by abrogating CRAC channel function. Nature. 2006;441:179–85.
8. Prakriya M, et al. Orai1 is an essential pore subunit of the CRAC channel. Nature. 2006;443:230–3.
9. McNally BA, Somasundaram A, Yamashita M, Prakriya M. Gated regulation of CRAC channel ion selectivity by STIM1. Nature. 2012;482:241–5.
10. Yang B, Gwozdz T, Dutko-Gwozdz J, Bolotina VM. Orai1 and Ca^{2+}-independent phospholipase A2 are required for store-operated Icat-SOC current, Ca^{2+} entry, and proliferation of primary vascular smooth muscle cells. Am J Physiol. 2012;302:C748–56.
11. Shuttleworth TJ. Orai channels—new insights, new ideas. J Physiol. 2012;590:4155–6.
12. Philipp S, et al. A novel capacitative calcium entry channel expressed in excitable cells. EMBO J. 1998;17:4274–82.
13. Beech DJ. TRPC1: store-operated channel and more. Pflugers Arch. 2005;451:53–60.
14. Beech DJ. Orai1 calcium channels in the vasculature. Pflugers Arch. 2012;463:635–47.
15. Damann N, Voets T, Nilius B. TRPs in our senses. Curr Biol. 2008;18:R880–9.
16. Bon RS, Beech DJ. In pursuit of small molecule chemistry for calcium-permeable non-selective TRPC channels—mirage or pot of gold? Br J Pharmacol. 2013;170:459–74.
17. Liao M, Cao E, Julius D, Cheng Y. Structure of the TRPV1 ion channel determined by electron cryo-microscopy. Nature. 2013;504:107–12.
18. Freichel M, et al. Lack of an endothelial store-operated Ca^{2+} current impairs agonist-dependent vasorelaxation in TRP4−/− mice. Nat Cell Biol. 2001;3:121–7.
19. Zeng F, et al. Human TRPC5 channel activated by a multiplicity of signals in a single cell. J Physiol. 2004;559:739–50.
20. Beech DJ. Harmony and discord in endothelial calcium entry. Circ Res. 2009;104:e22–3.
21. Xu SZ, et al. A sphingosine-1-phosphate-activated calcium channel controlling vascular smooth muscle cell motility. Circ Res. 2006;98:1381–9.
22. Xu SZ, et al. TRPC channel activation by extracellular thioredoxin. Nature. 2008;451:69–72.
23. Strubing C, Krapivinsky G, Krapivinsky L, Clapham DE. TRPC1 and TRPC5 form a novel cation channel in mammalian brain. Neuron. 2001;29:645–55.
24. Hou X, Pedi L, Diver MM, Long SB. Crystal structure of the calcium release-activated calcium channel Orai. Science. 2012;338:1308–13.

25. Thompson JL, Shuttleworth TJ. How many Orai's does it take to make a CRAC channel? Sci Rep. 2013;3:1961.
26. Grigoriev I, et al. STIM1 is a MT-plus-end-tracking protein involved in remodeling of the ER. Curr Biol. 2008;18:177–82.
27. Yuan JP, Zeng W, Huang GN, Worley PF, Muallem S. STIM1 heteromultimerizes TRPC channels to determine their function as store-operated channels. Nat Cell Biol. 2007;9:636–45.
28. Cheong A, et al. Potent suppression of vascular smooth muscle cell migration and human neointimal hyperplasia by $K_V1.3$ channel blockers. Cardiovasc Res. 2011;89:282–9.
29. Sweeney M, et al. Inhibition of endogenous TRP1 decreases capacitative Ca^{2+} entry and attenuates pulmonary artery smooth muscle cell proliferation. Am J Physiol. 2002;283:L144–55.
30. Bergdahl A, et al. Plasticity of TRPC expression in arterial smooth muscle: correlation with store-operated Ca^{2+} entry. Am J Physiol Cell Physiol. 2005;288:C872–80.
31. Kumar B, et al. Upregulated TRPC1 channel in vascular injury in vivo and its role in human neointimal hyperplasia. Circ Res. 2006;98:557–63.
32. Naylor J, Beech DJ. Generation of antibodies that are externally acting isoform-specific inhibitors of ion channels. Methods Mol Biol. 2013;998:245–56.
33. Xu SZ, Beech DJ. TrpC1 is a membrane-spanning subunit of store-operated Ca^{2+} channels in native vascular smooth muscle cells. Circ Res. 2001;88:84–7.
34. Xu SZ, et al. Generation of functional ion-channel tools by E3 targeting. Nat Biotechnol. 2005;23:1289–93.
35. Xu SZ, Boulay G, Flemming R, Beech DJ. E3-targeted anti-TRPC5 antibody inhibits store-operated calcium entry in freshly isolated pial arterioles. Am J Physiol. 2006;291:H2653–9.
36. Zeng W, et al. STIM1 gates TRPC channels, but not Orai1, by electrostatic interaction. Mol Cell. 2008;32:439–48.
37. Li J, et al. Interactions, functions, and independence of plasma membrane STIM1 and TRPC1 in vascular smooth muscle cells. Circ Res. 2008;103:e97–104.
38. Al-Shawaf E, et al. Short-term stimulation of calcium-permeable transient receptor potential canonical 5-containing channels by oxidized phospholipids. Arterioscler Thromb Vasc Biol. 2010;30:1453–9.
39. Berliner JA, Watson AD. A role for oxidized phospholipids in atherosclerosis. N Engl J Med. 2005;353:9–11.
40. Li J, et al. Nanomolar potency and selectivity of a Ca^{2+} release-activated Ca^{2+} channel inhibitor against store-operated Ca^{2+} entry and migration of vascular smooth muscle cells. Br J Pharmacol. 2011;164:382–93.
41. Cahalan MD. STIMulating store-operated Ca^{2+} entry. Nat Cell Biol. 2009;11:669–77.
42. Li J, et al. Orai1 and CRAC channel dependence of VEGF-activated Ca^{2+} entry and endothelial tube formation. Circ Res. 2011;108:1190–8.
43. Zhang W, et al. Orai1-mediated I_{CRAC} is essential for neointima formation after vascular injury. Circ Res. 2011;109:534–42.
44. Groschner K, et al. TRP proteins form store-operated cation channels in human vascular endothelial cells. FEBS Lett. 1998;437:101–6.
45. Tiruppathi C, et al. Impairment of store-operated Ca^{2+} entry in TRPC4(–/–) mice interferes with increase in lung microvascular permeability. Circ Res. 2002;91:70–6.
46. Cioffi DL, et al. Activation of the endothelial store-operated I_{SOC} Ca^{2+} channel requires interaction of protein 4.1 with TRPC4. Circ Res. 2005;97:1164–72.
47. Ma X, et al. Heteromeric TRPV4-C1 channels contribute to store-operated Ca^{2+} entry in vascular endothelial cells. Cell Calcium. 2011;50:502–9.
48. Sundivakkam PC, et al. The Ca^{2+} sensor stromal interaction molecule 1 (STIM1) is necessary and sufficient for the store-operated Ca^{2+} entry function of transient receptor potential canonical (TRPC) 1 and 4 channels in endothelial cells. Mol Pharmacol. 2012;81:510–26.
49. Hamdollah Zadeh MA, Glass CA, Magnussen A, Hancox JC, Bates DO. VEGF-mediated elevated intracellular calcium and angiogenesis in human microvascular endothelial cells in vitro are inhibited by dominant negative TRPC6. Microcirculation. 2008;15:605–14.

50. Ge R, et al. Critical role of TRPC6 channels in VEGF-mediated angiogenesis. Cancer Lett. 2009;283:43–51.
51. Yu PC, Gu SY, Bu JW, Du JL. TRPC1 is essential for in vivo angiogenesis in zebrafish. Circ Res. 2010;106:1221–32.
52. Antigny F, Girardin N, Frieden M. Transient receptor potential canonical channels are required for in vitro endothelial tube formation. J Biol Chem. 2012;287:5917–27.
53. Malli R, Naghdi S, Romanin C, Graier WF. Cytosolic Ca^{2+} prevents the subplasmalemmal clustering of STIM1: an intrinsic mechanism to avoid Ca^{2+} overload. J Cell Sci. 2008;121:3133–9.
54. Cheng KT, Liu X, Ong HL, Swaim W, Ambudkar IS. Local Ca^{2+} entry via Orai1 regulates plasma membrane recruitment of TRPC1 and controls cytosolic Ca^{2+} signals required for specific cell functions. PLoS Biol. 2011;9:e1001025.
55. Rosenbaum MA, Chaudhuri P, Graham LM. Hypercholesterolemia inhibits re-endothelialization of arterial injuries by TRPC channel activation. J Vasc Surg. 2014;62(4):1040–7. S0741-5214(14)00834-9.

Chapter 14
Smooth Muscle Cell Ion Channels in Pulmonary Arterial Hypertension: Pathogenic Role in Pulmonary Vasoconstriction and Vascular Remodeling

Ramon J. Ayon, Haiyang Tang, Ruby A. Fernandez, Ayako Makino, and Jason X.-J. Yuan

Abstract Pulmonary arterial hypertension (PAH) is a progressive hemodynamic disease that impacts right heart function ultimately resulting in mortality due to right heart failure. Increased pulmonary vascular resistance (PVR) and pulmonary arterial pressure (PAP) observed in PAH patients can be attributed in part to sustained vasoconstriction and excessive remodeling of the distal pulmonary arteries. Pulmonary vasoconstriction is a result of pulmonary artery smooth muscle cell (PASMC) contraction while pulmonary vascular remodeling is associated with increased cell proliferation and decreased apoptosis. Spanning the plasma membrane of PASMC are macromolecular pore-forming proteins known as ion channels that allow for passive distribution of ion across the membrane when open. Ion channel activity is crucial for driving critical physiological functions. Posttranscriptional regulation of ion channel expression and ion channel dysfunction in PASMCs have been linked to changes in vascular tone and the initiation of vascular remodeling. This chapter will introduce smooth muscle cell ion channels

A. Makino • J.X.-J. Yuan, M.D., Ph.D. (✉)
Division of Translational and Regenerative Medicine, The University of Arizona College of Medicine, 1295 North Martin Avenue, PO Box 210202, Tucson, AZ 85721, USA

Department of Physiology, The University of Arizona College of Medicine, Tucson, AZ 85721, USA
e-mail: jasonyuan@email.arizona.edu

R.J. Ayon • H. Tang • R.A. Fernandez
Division of Translational and Regenerative Medicine, The University of Arizona College of Medicine, 1295 North Martin Avenue, PO Box 210202, Tucson, AZ 85721, USA

Department of Medicine, The University of Arizona College of Medicine, Tucson, AZ 85721, USA

I. Levitan, A.M. Dopico (eds.), *Vascular Ion Channels in Physiology and Disease*, DOI 10.1007/978-3-319-29635-7_14

that regulate physiological function in the pulmonary vasculature and summarize their potential pathogenic contribution to the development of pulmonary vascular diseases such as PAH.

Keywords Ion channels • Pulmonary arterial hypertension • Ca^{2+} channels • Vascular smooth muscle cells • K^+ channels • Vasoconstriction • Vascular remodeling • Cell proliferation and apoptosis • Pulmonary artery

Introduction

Pulmonary Arterial Hypertension (PAH) is a progressive microvascular disease characterized by sustained elevation of PVR and PAP, which directly impacts right heart function. Conservative estimates indicate incidence of group 1 PAH is approximately 2.3 cases per million of the adult population per year in the United States [42]. Epidemiological studies have revealed a higher incident of PAH in females compared to males (~4:1 depending on the underlying disease pathology) [8, 42, 79]. In PAH, narrowing of distal pulmonary arteries (PA) increases PVR, restricting blood flow, which forces right heart compensation by increasing afterload. If left untreated, these factors can lead to right ventricular hypertrophy, right heart failure and ultimately death. Despite our ever increasing knowledge of the underlying mechanisms, and available treatment options to improve quality of life, it remains a severe disease with a poor long-term outlook for patients [60, 61]. To date there are no known therapies available that can cure PAH, therefore it is imperative to delve deeper into the pathophysiology of the disease, in order to identify potentially novel therapeutic strategies.

PAH is clinically defined by an elevated mean pulmonary arterial pressure (mPAP) of ≥25 mmHg which is diagnosed by the invasive method of right heart catheterization [55]. While noninvasive methods such as echocardiography are used to estimate PA pressure in patients with suspected PAH, pulmonary catheters can provide direct measurements of vascular pressure, CO, and calculations of PVR, thus right heart catheterization remains the gold standard for diagnosis of PAH. Measurement of pulmonary occlusion pressure or pulmonary capillary wedge pressure (≤15 mmHg) and in some cases PVR (≥3 Wood units) are also used as diagnostic tools to distinguish PAH from other forms of pulmonary hypertension [7]. PAH is divided into several categories including idiopathic PAH (IPAH), heritable PAH (HPAH), drug and toxin induced PAH, and PAH associated with other diseases (APAH). Although it is unclear whether all forms of PAH share a common genetic or pathogenic trigger, elevated PVR observed in PAH patients can be linked to four major contributors: (1) sustained vasoconstriction, (2) pulmonary vascular remodeling, (3) *in situ* thrombosis and (4) increased arterial wall stiffening.

Ion channels, which are macromolecular pore-forming proteins in the plasma membrane, play an important role in the pulmonary vasculature [97]. They allow for passive movement of ions across the membrane or between intracellular compartments to drive critical physiological functions such as maintenance and regulation of vascular tone. Ion channels also participate in key cellular processes including proliferation and apoptosis. Dysfunctional ion channel activity is associated with the development of vascular diseases such as PAH. For example, attenuated K^+ channel function was demonstrated in pulmonary artery smooth muscle cells (PASMC) from patients with PAH [154]. This chapter will provide an overview of ion channels in the pulmonary vasculature as well as examine the role of smooth muscle ion channel expression and function, with particular emphasis on their contribution to vasoconstriction and vascular remodeling, in the development of PAH.

Pathogenic Mechanisms of PAH

In PAH, blood flow through the pulmonary artery vessels become restricted. The right side of the heart has to compensate by increasing right ventricular afterload to force blood through the small arteries and arterioles of the lungs [37]. Eventually these counterbalancing mechanisms break down, causing right heart failure and death. Right heart dysfunction is usually the endpoint for patients with PAH, however the pathogenesis of the disease appears to originate from within the pulmonary vasculature. The pulmonary vascular wall is comprised of three layers: adventitia (fibroblasts and extracellular matrix), media (smooth muscle cells), and intima (endothelial cells). As mentioned previously, sustained vasoconstriction, pulmonary vascular remodeling, luminal obstruction due to *in situ* thrombosis, and increased arterial wall stiffening all contribute to the elevated PVR and decreased vessel wall compliance observed in PAH patients [59]. The process of pulmonary vascular remodeling involves excessive cell proliferation along with decreased cell apoptosis across all three layers of the pulmonary arterial wall, which leads to luminal obstruction. Sustained pulmonary vasoconstriction is believed to be an early component of the pulmonary hypertensive process [59]. Like pulmonary vascular remodeling, sustained vasoconstriction is also related to cellular dysfunction. In particular, dysfunctional endothelial cells lead to severely impaired production of vasodilators such as nitric oxide (NO), vasoactive intestinal polypeptide (VIP), and prostacyclin, in addition to enhanced production of vasoconstricting factors such as endothelin [52]. Many studies have identified several cellular and molecular mechanisms that play a pathogenic role in the development and progression of PAH [19, 48, 59, 106, 133]. There is a continuously growing body of evidence implicating ion channels among these signaling pathways in the pathogenesis of PAH.

Pulmonary Vascular Remodeling

Under normal circumstances, pulmonary arterial wall thickness is maintained by a delicate balance between cell proliferation and apoptosis of the intimal, medial and adventitial layers in the vasculature. If this balance is disrupted and the scale tips toward cell proliferation, initiation of pulmonary vascular remodeling will occur. Pulmonary vascular remodeling is a result of enhanced proliferation, decreased apoptosis, and hypertrophy of pulmonary artery smooth muscle cells (PASMC), endothelial cells (PAEC), and fibroblasts that ultimately cause occlusion of the pulmonary arterial lumen [53, 59, 114]. Medial and intimal thickening is mostly seen in medium- and small-sized arteries. Medial hypertrophy is consistently the most prominent pathologic finding in PAH. Smooth muscle cell growth as well as myointimal hypertrophy due to PASMC migration and PAEC proliferation are also implicated in this process [106]. Plexiform lesions are another example of pathological changes present in PAH. Plexiform lesions, which occur in very small arteries and arterioles, are clusters of cellular capillary formation resembling a vascular plexus present within the lumen of dilated thin-walled arteries [81]. All together these complex processes promote structural changes in the vasculature, leading to an increased PVR and PAP in PAH.

Pulmonary Vasoconstriction

Pulmonary vasoconstriction is triggered by a rise in intracellular free Ca^{2+} ($[Ca^{2+}]_i$) in PASMC. When $[Ca^{2+}]_i$ rises above basal levels, Ca^{2+} binds to calmodulin (CaM) forming the Ca^{2+}/CaM complex. This complex goes on to activate myosin light chain kinase (MLCK) which in turn phosphorylates myosin light chain. This phosphorylation increases the activity of myosin ATPase that hydrolyses ATP to release energy for subsequent cycling of the myosin cross-bridges with the actin filaments thus promoting contraction [120]. Activation of RhoA/Rho kinase (ROK) can also cause smooth muscle contraction by directly and indirectly inhibiting myosin light chain phosphatase (MLCP) activity. When MLCP is rendered inactive, dephosphorylation of myosin light chain is prevented which in turn prevents smooth muscle relaxation. In the pulmonary vasculature, hypoxia is an important regulator of vasoconstriction. Physiologically, hypoxic pulmonary vasoconstriction (HPV) is a process that functions to divert blood flow from poorly ventilated areas of the lung in order to optimize ventilated-perfusion and pulmonary gas exchange [140]. However, this process can have detrimental effects on PAP, particularly in pathological conditions. Sustained vasoconstriction is believed to be the initial trigger in pulmonary vascular diseases such as PAH. Vasoconstriction and cellular proliferation may both involve signaling processes that result in parallel intracellular events in vascular remodeling and in the development of PAH. Vasoconstriction induces elevated PAP and elastic stretch of the smooth muscle cell membrane both of which increase PASMC growth and cellular synthetic activity, leading to smooth muscle cell hypertrophy and hyperplasia [80, 97].

Contribution of Ion Flux to PASMC Membrane Potential

PASMC have been shown to actively maintain an elevated cytoplasmic K^+ concentration ($[K^+]_i \approx 140$ mM), a low cytoplasmic Na^+ concentration ($[Na^+]_i \approx 10$ mM), and a resting cytoplasmic Cl^- concentration ($[Cl^-]_i$) of approximately 50 mM. Under resting conditions, PASMC maintains a membrane potential (E_m) of −40 to −60 mV due to the relatively high permeability to K^+. However PASMC maintain a basal permeability to Cl^- and Na^+ which can explain why resting E_m is substantially more depolarized than predicted by the equilibrium potential for K^+ ($E_K \approx -85$ mV). In summary, E_m in PASMC is a function of the Na^+, K^+ and Cl^- concentration gradients across the plasma membrane and the relative ion permeability.

Functional and Molecular Classifications of Ion Channels in PASMC

Ionic flux across the membrane is not only important for cellular homeostasis but also necessary for critical physiological functions including vascular tone [38]. Ion channels are membrane-bound proteins that assemble into pore forming structures that control ion flux across the plasma membrane. Functional channels are composed of pore forming subunit(s) which are responsible for ion conductance and regulatory subunit(s) which alter activation and inactivation kinetics. Ion channels are classified according to the mechanisms that regulate gating. For example voltage gated ion channels are activated by changes in the voltage gradient across the membrane. Other classifications include ligand gated, temperature gated and mechanosensitive ion channels.

Ion channels fall within two major categories of channel based on their permeability and selectivity to charged ions: cation and anion channels. Cation channels are composed of Na^+, Ca^{2+} and K^+ channels and anion channels consist of Cl^- channels and HCO_3^- channels. This section will focus on characterizing ion channels in PASMC, in addition to describing their pathogenic role in sustained vasoconstriction and pulmonary vascular remodeling, two key mechanisms involved in the development and progression of PAH.

K^+ Channels

The membrane potential (E_m) of PASMC is an important regulator of cellular excitability, thus it plays a critical role in regulating pulmonary vascular tone. *In vitro* studies indicate that the resting membrane potential of human PASMC is approximately −40 to −60 mV [82, 103], which is close to the predicted equilibrium potential for K^+ ions (~−85 mV) [92]. The high intracellular and low extracellular K^+ concentration results in passive efflux of K^+ ions after activation of the channels, leading to membrane hyperpolarization, inhibition of VDCCs, and vasodilation. A small change in K^+ channel activity can have a substantial effect on membrane potential and thus factors targeting K^+ channels can modulate vessel tone [23]. Various K^+ channels have been identified based on their biophysical and pharmacological properties as well as their molecular correlates. The main families of K^+ channels identified in the pulmonary vasculature are (1) voltage-gated K^+ channels (K_v), (2) large-, intermediate-, and small-conductance Ca^{2+}-activated K^+ channels (BK_{Ca}, IK_{Ca}, and SK_{Ca}, respectively), (3) Inward rectifier K^+ channels (K_{ir}), (4) ATP-sensitive K^+ channels (K_{ATP}) and (5) two-pore-domain K^+ channels (K_{2P}).

Voltage-Gated K^+ Channels (K_v)

Voltage-gated K^+ channels, or K_v channels, are the most diverse and best characterized group of K^+ channels ubiquitously expressed in VSMCs [20, 101]. There are currently over 40 known K_v channel members grouped into 12 subfamilies with more than 20 K_v channel members identified in the pulmonary vasculature [51]. In PASMCs, the assembly of different α and β subunits into functional K_v channels gives rise to K_v currents with heterogeneous biophysical properties [103, 118]. Four distinct K_v channel currents have been recorded in PASMC using the whole-cell patch-clamp method: (1) rapidly activating and slowly inactivating K_v current, (2) rapidly activating and non-inactivating K_v current, (3) slowly activating and non-inactivating K_v current, and (4) rapidly activating and rapidly inactivating K_v current. K_v channels are ubiquitously expressed in animal and human PASMC (Fig. 14.1). K_v channel α- and β-subunits are found in greater abundance in resistance arteries compared to conduit arteries [3, 159]. K_v channel activity has been shown to be very important for controlling E_m and vessel tone. Pharmacological blockade of K_v channels with 4-aminopyridine (4-AP) induced vasoconstriction in isolated PA ring segments [30]. In freshly dissociated and cultured PASMCs, application of 4-AP reversibly attenuates K_v currents, causes membrane depolarization, and induces Ca^{2+} influx which leads to increased $[Ca^{2+}]_i$ [157–159]. K_v channels have also been shown to have an O_2 sensitive component as hypoxia attenuates K_v current in PASMCs [155]. O_2-induced changes in K_v current regulates vessel tone suggesting that K_v channel activity plays a crucial role in HPV [5].

Ca^{2+}-Activated K^+ Channels (K_{Ca})

Ca^{2+}-activated K^+ (K_{Ca}) channels are subcategorized on the basis of their conductance; large (BK), intermediate (IK), and small (SK). Although all three groups are activated by intracellular $[Ca^{2+}]_i$, SK and IK display greater sensitivity to $[Ca^{2+}]_i$ than BK channels. Another distinguishing feature between groups is that BK channels are highly voltage sensitive, while SK and IK are predominately voltage-independent.

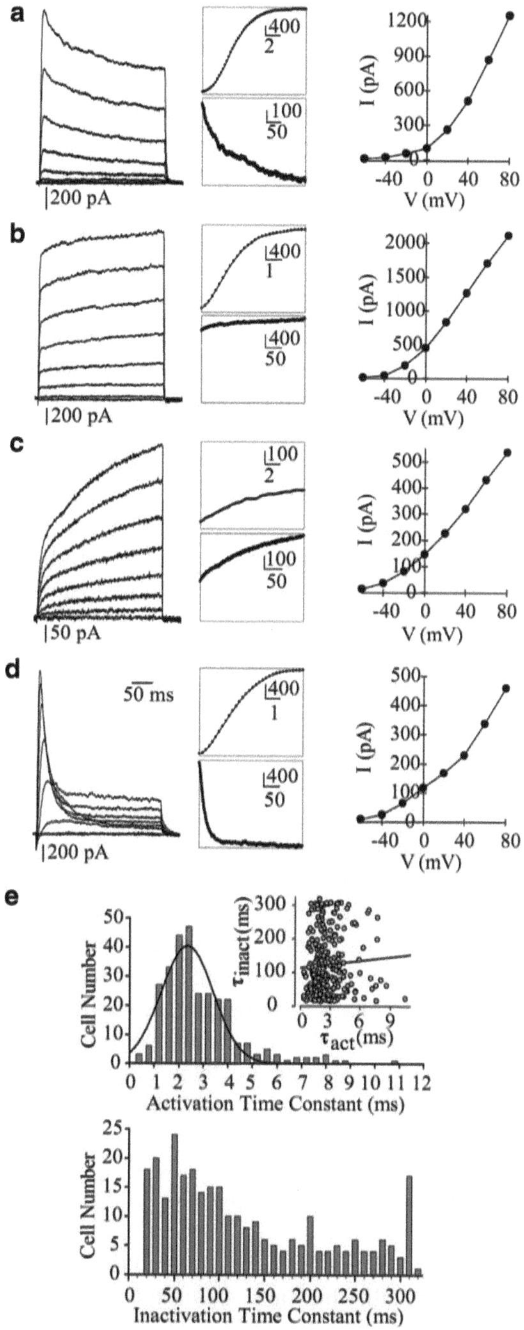

Fig. 14.1 Whole cell voltage-gated K^+ (Kv) currents in human pulmonary artery smooth muscle cells (PASMC). Four different types of K_v currents were elicited by step depolarizations from a holding potential of −70 mV to test potentials between −80 mV and +80 mV in 20 mV increments (**a–d**). Representative families of currents (*left* panels), enlarged trace segments showing steady-state activation (*middle*, *top* panels) and inactivation (*middle*, *bottom* panels) and I-V curves are presented for each type of current. Activation (*top*) and inactivation (*bottom*) time constants plotted as a function of cell number. The majority of currents are rapidly activated whereas the range of inactivation displays wider variation which is reflective of the diversity of current types found in PASMCs (**e**). Reproduced from Firth et al. 2011 [38]

Similar to K_v channels, BK channels have a transmembrane voltage sensing domain (S1–S4). Unique to the BK channel voltage sensing domain is the inclusion of an additional transmembrane segment (S0) [146]. The BK_{Ca} channel α subunit, which is encoded by a single gene KCNMA1, is ubiquitously expressed in mammalian VSMCs. The channel is regulated by the β subunit, which comprises two membrane-spanning domains separated by an extracellular loop. The Ca^{2+} sensitivity of BK_{Ca} arises from a pair of domains located on the C-terminus that regulate the conductance of K^+, known as RCK1 and RCK2 [64]. BK_{Ca} channels may act as a negative-feedback mechanism in response to vasoconstriction. Membrane depolarization leads to increased Ca^{2+} entry via VDCC, which in turn activates BK_{Ca} channels resulting in net K^+ efflux and thus attenuates vasoconstriction. Additionally, BK_{Ca} channels can be regulated by endothelial derived relaxing factor (EDRF) and endothelial derived hyperpolarizing factors (EDHF) which would suggest that BK_{Ca} channels play a role in endothelium dependent vasorelaxation [65, 164].

Although BK_{Ca} channels have been identified in pulmonary artery endothelial cells (PAEC), the physiological significance of these channels in this cell type remains unclear. However, IK_{Ca} and SK_{Ca} channels are highly expressed and have been shown to contribute to PAEC membrane hyperpolarization which in turn evokes NO synthesis, mediating NO-dependent vasorelaxation [117].

Inwardly Rectifying K^+ Channels (K_{ir})

Inwardly rectifying K^+ channels (K_{ir}) are appropriately named due to their ability to conduct inward K^+ current instead of outward K^+ current at more hyperpolarized potential than E_K. K_{ir} channels are expressed in PASMC [105, 129] and can be identified pharmacologically by their sensitivity to external Ba^{2+}, and insensitivity to several K^+ channel inhibitors (i.e. TEA, IBTX) and glibenclamide [82]. Homo- and Heterotetrameric combinations of these channel subunits form distinct functional K_{ir} channels. Generally, multimeric formations occur within the same subfamily, for example the G protein-gated K_{ir} channels (Kir3.x), Kir3.1 forms complexes with either Kir3.2, Kir3.3, or Kir3.4, although some exceptions have been reported [22, 128]. Physiologically, K_{ir} channels are thought to be involved in setting the resting E_m, preventing membrane hyperpolarization, mediating K^+-induced vasodilation and minimizing loss of intracellular K^+ [80].

ATP-Sensitive K^+ Channels (K_{ATP})

While K_{ATP} channels are considered to be a functionally distinct family of K^+ channels, in actuality they are a subfamily of the K_{ir} channels. These channels show little to no voltage dependence and have low open probability under basal conditions. Pharmacologically, K_{ATP} channels can be inhibited by intracellular ATP and glibenclamide, and activated by ADP and cromakalim. The SUR subunit 2B (SUR2B) co-assembles with Kir6.1 or Kir6.2 to produce two distinct K_{ATP} channel families, K_{NDP} (so named reflecting a primary role for various nucleotide diphosphates in their activation) and K_{ATP} (predominately sensitive to ATP), respectively [11]. Expression of both Kir 6.1 and SUR2B have been detected in human and rat PASMCs [24, 76]. Currently, the physiological role of K_{ATP} channels in vasoconstriction and vascular remodeling remains ambiguous. It has been suggested that K_{ATP} channels are normally closed under basal conditions in the pulmonary arteries and are not activated by the levels of hypoxia that cause constriction [111]. So one may assume they play a minor role in regulating pulmonary vascular tone. However, there is some evidence suggesting K_{ATP} channels are involved in regulating vessel tone as pharmacological activation of these channels causes vasodilation in preconstricted PA [35]. Additionally, it has been implied that the K_{ATP} channel may contribute to regulating E_m [24], however further studies are necessary to elucidate these assertions.

Two-Pore Domain K^+ Channels (K_{2P})

Two-pore domain K^+ (K_{2P}) channels have a unique topology compared to other K^+ channel family members. Each K_{2P} channel subunit is composed of four transmembrane segments (denoted M1–M4), two pore-forming domains (P1 and P2), and cytoplasmic N- and C-termini. It is believed that two K_{2P} α subunits are necessary to form functionally active channels. K_{2P} channels are time- and voltage-independent, and constitutively active at basal E_m [50, 96]. Multiple K_{2P} channels have been identified and subcategorized into six structural groups. Out of these groups, TASK-1, TASK-2, TREK-1, TREK-2, and TWIK-2 have been detected in the pulmonary vasculature [43, 49, 67]. TASK-1 is of particular interest as it has been detected at the transcript and protein level in PASMCs from rats, rabbits and humans. Additionally, these groups identified a membrane conductance that displayed a pharmacological profile similar to heterologously expressed TASK-1 channels [50, 96]. Initially dismissed as simply a "leak" channel, K_{2P} channels are gaining some consideration as regulators of resting E_m. TASK-1 channels are believed to play a physiological role in E_m homeostasis. TASK-1 channels are insensitive to glibenclamide, TEA, and Ca^{2+} channel blockers and display high sensitivity to external pH and the endogenous neurotransmitter anandamide. Inhibition of TASK-1 current by acid or anandamide or knockdown of TASK-1 expression by siRNA leads to significant membrane depolarization in human PASMCs providing further evidence that TASK-1 channels are involved in the regulation of E_m [96]. It is also noteworthy that TASK-1, in addition to its acid sensitivity, exhibits sensitivity to hypoxia and as such these channels are proposed to be functionally important in mechanisms underlying HPV.

Ca^{2+} Channels in PASMC

Intracellular Ca^{2+} concentration plays a critical role in the maintenance and regulation of tone and vascular remodeling in the pulmonary vasculature. Basal $[Ca^{2+}]_i$ has been reported to be approximately 100 nM which is significantly less than extracellular Ca^{2+} concentration (~1.8 mM) in PASMCs. Ca^{2+} channels are important because they facilitate Ca^{2+} entry which is necessary for excitation-contraction (E-C) coupling. It has been established there are three type of Ca^{2+} channels; these are distinguished by their activation mechanisms and pharmacology, which play a significant physiological role in PASMC. These channels are (1) the voltage-dependent Ca^{2+} channel (VDCC), (2) the receptor-operated Ca^{2+} channel (ROC) and (3) the store-operated Ca^{2+} channel (SOC).

VDCC

The open probability of VDCC is increased during membrane depolarization, thereby mediating Ca^{2+} entry and elevating $[Ca^{2+}]_i$, which serve as second messengers of electrical signaling, initiating a variety of cellular events, including smooth muscle contraction and gene expression. Differences in Ca^{2+} channel density and oxygen sensitivity in PASMC are observed using specific VDCC inhibitors and direct measurement of VDCC currents [40]. The combination of $\alpha 1$ subunits with different accessory subunits gives rise to six functionally distinct VDCC subfamilies: the L-, N-, P-, Q-, R- (high-voltage activated; HVA) and T-type (low-voltage activated; LVA) channels [18].

There is convincing evidence that the L- and T-type channels are the major players involved in capacitive Ca^{2+} entry (CCE) necessary for E-C coupling [130] and cell proliferation [68]. During E-C coupling, voltage-dependent Ca^{2+} channels (VDCC) are activated when the membrane is depolarized beyond the activation threshold [69]. L-type (or "long lasting") channels, which have been substantially examined in PASMC, are important in elevating cytosolic Ca^{2+} concentration for the purpose of promoting vasoconstriction. L-type channels are characterized by high-voltage activation ($\geq$–30 mV), sensitivity to dihydropyridine agonists and a 25-pS single channel conductance. T-type (or "transient") Ca^{2+} channels display a low voltage activation threshold (~–70 mV), insensitivity to dihydropyridine antagonists and 8-pS single channel conductance. Although expression of T-type channels has been detected in PASMC [66], their physiological role in the vasculature remains unclear. It has been argued that T-type Ca^{2+} channels do not contribute to regulating vessel tone as they are partially inactive at E_m. Recently it has been shown that T-type channels regulate cell proliferation as expression of these channels are upregulated in proliferating PASMC [112].

ROCE and SOCE

G protein coupled receptors are activated by ligand binding, leading to activation of phospholipase C (PLC) which in turn cleaves phosphoinositol bisphosphate (PIP2) into diacylglycerol (DAG) and inositol triphosphate (IP3), two very important second messengers. DAG stimulates Ca^{2+} influx, a process known as receptor-operated Ca^{2+}

entry (ROCE) via activation of receptor-operated Ca^{2+} channels (ROC). IP3 activates IP3 receptors located on ER/SR membrane which leads to Ca^{2+} release from these intracellular stores. Depletion of ER/SR Ca^{2+} stores triggers an influx of Ca^{2+} across the plasma membrane. This phenomenon is known as store-operated Ca^{2+} entry (SOCE) or capacitive Ca^{2+} entry (CCE) and occurs via activation of channels commonly termed, "store-operated Ca^{2+} channels" (SOC). In the pulmonary vasculature, the transient receptor potential (TRP) channels are capable of forming ROCs and SOCs. In addition, Ca^{2+}-release-activated Ca^{2+} channels (CRACs) have been identified as a functional Ca^{2+} entry pathway in several cell types and display a close interaction with TRP channels in smooth muscle cells.

TRP Channels

Like VDCCs, TRP channels are also part of the superfamily of six transmembrane spanning cation channels. One exception is that TRP channel lack a voltage sensor at the S4 TMD. They function as voltage-independent, nonselective cation channels that predominantly conduct Ca^{2+} ion across the plasma membrane but are also permeable to Na^{+}, K^{+}, Cs^{+}, Li^{+} and Mg^{2+}. Several subtypes of TRP channels have been identified in the pulmonary vasculature on the basis of their activation stimuli and presence of regulatory domains in the N- and C-termini; these include the classical or canonical TRP (TRPC), vanilloid-receptor-related TRP (TRPV) and melastatin-related TRP (TRPM) channels. Other TRP subfamilies, which tend to be associated with specific genetic disorders, are the polycystins (TRPP), mucolipidins (TRPML), the ankyrins (TRPA) and the mechanoreceptor potential C (TRPN).

In PASMCs more than ten TRP isoforms have been identified. Distinct domains in the N- and C-termini reflect the capability to form specific protein-protein interactions. TRPC1-TRPC7 have been detected in SMC by way of RT-PCR, Western Blot and/or immunofluorescence [9, 29]. TRPC1 is ubiquitously expressed in the vasculature and it is proposed to be one of the pore-forming subunits comprising SOCs in VSMCs [93]. TRPC1 can be assembled in a homo- or heterotetrameric (with TRPC5, TRPC3, and TRPP2) fashion to form functional voltage-independent nonselective cation channels [10]. TRPC4 is also widely expressed in PASMC [162], however its functional role is unclear,whereas in the endothelium it appears to play a more prominent role in regulating vascular permeability and vasorelaxation [41, 132]. TRPM2-TRPM8 and TRPV1-TRPV4 have all been detected at the transcript and protein level in the pulmonary artery [139, 148]. There is ever-growing evidence that TRP channels contribute to the regulation of PASMC proliferation. For example, TRPC1, TRPC4, and TRPC6 are upregulated in proliferating PASMC, and inhibition of these upregulated channels was shown to attenuate PASMC proliferation [45, 70, 125, 153, 162]. TRP channels are also implicated in the regulation of pulmonary vasoconstriction as overexpression of TRPC1 leads to enhanced CCE which is associated with increased pulmonary vasoconstriction [70].

Ca^{2+}-Release-Activated Ca^{2+} Channels

While TRPC has long been thought to contribute to SOCE, recent evidence suggests a significant role for CRAC channels in voltage-independent Ca^{2+} influx that is postulated to be a major contributor to SOCs in PASMCs. CRAC channels are comprised of stromal interaction molecule (STIM) and ORAI calcium release-activated calcium modulator (Orai) subunits. Recent studies have demonstrated that Orai-1 is the fundamental CRAC pore-forming subunit in the plasma membrane [16, 100]. ER/SR Ca^{2+} concentration is sensed by the EF-hand domain located on the N-terminus of STIM-1. When Ca^{2+} is not bound to the EF-hand domain as the case during ER/SR Ca^{2+} depletion, STIM-1 undergoes a conformation change. After this occurrence, STIM-1 oligomerizes, and is translocated along the ER/SR membrane to the ER/SR-plasma membrane junction bringing it in close proximity to the Orai-1 tetramer. This STIM-Orai interaction activates SOC and stimulates SOCE [113, 163]. The precise functional association between STIM-1 and Orai-1 is currently unknown and the mechanism of Ca^{2+} influx mediated by Orai-1 requires further investigation.

Cl^- Channels

Chloride is the most abundant intracellular and extracellular anion in multiple cell types. In smooth muscle, the intracellular concentration of Cl^- ($[Cl^-]_i$) is estimated to be approximately 30–50 mM, which is considerably higher than calculated in other cell types [21]. Cl^- ions are actively accumulated in the cell via one of three uptake mechanisms: the Na^+-K^+-$2Cl^-$ cotransporter, the Cl^-/HCO_3^- exchanger or the relatively unknown "pump III" [15, 21]. According to the Nernst equation, the equilibrium potential for Cl^- (E_{Cl}) is estimated to be around −30 to −20 mV for SMC [15, 71]. Since E_{Cl} is more positive than E_K (~−85 mV) and resting E_m (~−60 to −40 mV) in vascular SMC [71, 92], activation of Cl^- channels results in a net efflux of Cl^- ions across the electrochemical gradient, resulting in membrane depolarization, VDCC activation, elevated $[Ca^{2+}]_i$ and contraction [15, 54]. Currently there are five known Cl^- channels family members: transmembrane protein 16 (TMEM16)/anoctamin (ANO), bestrophins, voltage-gated Cl^- channels (CLC), cystic fibrosis transmembrane conductance regulator (CFTR), and ligand-gated Cl^- channels [33]. This section will focus on the first four family members as other Cl^- channels in PASMC have not been fully characterized.

Ca^{2+}-Activated Cl^- Channels (TMEM16A/ANO1)

Ca^{2+}-activated Cl^- channel (Cl_{Ca}), which is the most extensively studied Cl^- channel is ubiquitously expressed in a variety of SMC types including PASMC. The Cl^- current propagated by Cl_{Ca} channels is initiated by an elevation of $[Ca^{2+}]_i$, via Ca^{2+}

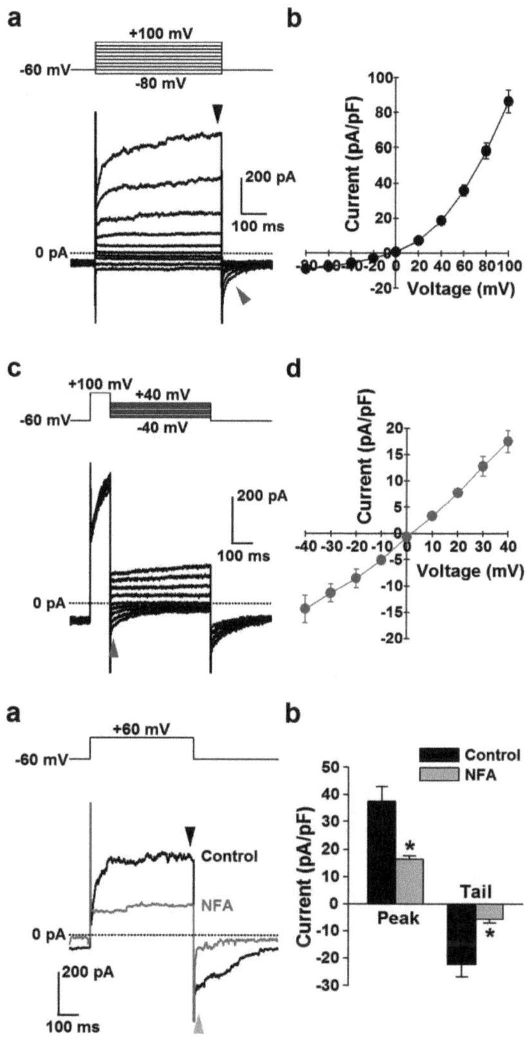

Fig. 14.2 Whole-cell Cl_{Ca} currents in human PASMCs. Cl_{Ca} currents were measured in human PASMCs. Representative outward currents (*black arrowhead*), elicited by depolarization from a holding potential of –60 mV to a series of test potentials (–80 to +100 mV) and inward tail currents (*gray arrowhead*), induced by repolarization to –60 mV in a human PASMC (**a**). I-V relationship at peak amplitude during depolarization. The current reverses at about 0 mV, the theoretical equilibrium potential of Cl^- under these experimental conditions (**b**). Representative tail currents (*gray arrowhead*) at a series of test potentials (–40 to +40 mV) for 500 ms after depolarization from a holding potential of –60 to +100 mV in a human PASMC (**c**). I-V relationship of tail currents. The reversal potential is close to 0 m V (**d**). Representative outward currents (*black arrowhead*), elicited by depolarization from a holding potential of –60 to +60 mV, and inward tail currents (*gray arrowhead*), induced by repolarization to –60 mV in the absence (*black line*) and presence (*gray line*) of 100 μM niflumic acid (NFA) in a human PASMC (**e**). Summarized data showing the effect of niflumic acid on outward and tail currents in human PASMCs (**f**). The number of cells examined is given in *parentheses*. Statistical significance versus control is indicated as $^*P<0.05$. Reproduced from Yamamura et al. 2011 [144]

influx through (1) VDCC [156], (2) reverse-mode Na^{+}/Ca^{2+} exchange (NCX) [72], (3) ROCE and SOCE [2, 144], or by agonist-induced Ca^{2+} release from intracellular stores (SR) [12, 156]. In the pulmonary artery of rabbits it has been shown that both norepinephrine and histamine can elicit Ca^{2+}-activated Cl^{-} currents ($I_{Cl(Ca)}$) [56, 138]. In rat and human PASMCs, depolarization or agonist-mediated elevation of $[Ca^{2+}]_i$ generates a time-dependent outwardly rectifying Cl^{-} current that reverses around 0 mV and is inhibited by putative Cl^{-} channel blockers, such as niflumic acid, flufenamic acid, and 4,4′-diisothiocyanatostilbene-2,2′-disulfonic acid (Fig. 14.2) [141, 144]. Cl_{Ca} channels are heterogeneous with several different characteristics; currently they can be divided into three distinct families: (1) "classical" Cl_{Ca} channels, (2) CaMKII-dependent Cl_{Ca} channels and (3) cyclic GMP-dependent Cl_{Ca} channels [58].

Although the biophysical and pharmacological properties of Cl_{Ca} channels have been extensively studied in human and animal PASMC [6, 46], the molecular correlate of Cl_{Ca} channels has proven to be challenging. Bestrophin, CLC, CLCA and Tweety have all been proposed as the Cl_{Ca} channel gene candidate. However studies with these recombinant channels failed to recapitulate the kinetic properties of the "native" SMC Cl_{Ca} current [71, 123]. Recently, three groups independently identified TMEM16A as a novel molecular candidate for Cl_{Ca} channels [17, 115, 149]. Among other vascular structures, TMEM16A channels are described at the transcript and protein level of rat and human PASMC [27, 83, 144]. In recombinant studies, TMEM16A elicits a current that displays similar Ca^{2+} dependence and I-V relationship to native Cl_{Ca} currents in vascular SMC [17, 83, 91, 116, 131].

Cl^{-} channel inhibition has also been shown to attenuate serotonin-induced membrane depolarization and agonist-induced pulmonary vasoconstriction [126, 156]. Most recently, a TMEM16A selective inhibitor was able to promote vasorelaxation in isolated blood vessels from mouse and humans [28]. These data suggest TMEM16A/Cl_{Ca} channels are involved in modulating E_m and regulating vasoconstriction. Furthermore, DIDS, a nonselective Cl^{-} channel inhibitor significantly attenuates ET-1-induced proliferation in cultured rat VSMC, while NFA and IAA-94 inhibited proliferation of PASMC from chronically hypoxic rats [142, 150], suggesting that Cl^{-} currents are important mediators of cell proliferation.

cGMP-Dependent Cl_{Ca} Channel (Bestrophin-3)

A cGMP-dependent Cl_{Ca} current has been described in a number of VSMC [84, 85]. These currents are distinct from classical Cl_{Ca} currents not only because they require cGMP for activation but also for their voltage independence and lower $[Ca^{2+}]_i$ sensitivity [85]. Additionally, they are relatively insensitive to classic Cl_{Ca} channel blockers [84]. Although the molecular identity of cGMP-dependent Cl_{Ca} channels is unclear, it is believed that members of the Bestrophin family contribute to this Cl^{-} conductance. Bestrophins are composed of six hydrophobic segments with both intracellular N- and C-termini. Bestrophins are ubiquitously expressed in a variety of tissue types, but it is Bestrophin-3 that garners the most interest. This is partially because mRNA and protein expression of Bestrophin-3 is significantly stronger in VSMC than other family member expression [86]. Bestrophin-3 has been reported to regulate rhythmic contraction and H_2O_2-induced apoptosis in rat mesenteric artery and basilar artery SMC, respectively [14, 63]. To date the physiological role of Bestrophin-3 in pulmonary artery or PASMC is unknown, therefore further studies are necessary to determine whether reported functions are shared across all vascular beds or to identify novel physiological functions in the pulmonary vasculature.

Volume-Sensitive Cl^- Channels (ClC-3)

Cell volume regulation is an important compensatory mechanism for many cellular processes including cell proliferation, differentiation and apoptosis [47]. Volume-sensitive chloride channels are ubiquitously expressed in VSMC and are believed to contribute to cell volume regulation. These channels are activated in response to cell swelling leading to Cl^- efflux, which is accompanied by passive water transport out of the cell followed by regulatory volume decrease (RVD). Yamazaki et al. demonstrated that functional volume-sensitive chloride channels were present in canine PASMC [145]. Perfusion of cells with hypotonic solutions activated an outwardly rectifying Cl^- conductance that is sensitive to inhibition by DIDS, ATP, and tamoxifen [31, 145]. ClC-3 is the only known molecular candidate for volume-sensitive Cl^- channels in SMC [15]. The structure of ClC-3 has a total of 18 α-helixes that are divided into two identical subunits which assemble homodimerically in an anti-parallel fashion [34]. There is some evidence suggesting that ClC-3 channels are involved in controlling SMC function. For example, ClC-3 knockdown inhibits ET-1-induced proliferation in rat aorta and basilar artery SMC [134]. Furthermore, ClC-3 overexpression in canine PASMC was shown to increase cell proliferation by inhibiting apoptosis [26]. While there is clear evidence that changes in ClC-3 expression alter physiological function in SMC, they are surrounded by a cloud of controversy as their contribution to the volume-sensitive Cl^- conductance has been called into question. Yamamoto-Mizuma et al. found that PASMC from ClC-3 knockout mice still produced volume-sensitive Cl^- currents [143]. More recently it has been shown that both volume-sensitive Cl^- currents and ClC-3 expression are involved in proliferation of human PASMC, however ClC-3 channels are intracellularly localized and do not contribute to the volume sensitive Cl^- flux [73].

Cystic Fibrosis Transmembrane Conductance Regulator (CFTR)

The cystic fibrosis transmembrane conductance regulator (CFTR) is a cAMP-activated Cl^- channel that was first identified in epithelial cells. CFTR consist of two membrane spanning domains each of which is composed of six transmembrane segments. CFTR expression has been detected in VSMC in rodents including rat PASMC [109, 110]. Disruption of CFTR Cl^- channel expression by homozygous knockout mice leads to increased vasoconstriction of isolated aortic rings compared to WT mice when stimulated by vasoactive agents [108]. Furthermore, pharmacological activation of CFTR induced vasorelaxation in preconstricted aortic rings from WT mice but not in KO mice [108], suggesting for the first time that CFTR Cl^- channels regulate vascular reactivity. CFTR activation is also thought to be a critical regulator of Sphingosine-1-phosphate (S1P) signaling which in itself is a central regulator of vascular tone and cellular proliferation [88]. Whether CFTR plays a similar physiological role in PASMC is still unknown and will require extensive study to elucidate their function under normal and pathological conditions.

Pathological Function of Ion Channels in PAH

Ionic conductance in vascular smooth muscle cells (VSMC) is vital for controlling intracellular Ca^{2+}, vessel constriction and thus vascular resistance to blood flow. Dysfunction of ion channels is a major cause of elevated PVR and PAP in patients with PAH. PASMC from animal models of PAH and PAH patients display a depolarized resting membrane potential and higher $[Ca^{2+}]_i$ when compared to non-PAH PASMC. This membrane depolarization is due to a decrease in K^+ channel expression and function, including Kv1.2, Kv1.5 and Kv2.1, which leads to enhanced Ca^{2+} influx due to an increase in open state probability of voltage-gated Ca^{2+} channels. Enhanced cytosolic Ca^{2+} levels are instrumental for pulmonary vasoconstriction and an important stimulus for PASMC proliferation. Changes in the expression and function of many K^+, Ca^{2+}, and Cl^- channels have been identified in experimental models of PH or IPAH and are summarized in Table 14.1.

Expression and Function of K^+ Channels in PAH

Plasmalemmal K^+ channel activity is important to maintaining the membrane potential and is also an important regulator of vascular contractility and smooth muscle cell excitability [13, 78, 94, 102, 107, 157]. Increased efflux of K^+ ion through the K^+ channels causes membrane hyperpolarization, which inhibits VDCC and results in vasodilation. Defective and downregulated K^+ channels have been observed in animal models of PAH and IPAH patients [36, 154, 160]. As a result of this downregulation, whole cell K^+ currents are reduced, which attenuates K^+ efflux and membrane depolarization ensues. Membrane depolarization increases the open probability of VDCC, which leads to enhanced Ca^{2+} influx, thereby elevating $[Ca^{2+}]_i$ in PASMC [104, 154]. This increase in $[Ca^{2+}]_i$ not only causes pulmonary vasoconstriction, but it also stimulates PASMC proliferation, which is a major contributor to the development of pulmonary vascular remodeling.

Downregulation of K_v channel activity also leads to the development of pulmonary vascular remodeling through the inhibition of apoptotic volume decrease (AVD) and apoptosis. During AVD there is enhanced K^+ and Cl^- efflux through open K^+

Table 14.1 Reported changes of ion channel expression in IPAH patents and animal models of PH

Channels	Sample type	Expression	Reference
K^+ channels			
Kv1.1	Hypoxic rat PASMC	Downregulated	[82]
Kv1.2	Hypoxic rat PASMC	Downregulated	[57, 136]
Kv1.5	Hypoxic rat PASMC; IPAH PASMC; MCT/CH rats	Downregulated	[4, 57, 87, 89, 119, 136, 154, 160]
Kv2.1	Hypoxic rat PASMC; CH rats	Downregulated	[99]
BKCa	CH rat PASMC; newborn lambs PH	Downregulated	[32]
TASK-1	Loss of function mutation in IPAH	Downregulated	[77]
Ca^{2+} channels			
TRPC1	MCT/CH rat PASMC; cigarette smoke exposed rat PASMC	Upregulated	[75, 135, 137, 151]
TRPC3	CH rat PASMC; IPAH PASMC	Upregulated	[74, 151]
TRPC4	MCT rat PASMC; lungs of newborn lambs at high altitude	Upregulated	[75, 98]
TRPC6	CH rat PASMC; cigarette smoke exposed rat PASMC; IPAH PASMC	Upregulated	[74, 75, 135, 151]
TRPV1	CH human PASMC	Upregulated	[139]
TRPV4	MCT rat PASMC; IPAH PASMC	Upregulated	[25, 122, 147]
TRPM7	IPAH PASMC	Upregulated	[122]
STIM1	PDGF stimulated human PASMC	Upregulated	[95]
STIM2	IPAH PASMC	Upregulated	[121]
Orai1	PDGF stimulated human PASMC	Upregulated	[95]
ASIC1	CH rat PASMC	Upregulated	[62]
Cl^- channels			
TMEM16A/ANO1	MCT/CH rat PASMC	Upregulated	[39, 124]
ClC-3	MCT rat PASMC	Upregulated	[26]
CLIC4	Lungs from IPAH patients	Upregulated	[1]

and Cl^- channels along the membrane. This results in a passive movement of water across the membrane through aquaporin that lead to cell shrinkage. In IPAH PASMC, BMP-induced apoptosis is inhibited, whereas overexpression of KCNA5 (Kv1.5) in normal PASMC increased whole-cell K_v currents and enhanced apoptosis [13, 161].

More recent studies have suggested a direct role for TASK-1 channels in HPV or as a pathogenic mechanisms of PAH. It has been shown that endothelin-1, a vasoconstrictive agonist known to be upregulated in PAH, inhibits TASK-1 in human PASMCs [127]. Missense variants of KCNK3, which is the gene encoding TASK-1 channel, have been identified as rare novel disease causing mutations in both hereditary pulmonary arterial hypertension (HPAH) and IPAH [44, 77]. These mutations resulted in a loss of ion channel function at physiological pH as indicated by the decreased TASK-1 currents [77], which suggests KCNK3 dysfunction may contribute to the pathogenic mechanisms of PAH.

Expression and Function of Ca^{2+} Channels in PAH

Maintenance of cytosolic Ca^{2+} concentration $[Ca^{2+}]_i$ is critical for the initiation of a variety of physiological events including contraction, migration, development, proliferation, apoptosis, and gene transcription. Functionally, TRP channels are critically important for regulating $[Ca^{2+}]_i$. In normal PASMC, proliferation is associated with upregulation of TRPC1 and TRPC6 channels and enhanced SOCE compared to growth-arrested PASMC [45, 125, 152, 153]. Inhibition of TRPC channels was shown to attenuate SOCE and proliferation in PASMC [125].

In PASMC from IPAH patients, basal $[Ca^{2+}]_i$ is greater than observed levels in PASMC from normal subjects due to upregulated TRPC6 channel expression, which stimulates increased Ca^{2+} influx (Fig. 14.3). For example, the amplitude of cyclopiazonic acid (CPA)-induced SOCE was significantly greater in IPAH-PASMC compared to normal PASMC [151, 152]. Along with increased SOCE, mRNA and protein expression of TRPC3 and TRPC6 are upregulated in PASMC from patients with IPAH compared to control patients [45]. Additionally, isoforms for the stromal interacting molecule (Stim2) and SOC channel (Orai2) families are upregulated and contribute to enhanced SOCE in IPAH-PASMC [121]. Recent studies by our group have shown that the mechanosensitive channels TRPV4 and TRMP7 are upregulated and contribute to the enhanced $[Ca^{2+}]_i$ increase induced by shear stress in PASMC from IPAH patients [122]. These results indicate that upregulation of TRP channels in IPAH-PASMC promotes greater Ca^{2+} entry, leading to enhanced $[Ca^{2+}]_i$, which ultimately contributes to sustained vasoconstriction, increased cell proliferation and decreased apoptosis.

Expression and Function of Cl^- Channels in PAH

Chloride (Cl^-) is the most abundant anion in many cell types. Cytosolic Cl^- in vascular smooth muscle have been shown to be quite high falling within a range of 51–86 mM, which according to the Nernst equation, corresponds with a reversal

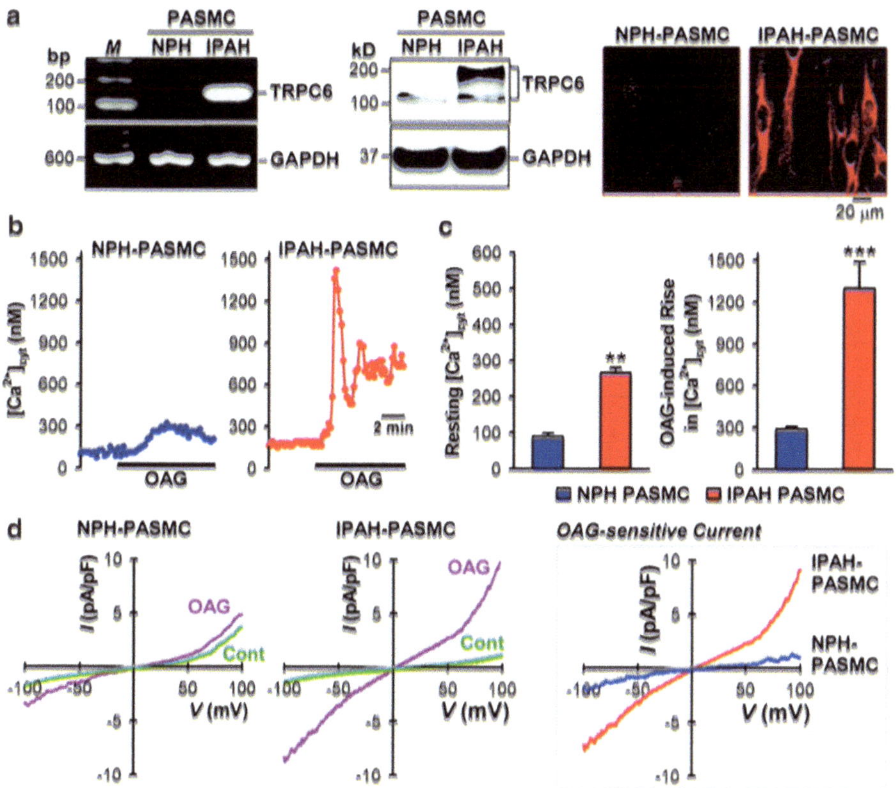

Fig. 14.3 Upregulated TRPC6 expression in PASMCs from IPAH patients increases the resting $[Ca^{2+}]_{cyt}$ and enhances agonist-mediated Ca^{2+} influx. mRNA and protein expression of TRPC6, determined by RT-PCR (*left*), Western blot (*middle*), and immunocytochemistry (*right*), respectively, is significantly higher in IPAH PASMCs than in control PASMCs from NPH patients (**a**). Representative records of $[Ca^{2+}]_{cyt}$ changes in response to OAG (100 μmol/L) in control (NPH) and IPAH PASMCs (**b**). Summarized data of the resting $[Ca^{2+}]_{cyt}$ and the amplitude of the OAG-induced increase in $[Ca^{2+}]_{cyt}$ in NPH and IPAH PASMCs. $^{**}P<0.01$, $^{***}P<0.001$ vs. NPH PASMCs (**c**). Representative currents recorded in NPH (*left*) and IPAH (*middle*) PASMCs before (Cont) and during (OAG) application with OAG (100 μmol/L). OAG-sensitive currents, generated by subtracting the currents recorded during OAG from the control currents, in NPH and IPAH PASMCs are shown in the *right* panel (**d**). Reproduced from Yu et al. 2009 [152]

potential of −25 to −12 mV. As previously mentioned, there is evidence to suggest that Cl^- efflux contributes to vasoconstriction and SMC proliferation.

In Monocrotaline (MCT)-induced PAH rats, the Cl^- channel blocker 4,4′-diisothiocyanatostilbene-2,2′-disulfonic acid (DIDS) was shown to dose-dependently inhibit the spontaneous active tone, which would suggest that this elevated muscular tone is mediated in part by activation of Cl^- channels [90]. TMEM16A transcript and protein expression along with Cl_{Ca} currents were

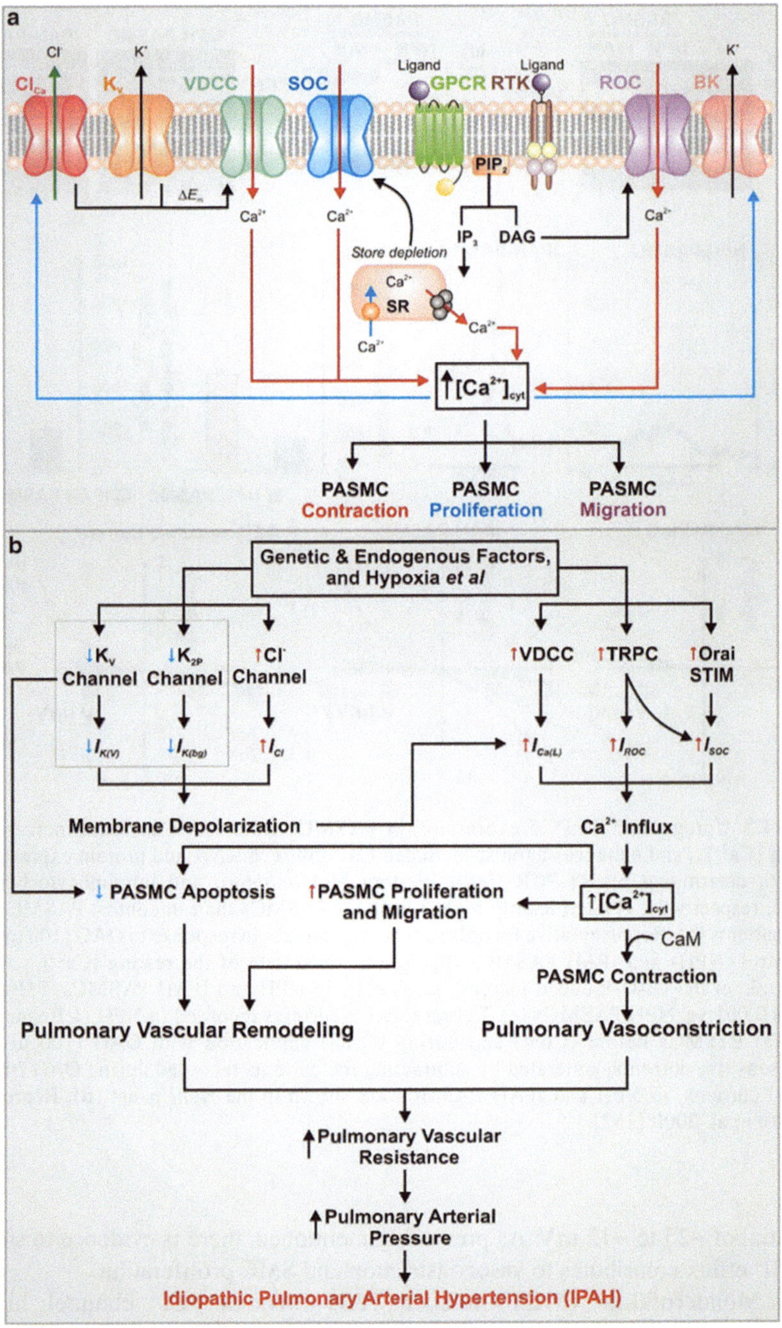

Fig. 14.4 Regulation of $[Ca^{2+}]_i$ in pulmonary artery smooth muscle cells (PASMC) and the pathogenic role of ion channels in the development of idiopathic pulmonary arterial hypertension (IPAH). Decreased K_v and increased Cl^- channel activity causes membrane depolarization which subsequently opens voltage-dependent Ca^{2+} channels (VDCC), increase Ca^{2+} influx, and

increased in PASMC from rats with chronic hypoxia and MCT-induced PH compared with their respective controls [39, 124]. Furthermore, agonist induced contractions of PA vessels from CH/MCT PH rats were enhanced versus control and shown to be sensitive to the putative Cl_{Ca} channel blockers niflumic acid and $T16A_{inh}$ suggesting increased TMEM16A/Cl_{Ca} channel activity contributes to increased and sustained pulmonary vasoconstriction in CH/MCT models of PH. In addition to TMEM16A, the volume regulated Cl^- channel ClC-3 was also found to be upregulated in PASMC from MCT-induced pulmonary hypertensive rats [26]. Overexpression of ClC-3 in PASMC inhibited apoptosis and improved resistance to reactive oxygen species, suggesting this volume-regulated Cl^- channel may also have a potential role in the pathogenesis of pulmonary hypertension [26]. While these data have provided some insight into the potential role of Cl^- channels in pulmonary vasoconstriction, further studies are required to determine whether pulmonary vascular remodeling can be regulated by Cl^- channel activity and if these findings reflect expressional and functional changes in IPAH patients.

Summary

Ion channels have a delicate role in controlling cellular homeostasis and cell excitability to ensure normal physiological function. Disruption of this balance due to changes in expression level or ion channel activity initiates pathogenic triggers such as sustained pulmonary vasoconstriction and excessive pulmonary vascular remodeling, due to increased cellular proliferation and decreased cellular apoptosis, which are key contributors to the development and progression of pulmonary vascular diseases such as PAH (Fig. 14.4). Thus, targeting ion channels may lead to potential pathways for treatment of PAH.

raises $[Ca^{2+}]_i$. Activation of G-protein coupled receptors (GPCR) and receptor tyrosine kinase (RTK) produces DAG and IP_3 which respectively activate receptor operated Ca^{2+} (ROC) and store operated Ca^{2+} (SOC) channels. Increased $[Ca^{2+}]_i$ is an important trigger for PASMC contraction, proliferation and migration (**a**). In response to unknown pathogenic and endogenous factors or hypoxia, K_V, and K_{2P} channels are downregulated while Cl_{Ca} channels are upregulated. The ensuing changes in expression causes changes to ion channel activity which leads to membrane depolarization. This membrane depolarization triggers VDCC and increases $[Ca^{2+}]_i$. Voltage dependent- and independent-Ca^{2+} channels are also upregulated and associated with enhanced Ca^{2+} entry. The augmentation of Ca^{2+} influx causes pulmonary vasoconstriction and pulmonary vascular remodeling. K_v channel also contribute to pulmonary vascular remodeling by inhibiting apoptotic volume decrease (AVD) and apoptosis in PASMC. Ultimately these changes in ion channel expression and activity can be associated with elevated pulmonary vascular resistance and pulmonary arterial pressure in patients with idiopathic pulmonary arterial hypertension (IPAH) (**b**)

References

1. Abdul-Salam VB, Wharton J, Cupitt J, Berryman M, Edwards RJ, Wilkins MR. Proteomic analysis of lung tissues from patients with pulmonary arterial hypertension. Circulation. 2010;122:2058–67.
2. Angermann JE, Forrest AS, Greenwood IA, Leblanc N. Activation of Ca^{2+}-activated Cl^- channels by store-operated Ca^{2+} entry in arterial smooth muscle cells does not require reverse-mode Na^+/Ca^{2+} exchange. Can J Physiol Pharmacol. 2012;90:903–21.
3. Archer SL, Huang JMC, Reeve HL, Hampl V, Tolarova S, Michelakis E, Weir EK. Differential distribution of electrophysiologically distinct myocytes in conduit and resistance arteries determines their response to nitric oxide and hypoxia. Circ Res. 1996;78:431–42.
4. Archer SL, Souil E, Dinh-Xuan AT, Schremmer B, Mercier JC, El Yaagoubi A, Nguyen-Huu L, Reeve HL, Hampl V. Molecular identification of the role of voltage-gated K^+ channels, Kv1.5 and Kv2.1, in hypoxic pulmonary vasoconstriction and control of resting membrane potential in rat pulmonary artery myocytes. J Clin Invest. 1998;101:2319–30.
5. Archer SL, Weir EK, Reeve HL, Michelakis E. Molecular identification of O_2 sensors and O_2-sensitive potassium channels in the pulmonary circulation. Adv Exp Med Biol. 2000;475:219–40.
6. Ayon R, Sones W, Forrest AS, Wiwchar M, Valencik ML, Sanguinetti AR, Perrino BA, Greenwood IA, Leblanc N. Complex phosphatase regulation of Ca^{2+}-activated Cl^- currents in pulmonary arterial smooth muscle cells. J Biol Chem. 2009;284:32507–21.
7. Badesch DB, Champion HC, Sanchez MA, Hoeper MM, Loyd JE, Manes A, McGoon M, Naeije R, Olschewski H, Oudiz RJ, Torbicki A. Diagnosis and assessment of pulmonary arterial hypertension. J Am Coll Cardiol. 2009;54:S55–66.
8. Badesch DB, Raskob GE, Elliott CG, Krichman AM, Farber HW, Frost AE, Barst RJ, Benza RL, Liou TG, Turner M, Giles S, Feldkircher K, Miller DP, McGoon MD. Pulmonary arterial hypertension: baseline characteristics from the REVEAL Registry. Chest. 2010;137:376–87.
9. Beech DJ, Muraki K, Flemming R. Non-selective cationic channels of smooth muscle and the mammalian homologues of Drosophila TRP. J Physiol. 2004;559:685–706.
10. Beech DJ, Xu SZ, McHugh D, Flemming R. TRPC1 store-operated cationic channel subunit. Cell Calcium. 2003;33:433–40.
11. Beech DJ, Zhang H, Nakao K, Bolton TB. K^+ channel activation by nucleotide diphosphates and its inhibition by glibenclamide in vascular smooth muscle cells. Br J Pharmacol. 1993;110:573–82.
12. Berg J, Yang H, Jan LY. Ca^{2+}-activated Cl^- channels at a glance. J Cell Sci. 2012;125:1367–71.
13. Brevnova EE, Platoshyn O, Zhang S, Yuan JX. Overexpression of human *KCNA5* increases $I_{K(V)}$ and enhances apoptosis. Am J Physiol Cell Physiol. 2004;287:C715–22.
14. Broegger T, Jacobsen JC, Secher Dam V, Boedtkjer DM, Kold-Petersen H, Pedersen FS, Aalkjaer C, Matchkov VV. Bestrophin is important for the rhythmic but not the tonic contraction in rat mesenteric small arteries. Cardiovasc Res. 2011;91:685–93.
15. Bulley S, Jaggar JH. Cl^- channels in smooth muscle cells. Pflugers Arch. 2014;466:861–72.
16. Cahalan MD. STIMulating store-operated Ca^{2+} entry. Nat Cell Biol. 2009;11:669–77.
17. Caputo A, Caci E, Ferrera L, Pedemonte N, Barsanti C, Sondo E, Pfeffer U, Ravazzolo R, Zegarra-Moran O, Galietta LJ. TMEM16A, a membrane protein associated with calcium-dependent chloride channel activity. Science. 2008;322:590–4.
18. Catterall WA, Perez-Reyes E, Snutch TP, Striessnig J. International Union of Pharmacology. XLVIII Nomenclature and structure-function relationships of voltage-gated calcium channels. Pharmacol Rev. 2005;57:411–25.
19. Chan SY, Loscalzo J. Pathogenic mechanisms of pulmonary arterial hypertension. J Mol Cell Cardiol. 2008;44:14–30.
20. Chen TT, Luykenaar KD, Walsh EJ, Walsh MP, Cole WC. Key role of Kv1 channels in vasoregulation. Circ Res. 2006;99:53–60.

21. Chipperfield AR, Harper AA. Chloride in smooth muscle. Prog Biophys Mol Biol. 2000;74:175–221.
22. Connors NC, Kofuji P. Dystrophin Dp71 is critical for the clustered localization of potassium channels in retinal glial cells. J Neurosci. 2002;22:4321–7.
23. Coppock EA, Martens JR, Tamkun MM. Molecular basis of hypoxia-induced pulmonary vasoconstriction: role of voltage-gated K^+ channels. Am J Physiol Lung Cell Mol Physiol. 2001;281:L1–12.
24. Cui Y, Tran S, Tinker A, Clapp LH. The molecular composition of K_{ATP} channels in human pulmonary artery smooth muscle cells and their modulation by growth. Am J Respir Cell Mol Biol. 2002;26:135–43.
25. Dahan D, Ducret T, Quignard JF, Marthan R, Savineau JP, Esteve E. Implication of the ryanodine receptor in TRPV4-induced calcium response in pulmonary arterial smooth muscle cells from normoxic and chronically hypoxic rats. Am J Physiol Lung Cell Mol Physiol. 2012;303:L824–33.
26. Dai YP, Bongalon S, Hatton WJ, Hume JR, Yamboliev IA. ClC-3 chloride channel is upregulated by hypertrophy and inflammation in rat and canine pulmonary artery. Br J Pharmacol. 2005;145:5–14.
27. Davis AJ, Forrest AS, Jepps TA, Valencik ML, Wiwchar M, Singer CA, Sones WR, Greenwood IA, Leblanc N. Expression profile and protein translation of TMEM16A in murine smooth muscle. Am J Physiol Cell Physiol. 2010;299:C948–59.
28. Davis AJ, Shi J, Pritchard HA, Chadha PS, Leblanc N, Vasilikostas G, Yao Z, Verkman AS, Albert AP, Greenwood IA. Potent vasorelaxant activity of the TMEM16A inhibitor $T16A_{inh}$-A01. Br J Pharmacol. 2013;168:773–84.
29. Dietrich A, Chubanov V, Kalwa H, Rost BR, Gudermann T. Cation channels of the transient receptor potential superfamily: their role in physiological and pathophysiological processes of smooth muscle cells. Pharmacol Ther. 2006;112:744–60.
30. Doi S, Damron DS, Ogawa K, Tanaka S, Horibe M, Murray PA. K^+ channel inhibition, calcium signaling, and vasomotor tone in canine pulmonary artery smooth muscle. Am J Physiol Lung Cell Mol Physiol. 2000;279:L242–51.
31. Duan D, Winter C, Cowley S, Hume JR, Horowitz B. Molecular identification of a volume-regulated chloride channel. Nature. 1997;390:417–21.
32. Dubuis E, Potier M, Wang R, Vandier C. Continuous inhalation of carbon monoxide attenuates hypoxic pulmonary hypertension development presumably through activation of BK_{Ca} channels. Cardiovasc Res. 2005;65:751–61.
33. Duran C, Thompson CH, Xiao Q, Hartzell HC. Chloride channels: often enigmatic, rarely predictable. Annu Rev Physiol. 2010;72:95–121.
34. Dutzler R, Campbell EB, Cadene M, Chait BT, MacKinnon R. X-ray structure of a ClC chloride channel at 3.0 Å reveals the molecular basis of anion selectivity. Nature. 2002;415:287–94.
35. Erdemli O, Hao Q, Cai B, Lippton HL, Hyman A. Influence of nifedipine, verapamil and diltiazem on pulmonary vascular resistance and vasoconstrictors in cats. Arch Int Pharmacodyn Ther. 1994;328:165–79.
36. Fike CD, Kaplowitz MR, Zhang Y, Madden JA. Voltage-gated K^+ channels at an early stage of chronic hypoxia-induced pulmonary hypertension in newborn piglets. Am J Physiol Lung Cell Mol Physiol. 2006;291:L1169–76.
37. Firth AL, Mandel J, Yuan JX. Idiopathic pulmonary arterial hypertension. Dis Model Mech. 2010;3:268–73.
38. Firth AL, Remillard CV, Platoshyn O, Fantozzi I, Ko EA, Yuan JX. Functional ion channels in human pulmonary artery smooth muscle cells: voltage-dependent cation channels. Pulm Circ. 2011;1:48–71.
39. Forrest AS, Joyce TC, Huebner ML, Ayon RJ, Wiwchar M, Joyce J, Freitas N, Davis AJ, Ye L, Duan DD, Singer CA, Valencik ML, Greenwood IA, Leblanc N. Increased TMEM16A-encoded calcium-activated chloride channel activity is associated with pulmonary hypertension. Am J Physiol Cell Physiol. 2012;303:C1229–43.

40. Franco-Obregon A, Lopez-Barneo J. Differential oxygen sensitivity of calcium channels in rabbit smooth muscle cells of conduit and resistance pulmonary arteries. J Physiol. 1996;491(Pt 2):511–8.
41. Freichel M, Suh SH, Pfeifer A, Schweig U, Trost C, Weiszgerber P, Biel M, Philipp S, Freise D, Droogmans G, Hofmann F, Flockerzi V, Nilius B. Lack of an endothelial store-operated Ca^{2+} current impairs agonist-dependent vasorelaxation in $TRP4^{-/-}$ mice. Nat Cell Biol. 2001;3:121–7.
42. Frost AE, Badesch DB, Barst RJ, Benza RL, Elliott CG, Farber HW, Krichman A, Liou TG, Raskob GE, Wason P, Feldkircher K, Turner M, McGoon MD. The changing picture of patients with pulmonary arterial hypertension in the United States: how REVEAL differs from historic and non-US Contemporary Registries. Chest. 2011;139:128–37.
43. Gardener MJ, Johnson IT, Burnham MP, Edwards G, Heagerty AM, Weston AH. Functional evidence of a role for two-pore domain potassium channels in rat mesenteric and pulmonary arteries. Br J Pharmacol. 2004;142:192–202.
44. Girerd B, Perros F, Antigny F, Humbert M, Montani D. KCNK3: new gene target for pulmonary hypertension? Expert Rev Respir Med. 2014;8:385–7.
45. Golovina VA, Platoshyn O, Bailey CL, Wang J, Limsuwan A, Sweeney M, Rubin LJ, Yuan JXJ. Upregulated TRP and enhanced capacitative Ca^{2+} entry in human pulmonary artery myocytes during proliferation. Am J Physiol Heart Circ Physiology. 2001;280:H746–55.
46. Greenwood IA, Ledoux J, Leblanc N. Differential regulation of Ca^{2+}-activated Cl^- currents in rabbit arterial and portal vein smooth muscle cells by Ca^{2+}-calmodulin-dependent kinase. J Physiol. 2001;534:395–408.
47. Guan YY, Wang GL, Zhou JG. The ClC-3 Cl^- channel in cell volume regulation, proliferation and apoptosis in vascular smooth muscle cells. Trends Pharmacol Sci. 2006;27:290–6.
48. Guignabert C, Dorfmuller P. Pathology and pathobiology of pulmonary hypertension. Semin Respir Crit Care Med. 2013;34:551–9.
49. Gurney AM, Osipenko ON, MacMillan D, Kempsill FE. Potassium channels underlying the resting potential of pulmonary artery smooth muscle cells. Clin Exp Pharmacol Physiol. 2002;29:330–3.
50. Gurney AM, Osipenko ON, MacMillan D, McFarlane KM, Tate RJ, Kempsill FE. Two-pore domain K channel, TASK-1, in pulmonary artery smooth muscle cells. Circ Res. 2003;93:957–64.
51. Gutman GA, Chandy KG, Grissmer S, Lazdunski M, McKinnon D, Pardo LA, Robertson GA, Rudy B, Sanguinetti MC, Stuhmer W, Wang X. International Union of Pharmacology. LIII Nomenclature and molecular relationships of voltage-gated potassium channels. Pharmacol Rev. 2005;57:473–508.
52. Handler CE, Coghlan G (2012) Pulmonary hypertension. In: Oxford specialist handbooks in cardiology. Oxford: Oxford University Press; 2012, p. xxv, 237 p.
53. Hassoun PM, Mouthon L, Barbera JA, Eddahibi S, Flores SC, Grimminger F, Jones PL, Maitland ML, Michelakis ED, Morrell NW, Newman JH, Rabinovitch M, Schermuly R, Stenmark KR, Voelkel NF, Yuan JX, Humbert M. Inflammation, growth factors, and pulmonary vascular remodeling. J Am Coll Cardiol. 2009;54:S10–9.
54. Hill MA, Zou H, Potocnik SJ, Meininger GA, Davis MJ. Signal transduction in smooth muscle—invited review: arteriolar smooth muscle mechanotransduction: Ca2+ signaling pathways underlying myogenic reactivity. J Appl Physiol. 2001;91:973–83.
55. Hoeper MM, Bogaard HJ, Condliffe R, Frantz R, Khanna D, Kurzyna M, Langleben D, Manes A, Satoh T, Torres F, Wilkins MR, Badesch DB. Definitions and diagnosis of pulmonary hypertension. J Am Coll Cardiol. 2013;62:D42–50.
56. Hogg RC, Wang Q, Helliwell RM, Large WA. Properties of spontaneous inward currents in rabbit pulmonary artery smooth muscle cells. Pflugers Arch. 1993;425:233–40.
57. Hong Z, Weir EK, Nelson DP, Olschewski A. Subacute hypoxia decreases voltage-activated potassium channel expression and function in pulmonary artery myocytes. Am J Respir Cell Mol Biol. 2004;31:337–43.
58. Huang F, Wong X, Jan LY. International Union of Basic and Clinical Pharmacology. LXXXV: calcium-activated chloride channels. Pharmacol Rev. 2012;64:1–15.

59. Humbert M, Morrell NW, Archer SL, Stenmark KR, MacLean MR, Lang IM, Christman BW, Weir EK, Eickelberg O, Voelkel NF, Rabinovitch M. Cellular and molecular pathobiology of pulmonary arterial hypertension. J Am Coll Cardiol. 2004;43:13S–24.
60. Humbert M, Sitbon O, Chaouat A, Bertocchi M, Habib G, Gressin V, Yaici A, Weitzenblum E, Cordier JF, Chabot F, Dromer C, Pison C, Reynaud-Gaubert M, Haloun A, Laurent M, Hachulla E, Cottin V, Degano B, Jais X, Montani D, Souza R, Simonneau G. Survival in patients with idiopathic, familial, and anorexigen-associated pulmonary arterial hypertension in the modern management era. Circulation. 2010;122:156–63.
61. Humbert M, Sitbon O, Yaïci A, Montani D, O'Callaghan DS, Jaïs X, Parent F, Savale L, Natali D, Günther S, Chaouat A, Chabot F, Cordier J-F, Habib G, Gressin V, Jing Z-C, Souza R, Simonneau G. Survival in incident and prevalent cohorts of patients with pulmonary arterial hypertension. Eur Respir J. 2010;36:549–55.
62. Jernigan NL, Herbert LM, Walker BR, Resta TC. Chronic hypoxia upregulates pulmonary arterial ASIC1: a novel mechanism of enhanced store-operated Ca^{2+} entry and receptor-dependent vasoconstriction. Am J Physiol Cell Physiol. 2012;302:C931–40.
63. Jiang L, Liu Y, Ma MM, Tang YB, Zhou JG, Guan YY. Mitochondria dependent pathway is involved in the protective effect of bestrophin-3 on hydrogen peroxide-induced apoptosis in basilar artery smooth muscle cells. Apoptosis. 2013;18:556–65.
64. Jiang Y, Pico A, Cadene M, Chait BT, MacKinnon R. Structure of the RCK domain from the E. coli K^+ channel and demonstration of its presence in the human BK channel. Neuron. 2001;29:593–601.
65. Kandilci HB, Gumusel B, Lippton H. Intermedin/adrenomedullin-2 (IMD/AM2) relaxes rat main pulmonary arterial rings via cGMP-dependent pathway: role of nitric oxide and large conductance calcium-activated potassium channels (BK(Ca)). Peptides. 2008;29:1321–8.
66. Ko EA, Wan J, Yamamura A, Zimnicka AM, Yamamura H, Yoo HY, Tang H, Smith KA, Sundivakkam PC, Zeifman A, Ayon RJ, Makino A, Yuan JX. Functional characterization of voltage-dependent Ca^{2+} channels in mouse pulmonary arterial smooth muscle cells: divergent effect of ROS. Am J Physiol Cell Physiol. 2013;304:C1042–52.
67. Koh SD, Monaghan K, Sergeant GP, Ro S, Walker RL, Sanders KM, Horowitz B. TREK-1 regulation by nitric oxide and cGMP-dependent protein kinase. An essential role in smooth muscle inhibitory neurotransmission. J Biol Chem. 2001;276:44338–46.
68. Kuga T, Kobayashi S, Hirakawa Y, Kanaide H, Takeshita A. Cell cycle-dependent expression of L- and T-type Ca^{2+} currents in rat aortic smooth muscle cells in primary culture. Circ Res. 1996;79:14–9.
69. Kuhr FK, Smith KA, Song MY, Levitan I, Yuan JX. New mechanisms of pulmonary arterial hypertension: role of Ca^{2+} signaling. Am J Physiol Heart Circ Physiol. 2012;302:H1546–62.
70. Kunichika N, Yu Y, Remillard CV, Platoshyn O, Zhang S, Yuan JX. Overexpression of TRPC1 enhances pulmonary vasoconstriction induced by capacitative Ca^{2+} entry. Am J Physiol Lung Cell Mol Physiol. 2004;287:L962–9.
71. Leblanc N, Ledoux J, Saleh S, Sanguinetti A, Angermann J, O'Driscoll K, Britton F, Perrino BA, Greenwood IA. Regulation of calcium-activated chloride channels in smooth muscle cells: a complex picture is emerging. Can J Physiol Pharmacol. 2005;83:541–56.
72. Leblanc N, Leung PM. Indirect stimulation of Ca^{2+}-activated Cl^- current by Na^+/Ca^{2+} exchange in rabbit portal vein smooth muscle. Am J Physiol Heart Circ Physiol. 1995;268:H1906–17.
73. Liang W, Huang L, Zhao D, He JZ, Sharma P, Liu J, Gramolini AO, Ward ME, Cho HC, Backx PH. Swelling-activated Cl^- currents and intracellular CLC-3 are involved in proliferation of human pulmonary artery smooth muscle cells. J Hypertens. 2014;32:318–30.
74. Lin MJ, Leung GP, Zhang WM, Yang XR, Yip KP, Tse CM, Sham JS. Chronic hypoxia-induced upregulation of store-operated and receptor-operated Ca^{2+} channels in pulmonary arterial smooth muscle cells: a novel mechanism of hypoxic pulmonary hypertension. Circ Res. 2004;95:496–505.
75. Liu XR, Zhang MF, Yang N, Liu Q, Wang RX, Cao YN, Yang XR, Sham JS, Lin MJ. Enhanced store-operated Ca^{2+} entry and TRPC channel expression in pulmonary arteries of monocrotaline-induced pulmonary hypertensive rats. Am J Physiol Cell Physiol. 2012;302:C77–87.

76. Liu Y, Zhang J, Yu L, Cao F, Rao J, Li J, Jiang C, Falck JR, Jacobs ER, Zhu D. A soluble epoxide hydrolase inhibitor—8-HUDE increases pulmonary vasoconstriction through inhibition of K_{ATP} channels. Pulm Pharmacol Ther. 2012;25:69–76.
77. Ma L, Roman-Campos D, Austin ED, Eyries M, Sampson KS, Soubrier F, Germain M, Tregouet DA, Borczuk A, Rosenzweig EB, Girerd B, Montani D, Humbert M, Loyd JE, Kass RS, Chung WK. A novel channelopathy in pulmonary arterial hypertension. N Engl J Med. 2013;369:351–61.
78. MacKinnon R. New insights into the structure and function of potassium channels. Curr Opin Neurobiol. 1991;1:14–9.
79. Mair KM, Johansen AK, Wright AF, Wallace E, MacLean MR. Pulmonary arterial hypertension: basis of sex differences in incidence and treatment response. Br J Pharmacol. 2014;171:567–79.
80. Makino A, Firth AL, Yuan JX. Endothelial and smooth muscle cell ion channels in pulmonary vasoconstriction and vascular remodeling. Compr Physiol. 2011;1:1555–602.
81. Mandegar M, Fung YC, Huang W, Remillard CV, Rubin LJ, Yuan JX. Cellular and molecular mechanisms of pulmonary vascular remodeling: role in the development of pulmonary hypertension. Microvasc Res. 2004;68:75–103.
82. Mandegar M, Remillard CV, Yuan JX. Ion channels in pulmonary arterial hypertension. Prog Cardiovasc Dis. 2002;45:81–114.
83. Manoury B, Tamuleviciute A, Tammaro P. TMEM16A/anoctamin 1 protein mediates calcium-activated chloride currents in pulmonary arterial smooth muscle cells. J Physiol. 2010;588:2305–14.
84. Matchkov VV, Aalkjaer C, Nilsson H. A cyclic GMP-dependent calcium-activated chloride current in smooth-muscle cells from rat mesenteric resistance arteries. J Gen Physiol. 2004;123:121–34.
85. Matchkov VV, Aalkjaer C, Nilsson H. Distribution of cGMP-dependent and cGMP-independent Ca^{2+}-activated Cl^- conductances in smooth muscle cells from different vascular beds and colon. Pflugers Arch. 2005;451:371–9.
86. Matchkov VV, Larsen P, Bouzinova EV, Rojek A, Boedtkjer DM, Golubinskaya V, Pedersen FS, Aalkjaer C, Nilsson H. Bestrophin-3 (vitelliform macular dystrophy 2-like 3 protein) is essential for the cGMP-dependent calcium-activated chloride conductance in vascular smooth muscle cells. Circ Res. 2008;103:864–72.
87. McMurtry MS, Bonnet S, Wu X, Dyck JR, Haromy A, Hashimoto K, Michelakis ED. Dichloroacetate prevents and reverses pulmonary hypertension by inducing pulmonary artery smooth muscle cell apoptosis. Circ Res. 2004;95:830–40.
88. Meissner A, Yang J, Kroetsch JT, Sauve M, Dax H, Momen A, Noyan-Ashraf MH, Heximer S, Husain M, Lidington D, Bolz SS. Tumor necrosis factor-alpha-mediated downregulation of the cystic fibrosis transmembrane conductance regulator drives pathological sphingosine-1-phosphate signaling in a mouse model of heart failure. Circulation. 2012;125:2739–50.
89. Michelakis ED, McMurtry MS, Wu XC, Dyck JR, Moudgil R, Hopkins TA, Lopaschuk GD, Puttagunta L, Waite R, Archer SL. Dichloroacetate, a metabolic modulator, prevents and reverses chronic hypoxic pulmonary hypertension in rats: role of increased expression and activity of voltage-gated potassium channels. Circulation. 2002;105:244–50.
90. Nakazawa H, Hori M, Murata T, Ozaki H, Karaki H. Contribution of chloride channel activation to the elevated muscular tone of the pulmonary artery in monocrotaline-induced pulmonary hypertensive rats. Jpn J Pharmacol. 2001;86:310–5.
91. Namkung W, Phuan PW, Verkman AS. TMEM16A inhibitors reveal TMEM16A as a minor component of calcium-activated chloride channel conductance in airway and intestinal epithelial cells. J Biol Chem. 2011;286:2365–74.
92. Nelson MT, Quayle JM. Physiological roles and properties of potassium channels in arterial smooth muscle. Am J Physiol. 1995;268:C799–822.
93. Ng L, Airey J, Hume J. The contribution of TRPC1 and STIM1 to capacitative Ca^{2+} entry in pulmonary artery. In: Yuan JXJ, Ward JPT, editors. Membrane receptors, channels and transporters in pulmonary circulation. New York: Humana Press; 2010. p. 123–35.
94. Noskov SY, Roux B. Ion selectivity in potassium channels. Biophys Chem. 2006;124:279–91.

95. Ogawa A, Firth AL, Smith KA, Maliakal MV, Yuan JX. PDGF enhances store-operated Ca^{2+} entry by upregulating STIM1/Orai1 via activation of Akt/mTOR in human pulmonary arterial smooth muscle cells. Am J Physiol Cell Physiol. 2012;302:C405–11.
96. Olschewski A, Li Y, Tang B, Hanze J, Eul B, Bohle RM, Wilhelm J, Morty RE, Brau ME, Weir EK, Kwapiszewska G, Klepetko W, Seeger W, Olschewski H. Impact of TASK-1 in human pulmonary artery smooth muscle cells. Circ Res. 2006;98:1072–80.
97. Olschewski A, Papp R, Nagaraj C, Olschewski H. Ion channels and transporters as therapeutic targets in the pulmonary circulation. Pharmacol Ther. 2014;144:349–68.
98. Parrau D, Ebensperger G, Herrera EA, Moraga F, Riquelme RA, Ulloa CE, Rojas RT, Silva P, Hernandez I, Ferrada J, Diaz M, Parer JT, Cabello G, Llanos AJ, Reyes RV. Store-operated channels in the pulmonary circulation of high- and low-altitude neonatal lambs. Am J Physiol Lung Cell Mol Physiol. 2013;304:L540–8.
99. Patel AJ, Lazdunski M, Honore E. Kv2.1/Kv9.3, a novel ATP-dependent delayed-rectifier K^+ channel in oxygen-sensitive pulmonary artery myocytes. EMBO J. 1997;16:6615–25.
100. Penna A, Demuro A, Yeromin AV, Zhang SL, Safrina O, Parker I, Cahalan MD. The CRAC channel consists of a tetramer formed by Stim-induced dimerization of Orai dimers. Nature. 2008;456:116–20.
101. Plane F, Johnson R, Kerr P, Wiehler W, Thorneloe K, Ishii K, Chen T, Cole W. Heteromultimeric Kv1 channels contribute to myogenic control of arterial diameter. Circ Res. 2005;96:216–24.
102. Platoshyn O, Brevnova EE, Burg ED, Yu Y, Remillard CV, Yuan JX. Acute hypoxia selectively inhibits KCNA5 channels in pulmonary artery smooth muscle cells. Am J Physiol Cell Physiol. 2006;290:C907–16.
103. Platoshyn O, Remillard CV, Fantozzi I, Mandegar M, Sison TT, Zhang S, Burg E, Yuan JX. Diversity of voltage-dependent K^+ channels in human pulmonary artery smooth muscle cells. Am J Physiol Lung Cell Mol Physiol. 2004;287:L226–38.
104. Platoshyn O, Yu Y, Golovina VA, McDaniel SS, Krick S, Li L, Wang JY, Rubin LJ, Yuan JX. Chronic hypoxia decreases K_V channel expression and function in pulmonary artery myocytes. Am J Physiol Lung Cell Mol Physiol. 2001;280:L801–12.
105. Quayle JM, Nelson MT, Standen NB. ATP-sensitive and inwardly rectifying potassium channels in smooth muscle. Physiol Rev. 1997;77:1165–232.
106. Rabinovitch M. Molecular pathogenesis of pulmonary arterial hypertension. J Clin Invest. 2012;122:4306–13.
107. Remillard CV, Tigno DD, Platoshyn O, Burg ED, Brevnova EE, Conger D, Nicholson A, Rana BK, Channick RN, Rubin LJ, O'Connor DT, Yuan JX. Function of Kv1.5 channels and genetic variations of KCNA5 in patients with idiopathic pulmonary arterial hypertension. Am J Physiol Cell Physiol. 2007;292:C1837–53.
108. Robert R, Norez C, Becq F. Disruption of CFTR chloride channel alters mechanical properties and cAMP-dependent Cl^- transport of mouse aortic smooth muscle cells. J Physiol. 2005;568:483–95.
109. Robert R, Savineau JP, Norez C, Becq F, Guibert C. Expression and function of cystic fibrosis transmembrane conductance regulator in rat intrapulmonary arteries. Eur Respir J. 2007;30:857–64.
110. Robert R, Thoreau V, Norez C, Cantereau A, Kitzis A, Mettey Y, Rogier C, Becq F. Regulation of the cystic fibrosis transmembrane conductance regulator channel by beta-adrenergic agonists and vasoactive intestinal peptide in rat smooth muscle cells and its role in vasorelaxation. J Biol Chem. 2004;279:21160–8.
111. Robertson BE, Kozlowski RZ, Nye PC. Opposing actions of tolbutamide and glibenclamide on hypoxic pulmonary vasoconstriction. Comp Biochem Physiol C. 1992;102:459–62.
112. Rodman DM, Reese K, Harral J, Fouty B, Wu S, West J, Hoedt-Miller M, Tada Y, Li KX, Cool C, Fagan K, Cribbs L. Low-voltage-activated (T-type) calcium channels control proliferation of human pulmonary artery myocytes. Circ Res. 2005;96:864–72.
113. Roos J, DiGregorio PJ, Yeromin AV, Ohlsen K, Lioudyno M, Zhang S, Safrina O, Kozak JA, Wagner SL, Cahalan MD, Velicelebi G, Stauderman KA. STIM1, an essential and conserved component of store-operated Ca^{2+} channel function. J Cell Biol. 2005;169:435–45.

114. Schermuly RT, Ghofrani HA, Wilkins MR, Grimminger F. Mechanisms of disease: pulmonary arterial hypertension. Nat Rev Cardiol. 2011;8:443–55.
115. Schroeder BC, Cheng T, Jan YN, Jan LY. Expression cloning of TMEM16A as a calcium-activated chloride channel subunit. Cell. 2008;134:1019–29.
116. Scudieri P, Sondo E, Ferrera L, Galietta LJ. The anoctamin family: TMEM16A and TMEM16B as calcium-activated chloride channels. Exp Physiol. 2012;97:177–83.
117. Sheng JZ, Braun AP. Small- and intermediate-conductance Ca^{2+}-activated K^+ channels directly control agonist-evoked nitric oxide synthesis in human vascular endothelial cells. Am J Physiol Cell Physiol. 2007;293:C458–67.
118. Smirnov SV, Beck R, Tammaro P, Ishii T, Aaronson PI. Electrophysiologically distinct smooth muscle cell subtypes in rat conduit and resistance pulmonary arteries. J Physiol. 2002;538:867–78.
119. Smirnov SV, Robertson TP, Ward JPT, Aaronson PI. Chronic hypoxia is associated with reduced delayed rectifier K^+ current in rat pulmonary artery muscle cells. Am J Physiol. 1994;266:H365–70.
120. Somlyo AP, Somlyo AV. Signal transduction and regulation in smooth muscle. Nature. 1994;372:231–6.
121. Song MY, Makino A, Yuan JX. STIM2 contributes to enhanced store-operated Ca^{2+} entry in pulmonary artery smooth muscle cells from patients with idiopathic pulmonary arterial hypertension. Pulm Circ. 2011;1:84–94.
122. Song S, Yamamura A, Yamamura H, Ayon RJ, Smith KA, Tang H, Makino A, Yuan JX. Flow shear stress enhances intracellular Ca^{2+} signaling in pulmonary artery smooth muscle cells from patients with pulmonary arterial hypertension. Am J Physiol Cell Physiol. 2014;307:C373–83.
123. Sun H, Tsunenari T, Yau KW, Nathans J. The vitelliform macular dystrophy protein defines a new family of chloride channels. Proc Natl Acad Sci U S A. 2002;99:4008–13.
124. Sun H, Xia Y, Paudel O, Yang XR, Sham JS. Chronic hypoxia-induced upregulation of Ca^{2+}-activated Cl^- channel in pulmonary arterial myocytes: a mechanism contributing to enhanced vasoreactivity. J Physiol. 2012;590:3507–21.
125. Sweeney M, Yu Y, Platoshyn O, Zhang S, McDaniel SS, Yuan JX. Inhibition of endogenous TRP1 decreases capacitative Ca^{2+} entry and attenuates pulmonary artery smooth muscle cell proliferation. Am J Physiol Lung Cell Mol Physiol. 2002;283:L144–55.
126. Takenaka T, Epstein M, Forster H, Landry DW, Iijima K, Goligorsky MS. Attenuation of endothelin effects by a chloride channel inhibitor, indanyloxyacetic acid. Am J Physiol. 1992;262:F799–806.
127. Tang B, Li Y, Nagaraj C, Morty RE, Gabor S, Stacher E, Voswinckel R, Weissmann N, Leithner K, Olschewski H, Olschewski A. Endothelin-1 inhibits background two-pore domain channel TASK-1 in primary human pulmonary artery smooth muscle cells. Am J Respir Cell Mol Biol. 2009;41:476–83.
128. Tang X, Schmidt TM, Perez-Leighton CE, Kofuji P. Inwardly rectifying potassium channel Kir4.1 is responsible for the native inward potassium conductance of satellite glial cells in sensory ganglia. Neuroscience. 2010;166:397–407.
129. Tennant BP, Cui Y, Tinker A, Clapp LH. Functional expression of inward rectifier potassium channels in cultured human pulmonary smooth muscle cells: evidence for a major role of Kir2.4 subunits. J Membr Biol. 2006;213:19–29.
130. Thakali KM, Kharade SV, Sonkusare SK, Rhee SW, Stimers JR, Rusch NJ. Intracellular Ca^{2+} silences L-type Ca^{2+} channels in mesenteric veins: mechanism of venous smooth muscle resistance to calcium channel blockers. Circ Res. 2010;106:739–47.
131. Thomas-Gatewood C, Neeb ZP, Bulley S, Adebiyi A, Bannister JP, Leo MD, Jaggar JH. TMEM16A channels generate Ca^{2+}-activated Cl^- currents in cerebral artery smooth muscle cells. Am J Physiol Heart Circ Physiol. 2011;301:H1819–27.
132. Tiruppathi C, Ahmmed GU, Vogel SM, Malik AB. Ca^{2+} signaling, TRP channels, and endothelial permeability. Microcirculation. 2006;13:693–708.

133. Tuder RM, Archer SL, Dorfmuller P, Erzurum SC, Guignabert C, Michelakis E, Rabinovitch M, Schermuly R, Stenmark KR, Morrell NW. Relevant issues in the pathology and pathobiology of pulmonary hypertension. J Am Coll Cardiol. 2013;62:D4–12.
134. Wang GL, Wang XR, Lin MJ, He H, Lan XJ, Guan YY. Deficiency in ClC-3 chloride channels prevents rat aortic smooth muscle cell proliferation. Circ Res. 2002;91:E28–32.
135. Wang J, Chen Y, Lin C, Jia J, Tian L, Yang K, Zhao L, Lai N, Jiang Q, Sun Y, Zhong N, Ran P, Lu W. Effects of chronic exposure to cigarette smoke on canonical transient receptor potential expression in rat pulmonary arterial smooth muscle. Am J Physiol Cell Physiol. 2014;306:C364–73.
136. Wang J, Juhaszova M, Rubin LJ, Yuan XJ. Hypoxia inhibits gene expression of voltage-gated K+ channel alpha subunits in pulmonary artery smooth muscle cells. J Clin Invest. 1997;100:2347–53.
137. Wang J, Weigand L, Lu W, Sylvester JT, Semenza GL, Shimoda LA. Hypoxia inducible factor 1 mediates hypoxia-induced TRPC expression and elevated intracellular Ca^{2+} in pulmonary arterial smooth muscle cells. Circ Res. 2006;98:1528–37.
138. Wang Q, Large WA. Action of histamine on single smooth muscle cells dispersed from the rabbit pulmonary artery. J Physiol. 1993;468:125–39.
139. Wang YX, Wang J, Wang C, Liu J, Shi LP, Xu M, Wang C. Functional expression of transient receptor potential vanilloid-related channels in chronically hypoxic human pulmonary arterial smooth muscle cells. J Membr Biol. 2008;223:151–9.
140. Ward JP, McMurtry IF. Mechanisms of hypoxic pulmonary vasoconstriction and their roles in pulmonary hypertension: new findings for an old problem. Curr Opin Pharmacol. 2009;9:287–96.
141. Wiwchar M, Ayon R, Greenwood IA, Leblanc N. Phosphorylation alters the pharmacology of Ca^{2+}-activated Cl^- channels in rabbit pulmonary arterial smooth muscle cells. Br J Pharmacol. 2009;158:1356–65.
142. Xiao GN, Guan YY, He H. Effects of Cl^- channel blockers on endothelin-1-induced proliferation of rat vascular smooth muscle cells. Life Sci. 2002;70:2233–41.
143. Yamamoto-Mizuma S, Wang GX, Liu LL, Schegg K, Hatton WJ, Duan D, Horowitz TL, Lamb FS, Hume JR. Altered properties of volume-sensitive osmolyte and anion channels (VSOACs) and membrane protein expression in cardiac and smooth muscle myocytes from $Clcn3^{-/-}$ mice. J Physiol. 2004;557:439–56.
144. Yamamura A, Yamamura H, Zeifman A, Yuan JX. Activity of Ca^{2+}-activated Cl^- channels contributes to regulating receptor- and store-operated Ca entry in human pulmonary artery smooth muscle cells. Pulm Circ. 2011;1:269–79.
145. Yamazaki J, Duan D, Janiak R, Kuenzli K, Horowitz B, Hume JR. Functional and molecular expression of volume-regulated chloride channels in canine vascular smooth muscle cells. J Physiol. 1998;507(Pt 3):729–36.
146. Yang H, Zhang G, Cui J. BK channels: multiple sensors, one activation gate. Front Physiol. 2015;6:29.
147. Yang XR, Lin AH, Hughes JM, Flavahan NA, Cao YN, Liedtke W, Sham JS. Upregulation of osmo-mechanosensitive TRPV4 channel facilitates chronic hypoxia-induced myogenic tone and pulmonary hypertension. Am J Physiol Lung Cell Mol Physiol. 2012;302:L555–68.
148. Yang XR, Lin MJ, McIntosh LS, Sham JS. Functional expression of transient receptor potential melastatin- and vanilloid-related channels in pulmonary arterial and aortic smooth muscle. Am J Physiol Lung Cell Mol Physiol. 2006;290:L1267–76.
149. Yang YD, Cho H, Koo JY, Tak MH, Cho Y, Shim WS, Park SP, Lee J, Lee B, Kim BM, Raouf R, Shin YK, Oh U. TMEM16A confers receptor-activated calcium-dependent chloride conductance. Nature. 2008;455:1210–5.
150. Yang Z, Zhang Z, Xu Y, Wang T, Ma D, Ye T. Effects of calcium-activated chloride channels on proliferation of pulmonary artery smooth muscle cells in rats under chronic hypoxic condition. J Biomed Res. 2008;22:39–43.

151. Yu Y, Fantozzi I, Remillard CV, Landsberg JW, Kunichika N, Platoshyn O, Tigno DD, Thistlethwaite PA, Rubin LJ, Yuan JX. Enhanced expression of transient receptor potential channels in idiopathic pulmonary arterial hypertension. Proc Natl Acad Sci U S A. 2004;101:13861–6.
152. Yu Y, Keller SH, Remillard CV, Safrina O, Nicholson A, Zhang SL, Jiang W, Vangala N, Landsberg JW, Wang JY, Thistlethwaite PA, Channick RN, Robbins IM, Loyd JE, Ghofrani HA, Grimminger F, Schermuly RT, Cahalan MD, Rubin LJ, Yuan JX. A functional single-nucleotide polymorphism in the TRPC6 gene promoter associated with idiopathic pulmonary arterial hypertension. Circulation. 2009;119:2313–22.
153. Yu Y, Sweeney M, Zhang S, Platoshyn O, Landsberg J, Rothman A, Yuan JX. PDGF stimulates pulmonary vascular smooth muscle cell proliferation by upregulating TRPC6 expression. Am J Physiol Cell Physiol. 2003;284:C316–30.
154. Yuan JX, Aldinger AM, Juhaszova M, Wang J, Conte Jr JV, Gaine SP, Orens JB, Rubin LJ. Dysfunctional voltage-gated K^+ channels in pulmonary artery smooth muscle cells of patients with primary pulmonary hypertension. Circulation. 1998;98:1400–6.
155. Yuan JX, Goldman W, Tod M, Rubin L, Blaustein M. Hypoxia reduces potassium currents in cultured rat pulmonary but not mesenteric arterial myocytes. Am J Physiol Lung Cell Mol Physiol. 1993;264:L116–23.
156. Yuan XJ. Role of calcium-activated chloride current in regulating pulmonary vasomotor tone. Am J Physiol. 1997;272:L959–68.
157. Yuan XJ. Voltage-gated K^+ currents regulate resting membrane potential and $[Ca^{2+}]_i$ in pulmonary arterial myocytes. Circ Res. 1995;77:370–8.
158. Yuan XJ, Goldman WF, Tod ML, Rubin LJ, Blaustein MP. Ionic currents in rat pulmonary and mesenteric arterial myocytes in primary culture and subculture. Am J Physiol. 1993;264:L107–15.
159. Yuan XJ, Tod ML, Rubin LJ, Blaustein MP. Hypoxic and metabolic regulation of voltage-gated K^+ channels in rat pulmonary artery smooth muscle cells. Exp Physiol. 1995;80:803–13.
160. Yuan XJ, Wang J, Juhaszova M, Gaine SP, Rubin LJ. Attenuated K^+ channel gene transcription in primary pulmonary hypertension. Lancet. 1998;351:726–7.
161. Zhang S, Fantozzi I, Tigno DD, Yi ES, Platoshyn O, Thistlethwaite PA, Kriett JM, Yung G, Rubin LJ, Yuan JX. Bone morphogenetic proteins induce apoptosis in human pulmonary vascular smooth muscle cells. Am J Physiol Lung Cell Mol Physiol. 2003;285:L740–54.
162. Zhang S, Remillard CV, Fantozzi I, Yuan JX. ATP-induced mitogenesis is mediated by cyclic AMP response element-binding protein-enhanced TRPC4 expression and activity in human pulmonary artery smooth muscle cells. Am J Physiol Cell Physiol. 2004;287:C1192–201.
163. Zhang SYL, Yu Y, Roos J, Kozak JA, Deerinck TJ, Ellisman MH, Stauderman KA, Cahalan MD. STIM1 is a Ca^{2+} sensor that activates CRAC channels and migrates from the Ca^{2+} store to the plasma membrane. Nature. 2005;437:902–5.
164. Zhao YJ, Wang J, Rubin LJ, Yuan XJ. Inhibition of K_V and K_{Ca} channels antagonizes NO-induced relaxation in pulmonary artery. Am J Physiol Heart Circ Physiol. 1997;272:H904–12.

Part IV
Ion Channel Regulation by Lipids and Channel Modifications in Metabolic Disease

Chapter 15
Physiological Roles and Cholesterol Sensitivity of Endothelial Inwardly-Rectifying K⁺ Channels: Specific Cholesterol-Protein Interactions Through Non Annular Binding Sites

Irena Levitan, Sang Joon Ahn, Ibra Fancher, and Avia Rosenhouse-Dantsker

Abstract Inwardly-rectifying K^+ channels (Kir) have been implicated to play a major role in endothelial sensation of shear stress forces and suggested to constitute a primary flow sensor. The studies of our group focused on elucidating the impact of hypercholesterolemia on endothelial Kir channels and elucidating molecular, biophysical and structural basis of cholesterol-induced Kir suppression. In this chapter, we first review briefly what is known about expression of Kir channels in different types of endothelial cells and their role in endothelial function and then discuss in detail the mechanisms of cholesterol-Kir interactions. Briefly, endothelial Kir channels are suppressed by loading the cells with cholesterol and by exposing them to atherogenic lipoproteins *in vitro* and by plasma hypercholesterolemia *in vivo*. A series of studies revealed that cholesterol interacts with the channels directly stabilizing them in a long-lived closed "silent" state and that multiple structural features of the channels are essential for conferring their cholesterol sensitivity. There is also a significant cross-talk between cholesterol, caveolin-1 and a regulatory phospholipid $PI(4,5)P_2$ in the regulation of these channels. Further studies are needed to determine the impact of cholesterol-induced suppression of Kir on endothelial function.

Keywords Potassium channels • Cholesterol • Cholesterol binding motifs • Hypercholesterolemia • Endothelial cells

---

I. Levitan (✉) • A. Rosenhouse-Dantsker
Division of Pulmonary and Critical Care, Department of Medicine, University of Illinois at Chicago, 840 South Wood Street, Chicago, IL 60612-7323, USA

Department of Pharmacology, University of Illinois at Chicago, Chicago, IL, USA
e-mail: levitan@uic.edu

S.J. Ahn • I. Fancher
Division of Pulmonary and Critical Care, Department of Medicine, University of Illinois at Chicago, 840 South Wood Street, Chicago, IL 60612-7323, USA

I. Levitan, A.M. Dopico (eds.), *Vascular Ion Channels in Physiology and Disease*, DOI 10.1007/978-3-319-29635-7_15

Introduction

Multiple studies established that dyslipidemia-induced endothelial dysfunction plays a key role in the early stage of the development of atherosclerosis (e.g. [1–3]. The mechanisms, however, that underlie dyslipidemia-induced endothelial dysfunction are still poorly understood. Our studies demonstrated that an increase in cellular cholesterol *in vitro* and plasma hypercholesterolemia *in vivo* result in strong suppression of endothelial K^+ channels belonging to a class of inwardly-rectifying K^+ channels (Kir) [4, 5]. These channels are known to play a major role in regulating resting membrane potential and excitability in a variety of cell types [6, 7] and were shown to be sensitive to fluid shear stress, a frictional force generated by blood flow [5, 8]. Furthermore, since activation of Kir channels is one of the fastest responses of endothelial cells to shear stress, it was proposed that shear stress-sensitive Kir channels may constitute a primary shear stress sensor responsible for the ability of endothelial cells to respond to their hemodynamic environment and thus play a major role in endothelial physiology [9]. Our observations that endothelial Kir channels are sensitive to plasma dyslipidemia led us to hypothesize that cholesterol-induced suppression of these channels may play a significant role in endothelial dysfunction under hypercholesterolemic conditions. In this chapter, we summarize what is currently known about the role of Kir channels in endothelial physiology and provide an in depth discussion of cholesterol interaction with these channels.

Expression of Kir Channels in Endothelial Cells

In general, endothelial cells (ECs) express two major classes of K^+ channels; inwardly-rectifying K+ (Kir) channels, which as described above, maintain stable membrane potential and are sensitive to fluid shear stress [5, 8] and Ca^{2+}-activated K^+ (K_{Ca}) channels that are sensitive to the level of intracellular Ca^{2+} [10, 11]. Among Kir channels that are divided into seven sub-families (Kir1-7) that differ in their biophysical properties and sensitivities to different regulators [6], endothelial cells express Kir2 channels, strong rectifiers with high basal open probability [12–15] and ATP-dependent Kir6 channels [16–20]. Among Ca^{2+}-sensitive K^+ channels, endothelial cells were shown to express mainly small and intermediate conductance K_{Ca} channels (SK and IK, respectively), [21–23] with some evidence for the expression of large-conductance (BK) channels as well [24, 25].

Kir2 channels or strong rectifiers were found in both conduit and microvasculature endothelial cells across multiple species. Specifically, functional Kir2 channels are expressed in human [14], bovine [4] and porcine [14] aortic endothelial cells, as well as in bovine pulmonary endothelium [26, 27], but were not found in rabbit [28] and mouse [29] aortic endothelium. Kir2 channels were also found in endothelial cells isolated from rat brain microvasculature [15, 30] and from bovine cornea [13]. Two studies also looked at the differential distribution of Kir channels between the luminal and the abluminal surfaces of the endothelium using a "stamp method" that allows to isolate endothelial cells with their abluminal side up. The study by Manabe et al. [31] revealed that in freshly-isolated cardiac endothelium functional Kir channels are expressed exclusively on the luminal side of the endothelium with no detectable expression on the subluminal side. In contrast, in cultured bovine aortic endothelial cells, there was no difference in Kir expression between the basal and the apical endothelial surfaces [32]. It is possible that these differences may be related to the hemodynamic conditions.

The subfamily of Kir2 channels consists of four members (Kir2.1-2.4) [6] but only Kir2.1 and Kir2.2 were found to be functionally expressed in endothelial cells. Our study showed that while all four of Kir2 channels are expressed in human aortic endothelial cells on the molecular level (mRNA), single-channel analysis and differential expression of specific dominant-negative Kir2.x subunits (dnKir2.1-2.4) revealed that only Kir2.1 and Kir2.2 are functionally expressed with a strong contribution of Kir2.2 channels [14]. Furthermore, our observations suggest that Kir2.1 and Kir2.2 subunits may interact in aortic endothelium. Equivalent expression of Kir2.1 and Kir2.2 channels on the mRNA levels was also found in bovine and rat brain capillary ECs but no functional analysis was performed to discriminate between relative contributions of these two channels [15, 30].

Kir6 or ATP-dependent Kir channels that are regulated by an auxiliary subunit of ATP-binding cassette (ABC) sulfonylurea receptors and thus sensitive to the metabolic state of the cells are also expressed in multiple types of endothelial cells. The two members of the Kir6 family (Kir6.1 and Kir6.2) were also found in both conduit and microvascular endothelial cells: specifically Kir6.1 and 6.2 were found in the endothelium of heart capillaries with a predominant expression of Kir6.2 [16] and in coronary conduit arteries [33]. Kir6.2 were also found in microvascular pulmonary endothelial cells where these channels were also shown to be shear stress sensitive [17, 20]. Furthermore, Chatterjee et al. demonstrated that Kir6.2 channels play an important role in endothelial response to the cessation of flow that may occur during lung ischemia.

Other Kir channels: In addition, there is molecular/histological evidence for the expression of Kir1 and G protein-coupled Kir3 channels in pulmonary vein endothelial cells but their functional role in these cells has not been studied [34].

Other K^+ channels: There are also multiple studies demonstrating the expression and functional roles of Ca^{2+}-sensitive K^+ channels in endothelial cells, particularly small-conductance (SK) channels that are expressed most abundantly in caveolin-rich domains [35] and endothelial cell gap junctions [36], and intermediate-conductance (IK) channels that are localized along endothelial projections adjacent to myoendothelial gap junctions [36, 37]. The roles of SK and IK channels in endothelial function are described in detail in the chapter by Kohler et al.

Flow Sensitivity and Functional Roles of Kir Channels in Endothelial Cells

Flow-sensitivity of endothelial Kir channels was first discovered in bovine aortic endothelial cells and this response was shown to underlie flow-induced endothelial hyperpolarization [8, 38]. It was proposed, therefore, that these channels may play a major role in endothelial responses to their hemodynamic environment [39]. Furthermore, Kir2.1 channels were shown to be sensitive to a mechanical stimulus when over-expressed in *Xenopus* oocytes [40]. Similar flow-induced activation of Kir channels was also observed later in our studies in human aortic endothelial cells [5, 41]. We also identified Kir2 channels as the dominant K^+ conductance in aortic endothelial cells under resting conditions [14]. However, due to a lack of appropriate genetic models, the evidence for a direct role of Kir2 channels in flow-induced endothelial responses is still rather sparse.

Endothelial Kir channels in control of vascular tone: Flow-induced vasodilation is a hallmark of endothelial response to flow and one of the essential vascular functions. Cooke et al. [42] were first to demonstrate that endothelial K^+ channels contribute significantly to flow-induced vasodilation by exposing pressurized rabbit iliac arteries to an array of K^+ channel blockers including Ba^{2+} that blocks Kir channels [6], as well as charybdotoxin and iberiotoxin that block Ca^{2+}-sensitive K^+ channels [43]. All blockers were shown to inhibit flow-induced vasodilatation, and it was proposed that the channel that is responsible for the flow response is a Ca^{2+}-sensitive K^+ channel [42]. The role of Kir channels was not established. A further study addressed the role of Kir channels in flow- and acetylcholine-induced vasodilation of cerebral arteries as measured by isometric force recordings of arterial rings [44]. In this study, Ba^{2+} was shown to inhibit the relaxation induced by intraluminal perfusion but not by acetylcholine suggesting that Kir channels are important for flow-induced but not for muscarinic-dependent vasodilation. In both studies, the effect of Ba^{2+} was eliminated by vessel denudation. In contrast, a recent study demonstrated that endothelial Ba^{2+}-sensitive Kir currents contribute to acetylcholine-induced vasodilation in rat mesenteric artery [45]. It was also shown that application of acetylcholine resulted in an increase in Kir currents in these cells but it was not established what sub-types of Kir channels were responsible for this effect [45]. Kir channels were also implicated in the regulation of myogenic response in renal arterioles as demonstrated by blunting the myogenic response by Ba^{2+} [46]. Interestingly, blocking Kir channels with Ba^{2+} was also shown to inhibit flow-induced Ca^{2+} response and augment flow-induced downregulation of endothelin-1 expression [47] suggesting that Kir channels may facilitate vasodilation by inhibiting endothelin-1 synthesis.

Notably, Kir channels are also implicated in the control of blood flow in human subjects. First, Dawes et al. [48] showed that infusing $BaCl_2$ (4 μM/min) into the brachial artery via a catheter to create a local increase in Ba^{2+} concentration resulted in a significant decrease in resting flow rate in the forearm of healthy subjects. Moreover, using the same approach, it was also shown that Ba^{2+} infusion inhibits both reactive and exercise-induced hyperemia (increase in blood flow) [49, 50]. Taken together, these studies suggest that Kir channels may play an important role in flow-induced vasodilation.

Endothelial Kir channels in cell proliferation and migration. Kir channels were also shown to play a role in the regulation of endothelial proliferation by basic Fibroblast Growth Factor (bFGF), one of the well-known regulators of endothelial cell proliferation [51]. Specifically, bFGF was shown to significantly increase endothelial Kir current in human umbilical vein endothelial cells, whereas exposing cells to Ba^{2+} abrogated bFGF-induced proliferation. In addition, Ba^{2+} was also shown to inhibit bFGF-induced NO release. In terms of migration, the role of Kir was tested in the wound-healing model, also in human umbilical vein ECs (HUVECs) [52]. Several ion channel blockers, including Ba^{2+} were shown to significantly inhibit the velocity of cell migration but since the effect was not specific, it is hard to determine the role of Kir in this process. Clearly, a lack of genetic models for Kir2 channels, a major class of endothelial Kir channels, presents a major constraint in determining the physiological roles of these channels in endothelial function.

Endothelial Kir channels in lung ischemia. A series of studies from Chatterjee and colleagues provided significant insights into the roles of endothelial Kir, specifically Kir6 channels, in endothelial response to the cessation of flow that occurs during lung ischemia leading to membrane depolarization and production of reactive oxygen species [17, 20, 53]. This topic is described in detail in several excellent reviews [53, 54].

Suppression of Endothelial Kir Channels by Cholesterol **In Vitro *and* In Vivo**

Endothelial Kir: As described above, since plasma hypercholesterolemia is known to play a major role in causing endothelial dysfunction and since cholesterol is a major lipid component of the plasma membrane in all mammalian cells, our studies focused on determining the impact of cholesterol on mechanosensitive endothelial ion channels. Our studies showed that endothelial Kir channels are suppressed by an increase in cellular cholesterol (Fig. 15.1a, [4]), and by the exposure to elevated levels of pro-atherogenic very low density lipoproteins (VLDL) (Fig. 15.1b, [5]). In the first study [4], bovine aortic endothelial cells (BAECs) that were either depleted of or enriched with cholesterol using the cholesterol carrier methyl-β-cyclodextrin (MβCD) resulting in significant increase or decrease of endothelial Kir current without affecting cell capacitance or biophysical characteristics of the current. In a later study, we used human aortic endothelial cells (HAECs) that also express Kir channels even though typical Kir current densities in HAECs are significantly lower than those in BAECs. In fact, most of our studies in HAECs were performed in high K^+ extracellular solution to increase Kir current density. The effect of MβCD on Kir current in BAECs and HAECs were very similar. Exposure to 5–50 μg/ml VLDL

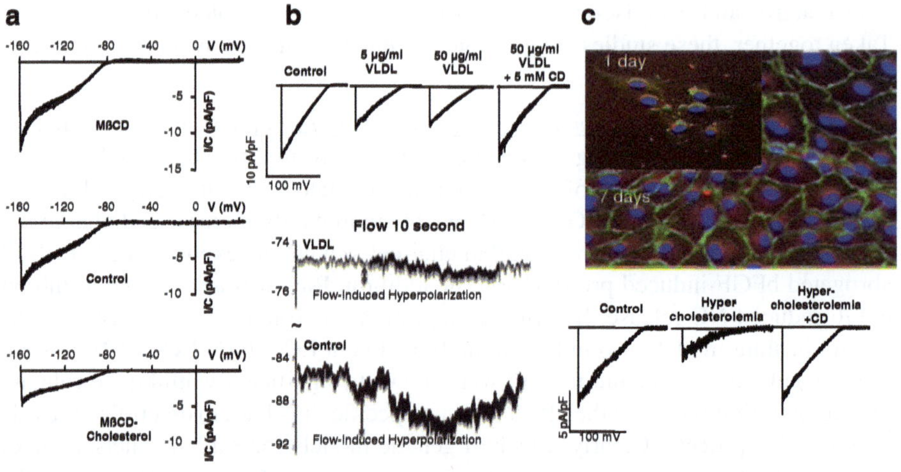

Fig. 15.1 Suppression of endothelial Kir channels by an increase in cellular cholesterol *in vitro* and *in vivo*. (**a**) Typical Kir currents recorded from BAECs exposed to 2.5 mM MβCD or MβCD-cholesterol for 1 h. Three superimposed traces are shown for each cell (recordings are performed in low K^+ solutions). **This research was originally published in Biophysical Journal** 83: p. 3211–3222. (**b**) Upper panel Typical Kir currents recorded from HAECs exposed to 5 or 50 μg/mL VLDL for 24 h or treated with 50 μg/mL VLDL for 24 h followed by 5 mM MβCD for 1 h (recordings are performed in high K^+ solutions). Lower panel Typical voltage traces recorded from a control and a VLDL-treated cells exposed to 2 dyn/cm^2 shear stress. (**c**) Upper panel Immunostaining of PAECs for PECAM (*Green*) and von-Willebrand Factor (*Red*). The staining was performed immediately after cell isolation, and repeated after cells were maintained in culture for 7 days to form a monolayer. Lower panel Representative recordings of Kir currents in freshly-isolated PAECs from control and hypercholesterolemic pigs, and in PAECs freshly-isolated from hypercholesterolemic pigs followed by a treatment of 5 mM MβCD for 1 h after isolation. **This research was originally published in Circulation Research** 2006. 98(8): p. 1064–1071

also resulted in current inhibition, which was completely reversible by MβCD, indicating that VLDL-induced suppression of Kir is mediated by an increase in the level of free cholesterol in endothelial membranes (Fig. 15.1b, upper panel). Similar effects were also observed for acetylated LDL (acLDL). Exposure to VLDL also significantly decreased flow-induced membrane hyperpolarization in aortic endothelium (Fig. 15.1b, lower panel). Most importantly, we showed that endothelial Kir channels were also suppressed by plasma dyslipidemia in a diet-induced porcine model of atherosclerosis. Endothelial cells were isolated by gentle mechanical scraping of porcine aortas immediately after animal sacrifice and harvesting of the aortas, and cells were identified by typical endothelial markers, PECAM and vWF (Fig. 15.1c, upper panel). Electrophysiological recordings were performed on freshly-isolated cells on the day of the sacrifice (Fig. 15.1c, lower panel). These recordings demonstrate that Kir currents recorded in aortic endothelial cells isolated from hypercholesterolemic animals were significantly lower than currents recorded in cells isolated from control animals. Furthermore, the effect was fully abrogated by exposing cells isolated from hypercholesterolemic animals to MβCD after the isolation [5]. These observations demonstrate that endothelial Kir channels are suppressed by plasma dyslipidemia *in vivo.*

Cholesterol sensitivity of Kir in progenitor cells and cardiomyocytes: Dyslipidemia-induced suppression of Kir channels was observed not only in mature endothelial cells but also in endothelial progenitor cells isolated from the bone marrow of hypercholesterolemic pigs [55]. This study was performed with side-population cells, a sub-type of progenitor cells that can differentiate into endothelial cells [56, 57]. These cells express strongly-rectifying K+ channels and the current density in freshly isolated cells from the bone marrow was almost tenfold higher than in mature endothelial cells, and the current was decreased upon cell differentiation [55]. Similarly to dyslipidemia-induced suppression of Kir channels in aortic endothelium, reduced Kir activity was observed in side-population cells freshly isolated from hypercholesterolemic pigs. Dyslipidemia-induced Kir suppression was also observed in cardiomyocytes isolated from hypercholesterolemic rats [58].

The effect of cholesterol on endothelial Kir currents is mediated by Kir2.1 and Kir2.2 channels which, as described above, constitute the dominant Kir conductance in aortic endothelium [14]. Indeed, both Kir2.1 and Kir2.2 are highly cholesterol sensitive as demonstrated by expressing the channels in cell lines that lack endogenous Kir channels, such as Chinese Hamster Ovary (CHO) cells or HEK293 cells [59, 60]. In contrast, Kir2.3 and Kir2.4 show less cholesterol sensitivity, which may result in differential effects of cholesterol on the membrane potential of different cell types. Cholesterol was also shown to suppress the activity of Kir6.2 [61], which as described briefly above plays an important role in microvascular endothelial cells. The effects of cholesterol on Kir channels in cardiomyocytes are complex: while basal currents that are underlined by Kir2 channels are suppressed by cholesterol, acetylcholine-sensitive Kir channels (GIRK or Kir3) are actually enhanced by cholesterol, an effect that is also confirmed in a heterologous expression system [58]. These opposite effects of cholesterol on basal and acetylcholine-sensitive currents in the same cells might be the basis of complex electrophysiological patterns under high cholesterol conditions, but this is beyond the topic of the current book chapter.

Biophysical Basis of Cholesterol Regulation of Kir Channels: The Silent Channel Hypothesis

Silencing of Kir by cholesterol: Earlier studies have shown that cholesterol may regulate Ca^{2+}-sensitive K^+ channels by decreasing their open probability, the ability of the channels to undergo the transformation between closed and open states [62]. This is not the case, however, for Kir2 channels: analysis of single channel activity of both endothelial Kir channels and Kir2.1 channels expressed in a null cell showed that enriching the cells with or depleting them of cholesterol results only in a small effect on the open probability of the channels (Fig. 15.2). In particular, following cholesterol enrichment of the cells, we observed only ~5 % decrease in the open probability of the channels [4], [59]. This small decrease in the open probability cannot account for the 2- to 3-fold decrease in the Kir current density observed in whole cell currents under the same experimental conditions. Moreover, since the basal open probability of Kir2.1 channels is very high as measured in the on cell configuration at negative voltages (−140 to −60 mV range) (>90 %), an increase in open probability would be impossible to account for a twofold elevation of the whole cell Kir current that is induced by cholesterol depletion. Changes in cellular cholesterol also had no effect on the unitary conductance of the channels indicating that cholesterol-Kir2 protein interaction does not affect the pore structure of the channels. An alternative possibility was that cholesterol might affect the expression of the channel protein or its trafficking to the plasma membrane. However, this was

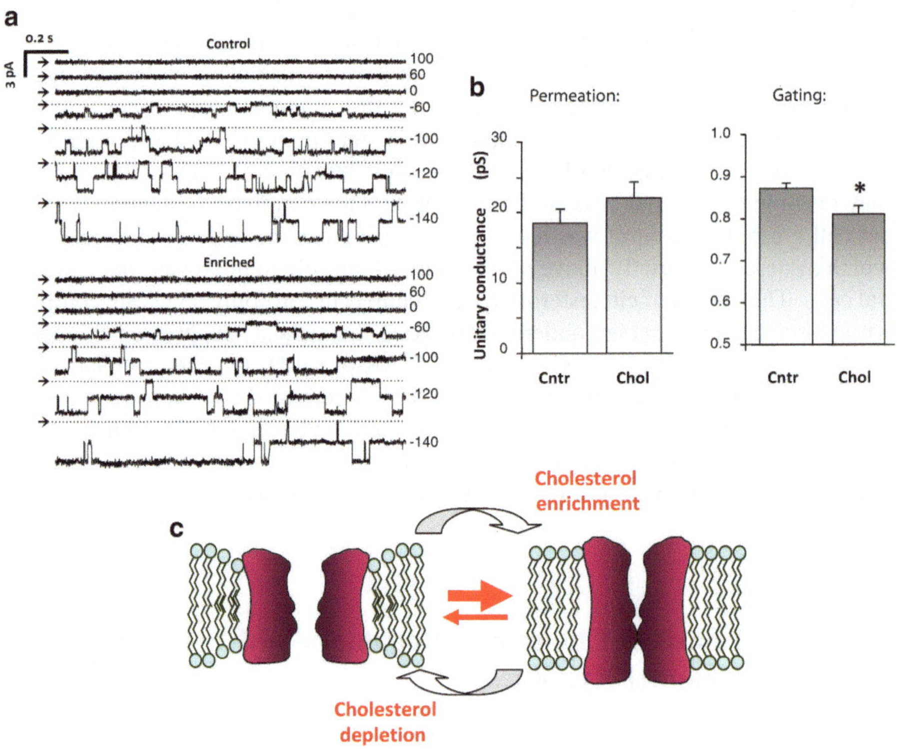

Fig. 15.2 Evidence for silent channel hypothesis. (**a**) Single channel recordings of Kir2.1 expressed in CHO cells under control and cholesterol-enriched conditions. (**b**) Average unitary conductance open probability for the two conditions. (**c**) An increase in membrane cholesterol shifts the conformation of the channels between "active" and "silent" states. **This research was originally published in Biophysical Journal**, 2004. 87: p. 3850–61

also not the case for Kir2 channels: there was no effect neither on the expression of the channels nor on their trafficking to the plasma membrane [59]. To resolve this apparent controversy, we proposed that an increase in membrane cholesterol stabilizes the channels in a long-lived closed "silent" state that decreases the number of "active" channels on the membrane and which cannot be therefore detected on a level of single channel.

Silencing of Kir by caveolin: More recently, the "silent channel" hypothesis was supported by demonstrating the same pattern of inhibition in caveolin-induced suppression of Kir current [60]. In this study, we showed that, Kir2.1 physically interact with cavelin-1 (Cav-1) (Fig. 15.3a) and that similarly to cholesterol, an increase in caveolin-1 (Cav-1) expression resulted in a significant decrease in Kir current density (Fig. 15.3b) without any detectable effect on the single channel properties of the channels or their expression on the plasma membrane [60]. Furthermore, a putative Cav-1 consensus binding motif that was previously defined as φXφXXXXφ, where φ is the aromatic amino acid Trp, Phe, or Tyr, [63], was identified at the interface between the outer transmembrane helix and the N-terminus of Kir2.1 channel (Fig. 15.3c, [60]. Comparative analysis of the closed and open conformation states of the channels based on the crystal structures reported earlier [64, 65] revealed that the putative caveolin-1 binding site of Kir2.1 channels would be more accessible to caveolin-1 when the channels are closed than when they are open and become partially obscured in the open state [60]. We proposed, therefore, that caveolin-1 binds preferentially to the closed conformation of Kir2 channels and stabilizes the

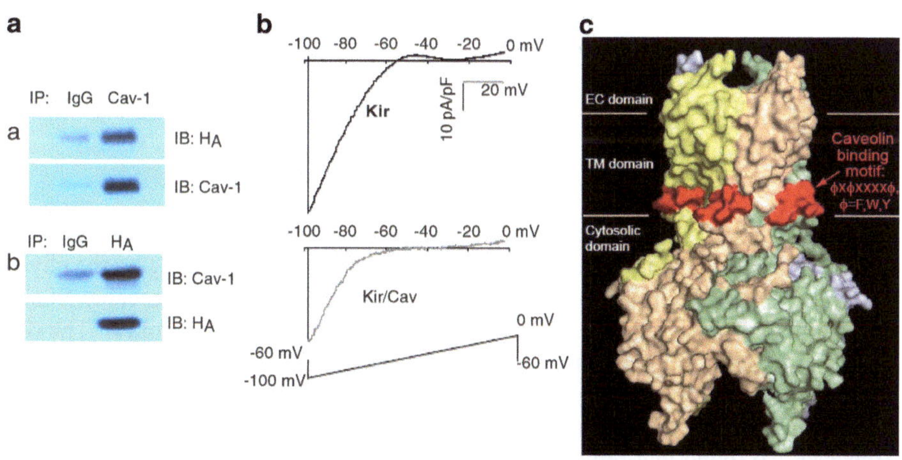

Fig. 15.3 Caveolin-1 is a negative regulator of Kir2.1 channels. (**a**) Co-immunoprecipitation of caveolin-1 with Kir2.1 channels. (**b**) Current traces for Kir2.1 with and without co-expression with Cav-1. (**c**) Location of the caveolin φXφXXXXφ binding motif (*red*) at the interface between the outer transmembrane helix and the N-terminus of the channels (residues 81–88 in Kir2.1) in a surface presentation of the crystal structure of Kir2.2. **This research was originally published in Journal of Physiology** 2014, 592:4025–38

channels in the closed "silent" state. As discussed in more detail below, however, our computational and site-directed mutagenesis analysis suggests that cholesterol and caveolin-1 do not compete for the same binding sites of Kir2 channels but regulate the channels through different binding sites with the two signals converging to the same intramolecular pathway mediated by common residues. These observations suggest that stabilization of Kir2 channels in a silent state is an important mechanism in the regulation of these channels by several effectors.

Cholesterol-rich domains: The partitioning of proteins into cholesterol-rich domains is believed to regulate their function by providing scaffolds to signaling platforms and protein-protein interactions. However, this process does not seem to be critical for cholesterol sensitivity of Kir2 channels. We found that Kir2 channels indeed do partition into cholesterol-rich membrane domains [59, 66] and co-precipitate with caveolin-1 [60], but there is only a small shift in the channel distributions between the domains upon cholesterol depletion or enrichment [66] suggesting that disengaging from these domains upon cholesterol depletion or a small increase in channel association with the domains upon cholesterol enrichment is unlikely to have a major impact on Kir2 function. Furthermore, the presence of caveolins is not a pre-requisite of cholesterol sensitivity of Kir channels, as demonstrated in cells devoid of caveolins. Specifically, we showed that while genetic deletion of caveolin-1 results in an increase in Kir currents in macrophages [60] and endothelial cells (not shown), the currents remain cholesterol sensitive in both cell types. Notably, since genetic deletion of caveolin-1 was shown to result in the loss of caveolae structure [67], preservation of cholesterol sensitivity of Kir2 channels in Cav-1 knock out mice indicates that neither the integrity of caveolae nor partitioning of the channels into these domains play a critical role in their sensitivity to cholesterol.

Direct Cholesterol-Kir Interactions: Identification of Cholesterol-Binding Kir2 Domains

Comparative sterol analysis: The first indication that cholesterol regulates Kir2 channels by specific cholesterol-protein interactions and not by changing the physical properties of the lipid bilayers, as was primarily believed earlier, came from a comparison between cholesterol and its chiral analogue epicholesterol [4]. The two sterols differ in an angle of a single OH group and are known to have similar though not identical effects on membrane fluidity [68, 69]. Cells cannot survive without cholesterol and it is also impossible to substitute it completely with epicholesterol. However, we found that removing ~50 % of the membrane cholesterol from aortic endothelial cells and substituting it with a similar amount of epicholesterol maintains the cells' viability and membrane integrity. We could, therefore, test how this substitution affects endothelial Kir channels. To our surprise, substituting endothelial cholesterol with epicholesterol resulted in a significant increase in Kir current that was even stronger than the increase induced by cholesterol depletion [4]. These observations indicated that cholesterol-induced regulation of endothelial Kir channels depends on specific cholesterol-protein interactions and suggested that the two sterols compete for a binding site either in the Kir protein itself or in an auxiliary protein that might mediate cholesterol effects on the channels. Interestingly, this is not the case for volume-activated anion channels (VRAC) in the same cells. VRAC are not sensitive to the chiral nature of cholesterol, and are regulated instead by changes in the physical properties of the membrane [70, 71].

Insights from purified channels: To discriminate between direct and indirect effects of cholesterol on Kir channels, we reconstituted a purified bacterial analogue of Kir2 cannels, KirBac1.1, into lipid vesicles and found that cholesterol can regulate Kir channels in a purified system without any intermediates. We also found that there is no correlation between KirBac1.1 function and membrane fluidity when the two parameters were compared for an array of different sterols [69]. Importantly, the same conclusion was reached in a later study that used purified mammalian Kir2.1 channels which were reconstituted in lipid vesicles that contained either cholesterol or ent-cholesterol, another chiral analogue of cholesterol whose physical properties are virtually identical to that of cholesterol [72]. Furthermore, we also demonstrated that cholesterol binds to KirBac1.1 channels in a specific and saturable way, and that inhibition of cholesterol-KirBac1.1 binding abrogates the inhibitory effect of cholesterol on KirBac1.1 function [73]. As expected from our studies on cholesterol/epicholesterol substitution in endothelial cells, epicholesterol was shown to compete with cholesterol for KirBac1.1 binding. Thus, clearly Kir2 channels are regulated by specific and direct cholesterol-protein interactions.

First insights into the structural determinants of cholesterol sensitivity of Kir channels: Our quest to identify the structural determinants of cholesterol sensitivity in Kir2 channels started [74] with testing the residues on the protein-lipid interface of the transmembrane domains [75] and on the interface between the transmembrane and cytoplasmic domains, specifically focusing on the residues that were identified earlier to confer the sensitivity of the channels to a regulatory phospholipid PI(4,5) P_2 [76]. These studies led to the identification of the first residues that are essential for cholesterol sensitivity of Kir2.1, which surprisingly were found on the C-terminus of the cytosolic domain of the channels and not on the lipid-protein interface between the channel and the lipid bilayer [74]. More specifically, we found that cholesterol sensitivity of Kir2.1 critically depends on a set of specific residues within the CD loop of the C-terminus with one of the residues, leucine 222 (L222) having the most profound effect (see Fig. 15.2 for the position of L222 in the channel). L222 was also found earlier to play a critical role in the sensitivity of Kir2.1 channels to PI(4,5)P_2 [76]. A loss of cholesterol sensitivity was not associated with a loss of the ability of the channels to partition into cholesterol-rich lipid domains [74]. Importantly, identifying a residue that confers cholesterol sensitivity to the channels does not mean that this residue is part of a cholesterol-binding site. Indeed, we proposed that the CD loop may be important for maintaining the channels in a conformation susceptible to cholesterol modulation.

Cholesterol-sensitivity belt: Further analysis based on the differential cholesterol sensitivities of Kir2.1 and Kir2.3 channels revealed that the cytosolic CD loop is part of a regulatory belt of residues that surrounds the cytosolic pore of the channels in proximity of the inner leaflet of the membrane (Fig. 15.4a shows the side and top of view [77], Fig. 15.4b shows the side view of two opposite subunits indicating the position of the CD loop). However, docking analysis of the cholesterol molecule to the channel protein [78–80] showed that cholesterol does not dock to the residues that comprise the cholesterol sensitivity belt, suggesting that this structure does not represent a cholesterol-binding site of Kir2 channels. Instead, based on a database of crystallographic structures of Kir channels we found that the residues of the cholesterol sensitivity belt were correlated with the residues known to be critical for gating of these channels suggesting that the cholesterol sensitivity belt constitutes a regulatory site that couples cholesterol-binding to channel gating [77].

Two-way molecular switch: Surprisingly, we found that the cholesterol sensitivity of Kir2 channels is also regulated by a two-way molecular switch that is comprised of two distant cytosolic residues, L222 of the cholesterol sensitivity belt that is located close to the interface between the C-terminus and the transmembrane domain, and N251 that is located further away from the transmembrane domain in the EF loop (Fig. 15.4b, c [81]). Specifically, the L222I and N251D mutations each render the channels to be cholesterol insensitive, but together they cancel each other, and the double mutants are cholesterol sensitive. Moreover, we showed that the same switch also regulates the sensitivity of the channels to PI(4,5)P_2. Again, each one of these two mutations alone reduced the strength of the interaction between the channel and PI(4,5)P_2, but together they cancel each other. Based on the crystal structure of the cytosolic domain of the channel, the distance between the two residues is ~24 Å excluding the possibility that these two residues interact directly with each other. Instead, molecular dynamics simulations of the Kir2.1-WT, L222I-Kir2.1 mutant and the L222-N251D-Kir2.1 double mutant showed that the effects of these mutations span extensive regions of the C-terminus suggesting that the two residues are connected through a reversal-residue-chain that seems to serve as the wiring between the two residues (Fig. 15.4c, lower panel, [81]). In addition, our data suggest that functional links between the N- and the C-termini that couple the intracellular domains of the four subunits of the channels during gating also play an important role in cholesterol sensitivity of Kir2 channels [82]. Accordingly, multiple cytosolic structures in Kir2 channels play a critical role in their sensitivity to cholesterol. Yet, neither of these structures has the characteristics of a cholesterol-binding site.

Identification of novel cholesterol-binding sites. Our next strategy was to test whether Kir2 channels have regions homologous with the known cholesterol-binding motifs: the cholesterol consensus motif (CCM) and the cholesterol recognition amino acid consensus motif (CRAC), two well-established cholesterol-binding motifs that have been found and characterized in other proteins. Indeed, recent studies showed that CRAC is found in TRPV1 channels [83] and in BK channels [84]. A variation of CRAC motif, CARC, an inverted sequence of CRAC was suggested

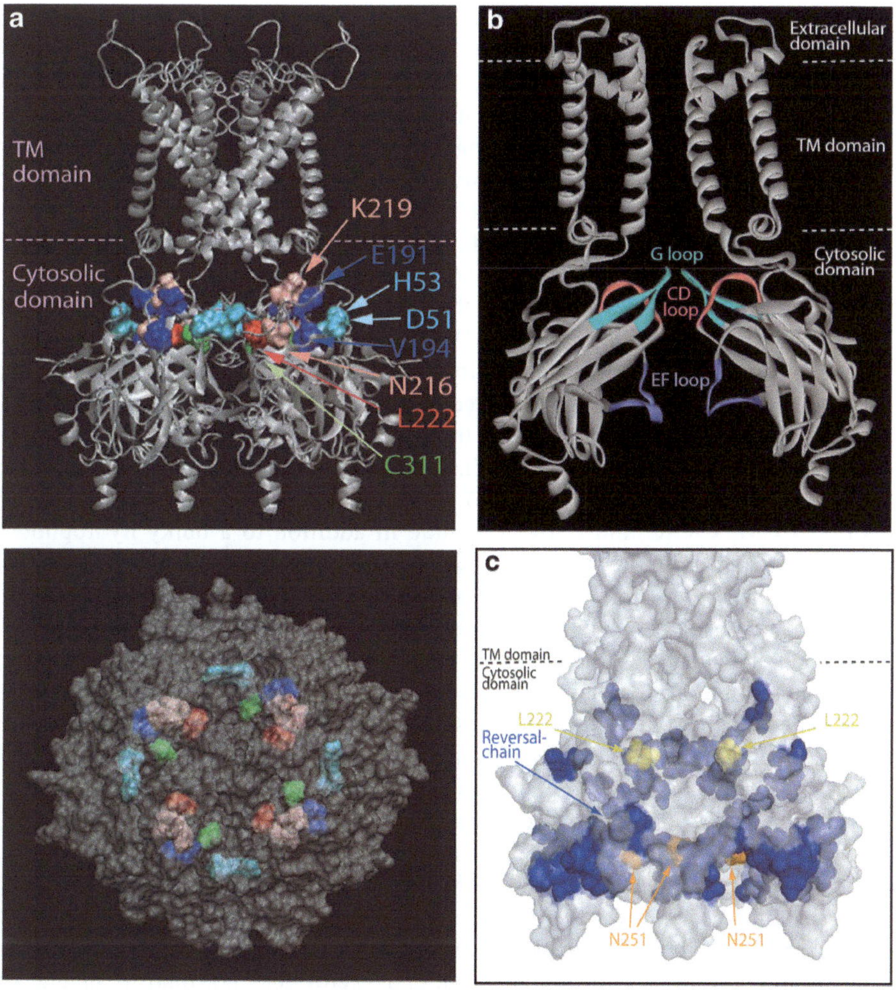

Fig. 15.4 Structural features of cholesterol sensitivity of Kir2.1 channels. (**a**) Upper panel Side view of a model of Kir2.1 that includes all four subunits. Shown in the model are the residues whose mutation affects cholesterol sensitivity: D51 and H53 (*cyan*), E191 and V194 (*blue*), N216 and K219 (*pink*), L222 (*red*) and C311 (*green*). Lower panel. Top view of the model of Kir2.1 from the membrane showing the cholesterol sensitivity belt formed by the residues whose mutation affects the cholesterol sensitivity of the channel. **This research was originally published in Biophysical Journal**, 2011:100:381–9. (**b**) Model of two opposite facing subunits of Kir2.1 showing the cytosolic G-loop, CD-loop and EF-loop; (**c**) Surface presentation of the cytosolic domain of Kir2.1 showing the reversal residue chain (*blue*), L222I (*yellow*), and N251 (*orange*). **This research was originally published in Journal of Biological Chemistry**. Rosenhouse-Dantsker, A., et al., *Distant Cytosolic Residues Mediate a Two-way Molecular Switch That Controls the Modulation of Inwardly Rectifying Potassium (Kir) Channels by Cholesterol and Phosphatidylinositol 4,5-Bisphosphate (PI(4,5)P2)*. Journal of Biological Chemistry, 2012. 287(48): p. 40266-40278. © **the American Society for Biochemistry and Molecular Biology**

to underlie cholesterol sensitivity of AChR [85]. However, based on the sequence analysis, Kir2.1 channels have no CCM motif and no appropriate CRAC or CARC motifs that would be energetically favorable to bind a cholesterol molecule [86]. Therefore, we turned to an alternative approach of identifying cholesterol-binding regions in Kir2 channels using a combination of molecular docking, molecular dynamics (MD) simulations and site-directed mutagenesis that is not biased or limited only to the known cholesterol-binding motifs.

These studies led to the identification of two novel non-annular hydrophobic cholesterol-binding regions in Kir2.1 [86] (See Fig. 15.5a, b).

One binding region was located in the center of the transmembrane domain of the channel (region 1, Fig. 15.5d) [86]. Within this region, we identified six residues whose mutation abrogated the sensitivity of the channel to cholesterol. In addition, mutations of two other residues resulted in a significant decrease in the channel's cholesterol sensitivity. The majority of these eight residues were bulky hydrophobic residues (isoleucine, leucine and valine), and the remaining two residues were smaller residues (alanine and serine). Notably, the known cholesterol-binding motifs (CCM, CRAC and CARC) include in addition to a bulky hydrophobic residue also an aromatic residue (tyrosine, tryptophan or phenylalanine) and a positively charged residue (arginine or lysine). However, mutations of several aromatic and positively charged residues located in region 1 of Kir2.1 resulted in a non-functional channel or did not affect the sensitivity of the channel to cholesterol. Since mild mutations may not affect the interaction of the channel with the cholesterol molecule, the role of these residues in cholesterol binding cannot be excluded.

The second binding region was located at the interface between the transmembrane and cytosolic domains of the channel (region 2) [86]. Within this region, we identified five hydrophobic residues (alanine, leucine, valine and methionine) that affected the sensitivity of the channel to cholesterol. Among these, three mutations abrogated the sensitivity of the channel to cholesterol. As noted above, our earlier studies have shown that mutations of residues located at the lipid-protein interface between the channel and the lipid bilayer do not affect the sensitivity of the channel to cholesterol [74] implying that cholesterol does not bind to annular or boundary sites located on the transmembrane surface of the channel protein. In contrast, in both of the putative cholesterol-binding regions described above, the residues whose mutation affected the sensitivity of the channel to cholesterol were distributed among α-helices of two adjacent subunits of the channel suggesting that the cholesterol molecule would bind in between the α-helices. Furthermore, MD simulations demonstrated that due to the matching between the cholesterol molecule and the hydrophobic and aromatic moieties in the two putative cholesterol-binding regions, these regions prefer cholesterol to phospholipids. Together, this suggests that these two cholesterol-binding regions are non-annular surfaces that are occluded from phospholipid binding (Fig. 15.3c) [86].

In order to compare the strength of the interactions between the cholesterol molecule and the channel in each of the two putative cholesterol-binding regions we assessed the binding energy and binding affinity of cholesterol to the channel.

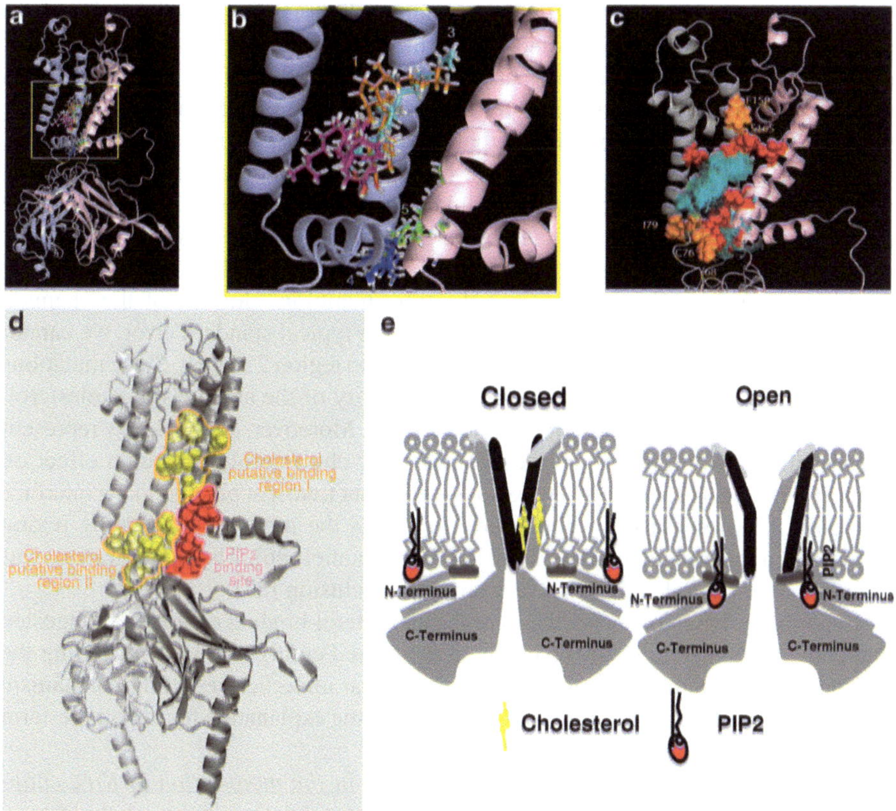

Fig. 15.5 Proximity of the cholesterol and PI(4,5)P_2 binding sites in Kir2 channels. (**a**) Location of two putative cholesterol-binding regions in Kir2.1 that were obtained from 50-ns all-atom full-membrane MD simulations. The starting points of the simulations were the centers of five clusters that were obtained from docking analysis. (**b**) E*nlargement* of the channel region that includes the two putative cholesterol-binding regions depicted as a *yellow rectangle* in **a**. In both **a** and **b**, the two adjacent subunits of the channel that interact with the cholesterol molecule are shown in a *ribbon representation*. (**c**) *Ribbon representation* of the transmembrane and extracellular domains of two adjacent subunits of Kir2.1 (in *gray* and *pink*) depicting the locations of primary (*red*, directly interacting residues) and secondary (*orange*, within 4 Å from the primary residues) cholesterol sensitive residues. Also shown are the locations of the cholesterol molecules at the two putative cholesterol-binding regions (*cyan sticks* and *surface representations*). (**a–c**) **This research was originally published in Journal of Biological Chemistry**. Rosenhouse-Dantskeret al., *Identification of Novel Cholesterol-binding Regions in Kir2 Channels*. Journal of Biological Chemistry, 2013. 288(43): p. 31154–31164 **© the American Society for Biochemistry and Molecular Biology**. (**d**) A ribbon presentation of two adjacent subunits (*Gray* and *light gray*) of the crystal structure of Kir2.2 (PDB ID 3spi) showing the binding site of PI(4,5)P_2 in the channel (*red balls*). Also shown are the corresponding Kir2.2 residues to Kir2.1 residues that form two putative cholesterol-binding regions in Kir2.1 based on functional data and molecular modeling (*yellow balls*—direct interaction; *light yellow balls*—secondary effect). (**e**) Schematic model of cholesterol and PI(4,5)P_2 interactions with Kir2 channels. (**d–e**) **This research was originally published in Computational Structural Biotechnology J**. Rosenhouse-Dantsker, A., Y. Epshtein, and I. Levitan, *Interplay Between Lipid Modulators of Kir2 Channels: Cholesterol and PIP2*. Comput Struct Biotechnol J, 2014. 11: p. 131–7

To that end, we calculated the binding enthalpy, which is a quantitative indicator of changes in the binding energy, and the equilibrium free energy of the process that enables to assess binding affinity and stability. Our calculations showed that the average binding enthalpy of cholesterol at region 1 (-6 ± 2 kcal/mol) was slightly stronger than the average binding enthalpy to the cholesterol molecule at region 2 (-4 ± 2 kcal/mol). Our calculations also indicated that the equilibrium free energy was favorable in region 1 and unfavorable in region 2. Yet, even for the favorable region 1, the absolute value of the free energy was small suggesting that the binding affinity of cholesterol to this region is weak. Overall, these results suggest preference for binding region 1 in the center of the transmembrane domain of the channel. However, due to the small absolute values and the typical standard error, we cannot rule out the possibility of weak cholesterol binding to region 2 as well. Since mutations of residues in both regions abrogated the sensitivity of the channels to cholesterol, both regions may form cholesterolophilic surfaces. Moreover, region 2 may represent a transient cholesterol-binding site necessary for cholesterol to have an effect on channel function. Alternatively, it is also possible that transient binding to the interface of region 2 is necessary for cholesterol to access the more stable binding region (region 1) in the transmembrane domain of the channel [86]. The same strategy was used in a later study leading to a very similar conclusion [87].

Importantly, based on the locations of the cholesterol-binding regions we suggested that cholesterol stabilizes the channels in the closed configuration by opposing the hinging motion of the transmembrane domain that underlie the gating mechanism of the channels, thereby providing the first mechanistic explanation for how cholesterol shifts Kir channels into the "silent state" [86].

Interplay between cholesterol and PI(4,5)P2 in the modulation of Kir2 channels. As noted above, our first studies of the structural determinants of cholesterol sensitivity in Kir2 channels led to cytosolic residues [74] that have been previously shown to affect the sensitivity of the channels to a regulatory phospholipid PI(4,5)P_2 [76]. Furthermore, our subsequent studies [81] resulted in the identification of a two-way molecular cytosolic switch that regulates the sensitivity of Kir2.1 to both cholesterol and PI(4,5)P_2. We thus explored whether cholesterol regulates Kir channels by regulating their access to PI(4,5)P_2. This, however, was not the case because sequestering PI(4,5)P_2 had no effect on cholesterol sensitivity of Kir2.1 channels [74]. Moreover, examining the role of cholesterol in channel-PI(4,5)P_2 interactions, we then showed that whereas cholesterol depletion resulted in strengthening of Kir2-PI(4,5)P_2 interactions, cholesterol enrichment had no effect on Kir2-PI(4,5)P_2 interactions [88]. Taken together, these data suggest that cholesterol and PI(4,5)P_2 act through overlapping regions of the channel. However, since residues that have been shown to bind directly to PI(4,5)P_2 do not affect the sensitivity of the channel to cholesterol [74], it is likely that the two lipids bind to distinct binding sites. This notion is further corroborated by comparing the locations of the two putative cholesterol binding regions described above [86] with the binding site of PI(4,5)P_2 in Kir2.2 [65, 88]. Accordingly, these two lipids bind to different adjacent non-overlapping sites in the channel (Fig. 15.5d). This structural proximity between the sites gives rise to the possibility that in addition to the common mechanism that the two lipids share, functional interplay between cholesterol and PI(4,5)P_2 may also originate from interactions between the binding sites. Figure 15.5e summarizes our current working hypothesis of the mechanism that underlies cholesterol regulation of Kir channel. Cholesterol binding in between the transmembrane helices of the channel stabilizes the closed state of the channel by interfering with the hinging motion of the inner transmembrane of the channel. Removal of cholesterol strengthens channel-PI(4,5)P_2 interactions and stabilizes the channel in its open state.

Conclusions: What Is Now Known and What Is Unknown

It is well established today that Kir channels are expressed in several types of endothelial cells and that their activity is strongly suppressed by the elevation of cellular cholesterol *in vitro* and *in vivo*. There is also no doubt that cholesterol interacts with the channels directly via specific sterol-protein interactions. Moreover, the preponderance of evidence suggests that cholesterol binds to the channels not at the protein-membrane interface but at non-annular sites in hydrophobic pockets formed by the helices of the transmembrane domains of the channels. The binding of cholesterol to these sites affects the gating machinery of the cytosolic pore. What is not clear, however, is how binding of the cholesterol molecule to a hydrophobic pocket in the transmembrane domain of the channels is transduced to the cytosolic domain to affect channel gating. This question should be addressed by a combination of molecular dynamic simulations, site-directed mutagenesis and biophysical analysis of the channel activities. The physiological impact of cholesterol-induced suppression of Kir on endothelial function remains to be explored. The only evidence so far that suppression of Kir plays a significant role in endothelial function is a correlation between the loss of channel activity and impairment of flow-induced vasodilation in a porcine model of diet-induced hypercholesterolemia. Clearly, this is not sufficient to establish the role of Kir. Structural studies provide novel and unique tools to address this question. Specifically, our studies identified multiple mutants of Kir2.1 channels that lose their cholesterol sensitivity. Creating transgenic models that substitute native Kir channels with their cholesterol-insensitive mutants will allow to establish the impact of cholesterol sensitivity of these channels on vascular function.

Acknowledgements We thank Mr. Gregory Kowalsky for his help in the layout and formatting of the Figures. The work was supported by National Institutes of Health grants HL073965 and HL083298 (to I. Levitan) and a Scientist Development Grant (11SDG5190025) from the American Heart Association (to A. Rosenhouse-Dantsker).

References

1. Ross R. Atherosclerosis—an inflammatory disease. N Engl J Med. 1999;340:115–26.
2. Kinlay S, Libby P, Ganz P. Endothelial function and coronary artery disease. Curr Opin Lipidol. 2001;12:383–9.
3. Kruth HS. Lipoprotein cholesterol and atherosclerosis. Curr Mol Med. 2001;1:633–53.
4. Romanenko VG, Rothblat GH, Levitan I. Modulation of endothelial inward-rectifier K+ current by optical isomers of cholesterol. Biophys J. 2002;83(6):3211–22.
5. Fang Y, et al. Hypercholesterolemia suppresses inwardly rectifying K+ channels in aortic endothelium in vitro and in vivo. Circ Res. 2006;98(8):1064–71.
6. Kubo Y, et al. International Union of Pharmacology. LIV. Nomenclature and molecular relationships of inwardly rectifying potassium channels. Pharmacol Rev. 2005;57(4):509–26.
7. Hibino H, et al. Inwardly rectifying potassium channels: their structure, function, and physiological roles. Physiol Rev. 2010;90(1):291–366.
8. Olesen S-P, Clapham DE, Davies PF. Hemodynamic shear stress activates a K+ current in vascular endothelial cells. Nature. 1988;331(6152):168–70.
9. Davies PF, Dull RO. Hemodynamic forces in relation to mechanosensitive ion channels in endothelial cells. In: Frangos JA, editor. Physical forces and the mammalian cell. San Diego: Academic Press; 1993. p. 125–38.
10. Ungvari Z, Csiszar A, Kohler A. Increases in endothelial Ca2+ activate KCa channels and elicit EDHF-type arteriolar dilation via gap junctions. Am J Physiol Heart Circ Physiol. 2002;282:H1760–7.
11. Wulff H, Köhler R. Endothelial small-conductance and intermediate-conductance KCa channels: an update on their pharmacology and usefulness as cardiovascular targets. J Cardiovasc Pharmacol. 2013;61:102–12.
12. Forsyth SE, Hoger A, Hoger JH. Molecular cloning and expression of a bovine endothelial inward rectifier potassium channel. FEBS Lett. 1997;409:277–82.
13. Yang D, et al. Expression of the inwardly rectifying K+ channel Kir2.1 in native bovine corneal endothelial cells. Invest Ophthalmol Vis Sci. 2003;44:3511–9.
14. Fang Y, et al. Functional expression of Kir2.x in human aortic endothelial cells: the dominant role of Kir2.2. Am J Physiol Cell Physiol. 2005;289(5):C1134–44.
15. Millar ID, et al. Kv1 and Kir2 potassium channels are expressed in rat brain endothelial cells. Pflugers Arch. 2008;456(2):379–91.
16. Mederos Y, et al. ATP-sensitive potassium channels in capillaries isolated from guinea-pig heart. J Physiol. 2000;525:307–17.
17. Chatterjee S, et al. Shear stress increases expression of a KATP channel in rat and bovine pulmonary vascular endothelial cells. Am J Physiol Cell Physiol. 2003;285(4):C959–67.
18. Yoshida H, et al. K ATP channels of primary human coronary artery endothelial cells consist of a heteromultimeric complex of Kir6.1, Kir6.2, and SUR2B subunits. J Mol Cell Cardiol. 2004;37(4):857–69.
19. Morrissey A, et al. Immunolocalization of KATP channel subunits in mouse and rat cardiac myocytes and the coronary vasculature. BMC Physiol. 2005;5(1):1.
20. Chatterjee S, Levitan I, Fisher AB. The KATP channel is an important component of flow sensing in the pulmonary microvasculature. Microcirculation. 2006;13:633–44.
21. Kohler R, et al. Expression and function of endothelial Ca2+-activated K+ channels in human mesenteric artery: a single-cell reverse transcriptase-polymerase chain reaction and electrophysiological study in situ. Circ Res. 2000;87(6):496–503.
22. Bychkov R, et al. Characterization of a charybdotoxin-sensitive intermediate conductance Ca2+-activated K+ channel in porcine coronary endothelium: relevance to EDHF. Br J Pharmacol. 2002;137(8):1346.
23. Kohler R, Hoyer J. The endothelium-derived hyperpolarizing factor: insights from genetic animal models. Kidney Int. 2007;72(2):145.

24. Papassotiriou J, et al. Endothelial K+ channel lacks the Ca2+ sensitivity-regulating β subunit. FASEB J. 2000;14(7):885–94.
25. Wang X-L, et al. Caveolae targeting and regulation of large conductance Ca2+-activated K+ channels in vascular endothelial cells. J Biol Chem. 2005;280(12):11656–64.
26. Kamouchi M, et al. Modulation of inwardly rectifying potassium channels in cultured bovine pulmonary artery endothelial cells. J Physiol. 1997;504:545–56.
27. Leung YM, Kwan CY, Daniel EE. Block of inwardly rectifying K+ currents by extracellular Mg2+ and Ba2+ in bovine pulmonary artery endothelial cells. Can J Physiol Pharmacol. 2000;78(9):751–6.
28. Rusko J, et al. Calcium-activated potassium channels in native endothelial cells from rabbit aorta: conductance, Ca2+ sensitivity and block. J Physiol. 1992;455:601–21.
29. Ledoux J, Bonev AD, Nelson MT. Ca2+-activated K+ channels in murine endothelial cells: block by intracellular calcium and magnesium. J Gen Physiol. 2008;131(2):125–35.
30. Yamazaki D, et al. Novel functions of small conductance Ca2+-activated K+ channel in enhanced cell proliferation by ATP in brain endothelial cells. J Biol Chem. 2006;281(50):38430–9.
31. Manabe K, et al. Classification of ion channels in the luminal and abluminal membranes of guinea-pig endocardial endothelial cells. J Physiol. 1995;484(1):41.
32. Colden-Stanfield M, Cramer EB, Gallin EK. Comparison of apical and basal surfaces of confluent endothelial cells: patch-clamp and viral studies. AM J Physiol. 1992;263:C573–83.
33. Malester B, et al. Transgenic expression of a dominant negative K(ATP) channel subunit in the mouse endothelium: effects on coronary flow and endothelin-1 secretion. FASEB J. 2007;21(9):2162–72.
34. Michelakis ED, et al. Potassium channels regulate tone in rat pulmonary veins. Am J Physiol Lung Cell Mol Physiol. 2001;280:L1138–47.
35. Absi M, et al. Effects of methyl beta-cyclodextrin on EDHF responses in pig and rat arteries; association between SK(Ca) channels and caveolin-rich domains. Br J Pharmacol. 2007;151(3):332–40.
36. Sandow SL, et al. Spatial separation of endothelial small- and intermediate-conductance calcium-activated potassium channels (K(Ca)) and connexins: possible relationship to vasodilator function? J Anat. 2006;209(5):689–98.
37. Ledoux J, et al. Functional architecture of inositol 1,4,5-trisphosphate signaling in restricted spaces of myoendothelial projections. Proc Natl Acad Sci U S A. 2008;105(28):9627–32.
38. Jacobs ER, et al. Shear activated channels in cell-attached patches of cultured bovine aortic endothelial cells. Pflugers Arch. 1995;431:129–31.
39. Davies PF, Spaan JA, Krams R. Shear stress biology of the endothelium. Ann Biomed Eng. 2005;V33(12):1714.
40. Hoger JH, et al. Shear stress regulates the endothelial Kir2.1 ion channel. Proc Natl Acad Sci U S A. 2002;99(11):7780–5.
41. Fang Y, et al. Flow sensitivity of inwardly-rectifying K channels in human aortic endothelium. Third IASTED International Conference on Biomechanics. Benidorm, Spain: ACTA Press; 2005.
42. Cooke JP, et al. Flow activates an endothelial potassium channel to release an endogenous nitrovasodilator. J Clin Invest. 1991;88:1663–71.
43. Wei AD, et al. International Union of Pharmacology. LII. Nomenclature and molecular relationships of calcium-activated potassium channels. Pharmacol Rev. 2005;57(4):463–72.
44. Wellman GC, Bevan JA. Barium inhibits the endothelium-dependent component of flow but not acetylcholine-induced relaxation in isolated rabbit cerebral arteries. J Pharmacol Exp Ther. 1995;274:47–53.
45. Climent B, et al. Intact rat superior mesenteric artery endothelium is an electrical syncytium and expresses strong inward rectifier K+ conductance. Biochem Biophys Res Commun. 2011;410(3):501–7.
46. Chilton L, Loutzenhiser R. Functional evidence for an inward rectifier potassium current in rat renal afferent arterioles. Circ Res. 2001;88(2):152–8.

47. Malek AM, et al. Endothelin-1 gene suppression by shear stress: pharmacological evaluation of the role of tyrosine kinase, intracellular calcium, cytoskeleton, and mechanosensitive channels. J Mol Cell Cardiol. 1999;31(2):387–99.
48. Dawes M, et al. Barium reduces resting blood flow and inhibits potassium-induced vasodilation in the human forearm. Circulation. 2002;105(11):1323–8.
49. Crecelius AR, et al. Reactive hyperemia occurs via activation of inwardly rectifying potassium channels and Na+/K+-ATPase in humans. Circ Res. 2013;113(8):1023–32.
50. Crecelius AR, et al. KIR channel activation contributes to onset and steady-state exercise hyperemia in humans. Am J Physiol Heart Circ Physiol. 2014;307(5):H782–91.
51. Scharbrodt W, et al. Basic fibroblast growth factor-induced endothelial proliferation and NO synthesis involves inward rectifier K+ current. Arterioscler Thromb Vasc Biol. 2004;24(7):1229–33.
52. Gojova A, Barakat AI. Vascular endothelial wound closure under shear stress: role of membrane fluidity and flow-sensitive ion channels. J Appl Physiol (1985). 2005;98(6):2355–62.
53. Chatterjee S, Chapman K, Fisher A. Lung ischemia: a model for endothelial mechanotransduction. Cell Biochem Biophys. 2008;52(3):125–38.
54. Chatterjee S, Fisher AB. Mechanotransduction in the endothelium: role of membrane proteins and reactive oxygen species in sensing, transduction, and transmission of the signal with altered blood flow. Antioxid Redox Signal. 2014;20:899–913.
55. Mohler Iii ER, et al. Hypercholesterolemia suppresses Kir channels in porcine bone marrow progenitor cells in vivo. Biochem Biophys Res Commun. 2007;358(1):317–24.
56. Goodell MA, et al. Isolation and functional properties of murine hematopoietic stem cells that are replicating in vivo. J Exp Med. 1996;183(4):1797–806.
57. Jackson KA, et al. Regeneration of ischemic cardiac muscle and vascular endothelium by adult stem cells. J Clin Invest. 2001;107(11):1395–402.
58. Deng W, et al. Hypercholesterolemia induces up-regulation of KACh cardiac currents via a mechanism independent of phosphatidylinositol 4,5-bisphosphate and Gβγ. J Biol Chem. 2012;287(7):4925–35.
59. Romanenko VG, et al. Cholesterol sensitivity and lipid raft targeting of Kir2.1 channels. Biophys J. 2004;87:3850–61.
60. Han H, et al. Silencing of Kir2 channels by caveolin-1: cross-talk with cholesterol. J Physiol. 2014;592:4025–38.
61. Rosenhouse-Dantsker A, et al. Comparative analysis of cholesterol sensitivity of Kir channels: role of the CD loop. Channels (Austin). 2010;4:63–6.
62. Chang HM, et al. Attenuation of channel kinetics and conductance by cholesterol: an interpretation using structural stress as a unifying concept. J Membr Biol. 1995;143:51–63.
63. Couet J, et al. Identification of peptide and protein ligands for the caveolin-scaffolding domain. J Biol Chem. 1997;272(10):6525–33.
64. Tao X, et al. Crystal structure of the eukaryotic strong inward-rectifier K+ channel Kir2.2 at 3.1 Ã… resolution. Science. 2009;326(5960):1668–74.
65. Hansen SB, Tao X, MacKinnon R. Structural basis of PIP2 activation of the classical inward rectifier K+ channel Kir2.2. Nature. 2011;477(7365):495.
66. Tikku S, et al. Relationship between Kir2.1/Kir2.3 activity and their distribution between cholesterol-rich and cholesterol-poor membrane domains. Am J Physiol Cell Physiol. 2007;293:C440–50.
67. Drab M, et al. Loss of caveolae, vascular dysfunction, and pulmonary defects in caveolin-1 gene-disrupted mice. Science. 2001;293:2449–52.
68. Gimpl G, Burger K, Fahrenholz F. Cholesterol as modulator of receptor function. Biochemistry. 1997;36:10959–74.
69. Singh DK, et al. Direct regulation of prokaryotic Kir channel by cholesterol. J Biol Chem. 2009;284(44):30727–36.
70. Romanenko VG, Rothblat GH, Levitan I. Sensitivity of volume-regulated anion current to cholesterol structural analogues. J Gen Physiol. 2004;123(1):77–88.

71. Byfield FJ, et al. Evidence for the role of cell stiffness in modulation of volume-regulated anion channels. Acta Physiol. 2006;187(1–2):285–94.
72. D'Avanzo N, et al. Enantioselective protein-sterol interactions mediate regulation of both prokaryotic and eukaryotic inward rectifier K^+ channels by cholesterol. PLoS One. 2011; 6(4):e19393.
73. Singh DK, et al. Cholesterol regulates prokaryotic Kir channel by direct binding to channel protein. Biochim Biophys Acta (BBA) Biomembr. 2011;1808(10):2527.
74. Epshtein Y, et al. Identification of a C-terminus domain critical for the sensitivity of Kir2.1 channels to cholesterol. PNAS. 2009;106:8055–60.
75. Minor DL, et al. Transmembrane structure of an inwardly rectifying potassium channel. Cell. 1999;96:879–91.
76. Logothetis DE, et al. Phosphoinositide-mediated gating of inwardly rectifying K(+) channels. Pflugers Arch. 2007;455(1):83–95.
77. Rosenhouse-Dantsker A, Logothetis DE, Levitan I. Cholesterol sensitivity of KIR2.1 is controlled by a belt of residues around the cytosolic pore. Biophys J. 2011;100(2):381.
78. Pegan S, et al. Cytoplasmic domain structures of Kir2.1 and Kir3.1 show sites for modulating gating and rectification. Nat Neurosci. 2005;8:279–87.
79. Nishida M, MacKinnon R. Structural basis of inward rectification: cytoplasmic pore of the G protein-gated inward rectifier GIRK1 at 1.8 A resolution. Cell. 2002;111(7):957.
80. Nishida M, et al. Crystal structure of a Kir3.1-prokaryotic Kir channel chimera. EMBO J. 2007;26:4005–15.
81. Rosenhouse-Dantsker A, et al. Distant cytosolic residues mediate a two-way molecular switch that controls the modulation of inwardly rectifying potassium (Kir) channels by cholesterol and phosphatidylinositol 4,5-bisphosphate (PI(4,5)P2). J Biol Chem. 2012; 287(48):40266–78.
82. Rosenhouse-Dantsker A, et al. Cholesterol sensitivity of KIR2.1 depends on functional inter-links between the N and C termini. Channels (Austin). 2013;7:303–12.
83. Picazo-Juarez G, et al. Identification of a binding motif in the S5 helix that confers cholesterol-sensitivity to TRPV1. J Biol Chem. 2011;286(28):24966–76.
84. Singh AK, et al. Multiple cholesterol recognition/interaction amino acid consensus (CRAC) motifs in cytosolic C tail of Slo1 subunit determine cholesterol sensitivity of Ca2+- and voltage-gated K+ (BK) channels. J Biol Chem. 2012;287(24):20509–21.
85. Fantini J, Barrantes FJ. How cholesterol interacts with membrane proteins: an exploration of cholesterol-binding sites including CRAC, CARC and tilted domains. Front Physiol. 2013;4:31.
86. Rosenhouse-Dantsker A, et al. Identification of novel cholesterol-binding regions in Kir2 channels. J Biol Chem. 2013;288(43):31154–64.
87. Fürst O, et al. Identification of a cholesterol-binding pocket in inward rectifier k(+) (kir) channels. Biophys J. 2014;107:2786–96.
88. Rosenhouse-Dantsker A, Epshtein Y, Levitan I. Interplay between lipid modulators of Kir2 channels: cholesterol and PIP2. Comput Struct Biotechnol J. 2014;11:131–7.

79. Schumacher MA, Rivard AF, [illegible] Structure of the gating domain of a Ca^{2+}-activated K^+ channel complexed with Ca^{2+}/calmodulin. Nature. 2001;410(6832):1120–4.
80. Keen JE, Khawaled R, [illegible] Domains responsible for constitutive and Ca^{2+}-dependent interactions between calmodulin and small conductance Ca^{2+}-activated potassium channels. J Neurosci. 1999;19(20):8830–8.
81. [illegible] Maylie J, [illegible] Adelman JP. [illegible] EMBO J. [illegible]
89. [illegible]

Chapter 16
Membrane Lipids and Modulation of Vascular Smooth Muscle Ion Channels

Alex M. Dopico, Anna N. Bukiya, and Guruprasad Kuntamallappanavar

Abstract In vascular smooth muscle (VSM), lipids regulate the function of many cellular proteins, including ion channels. Lipids can directly target ion channels or regulate ion channels indirectly, that is, through interactions with cell signalling molecules that eventually target the ion channel. We summarize current knowledge on VSM ion channel regulation by lipids, whether lipids are originated in the smooth muscle itself, vascular endothelium, or paracrine or circulation sources, with a focus on possible direct interaction between lipid molecules and the proteins that constitute the VSM ion channel complex. We also describe the contribution of such lipid-VSM ion channel interaction to VSM physiology and pathophysiology. However, more than one mechanism or type of interaction may be involved in lipid-induced regulation of VSM ion channels and some mechanisms are not mutually exclusive, which adds significant complexity to the field. Understanding the mechanisms that underlie lipid modulation of VSM ion channels, however, is necessary for developing novel therapeutic approaches to improve vascular function and prevent or counteract widespread cardio/cerebrovascular disorders.

Keywords Ion channels • Membrane lipids • Lipid signaling • Smooth muscle • Lipid-recognition protein site

Introduction

In contrast to other major classes of biologically-relevant organic compounds (nucleotides, amino acids and carbohydrates), which can be grouped around a rather precise chemical definition, the term lipid engulfs a vast number of molecules that merely share "a much better solubility in organic solvents than in water" [1].

A.M. Dopico, M.D., Ph.D. (✉) • A.N. Bukiya • G. Kuntamallappanavar
Department of Pharmacology, College of Medicine, The University of Tennessee Health Science Center, 874 Union Ave., Memphis, TN 38163, USA
e-mail: adopico@uthsc.edu

I. Levitan, A.M. Dopico (eds.), *Vascular Ion Channels in Physiology and Disease*, DOI 10.1007/978-3-319-29635-7_16

Thus, lipids can be found as structural and dynamic components of cell membranes and organelles, cytosolic and inter-organelle signaling molecules, local and paracrine mediators, hormones, and even neurotransmitters (e.g., bile acids in the olfactory systems of sea lamprey and salmon) [2, 3]. Within this myriad of compounds, many regulate key aspects of vascular smooth muscle (VSM) biology, from tissue proliferation to control of myogenic tone. As expected for an excitable tissue, lipid regulation of VSM function often entangles ion channel proteins, whether involving direct interaction between lipid and channel complex subunits or regulating channel function or expression *via* lipid targeting of receptors other than the ion channel complex, such receptors targeting the ion channel *via* protein-protein interactions or through additional cell signaling molecules. The focus of this review, however, is the regulation of VSM ion channels by lipids that are generated in the myocyte itself or its immediate vicinity (i.e., vascular endothelium), with special emphasis on possible direct interactions between lipid molecule and SM channel protein and eventual modification of channel function (Fig. 16.1). Recognizing some unavoidable overlapping, we present the information in this sequence: lipid regulation of voltage-gated, ligand-gated and mechano-gated ion channels, starting with channels that predominate in the plasmalemma and ending with intracellular ion channels of established significance in SM function. For each channel, we will

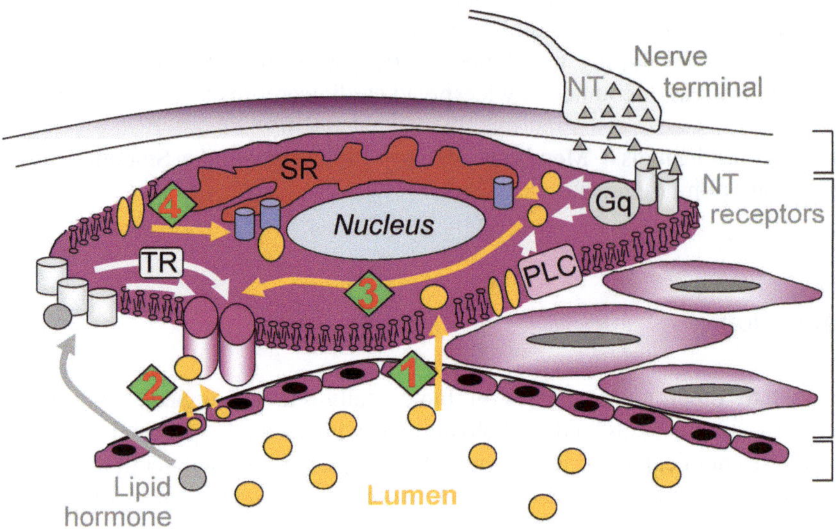

Fig. 16.1 Major pathways by which lipids target ion channels in vascular smooth muscle. Orange arrows correspond to pathways including a direct interaction between lipid and ion channel proteins. Origin of the lipid: (1) from circulation; (2) in the endothelium and accessing smooth muscle effectors; (3) within the myocyte, functioning as messenger/signaling molecules; (4) released from the plasma membrane lipid bilayer. *A* adventitia, *E* endothelium, *Gq G protein-coupled to protein kinase C*, *NT* neurotransmitter, *PLC* phospholipase C, *SM* smooth muscle, *SR* sarcoplasmic reticulum, *TR* transcription/translation regulators

follow, as much as possible, the following order: regulation by phosphoglycerides, phosphoinositides, lysophospholipids, diacylglycerol and analogs, fatty acids (FA), prostaglandins and eicosanoids, sphingolipids, and steroids.

Inwardly Rectifying K^+ Channels (Kir)

Like their counterparts in other excitable tissues, Kir in VSM are main determinants of the myocyte resting potential because their inward rectification allows them to pass significant current at the cell resting potential levels (−120 to −40 mV) [4]. Kir channels are tetramers with a symmetric topology around the pore [5]. Each monomer is comprised of a transmembrane (TM) "core" and a cytosolic domain, with each core having two TM segments, an outer helix and a pore-lining helix. There are seven Kir subfamilies (Kir1-7), with several subfamilies having multiple members (Kir 4.1 and Kir 4.2, etc.).

Phospholipids. Reconstitution of human Kir2.1 and Kir2.2 channels into liposomes of known lipid composition allowed researchers to screen the effects of membrane phospholipids on Kir activity [6]: normal gating of these channels required phosphatidylinositol 4,5 bisphosphate (PIP_2) and *bilayer-forming* phospholipids with anionic headgroups. In particular, 1-palmitoyl 2-oleoyl phosphatidylglycerol increased Kir open probability (Po) and unitary conductance (γ) whereas neither neutral nor cationic phospholipids activated the channels. Oleoyl CoA inhibited Kir2.1 by antagonizing channel's activation by PIP_2. Thus, Kir channels are exquisite sensors of the lipid membrane microenvironment. Using Kir6.2 channel mutants, researchers have narrowed down the phospholipid-sensing residues to Glu308, Ile309, Trp311, and Phe315, which are residues lying on one side of the α helix within the channel C-terminus [7].

Phosphoinositides. Kir channel activation by PIP_2 has been documented by several groups [7–9]. Activation of Kir2.1 and Kir2.2 channels by PIP_2 is highly selective: PIP_2 cannot be substituted by PI(4)P or PI(5)P while phosphatidylinositol 3,4,5-trisphosphate (PIP_3) has minimal activity on the channel [9]. Amino acid motifs and bonds involved in PIP_2 sensitivity of Kir channels include residues 176–222 and 301–312 in the K_{ATP} channel (Kir6.2) C-terminus [8], and K188,R189 in the Kir2.1 protein. In G protein-gated Kir (GIRK) channels, the ionic bond between His69 and Asp228 in the CD loop decreased PIP_2 sensitivity [10]. In all instances, proposed PIP_2-interaction sites include positively charged amino acids.

Ceramides and sphingolipids. Sphingosine 1-phosphate (SP1) slowed the spontaneous pacemaker activity and hyperpolarized maximal diastolic potential in rabbit sino-atrial node cells [11]. SP1 activated a pertussis toxin-sensitive Kir (IK_{Ach}) current and reversed the increase in pacing rate evoked by treatment of the cells with isoproterenol. The mechanism of Kir activation by SP1 in sino-atrial myocytes remains elusive. To our knowledge, there is no report on sphingolipid modulation of VSM Kir.

Fatty acids. Arachidonic acid (AA) has been shown to potentiate currents conducted by human Kir2.3 channels [12]. Comparison of AA-induced potentiation in whole-cell current vs. inside-out patches pointed at a decreased inward rectification as a mechanism that could underlie the AA effect. Whether interference of channel rectification by AA involves direct interaction between AA and Kir channel protein remains unknown.

Prostacyclin. Cicaprost (a synthetic prostacyclin analog) dilation of rat tail arteries pre-constricted with phenylephrine has been linked to modulation of Kir channels. [13]. Cicaprost-induced vasodilation was attenuated by Ba^{2+} block of Kir but unaffected by blockers of Gi/Go, K_{ATP} or GIRK channels. Vasodilation by cicaprost was endothelium-dependent, yet did not require NO• [13]. A direct interaction(s) between prostacyclin and Kir proteins cannot be ruled out.

Cholesterol. Cholesterol (CLR) modulation of Kir channels is diverse, with both CLR-induced channel inhibition and activation being reported for different Kir members. Experiments on human Kir2.1 incorporated into liposomes of controlled lipid composition documented inhibition of Kir2.1 channel by CLR [14]. Remarkably, CLR enantiomer failed to do so, which led to the conclusion that a selective CLR-Kir2.1 protein interaction likely mediated CLR action. Data from CLR optical isomers on Kir currents in bovine aortic endothelial cells rendered a similar conclusion [15]. While ion current mediated by other Kir channels, such as Kir1.1, Kir4.1 and Kir6.2Delta36, were suppressed by increased membrane CLR, the activities of Kir3.4* (K3.4 S143T) and Kir7.1 were enhanced by CLR enrichment [16]. Despite the opposite effects of CLR on the function of Kir2.1 and Kir3.4*, substitutions at corresponding positions within the CD loop inhibited CLR actions. Further work revealed that the CD loop was a part of a regulatory site that also included residues in the G-loop, the N-terminus and the connecting segment between the C-terminus and the inner TM helix [17]. Altogether, these findings point at the existence of a "cytosolic belt" that surrounds the channel pore and modulates Kir's CLR sensitivity. A detailed description of CLR regulation of vascular endothelium and recombinant IR channels is presented in this book by *Levitan et al.*

Modulation of Kir channel function by CLR was also studied by placing rats on a high-CLR diet and evaluating the effect of this diet on Kir current using patch-clamp on freshly isolated atrial cardiomyocytes. Unlike many other members of the Kir family, IK_{Ach} was up-regulated by CLR [21]. Two-electrode voltage-clamp recordings from *Xenopus laevis* oocytes expressing Kir3.1/Kir3.4 heteromers (e.g. subunits that underlie cardiac IRK currents) revealed enhanced IKAch current following CLR enrichment. CLR modulation of IK_{Ach} current was independent of endogenous regulators of K_{Ach} function, such as PIP_2 and Gβγ [18]. Thus, a direct interaction between K_{Ach} and CLR may be hypothesized.

Voltage-Gated Ion Channels of the TM6 Superfamily

Ion channels of the TM6 superfamily result from symmetric tetrameric assembly around the pore, which can be highly selective for K^+, Na^+ or Ca^{2+}, rendering different families. In all cases, activity increases with more positive voltages, the gating charges being located in S4 of the TM6 structure. The TM6 structure of "primarily" voltage-gated K^+ (K_V) channels results from association of four homomeric polypeptides while those of voltage-gated Na^+ (Na_V) and voltage-dependent Ca^{2+} channels (VDCC) results from association of equivalent repeats (I to IV) within a single polypeptide, with each repeat corresponding to a K_V monomer [19–21].

Voltage-Gated Ca^{2+} Channels (VDCC)

Upon plasma membrane depolarization, VDCC activation generates Ca^{2+} influx and thus, initiation of Ca^{2+}-dependent signal transduction. There are three VDCC subfamilies (Ca_V1-3) with each family having multiple members. According to biophysical and pharmacological properties, VDCC-mediated Ca^{2+} currents have been grouped into L-, N-, P/Q-, R-, and T-types [20]. Although L-, T-, and P/Q-types have all been detected in VSM [22, 23], L-type channels seem to prevail as regulator of VSM contractility. These channels are characterized by high voltage of activation, large unitary conductance and slow voltage-dependent inactivation [20]. L-type VDCC contains $\alpha 1$, $\alpha 2\delta$, β and γ subunits. The $\alpha 1$ subunit contains distinct recognition sites for VDCC selective blockers: 1,4-dihydropyridines, phenylalkylamines and benzothiazepines. These agents have vasoregulatory properties with the former being widely used as effective vasodilating therapy [24].

Phosphoinositides. L-type Ca^{2+}-channel activation by PIP_3 was documented in rat portal vein myocytes [25]. Moreover, authors postulated this mechanism as mediator of angiotensin II-induced Ca^{2+} influx into VSM.

Lysophospholipids. Lysophosphatidic acid (LPA) produces a diverse range of effects in the cardiovascular system [26]. In isolated rabbit ventricular myocytes, LPA increased action potential duration at 90 % repolarization and Ca^{2+} current peak current. However, current potentiation required Gi proteins [27]. Thus, the mechanism of LPA-induced VDCC activation seems to be mostly (if not fully) indirect. Whether a direct LPA-VDCC interaction takes place in VSM remains unknown.

1,2-Dioctanoyl-glycerol. 1,2-Dioctanoyl-glycerol (DiC8) is an analog of the naturally produced diacylglycerol (DAG) and a potent inhibitor of L-type VDCC [28]. DiC8 effect was related to a decrease in channel slope conductance and a hyperpolarizing shift in steady-state inactivation. Thus, as another VDCC inhibitor (AA; see above), DiC8 facilitates current inactivation.

Palmitoyl-l-carnitine. Palmitoyl-L-carnitine (L-PC) is an intermediate product in mitochondrial FA oxidation. L-PC increased the amplitude of guinea pig ventricular myocyte contractions [29]. Following channel reconstitution into lipid bilayers, 1 μM L-PC increased VDCC Po within the first minute of application. However, a decrease in Po was detected with protracted L-PC exposure [30]. In addition, higher L-PC concentrations (10 μM) decreased unitary conductance (γ). Considering that artificial planar lipid bilayers include a small number of second messengers, if any, it is conceivable that L-PC directly interacts with VDCC proteins and/or their immediate lipid microenvironment.

Fatty acids. An early study documented AA-induced inhibition of L-type VDCC in frog ventricular myocytes. This effect was evident from a negative shift in the inactivation curve, with peak conductance remaining unchanged. The study also revealed that increased protein phosphatase activity participated in AA action on L-type VDCC [31]. Inhibition of VDCC by AA was also reported in rat ventricular myocytes [32]. Interestingly, application of AA to the intracellular side of the myocyte membrane in the whole-cell patch clamp configuration did not affect current density nor modified the decrease in L-type Ca^{2+} current evoked by application of AA to the extracellular side of the membrane. Authors suggested that AA was targeting sites located at the extracellular side of the plasma membrane [42]. The study [32] also demonstrated that low μM levels of AA progressively decreased cell shortening. Thus, AA-induced block of L-type VDCC may have physiological consequences at the vessel level.

Location of AA-recognition sites in VDCC proteins has remained elusive. AA and anandamide inhibited binding of 1,4-dihydropyridine while increasing binding of 1,5-benzothiazepine and phenylalkylamine to rabbit VDCC channels [33]. This raised speculation that AA might directly interact with a site(s) located in the VDCC α1 subunit. Data from recombinant VDCC expressed in HEK cells linked the inhibitory effect of AA to the stabilization of the channel in a "deep closed-state" conformation, with palmitoyl groups of the β2a subunit interfering with channel inhibition by AA [34]. Whether β2a subunits . provide additional AA-recognition sites or contribute to destabilize the putative α1-AA interaction remains unknown.

Cytochrome P450-mediated metabolism of AA is an important pathway for the formation of lipid mediators, with omega-hydroxylation of AA taking place in the vasculature and resulting in release of 20-hydroxyeicosatetraenoic acid (20-HETE) [35]. In pig airway SM 20-HETE-induced inotropic effect was lost in presence of nifedipine [36]. However, patch-clamp recordings demonstrated that 20-HETE activated a nonselective cationic inward current (presumably mediated by a TRP-type channel). Therefore, 20-HETE-induced membrane depolarization and the resulting positive inotropic effect have been attributed to activation of nonselective cation influx into SM by 20-HETE rather than to direct activation of VDCC by this lipid. Subsequent attempts to study a possible effect of 20-HETE on L-type VDCC, however, showed that 20-HETE-induced VDCC current activation in rat isolated cardiomyocytes [37]. This study revealed that 20-HETE-induced NADPH-oxidase-derived superoxide production, which in turn activated L-type VDCC *via* protein kinase C (PKC).

Direct interaction between cytochrome P450 metabolites of AA (epoxyeicosatrienoic acids; EETs) and porcine L-type VDCC was reported in artificial lipid bilayers: EET (nM) reduced VDCC Po and γ, and accelerated current inactivation [38]. This inhibition of current remained in presence of protein phosphatases, suggesting that dephosphorylation was not the mechanism underlying EET-induced inhibition of VDCC. Interestingly, restricting 11,12-EET to the hydrophobic phase of the bilayer did not affect channel inhibition. In addition, there was no selectivity among different EET isomers to inhibit VDCC. Altogether, these findings raised the speculation that EETs affected VDCC activity *via* bilayer mechanisms rather than by selective lipid-protein interaction(s).

A key question that deserves consideration is the physiological relevance of bilayer findings to tissue and organ physiology. EET-induced channel inhibition reported in artificial bilayers seems to fit with the well-known vasodilatory properties of this group of AA metabolites. However, 14,15-EET evoked constriction of porcine coronary artery rings *via* VDCC. While 14,15-EET produced relaxation of arterial rings preconstricted with a thromboxane mimetic, application of 14,15-EET to rings preconstricted with acetylcholine or KCl resulted in additional constriction. The latter was mediated by 14,15-EET-induced Ca^{2+} influx in VSM *via* VDCC [39]. Authors suggested that EET-induced increase in intracellular Ca^{2+} (Ca^{2+}_i) could favor vasoconstriction while opposing vasodilation. In synthesis, it is conceivable that many mechanisms underlie differential effects of EETs on VDCC and thus, vessel diameter, including bilayer-mediated mechanisms, selective protein-lipid interactions or involvement of signaling cascades.

Finally, the lipoxygenase metabolite 12(S)-hydroxyeicosatetraenoic acid (12(S)-HETE) has been shown to constrict renal vessels [40]. 12(S)-HETE increased Ca^{2+}_i in renal microvascular myocytes, this increase being attenuated by VDCC blockade. The molecular mechanisms involved in VDCC-dependent 12(S)-HETE vascular effects remain unknown, and a direct interaction(s) between VDCC proteins and 12(S)-HETE cannot be ruled out.

Polyunsaturated fatty acids. Polyunsaturated FAs (PUFAs) have hypotensive and antiarrhythmic effects [41]. Moreover, PUFAs inhibited L-type VDCC activity in isolated rat ventricular myocytes [42], guinea pig tracheal SM [43] and ventricular myocytes [44], and a rat aortic SM cell line [45]. The mechanisms underlying this PUFA action remain unknown.

Sphingosine and sphingosine-1-phosphate. Sphingosine (SPH) and sphingosine-1-phosphate (S1P) are important signaling molecules. S1P evoked vasoconstriction in several vascular beds, including renal afferent microvasculature [46]. Vasoconstriction by S1P was abolished by L-type VDCC blockers [46]. Whole-cell patch-clamp recordings revealed block of L-type VDCC-mediated currents by (−) SPH, an S1P-related compound [47]. Whether the effects of SPH and S1P on VDCC involve selective lipid-protein interaction(s) remains to be determined.

Ceramides. Ceramides consist of SPH and a FA. C2, a membrane permeable ceramide analog, inhibited L-type VDCC in rat ventricular myocytes [48]. C2 increased VDCC inactivation without altering other current properties. The mechanism(s) of this increase remains unknown, although C2 and interleukin-1 may share a pathway for regulating VDCC: C2 had no effect on VDCC current properties in presence of maximally effective concentration of interleukin-1 that inhibited VDCC current [48].

Cholesterol. Exposure of rat pulmonary artery myocytes to CLR-rich liposomes increased VDCC current by inducing a right shift in the inactivation potential [49]. CLR-induced increase in L-type current was accompanied by decreased sensitivity to (–,R)-PN-202-791, a dihydropyridine antagonist. Interestingly, no change in peak current was detected. Authors proposed that CLR effect on L-type VDCC was likely due to alterations in channel function without modification in channel expression [49].

Lipophilic vitamins. Inhibition of L-type VDCC by 24R,25-dihydroxyvitamin D3 was reported in myocytes isolated from rat tail artery [50], although the mechanism underlying this action remains unknown.

Voltage-Gated Na^+ Channels (VGSC)

Upon activation by plasma membrane depolarization, VGSC generate inward Na^+ currents that depolarize the membrane even further [21]. Thus, VGSCs play a central role in cardiac contractility. Moreover, both loss- and gain-of-function mutations of cardiac VGSCs are associated with cardiac arrhythmias [51]. As other members of the TM6 superfamily, VGSC may be associated with non-pore-forming proteins, such as β subunits [52].

Fatty acids. Following human cardiac VGSC (SCN5A and DeltaKPQ-mutants) expression in HEK 293 cells, acute application of docosahexaenoic or eicosapentaenoic acid decreased the peak of the sustained component of the rapidly inactivating Na^+ current and shifted steady-state inactivation toward hyperpolarizing potentials [53]. PUFA-induced VGSC inhibition was concentration-dependent. However, the mechanism(s) of VGSC inhibition by PUFA remains elusive.

Ceramides. In cultured rat myoblasts, C(6)-ceramide reduced the peak amplitude of VGSC-mediated current and shifted the inactivation curve toward hyperpolarizing potentials. This inhibition was attenuated by inositol 1,4,5-trisphosphate (IP3) receptor blockers [54]. Thus, ceramide inhibition of VGSC likely involves secondary messenger pathway(s), yet a direct ceramide-VGSC interaction cannot be ruled out.

Voltage-Gated Potassium Channels (K_V)

Upon membrane depolarization, K_V channel activation leads to outward current that results in decreased depolarization and return of plasma membrane potential towards resting levels. Thus, K_V plays a key physiological role, particularly in excitable tissues including VSM [55]. Data on membrane lipid modulation of K_V channels are scarce, yet they reveal an important role of such regulation in common pathologies.

Eicosanoids. 15-Hydroxyeicosatetraenoic acid (15-HETE) production is activated by chronic hypoxia, with newly produced 15-HETE playing a critical role in hypoxia-induced pulmonary vasoconstriction. Perfusion of pulmonary artery rings with 15-HETE-containing solution resulted in vasoconstriction, which was abolished by K_V blockade but remained in presence of K_{ATP} or large conductance, voltage- and

Ca^{2+}-gated K^+ (BK) channel blockade [56]. Moreover, 15-HETE inhibited whole-cell K_V currents in freshly dispersed rabbit pulmonary artery myocytes [56]. Direct 15-HETE-K_V channel interactions remain unexplored.

Ceramides and sphingolipids. C6-ceramide inhibited 4-aminopyridine-sensitive K_V currents in rat pulmonary but not mesenteric artery myocytes [57], raising speculation that ceramide's vasoactive properties are selective towards particular vascular territories. Ceramide-induced inhibition of K_V current in rat pulmonary artery myocytes was significantly reduced by selective block of either $K_V2.1$ or $K_V1.5$ channels, suggesting that ceramide inhibited both major components of K_V current in pulmonary artery myocytes. A PKC-ζ pseudosubstrate inhibitor prevented ceramide-induced K_V current inhibition and rat pulmonary artery constriction. A most interesting finding, however, is the description of K_V current inhibition by *Bacillus cereus* bacterial sphingomyelinase (SMase) in both rat and human pulmonary artery myocytes. SMase application could result in increase of endogenous ceramide production, the latter eventually playing a critical role in K_V current regulation. In addition, SMase constricted human pulmonary arteries *in vitro*, underscoring the vasoactive properties of endogenously produced ceramide [57].

Ca^{2+}-Activated Ion Channels

Under this subheading, we include several ion channels that are heterogenous from both phylogenetic and phenotypic considerations. However, they all gate in response to Ca^{2+}_i within this divalent physiological concentration range, which is critical to understand the role of these channels in VSM physiology, pathology, and their modulation by lipids.

Ca^{2+} and Voltage-Gated, Large Conductance K^+ (BK; slo1) Channels

BK channels, encoded by *KCNMA1* or *slo1* in mammals, are widely expressed and participate in several physiological processes [58]. BK channels are homotetramers around a central pore that shows high conductance for K^+ and exquisite selectivity over other monovalents. BK channel-forming (slo1) proteins share with K_V channels the S1-S6 TM core [59] and thus, are usually included in the TM6 superfamily. Slo1 channel's voltage-dependence of gating, however, is shallower than that of K_V [59]. Thus, at zero or submicromolar (resting) Ca^{2+}_i, BK activity within the membrane potentials found in VSM (−60 mV at rest to −20 mV during depolarization) is rather low [60]. When compared with their K_V counterparts, however, slo1 proteins have an additional S0, which leads to an exofacial N-end, and a long (>800 residues) cytosolic tail domain (CTD), which is responsible for sensing changes in $Ca^{2+}{}_i$ within physiological levels. In VSM, an increase in $Ca^{2+}{}_i$ near the CTD Ca^{2+}-sensors due to VDCC-mediated Ca^{2+} influx or Ca^{2+} release from internal stores (in particular, *via* Ca^{2+} "sparks" from RyR activity) [61], leads to BK channel activation. This generates outward currents that drive the membrane towards more negative potentials and thus, provide a negative feedback on Ca^{2+} entry *via* VDCC, thereby blunting increase in $Ca^{2+}{}_i$, opposing SM contraction while favoring relaxation and vasodilation [62, 63].

Abundant literature documents that lipids such as phosphoinositides, FA, prostanoids, EETs, leukotrienes (LT), CLR and other steroids modulate SM BK channel currents. In a few cases, we refer the reader to previous reviews on BK channel regulation by specific lipids. Collectively, the majority of studies report that lipids potentiate the activity of BK channels from SM cells, cholesterol being a glaring exception.

Phosphoinositides. BK channel activation by PIP_2 and other phosphoinositides has been reported in cerebral artery myocytes and in *Xenopus* oocytes expressing slo1 cloned from cerebral artery SM (cbv1 channels) [64]. PIP_2-induced BK current potentiation resulted from increased Po without modification in γ. Further, PIP_2 action required $Ca^{2+}{}_i$ while being independent of voltage [64].

PIP_2-induced BK activation required negative charge and the inositol moiety within the phospholipid head group. In addition, PIP_2 was ~10 times more effective in activating BK channels than more water-soluble analogues such as diC4 and diC8, a difference that could be explained by the shorter analogs having lower affinity for a site(s) and/or poorer partitioning into the membrane hydrophobic core. Moreover, phosphoinositides were more effective when applied to the intracellular side of the membrane. Thus, if membrane partitioning is required for PI to access a site of action in the slo1 channel, this site should be located in the TM and/or near the intracellular region of the protein. Indeed, R334, K335, K336, identified as integral parts of a "PIP_2 sensor", are located in the S6-S7 linker of the cbv1 subunit [64].

PIP_2-induced increase in BK activity was more pronounced in cerebral artery SM (where β1 subunit is the most highly expressed BK auxiliary subunit) than in skeletal muscle cells (where β3 prevails). Moreover, β1 amplified PIP_2 action on cbv1 channels. Collectively, data strongly suggest a role for β1 in amplifying PIP_2 action on VSM BK channels [64]. Whether this amplification involves distinct PIP_2-recognition sites in β1 or allosteric coupling between β1 and PIP_2-bound cbv1 is under investigation. Pharmacological (Fig. 16.2) and other manipulations showed that PIP_2 direct activation of VSM BK channels led to cerebral artery dilation, and *endogenous* PIP_2 regulated myogenic tone and cerebral artery diameter independently of circulating, neural and endothelial factors [64] (Fig. 16.3).

Fatty Acids. Activation of BK channels by low μM FA has been reported in native myocytes from rabbit coronary [65] pulmonary [66, 67]; rat [68]; bovine [69] and porcine coronary arteries [70]; mouse aorta [71]; human umbilical artery [72] and rat pulmonary arteries [73]. FAs increased BK channel activity in a dose-dependent manner [65, 73], this FA action persisting over a wide range of $[Ca^{2+}]_i$ and $[Mg^{2+}]_i$ [65]. FA-induced activation of BK channel was found in the virtual absence of internal Ca^{2+} [71] and with the voltage sensors in resting state [74]. Thus, it was suggested that FA, i.e., docosahexanoic acid (DHA) activation of BK

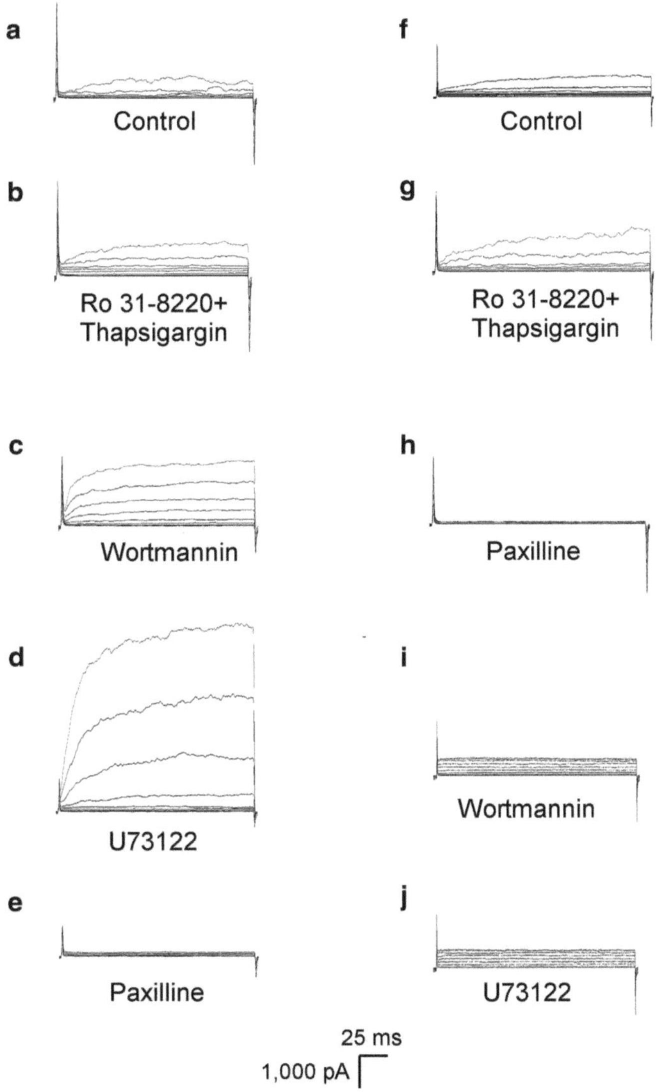

Fig. 16.2 Endogenous PIP2 activates native BK currents in the presence of blockers of PLC-mediated PIP2 downstream products. Perforated patch-recordings from two (**a–e** and **f–j**) freshly isolated cerebral artery myocytes bathed in physiological saline solution. Total outward K^+ currents were recorded in continuous presence of 5 mM 4-AP and 0.1 mM niflumic acid. Bath application of 2 μM Ro31-8220 and 0.2 μM thapsigargin increases current by 95 % (**b** vs. **a**; **g** vs. **f**). Subsequent inhibition of PI3 kinase by 5 nM wortmannin further increases current (**c**). Inhibition of PLC by 25 μM U73122 drastically increases current (**d**), likely due to PIP2 buildup in the membrane. The current is blocked by 0.3 μM paxilline (**e, h**), indicating it is mediated by BK channels. Pre-application of paxilline prevents both wortmannin (**i**) and U73122 (**j**) actions. ©Vaithianathan et al., 2008. Originally published in Journal of General Physiology. doi:10.1085/jgp.200709913

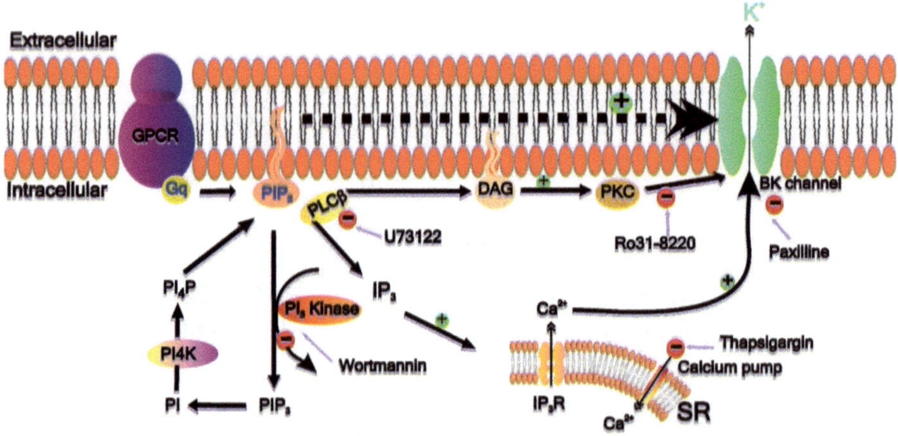

Fig. 16.3 Indirect and direct mechanisms of PIP_2 action on BK currents. PLC-mediated cleavage of PIP_2 results in PIP_3, IP3 and diacylglycerol (DAG). IP3 releases Ca^{2+} from SR, which raises Ca^{2+}_i, whereas DAG activates PKC. BK channel modulation by PKC and Ca^{2+}_i controls degree of vascular myocyte contraction. However, PIP_2 may directly activate BK channels (*bold dashed black arrow*), which involves the negative charges of the phosphoinositide headgroup and the sequence of positive residues RKK in the channel S6-S7 cytosolic linker. In intact myocytes, inhibition of PLC by U73122 increases BK current in presence of a PI_3 kinase blocker (wortmannin), PKC inhibitor (Ro 31-8220), and a selective blocker of the SR Ca^{2+}-ATPase (thapsigargin). Under these conditions, PIP_2 level is expected to rise, which renders BK channel activation. GPCR G *protein-coupled receptor* ©Vaithianathan et al., 2008. Originally published in *Journal of General Physiology*. doi:10.1085/jgp.200709913

channels was due to destabilization of the closed conformation of the pore-gate "domain" [71, 72, 75].

FA action persisted when the lipid was applied by bath perfusion [65, 69, 72, 75], indicating that the FA-recognition site(s) was accessible from the inner membrane leaflet. In addition, SAR studies revealed that long-chain FAs were more effective than their short-chain counterparts in activating VSM BK channels [65]. There was no significant difference, however, between the effects of saturated and unsaturated FAs [65]. The mechanisms and targets underlying differential modulation of BK currents by FA of variant structure remain unidentified. BK β1 subunits, however, played a critical role in DHA potentiation of BK current [71, 72]. Furthermore, β1 residues R11 and C18 were involved in DHA action on β1-containing BK channels [74]. Whether this structural feature applies to FA other than DHA remains to be determined.

The physiological and pathophysiological consequences of BK channel modulation by FAs are under active investigation. DHA activated BK . channels and dilated isolated coronary arteries. These DHA actions have been associated with DHA's beneficial effects on the cardiovascular system [68, 70]. Likewise, this FA lowered blood pressure by directly activating BK channels in aortic SM [71]. In turn, AA-induced relaxation of pulmonary artery was mediated by BK current modulation [76], yet the exact contribution of a direct FA-BK interaction to AA-induced vasodilation remains unclear.

Prostanoids. PGI_2 and PGE_2 have been shown to activate BK currents in rat tail [77], human coronary [78], and porcine coronary [79] artery myocytes. In contrast with these findings in VSM, PGE_2 inhibited Spontaneous Transient Outward Currents (STOCs; determined by BK activation) in guinea pig detrusor myocytes [80].

PGI_2-induced BK channel activation required activation of cAMP-dependent PKA [77]. However, in other VSM, PGI_2-induced activation of BK current involved a cAMP-independent, Gs protein-dependent component [81]. Interestingly, PGE_2-induced BK activation required cAMP-stimulated cross-activation of PKG, but not of PKA [78]. On the other hand, studies on guinea pig detrusor SM show that PGE_2-induced *inhibition* of Spontaneous Transient Outward Currents (STOCs) was due to activation of PKC *via* EP1 . receptors [80]. Thus, direct activation of BK channels by prostanoids remains to be established. It is possible to affirm, however, that prostanoid-induced activation of BK channels contributes to VSM relaxation and vasodilation [78, 79, 81].

Epoxyeicosatrienoic acids. EET and derivatives activate BK channels from rat renal [82]; rat coronary [83]; rat cerebral [84]; rat mesenteric [85]; mouse pulmonary artery [86]; bovine [87]; canine and porcine coronary [88]; human mammary [89] arteries. EET-related epoxyicosatetraeonic acids and 5-oxo-eicosatetraenoic acid potentiated BK current in human pulmonary artery and distal bronchi [90, 91], and in cerebral and mesenteric SM [92], with effective concentrations of EET ranging from nM to low μM [84, 91–94].

EET-induced activation of BK current was explained by increased Po without increase in γ [93, 94]. EET and 11,12-dihydroxyeicosatrienoic acid (DHET) potentiation of Po was related to several actions: dose-dependent increase in channel open-times, shortening of closed-times, and reduced open-close transitions [83, 84, 90–92, 94, 95].

EET-induced BK activation was detected in inside-out patches [83, 85] and following recombinant channel reconstitution into artificial lipid bilayers [94], suggesting that EET action did not require internal organelles, freely diffusing cytosolic signals or complex proteo-lipid domains. Moreover, EET-induced activation of BK channels was observed in cerebral and mesenteric SM lacking BK $\beta 1$ subunits [92]. Consistently, studies on HEK293 cells transfected with recombinant channels showed that EET's action on BK channels did not require $\beta 1$ subunits [96].

EET-induced BK activation led to SMC hyperpolarization and dilation of renal [82]; coronary [88], internal mammary [89], pulmonary [90] and mesenteric arteries [85]. Likewise, EETs and 20-hydroxyeicosatetraenoic acid (20-HETE) caused membrane hyperpolarization and relaxation of human distal bronchi [110, 111]. However, EET-stimulated. physical association of BK α and $\beta 1$ subunits in mitochondria could lead to reduced mitochondrial membrane potential and thus, depolarization, which could counteract SM relaxation, as reported in pulmonary VSM [86].

Leukotrienes. Among all LT tested (LTA_4, LTB_4, LTC_4, LTD_4, LTE_4), only LTB_4 directly activated BK channels cloned from cerebral artery SM after expression in *Xenopus* oocytes [97]. This activation required BK β1 but not other BK β (2–4) subunits. Computational modeling predicted a direct interaction between LTB_4 both the cholane steroid-sensing site (Thr-169, Leu-172, Leu-173) and two additional residues (Ala-176 and Lys-179) located in the β1 TM2 domain [97, 98]. Therefore, two lipids from very different families (LT and cholane steroids) can share a common site in a BK subunit. Moreover, their joint application failed to result in synergism on channel activation (Fig. 16.4).

Cholesterol. BK current sensitivity to various treatments that modified membrane CLR levels was first reported in cultured myocytes from human and rabbit aortas [99, 100] and, later on, extended to vascular and nonvascular myocytes from a wide variety of species [101]. Collectively, these studies show that in most cases increases in CLR decrease BK activity whereas decreases in CLR increase channel activity (reviewed by Dopico et al. [58]).

CLR reduced cbv1 Po following channel reconstitution into POPE:POPS (3:1 w/w) bilayers [102]. These results underscored that neither the complex cytoarchitecture of cell membranes nor cell signaling pathways was required for CLR inhibition of BK channels. Rather suggesting that channel-forming subunits were sufficient to support CLR action. In the same bilayer type, β1 subunits cloned from the same origin (rat cerebral artery myocyte) apparently failed to modify CLR action [102]. Bilayer data showing lack of a critical role for BK β1 in regulating CLR action on BK channels were validated by patch-clamp electrophysiology on native BK channels in cerebral artery myocyte membranes: CLR-depletion by methyl-β-cyclodextrin (MβCD) led to a similar increase in Po in C57BL/6 and *KCNMB1* K/O mouse [102]. Collectively, these results from recombinant and native cerebrovascular BK channels suggest that β1 do not drastically modify CLR action on channel steady-state activity.

The similarities in dwell-time changes and sterol potency (CLR $IC_{90} \approx 33$ mol%) reported in native membranes *vs.* artificial lipid bilayers strongly suggested that CLR action on BK channels was mediated by a recognition site(s) common to the different experimental systems: the two phospholipid species, the slo1 proteins, and/or some proteolipid product imported from the cell membrane preparation where the channel was originally expressed. Regarding slo1 proteins, we mapped seven CLR Recognition Amino acid Consensus (CRAC) motifs to the cbv1 CTD, with CRAC4 (which is adjacent to the inner membrane leaflet) playing a major role in CLR inhibition of BK channels [103] (Fig. 16.5). The relative role of the different CTD CRACs and the structural bases of CLR-sensing by BK channels is under current investigation in our laboratory [104].

Changes in BK current associated with hypercholesterolemia have been documented in various animal models [105]. Hypercholesterolemia blunted endothelium-independent, BK-mediated dilation of rabbit carotid [106, 107] and hindlimb arteries [108], and rat cerebral arteries [102]. This effect could contribute to vascular pathology associated with dyslipidemia, systemic hypertension, vas-

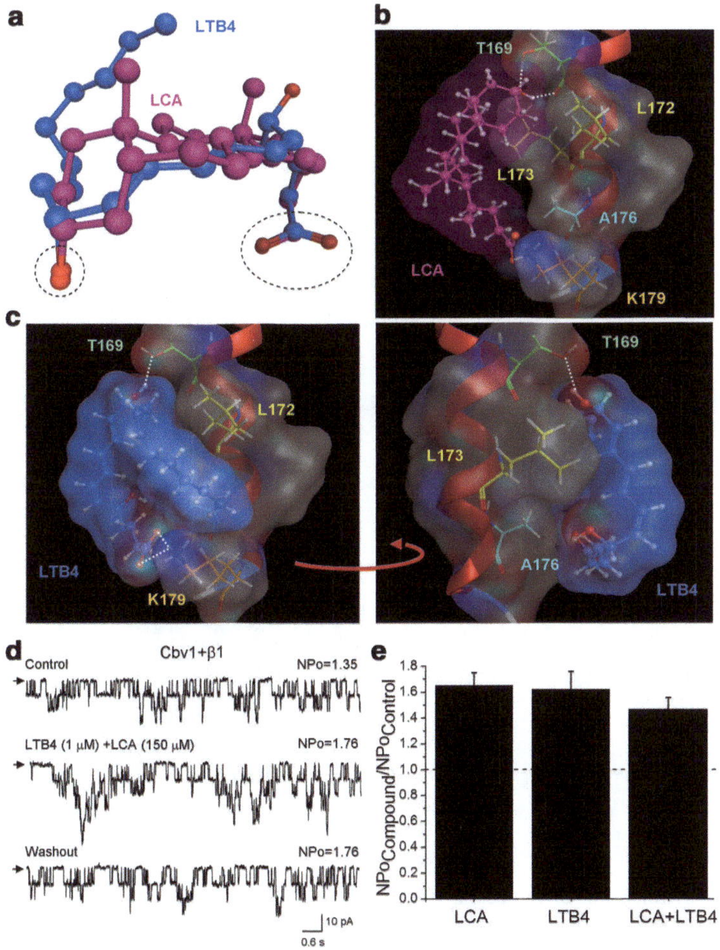

Fig. 16.4 LTB4 docking on the steroid-sensing site in the BK β1 subunit TM2. (**a**) Three-dimensional superposition of lithocholic acid (LCA; in *pink*) and LTB4 (in *blue*). *Black ovals* underscore spatial overlap between carboxyl and oxygen in LCA or LTB4 molecules. (**b**) Model of LCA docking on BK β1 TM2. LCA steroidal nucleus interacts with T169, L172 and L173, while the ionized carboxylate in LCA side chain resides near K179. (**c**) Model of LTB4 docking on the LCA-sensing site. In addition to tight steric interaction between T169, L172, L173 and the LTB4 molecule, A176 and K179 play a critical role in LTB4 recognition and retention. In **b** and **c**, the BK β1 TM2 α helix is in *red*; *dotted white lines* indicate hydrogen bonding. (**d**) Single-channel recordings from I/O patches excised from *Xenopus* oocytes expressing cbv1 + β1 subunits and probed with 1 μM LTB4 + 150 μM LCA. Vm = −40 mV, $[Ca^{2+}]_i = 10$ μM. (**e**) Averaged data showing similar increase in BK NPo by either single activator (LTB4, LCA) or their combination. @Bukiya et al., 2014. Originally published in *Journal of Biological Chemistry* 2014; 289: 35314–35325. ©the American Society for Biochemistry and Molecular Biology

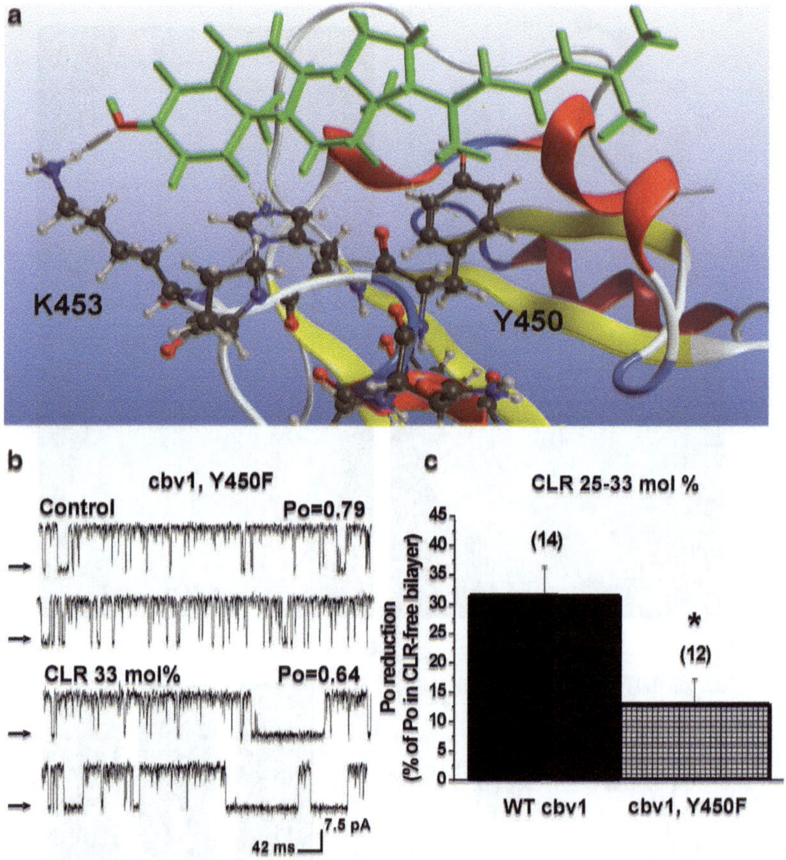

Fig. 16.5 (**a**) Snapshot of molecular dynamics simulations of CLR interaction with CRAC4 domain in the BK channel-forming α subunit CTD. CTD is composed of *yellow*, *red*, *grey* and *blue* ribbon structures; *green*: CLR. CLR hydroxyl hydrogen bonds with K453 while CLR hydrophobic nucleus resides near Y450. For visual clarity, part of the CTD downstream CRAC4 was hidden. (**b**) Single channel records from full-length cbv1, Y450F channels in CLR-free and 33 mol% CLR-containing bilayers. (**c**) Averaged inhibition of *wild type* cbv1 *vs*. cbv1 Y450F channel activity by 33 mol% CLR; *P = 0.0244. @Singh et al., 2012. Originally published in *Journal of Biological Chemistry* 2012; 287: 20509–20521. ©the American Society for Biochemistry and Molecular Biology

cular peripheral disease, cerebrovascular constriction and stroke. In particular, alcohol and CLR synergistically reduced BK current in cerebral artery SM. Remarkably, hypercholesterolemia and moderate–heavy alcohol intake both constitute risk factors for cerebrovascular disease [102].

Other Steroids. Among 21 C derivatives, *allotetrahydrocorticosterone* activated BK channels in guinea pig bronchial SM [109]. This action required cell integrity and subcellular pathways *via* pertussis toxin sensitive G-proteins [109]. Likewise, *progesterone* potentiated BK channels in porcine coronary [110] and sheep uterine artery [111] SMC. Progesterone action was mainly due to an increase in subsarcolemmal Ca^{2+} evoked by stimulation of bicuculline-sensitive GABA-A receptors [110]. Progesterone also upregulated BK β1 subunits and eventually decreased myogenic tone of sheep uterine arteries [111].

Testosterone (19 C) potentiated BK channels in pig coronary [112], rat mesenteric [113] and human umbilical [114] arteries. Testosterone-induced activation of BK channels in coronary arteries was attributed to increased cGMP synthesis and PKG activation [112].

Direct activation of BK channels by *17β-estradiol* (18 C) and *diosgenin* (phytoestrogen) was demonstrated in human [115] coronary, and human mesenteric, and sheep uterine [111] artery myocytes. Furthermore, estradiol increased the activity of slo1 + β1 channels but failed to activate slo1 channels [116]. Thus, BK β1 subunit seemed to be required for estradiol's direct action on BK channels. Studies from rat coronary [117], ovine uterine [118] and human coronary [119, 120] artery myocytes, however, showed that estrogen might modulate BK channel activity *via* indirect pathways. Indirect BK activation by estrogen involved the NO•/cGMP signaling pathway [117–119] and/or G-protein-coupled estrogen receptor activation [120]. Whether by direct, indirect or combined mechanisms, BK channels were involved in estradiol-induced dilation of rat mesenteric arteries, as this effect of the estrogen was significantly blocked by iberiotoxin (Ibtx) [121–123]. Finally, estrogen regulated gene and protein expression levels of slo1 and β1 subunits [111, 123, 124].

Xenoestrogens such as tamoxifen and analogues had dual action on BK channel steady-state, as reported in human jejunum, canine mesenteric and colonic [125–127], and mouse colonic and cerebral artery myocytes [128]. BK β1 was required for tamoxifen effect on colonic myocytes [127] while this subunit was unnecessary for tamoxifen to modulate BK channels in cerebral artery myocytes [128]. The tamoxifen analogue *Quat DME-oestradiol* activated BK channels *via* NO• signaling and thus dilated rat aorta [129].

Ca^{2+}-Activated K^+ Channels of Small (SK) and Intermediate (IK) Conductance

SK and IK channels are expressed both in endothelial and VSM cells [61]. In endothelial cells, activation of IK and SK channels leads to membrane hyperpolarization that increases the driving force for Ca^{2+} entry through non-voltage-gated Ca^{2+} channels. Thus, these channels contribute to endothelium-dependent relaxation by the Ca^{2+}-dependent production of NO• and/or some other hyperpolarizing factor(s) [130]. Furthermore, endothelial hyperpolarization is transmitted to underlying myocytes causing endothelium-dependent, VSM hyperpolarization and relaxation [131].

Phosphoinositides. Following transfection of CHO cells with IK, phosphatidylinositol 3-phosphate indirectly increased activity, this action involving a stretch of 14 amino acids in the IK carboxy terminus [132, 133]. To our knowledge, there are no studies reporting phosphoinositide modulation of IK or SK in VSM.

Lysophospholipids. Lysophospholipids directly activated SK and IK channels in human umbilical vein endothelial [134] and microglia cells [135]. Physiological consequences of this activation remain unknown.

Sphingolipids. The glycosphingolipid monosialotetrahexosylganglioside (GM1) caused endothelium-dependent relaxation of rat mesenteric arteries. This GM1-induced vasodilation involved IK channels while remaining unaffected by selective SK channel blockade [136]. Direct *vs.* indirect mechanisms in sphingolipid action on SK/IK channels remain to be addressed.

Epoxyeicosatrienoic acids. Although there is no report on direct action of EETs on SM SK or IK channels, there is some evidence that bradykinin-induced dilation of porcine coronary artery is mediated by EET. Bradykinin-induced hyperpolarization of porcine coronary myocytes was due to production of endothelial EET which, in turn, activated endothelial SK and IK channels [137].

Ca^{2+}-Activated Cl^- Channels (Cl^-_{Ca})

Cl^-_{Ca} is expressed in VSM where contributes to regulate vascular tone [138]. Because these channels carry an anionic current, Cl^-_{Ca} activation leads to depolarization, increased VSM tone and vasoconstriction [138].

Phosphoinositides. In rat pulmonary artery myocytes and HEK cells transfected with TMEM16A-encoded Cl^-_{ca} channels, PIP_2 inhibited Cl^-_{Ca} current. This modulation was likely due to direct binding between PIP_2 and Cl^-_{Ca} channel proteins [139]. Putative binding sites for PIP_2 on Cl^-_{Ca} are still under investigation.

Sphingolipids. Cerebrosides served as potential blocker of Cl^-_{Ca} in rat pulmonary artery SMC. Moreover, cerebroside slowed down activation of outward current and accelerated the decay of inward tail current in rat pulmonary artery myocytes [140]. Possible contributions of cerebroside modulation of SM Cl^-_{Ca} to vascular function remain unknown.

Cholesterol. Exposure of mouse portal vein myocytes to MβCD increased Cl^-_{Ca} Po. In contrast with data from BK channels [58], CLR-bound MβCD failed to modulate Cl^-_{Ca} Po. Thus, Cl^-_{Ca} seems to be sensitive to CLR depletion but not CLR enrichment. Modulation of Cl^-_{Ca} current to CLR depletion occurred without any change in reversal potential or kinetics [141].

Mechano-Gated Ion Channels and Transient Receptor Potential (TRP) Channels

TRP channels constitute a heterogeneous group of cationic channels that are primarily located in the plasma membrane. They have been sub-grouped into six subfamilies: TRPV (V for vanilloid), TRPC (C for canonical), TRPM (M for melastatin), TRPA (A for ankyrin), TRPP (P for polycystic) and TRPML (ML for mucolipin), with a subsequent number designating each specific member within subfamily [142]. TRPs have been reviewed elsewhere [143–145]. Given their activation by a wide variety of physiological stimuli, including lipid signals, TRPs as a whole participate in several key processes in the cardiovascular system, such as cell proliferation, differentiation and apoptosis [146]. In VSM, TRPV4, TRPV6, TRPC1, TRPC3, TRPC4, TRPC6 and TRPM4 constitute the most highly expressed members [147, 148] and, in consonance with endothelial TRPs, serve to control myogenic tone and vascular reactivity [146, 147, 149–151]. Most studies in VSM fall short from straightforwardly documenting direct, lipid-TRP protein interactions.

Phosphoinositides. Endothelin-activation of TRPC-like cationic currents in rabbit coronary artery myocytes is mediated by PIP_2 [181]. Moreover, authors showed that current stimulation by diC8-PIP_3 was blocked by anti-TRPC1 antibodies. On the other hand, endothelin receptor-mediated TRPC1 activity was inhibited by tPI-phospholipase C block. In addition, 1-oleoyl-2-acetyl-sn-glycerol (OAG), an analogue of the PI-PLC product diacylglycerol (DAG), increased PKC-dependent TRPC1 activity, which was prevented by PIP_2 depletion. These findings led to the conclusion that stimulation of endothelin receptors (type B) increased TRPC1 activity in a PKC-dependent manner, which involved both PI-PLC-mediated generation of DAG and a facilitating role by PIP_2 itself [152]. Likewise, PIP_3 was reported to facilitate TRPC6/7 activation by OAG in rabbit portal vein, yet not in mesenteric artery myocytes. PIP_3 effect was attributed to removal of PIP_2-mediated tonic inhibition [153]. Likewise, removal of PIP_2-mediated tonic inhibition appeared responsible for TRPC6-related current potentiation by angiotensin II in rabbit mesenteric artery myocytes. This potentiation was linked to DAG generation, yet independent of PKC. Thus, TRPC6 channel potentiation was interpreted as resulting from antagonism between PIP_2 and DAG [148]. Whether PIP_2 or PIP_3 directly binds to TRPC channels, as well as the structural bases of TRPC-phosphoinositide interactions remain undetermined.

Lysophospholipids. Structure Activity Relationship (SAR) studies on lysophospholipid modulation of TRPV1 expressed in HEK cells complemented with computational modeling revealed that lysophosphatidic acid (LPA) and analogs might directly interact with TRPV1 proteins and thus, modify channel function. Following development of a pharmacophore, a model was proposed to describe how these lipids could induce a conformational change that activated TRPV1 [154]. Whether these conclusions can be applied to TRPV1 or other TRP when evaluated in the actual environment of VSM remains unknown.

Remarkably, caveolin1-expressing VSM were more susceptible to oxidized low-density lipoprotein (oxLDL)-induced apoptosis, which was related to increased Ca^{2+} influx into the SM [155]. Lysophosphatidylcholine (LPC), a primary component of oxLDL, has been shown to activate nonselective cationic channels [156], but it remains unclear whether such activation involves direct targeting of TRP by LPC and, if so, whether such activation could participate in oxLDL-induced apoptosis of vascular myocytes.

Other oxidized lipids. 1-Palmitoyl-2-glutaroyl-phosphatidylcholine (PGPC) and 1-palmitoyl-2-oxovaleroyl-phosphatidylcholine (POVPC) are LPC analogs. At low μM levels, PGPC and POVPC potentiated TRPC5-mediated currents after channel expression in HEK293 cells. Authors further showed that the oxidized phospholipids activated TRPC5 channels in VSM, which could be linked to cell migration [187]. In contrast to LPC action on SM endogenous cationic current (see previous subheading), PGPC and POVPC did not evoke Ca^{2+} mobilization, and their action on TRPC5 depended almost completely on Gi/o protein signaling. Previously identified G protein-coupled receptors (GPCR) for oxidized phospholipids were not involved in PGPC and POVPC action, suggesting that the effects of these oxidized phospholipids on TRPC occurred through a previously unrecognized receptor or independently of receptors [157], possibly involving direct TRPC-PGPC/POVPC interactions.

A recent study in rat cerebral arteries also points to a functional interaction between lipid peroxides and endothelial TRP (in this case, TRPA1) channels. Authors proposed that SM relaxation resulted from electronic spread of endothelium-generated hyperpolarization at myo-endothelial gap junctions. In turn, endothelial hyperpolarization was the result of IK channel activation by TRPA1-generated "Ca^{2+}-sparklets". These localized Ca^{2+} signals resulted from the activation of a few, clustered TRPA1 channels, this activation being evoked by lipid peroxides originated from membrane lipids following increased endothelial levels of H_2O_2 and/or OH• [158].

Diacylglycerol. In rat cerebellar and posterior cerebral artery myocytes, the membrane-permeant DAG analog 1,2-dioctanoyl-sn-glycerol (DOG) stimulated cationic, TRP-like currents. DOG action, however, was ablated by PKC blockade [159] suggesting that DAG effect on current was largely mediated by PKC activation rather than by direct DAG-TRP interaction. In rabbit portal vein myocytes, native currents with a phenotype identical to that of recombinant TRPC6 channel-mediated currents were potentiated by bath application of DOG and 1-oleoyl-2-acetyl-sn-glycerol (OAG). DAG is the likely mediator of phenylephrine potentiation of such native currents *via* Gq-coupled alpha-1 adrenoceptors and eventual phospholipase-C stimulation [160]. Furthermore, currents mediated by heterologously expressed TRP6 cloned from mouse brain were resistant to PKC block and activated by the DAG analogs. Possible targeting of PKC in lipid action, however, remains inconclusive from these data. In the A7r5 aortic SM cell line, a current with a phenotype similar to channels described in rabbit portal vein myocytes [190] was modulated by membrane permeable DAG analogs, an effect that was PKC-independent [161]. Finally, depolarization of rat cerebral artery

SM and consequent vasoconstriction in response to PKC activation was blunted by antisense directed against TRPM4 [151]. Whether lipid (DAG in particular) modulation of PKC is involved in TRPM4-mediated control of cerebral artery tone remains to be established.

11,12 epoxyeicosatrienoic acid (11,12 EET). Endothelial-dependent dilation of cerebral arteries involves functional targeting of myocyte TRPV4 by 11,12 EET: the released lipid from endothelial cells activated TRPV4 channels in the arterial myocyte. Ca^{2+} influx through TRPV4 permeation led to increased RyR-mediated Ca^{2+} spark frequency and thus, increased BK channel-mediated STOCs and myocyte hyperpolarization, favoring SM relaxation and vasodilation. Indeed, sRNAi suppression of TRPV4 in intact arteries prevented 11,12 EET-induced SM hyperpolarization and cerebral artery dilation [162]. Likewise, 5,6 EET activated TRPV4 in a membrane-delimited manner, leading to Ca^{2+} influx *via* TRPV4-like channels in vascular endothelium. EET can be generated from AA and anandamide and thus, TRPV4 activation in vascular endothelial cells might therefore contribute to SM relaxation by endocannabinoids [163].

Internal Ca^{2+} Release Channels

Internal Ca^{2+} release channels serve fundamental roles in VSM physiology, including myogenic tone regulation, vascular reactivity, cell migration and apoptosis, as their coordinated function and integration with different Ca^{2+} influx pathways contribute not only to regulate overall cytosolic Ca^{2+} levels but also, and mainly, to generate a variety of Ca^{2+} signals ("waves", "sparks", "sparklets", etc.,), each having distinct tempo-spatial characteristics and physiological role [164–166]. With exception of the specific gating of Inositol 1,4,5-triphosphate (IP3) receptors by IP3 and analogs [167, 168], the modulation of internal release Ca^{2+} channels by lipids has not received much attention. Thus, modulation of each major ion channel family by several lipid species is presented within a single subheading.

Ryanodine Receptors (RyR)

RyR are intracellular Ca^{2+}-release channels found in the endoplasmic and sarcoplasmic reticulum (SR) membranes of most eukaryotes. Three RyR isoforms have been isolated: RyR1, which predominates in skeletal muscle; RyR2, which predominates in cardiac muscle, and RyR3, which is widely expressed, yet particularly abundant in brain neurons [169, 170]. Each isoform consists of a homotetramer of one RyR gene product, although native receptors are heterooligomers with subunit composition of $(RyR_{monomer})_4(FKBP)_4$ where FKBP refers to the FK506 Binding Protein [170, 171]. In VSM, RyR2 and RyR3 predominate. These RyRs may operate in orchestration with VDCC, BK, TRP channels and IP3 receptors to control a variety of Ca^{2+} signal modalities in the arterial myocyte and thus, control myogenic tone and vasomotion [147, 164, 172, 173]; (see Chap. 1 by McCarron et al. in this volume).

Phosphoinositide 3,5-bisphosphate (PI-3,5P_2) elevated $Ca^{2+}{}_i$ in VSM and thus contracted aortic rings. While the bulk of PI-3,5P_2 action remained intact in presence of IP3 receptor block, depletion of SR Ca^{2+} stores with thapsigargin or caffeine and/or ryanodine blunted the phosphoinositide-induced Ca^{2+} increase and greatly attenuated its associated contraction. Removal of extracellular Ca^{2+} or VDCC block reduced but did not totally suppress the Ca^{2+} or contractile responses to PI-3,5P_2, leading authors to advance that full and sustained aortic contractions to PI-3,5P_2 required the release of SR Ca^{2+} which likely involved RyR [174]. A direct RyR-phosphoinositide interaction remains to be probed.

Endothelium-derived *EETs* indirectly modulated SM RyR *via* SM TRPV4 channels: TRPV4-mediated Ca^{2+}-influx activated nearby RyR channels, with the resulting spark activity leading to BK current potentiation and, eventually, SM relaxation and vasodilation (reviewed in [175]). In contrast, RyR indirect activation by thromboxane analogs, i.e., *via* ligand interaction with thromboxane receptors, evoked constriction of renal arterioles [176].

Phosphatidylinositol (3,4,5)-Trisphosphate Receptors

The reader is referred to comprehensive reviews on the structural basis of IP3 receptor gating by IP3 [167, 168] and the receptor's contribution to VSM function [177, 178]. The interplay between RyR, IP3 and mitochondria in internal Ca^{2+} mobilization in SM is extensively reviewed by McCarron et al. in this volume.

Lipopolysaccharide (LPS) modulation of $Ca^{2+}{}_i$ oscillations has been linked to LPS-mediated activation of nuclear transcription factor NFκB in lung microvessel endothelial cells. LPC induced-enhancement of Ca^{2+} oscillation amplitude was blocked by xestospongin C, which pointed at IP3 receptors as mediators of LPS action [179]. A possible LPS-IP3 direct interaction remains to be explored.

Regarding *FA and analogs, prostaglandins and eicosanoids*, antagonism of IP3 receptors by 2-aminoethoxydiphenyl borate (2-APB) abolished oleoylethanolamide (OEA)-induced relaxation of rat small mesenteric arteries. OEA-induced relaxation involved Ca^{2+} released from IP3-sensitive endothelial stores by mechanisms involving RhoA kinase and phospholipase C [180], pointing at an indirect modulation of IP3 by OEA. Histamine stimulation of H1 receptors led to increased $Ca^{2+}{}_i$ *via* IP3 receptor activation in human aortic myocytes. Selective stimulation of prostaglandin receptors (EP2 or EP4) attenuated the histamine-evoked Ca^{2+} signal. However, *prostaglandin E2*-evoked reduction in Ca^{2+} signal was primarily mediated by EP2 receptors. Thus, IP3 receptors serve as common effectors of VSM actions of histamine and prostaglandin E2, two important inflammatory mediators [181]. On the other hand, prostaglandin F2α at sub-μM levels caused IP3-dependent Ca^{2+} release in rat intrapulmonary arteries, an action mediated by prostaglandin FP receptors [182]. Finally, in pulmonary artery myocytes, 15-hydroxyeicosatetraenoic acid

(15-HETE)-induced increase in Ca^{2+}_i was characterized by an initial rapid rise followed by a slow phase. While the slow phase was sensitive to VDCC and TRPC channel block, the initial rapid phase was reduced by the IP3 receptor antagonist 2-APB or depletion of RyR-operated stores by caffeine [183].

A few studies addressed the role of genetic, dietary or *in vitro* manipulation of steroid levels on IP3-generated signals in SM. Thus, IP3 receptor-mediated Ca^{2+}_i release was elevated in aortic SM from hypercholesterolemic, yet plaque-free, apoE−/− mice [184]. On the other hand, 24-h exposure of A7r5 aortic myocytes to *oxysterols* inhibited Ca^{2+} transients evoked by arginine vasopressin (AVP) and bradykinin. Moreover, 7β-hydroxycholesterol-induced reduction of AVP-stimulated Ca^{2+} signals were primarily due to modification of Ca^{2+} release from intracellular stores, with type 1 IP3 and RyR1 receptors being significantly downregulated by oxysterol treatment [185]. CLR-depleting *in vitro* treatment with methyl-β-cyclodextrin (MβCD) reduced IP3-induced cationic current and the associated constriction of pressurized, cerebral arteries in the mouse, actions that were reversed by CLR replenishment. IP3-induced vasoconstriction was reduced by shRNA directed against caveolin-1, a major structural determinant of caveolae integrity. Moreover, a cav-1 antibody co-immunoprecipitated caveolin-1, IP3 receptor type 1 and TRPC3, these proteins forming a localized unit in arterial SM, as revealed by immunoFRET. MβCD treatment disrupted such protein triad whereas CLR replenishment re-established IP3 receptor type 1 and TRPC3 channels spatial proximity. Thus, in cerebral artery SM, type 1 IP3 receptors in the SR co-localized with plasma membrane TRPC3 channels. This association required caveolin-1 and caveola integrity, the latter being ensured by the presence of CLR in the membrane(s) [186]. Whether any of these manipulations that perturbed steroid levels in VSM involve direct modulation of IP3 by the steroid remains to be determined.

ORAI and Stim1 Proteins

As in most excitable tissues, VSM presents the so-called store-operated Ca^{2+} (SOC) influx: Ca^{2+} depletion inside ER stores (mainly by IP3 receptor activation) leads to Ca^{2+} influx into the cell, which helps to refill the store with Ca^{2+} [187–189]. The emptying of the store is sensed by ER stromal interaction molecules (STIM1 and STIM2) *via* EF-hand motifs located in their N-termini, which face the ER lumen. In turn, STIM1 interacts with SOC channels constituted by tetra-(or hexa-)meric association of the Orai1 protein in the plasma membrane. Thus, regarding SOC influx, lipid modulation of STIM/Orai proteins is indirect, as conformational and functional changes in these proteins in response to store Ca^{2+} depletion are determined by *IP3* interaction with IP3 receptors, rather than with STIM/Orai themselves [189, 190]. Likewise, *S1P* indirectly modulated SOC influx in aortic SM isolated from C57BL/6 mouse; S1P effect was secondary to S1P targeting S1P type 2 and type 3 receptors [191]. A direct lipid-Orai interaction, however, had been hypothesized in cultured SM where STIM1, Orai1, and Orai3 acted as constitutive members of the thrombin-activated Ca^{2+} channel. Activation of the latter is *independent* of

IP3-mediated Ca^{2+} depletion or SOC influx [189, 192]. Rather, receptor activation led to generation of AA *via* PLA2. Eventually, *AA* is converted to *LTC4* [193], which was postulated to directly interact with Orai3. More recent electrophysiological data seemed to indicate that AA- and LTC4-activated currents were mediated by the same STIM1/Orai1/Orai3 channel complex [194]. Finally, up-regulation of this ion channel complex effector could lead to change in SM transcriptional regulation and neointimal formation [192].

Concluding Remarks

Lipid regulation of ion channel function in VSM constitutes a rapidly developing field of investigation, as mounting evidence indicates that such regulation participates not only in normal physiology but also in pathophysiology of cardiovascular disease. There is increasing evidence that modulation of ion channels in smooth muscle cells may result from direct recognition of lipids by the ion channel protein themselves. The sources of lipid species that seem to directly interact with ion channel proteins include the circulation, local nerves, endothelial cells and the vascular myocyte. The latter generates lipid species from cytosolic pathways, the plasma membrane or other organelles. Lack of structural biology data makes it difficult to discern whether direct recognition of a lipid species by an ion channel protein recognition site involves actual binding of the lipid to this site or, rather, the lipid binds to another, yet functionally-associated domain(s), while the identified region couples the actual binding site with regions that control channel gating, ion permeation and/or sorting and trafficking. Another complexity results from the quaternary nature of most ion channel proteins. One of the channel complexes where direct recognition of lipids by channel subunits has been best studied is the BK channel. Even for BK channels from a given source (e.g., cerebrovascular myocyte), direct lipid-channel interaction follows several modalities: direct lipid interaction with BK channel-forming subunits (e.g., cholesterol), direct recognition by channel-forming subunits with major amplification of lipid effect by the presence of regulatory β1 subunits (e.g., PIP2), and direct interaction with β1 subunits (cholane steroids and LTB4). Certainly, a more complex scenario waits ahead.

References

1. Dopico AM, Tigyi GJ. A glance at the structural and functional diversity of membrane lipids. Methods Mol Biol. 2007;400:1–13.
2. Lo YH, Bradley TM, Rhoads DE. High-affinity Ca^{2+}, Mg^{2+}-ATPase in plasma membrane-rich preparations from olfactory epithelium of Atlantic salmon. Biochim Biophys Acta. 1994;1192(2):153–8.
3. Li W, Sorensen PW, Gallaher DD. The olfactory system of migratory adult sea lamprey (Petromyzon marinus) is specifically and acutely sensitive to unique bile acids released by conspecific larvae. J Gen Physiol. 1995;105(5):569–87.

4. Smith PD, et al. KIR channels function as electrical amplifiers in rat vascular smooth muscle. J Physiol. 2008;586(4):1147–60.
5. Kuo A, et al. Crystal structure of the potassium channel KirBac1.1 in the closed state. Science. 2003;300(5627):1922–6.
6. Cheng WW, et al. Dual-mode phospholipid regulation of human inward rectifying potassium channels. Biophys J. 2011;100(3):620–8.
7. Cukras CA, Jeliazkova I, Nichols CG. Structural and functional determinants of conserved lipid interaction domains of inward rectifying Kir6.2 channels. J Gen Physiol. 2002;119(6):581–91.
8. Shyng SL, et al. Structural determinants of PIP(2) regulation of inward rectifier K(ATP) channels. J Gen Physiol. 2000;116(5):599–608.
9. D'Avanzo N, et al. Direct and specific activation of human inward rectifier K+ channels by membrane phosphatidylinositol 4,5-bisphosphate. J Biol Chem. 2010;285(48):37129–32.
10. Inanobe A, et al. A structural determinant for the control of PIP2 sensitivity in G protein-gated inward rectifier K^+ channels. J Biol Chem. 2010;285(49):38517–23.
11. Guo J, MacDonell KL, Giles WR. Effects of sphingosine 1-phosphate on pacemaker activity in rabbit sino-atrial node cells. Pflugers Arch. 1999;438(5):642–8.
12. Liu Y, Liu D, Krafte DS. Decrease of inward rectification as a mechanism for arachidonic acid-induced potentiation of hKir2.3. Eur Biophys J. 2002;31(7):497–503.
13. Orie NN, Fry CH, Clapp LH. Evidence that inward rectifier K^+ channels mediate relaxation by the PGI2 receptor agonist cicaprost via a cyclic AMP-independent mechanism. Cardiovasc Res. 2006;69(1):107–15.
14. D'Avanzo N, et al. Enantioselective protein-sterol interactions mediate regulation of both prokaryotic and eukaryotic inward rectifier K^+ channels by cholesterol. PLoS One. 2011; 6(4):e19393.
15. Romanenko VG, Rothblat GH, Levitan I. Modulation of endothelial inward-rectifier K^+ current by optical isomers of cholesterol. Biophys J. 2002;83(6):3211–22.
16. Rosenhouse-Dantsker A, et al. Comparative analysis of cholesterol sensitivity of Kir channels: role of the CD loop. Channels (Austin). 2010;4(1):63–6.
17. Rosenhouse-Dantsker A, Logothetis DE, Levitan I. Cholesterol sensitivity of KIR2.1 is controlled by a belt of residues around the cytosolic pore. Biophys J. 2011;100(2):381–9.
18. Deng W, et al. Hypercholesterolemia induces up-regulation of K_{ACh} cardiac currents via a mechanism independent of phosphatidylinositol 4,5-bisphosphate and Gbetagamma. J Biol Chem. 2012;287(7):4925–35.
19. Durell SR, Hao Y, Guy HR. Structural models of the transmembrane region of voltage-gated and other K^+ channels in open, closed, and inactivated conformations. J Struct Biol. 1998;121(2):263–84.
20. Catterall WA. Voltage-gated calcium channels. Cold Spring Harb Perspect Biol. 2011;3(8): a003947.
21. Peters CH, Ruben PC. Introduction to sodium channels. Handb Exp Pharmacol. 2014;221: 1–6.
22. Hansen PB. Functional and pharmacological consequences of the distribution of voltage-gated calcium channels in the renal blood vessels. Acta Physiol (Oxf). 2013;207(4):690–9.
23. Harraz OF, Welsh DG. T-type Ca^{2+} channels in cerebral arteries: approaches, hypotheses, and speculation. Microcirculation. 2013;20(4):299–306.
24. Glossmann H, Striessnig J. Molecular properties of calcium channels. Rev Physiol Biochem Pharmacol. 1990;114:1–105.
25. Le Blanc C, et al. Regulation of vascular L-type Ca^{2+} channels by phosphatidylinositol 3,4,5-trisphosphate. Circ Res. 2004;95(3):300–7.
26. Teo ST, et al. Lysophosphatidic acid in vascular development and disease. IUBMB Life. 2009;61(8):791–9.
27. Wei Y, et al. Lysophosphatidic acid increases the electrophysiological instability of adult rabbit ventricular myocardium by augmenting L-type calcium current. PLoS One. 2012;7(9):e45862.

28. Conforti L, Sumii K, Sperelakis N. Dioctanoyl-glycerol inhibits L-type calcium current in embryonic chick cardiomyocytes independent of protein kinase C activation. J Mol Cell Cardiol. 1995;27(5):1219–24.
29. Shen JB, Pappano AJ. Palmitoyl-L-carnitine acts like ouabain on voltage, current, and contraction in guinea pig ventricular cells. Am J Physiol. 1995;268(3 Pt 2):H1027–36.
30. Liu QY, Rosenberg RL. Activation and inhibition of reconstituted cardiac L-type calcium channels by palmitoyl-L-carnitine. Biochem Biophys Res Commun. 1996;228(2):252–8.
31. Petit-Jacques J, Hartzell HC. Effect of arachidonic acid on the L-type calcium current in frog cardiac myocytes. J Physiol. 1996;493(Pt 1):67–81.
32. Liu SJ. Inhibition of L-type Ca^{2+} channel current and negative inotropy induced by arachidonic acid in adult rat ventricular myocytes. Am J Physiol Cell Physiol. 2007;293(5):C1594–604.
33. Shimasue K, et al. Effects of anandamide and arachidonic acid on specific binding of (+) -PN200-110, diltiazem and (−) -desmethoxyverapamil to L-type Ca^{2+} channel. Eur J Pharmacol. 1996;296(3):347–50.
34. Roberts-Crowley ML, Rittenhouse AR. Arachidonic acid inhibition of L-type calcium (CaV1.3b) channels varies with accessory CaVbeta subunits. J Gen Physiol. 2009;133(4):387–403.
35. Kroetz DL, Xu F. Regulation and inhibition of arachidonic acid omega-hydroxylases and 20-HETE formation. Annu Rev Pharmacol Toxicol. 2005;45:413–38.
36. Cloutier M, et al. 20-HETE inotropic effects involve the activation of a nonselective cationic current in airway smooth muscle. Am J Physiol Lung Cell Mol Physiol. 2003;285(3): L560–8.
37. Zeng Q, et al. 20-HETE increases NADPH oxidase-derived ROS production and stimulates the L-type Ca^{2+} channel via a PKC-dependent mechanism in cardiomyocytes. Am J Physiol Heart Circ Physiol. 2010;299(4):H1109–17.
38. Chen J, et al. Inhibition of cardiac L-type calcium channels by epoxyeicosatrienoic acids. Mol Pharmacol. 1999;55(2):288–95.
39. Fang X, et al. Epoxyeicosatrienoic acids increase intracellular calcium concentration in vascular smooth muscle cells. Hypertension. 1999;34(6):1242–6.
40. Yiu SS, et al. 12-Hydroxyeicosatetraenoic acid participates in angiotensin II afferent arteriolar vasoconstriction by activating L-type calcium channels. J Lipid Res. 2003;44(12):2391–9.
41. DiNicolantonio JJ, et al. Omega-3s and cardiovascular health. Ochsner J. 2014;14(3): 399–412.
42. Xiao YF, et al. Suppression of voltage-gated L-type Ca^{2+} currents by polyunsaturated fatty acids in adult and neonatal rat ventricular myocytes. Proc Natl Acad Sci U S A. 1997;94(8):4182–7.
43. Hazama H, et al. Omega-3 polyunsaturated fatty acids—modulation of voltage-dependent L-type Ca^{2+} current in guinea-pig tracheal smooth muscle cells. Eur J Pharmacol. 1998; 355(2–3):257–66.
44. Ferrier GR, et al. Differential effects of docosahexaenoic acid on contractions and L-type Ca^{2+} current in adult cardiac myocytes. Cardiovasc Res. 2002;54(3):601–10.
45. Asano M, et al. Eicosapentaenoic acid inhibits vasopressin-activated Ca^{2+} influx and cell proliferation in rat aortic smooth muscle cell lines. Eur J Pharmacol. 1999;379(2–3):199–209.
46. Guan Z, et al. Sphingosine-1-phosphate evokes unique segment-specific vasoconstriction of the renal microvasculature. J Am Soc Nephrol. 2014;25(8):1774–85.
47. McDonough PM, et al. Control of cardiac Ca^{2+} levels. Inhibitory actions of sphingosine on Ca^{2+} transients and L-type Ca^{2+} channel conductance. Circ Res. 1994;75(6):981–9.
48. Schreur KD, Liu S. Involvement of ceramide in inhibitory effect of IL-1 beta on L-type Ca^{2+} current in adult rat ventricular myocytes. Am J Physiol. 1997;272(6 Pt 2):H2591–8.
49. Sen L, et al. Cholesterol increases the L-type voltage-sensitive calcium channel current in arterial smooth muscle cells. Circ Res. 1992;71(4):1008–14.
50. Shan JJ, et al. Inhibition of membrane L-type calcium channel activity and intracellular calcium concentration by 24R, 25-dihydroxyvitamin D3 in vascular smooth muscle. Steroids. 1996;61(11):657–63.
51. van Hoeijen DA, Blom MT, Tan HL. Cardiac sodium channels and inherited electrophysiological disorders: an update on the pharmacotherapy. Expert Opin Pharmacother. 2014;15(13): 1875–87.

52. Calhoun JD, Isom LL. The role of non-pore-forming beta subunits in physiology and pathophysiology of voltage-gated sodium channels. Handb Exp Pharmacol. 2014;221:51–89.
53. Pignier C, et al. Direct protective effects of poly-unsaturated fatty acids, DHA and EPA, against activation of cardiac late sodium current: a mechanism for ischemia selectivity. Basic Res Cardiol. 2007;102(6):553–64.
54. Liu Z, et al. C(6)-ceramide inhibited Na^+ currents by intracellular Ca^{2+} release in rat myoblasts. J Cell Physiol. 2007;213(1):151–60.
55. Hille B. Ionic channels: molecular pores of excitable membranes. Harvey Lect. 1986;82:47–69.
56. Han WN, et al. Effect of 15-HETE on potassium channels of rabbit pulmonary arterial smooth muscles during hypoxia. Sheng Li Xue Bao. 2004;56(6):717–22.
57. Moral-Sanz J, et al. Ceramide inhibits K_v currents and contributes to TP-receptor-induced vasoconstriction in rat and human pulmonary arteries. Am J Physiol Cell Physiol. 2011;301(1):C186–94.
58. Dopico AM, Bukiya AN, Singh AK. Large conductance, calcium- and voltage-gated potassium (BK) channels: regulation by cholesterol. Pharmacol Ther. 2012;135(2):133–50.
59. Yang H, Zhang G, Cui J. BK channels: multiple sensors, one activation gate. Front Physiol. 2015;6:29.
60. Knot HJ, Nelson MT. Regulation of arterial diameter and wall [Ca^{2+}] in cerebral arteries of rat by membrane potential and intravascular pressure. J Physiol. 1998;508(Pt 1):199–209.
61. Ledoux J, et al. Calcium-activated potassium channels and the regulation of vascular tone. Physiology (Bethesda). 2006;21:69–78.
62. Brayden JE, Nelson MT. Regulation of arterial tone by activation of calcium-dependent potassium channels. Science. 1992;256(5056):532–5.
63. Jaggar JH, et al. Ca2+ channels, ryanodine receptors and Ca(2+)-activated K+ channels: a functional unit for regulating arterial tone. Acta Physiol Scand. 1998;164(4):577–87.
64. Vaithianathan T, et al. Direct regulation of BK channels by phosphatidylinositol 4,5-bisphosphate as a novel signaling pathway. J Gen Physiol. 2008;132(1):13–28.
65. Ahn DS, et al. Fatty acids directly increase the activity of Ca^{2+}-activated K^+ channels in rabbit coronary smooth muscle cells. Yonsei Med J. 1994;35(1):10–24.
66. Kirber MT, et al. Both membrane stretch and fatty acids directly activate large conductance Ca^{2+}-activated K^+ channels in vascular smooth muscle cells. FEBS Lett. 1992;297(1–2):24–8.
67. Clarke AL, et al. Site of action of fatty acids and other charged lipids on BK_{Ca} channels from arterial smooth muscle cells. Am J Physiol Cell Physiol. 2003;284(3):C607–19.
68. Lai LH, et al. Effects of docosahexaenoic acid on large-conductance Ca^{2+}-activated K^+ channels and voltage-dependent K^+ channels in rat coronary artery smooth muscle cells. Acta Pharmacol Sin. 2009;30(3):314–20.
69. Li X, et al. Docosahexanoic acid-induced coronary arterial dilation: actions of 17S-hydroxy docosahexanoic acid on K^+ channel activity. J Pharmacol Exp Ther. 2011;336(3):891–9.
70. Ye D, et al. Cytochrome p-450 epoxygenase metabolites of docosahexaenoate potently dilate coronary arterioles by activating large-conductance calcium-activated potassium channels. J Pharmacol Exp Ther. 2002;303(2):768–76.
71. Hoshi T, et al. Omega-3 fatty acids lower blood pressure by directly activating large-conductance Ca^{2+}-dependent K^+ channels. Proc Natl Acad Sci U S A. 2013;110(12):4816–21.
72. Martin P, et al. Arachidonic acid activation of BK_{Ca} (Slo1) channels associated to the beta1-subunit in human vascular smooth muscle cells. Pflugers Arch. 2014;466(9):1779–92.
73. Yan J, et al. Docosahexaenoic acid attenuates hypoxic pulmonary vasoconstriction by activating the large conductance Ca^{2+}-activated K^+ currents in pulmonary artery smooth muscle cells. Pulm Pharmacol Ther. 2014;28(1):9–16.
74. Hoshi T, et al. A point mutation in the human Slo1 channel that impairs its sensitivity to omega-3 docosahexaenoic acid. J Gen Physiol. 2013;142(5):507–22.
75. Hoshi T, et al. Mechanism of the modulation of BK potassium channel complexes with different auxiliary subunit compositions by the omega-3 fatty acid DHA. Proc Natl Acad Sci U S A. 2013;110(12):4822–7.

76. Guerard P, et al. Arachidonic acid relaxes human pulmonary arteries through K^+ channels and nitric oxide pathways. Eur J Pharmacol. 2004;501(1-3):127–35.
77. Schubert R, et al. Iloprost activates K_{Ca} channels of vascular smooth muscle cells: role of cAMP-dependent protein kinase. Am J Physiol. 1996;271(4 Pt 1):C1203–11.
78. Zhu S, Han G, White RE. PGE2 action in human coronary artery smooth muscle: role of potassium channels and signaling cross-talk. J Vasc Res. 2002;39(6):477–88.
79. Klockgether-Radke AP, et al. Mechanisms involved in the relaxing effect of midazolam on coronary arteries. Eur J Anaesthesiol. 2005;22(2):135–9.
80. Parajuli SP, Provence A, Petkov GV. Prostaglandin E2 excitatory effects on guinea pig urinary bladder smooth muscle: a novel regulatory mechanism mediated by large-conductance voltage- and Ca^{2+}-activated K^+ channels. Eur J Pharmacol. 2014;738:179–85.
81. Tanaka Y, Koike K, Toro L. MaxiK channel roles in blood vessel relaxations induced by endothelium-derived relaxing factors and their molecular mechanisms. J Smooth Muscle Res. 2004;40(4-5):125–53.
82. Zou AP, et al. Stereospecific effects of epoxyeicosatrienoic acids on renal vascular tone and K^+-channel activity. Am J Physiol. 1996;270(5 Pt 2):F822–32.
83. Lu T, et al. Dihydroxyeicosatrienoic acids are potent activators of Ca^{2+}-activated K^+ channels in isolated rat coronary arterial myocytes. J Physiol. 2001;534(Pt 3):651–67.
84. Lauterbach B, et al. Cytochrome P450-dependent eicosapentaenoic acid metabolites are novel BK channel activators. Hypertension. 2002;39(2 Pt 2):609–13.
85. Dimitropoulou C, et al. Protein phosphatase 2A and Ca^{2+}-activated K^+ channels contribute to 11,12-epoxyeicosatrienoic acid analog mediated mesenteric arterial relaxation. Prostaglandins Other Lipid Mediat. 2007;83(1-2):50–61.
86. Loot AE, et al. 11,12-EET stimulates the association of BK channel alpha and $beta_1$ subunits in mitochondria to induce pulmonary vasoconstriction. PLoS One. 2012;7(9):e46065.
87. Li PL, Campbell WB. Epoxyeicosatrienoic acids activate K^+ channels in coronary smooth muscle through a guanine nucleotide binding protein. Circ Res. 1997;80(6):877–84.
88. Zhang Y, et al. EET homologs potently dilate coronary microvessels and activate BK_{Ca} channels. Am J Physiol Heart Circ Physiol. 2001;280(6):H2430–40.
89. Archer SL, et al. Endothelium-derived hyperpolarizing factor in human internal mammary artery is 11,12-epoxyeicosatrienoic acid and causes relaxation by activating smooth muscle BK(Ca) channels. Circulation. 2003;107(5):769–76.
90. Morin C, et al. Relaxing effects of 5-oxo-ETE on human bronchi involve BK_{Ca} channel activation. Prostaglandins Other Lipid Mediat. 2007;83(4):311–9.
91. Morin C, et al. Relaxing effects of 17(18)-EpETE on arterial and airway smooth muscles in human lung. Am J Physiol Lung Cell Mol Physiol. 2009;296(1):L130–9.
92. Hercule HC, et al. The vasodilator 17,18-epoxyeicosatetraenoic acid targets the pore-forming BK alpha channel subunit in rodents. Exp Physiol. 2007;92(6):1067–76.
93. Wu SN, Li HF, Chiang HT. Actions of epoxyeicosatrienoic acid on large-conductance Ca^{2+}-activated K^+ channels in pituitary GH_3 cells. Biochem Pharmacol. 2000;60(2):251–62.
94. Benoit C, et al. EETs relax airway smooth muscle via an EpDHF effect: BK_{Ca} channel activation and hyperpolarization. Am J Physiol Lung Cell Mol Physiol. 2001;280(5):L965–73.
95. Subramani J, et al. Atorvastatin restores the impaired vascular endothelium-dependent relaxations mediated by nitric oxide and endothelium-derived hyperpolarizing factors but not hypotension in sepsis. J Cardiovasc Pharmacol. 2009;54(6):526–34.
96. Fukao M, et al. Regulation of BK_{Ca} channels expressed in human embryonic kidney 293 cells by epoxyeicosatrienoic acid. Mol Pharmacol. 2001;59(1):16–23.
97. Bukiya AN, et al. Activation of calcium- and voltage-gated potassium channels of large conductance by leukotriene B4. J Biol Chem. 2014;289(51):35314–25.
98. Bukiya AN, et al. An alcohol-sensing site in the calcium- and voltage-gated, large conductance potassium (BK) channel. Proc Natl Acad Sci U S A. 2014;111(25):9313–8.
99. Bolotina V, et al. Variations of membrane cholesterol alter the kinetics of Ca^{2+}-dependent K^+ channels and membrane fluidity in vascular smooth muscle cells. Pflugers Arch. 1989; 415(3):262–8.

100. Bregestovski PD, Bolotina VM, Serebryakov VN. Fatty acid modifies Ca^{2+}-dependent potassium channel activity in smooth muscle cells from the human aorta. Proc R Soc Lond B Biol Sci. 1989;237(1288):259–66.
101. Brainard AM, et al. Maxi-K channels localize to caveolae in human myometrium: a role for an actin-channel-caveolin complex in the regulation of myometrial smooth muscle K+ current. Am J Physiol Cell Physiol. 2005;289(1):C49–57.
102. Bukiya AN, et al. Smooth muscle cholesterol enables BK beta1 subunit-mediated channel inhibition and subsequent vasoconstriction evoked by alcohol. Arterioscler Thromb Vasc Biol. 2011;31(11):2410–23.
103. Singh AK, et al. Multiple cholesterol recognition/interaction amino acid consensus (CRAC) motifs in cytosolic C tail of Slo1 subunit determine cholesterol sensitivity of Ca^{2+}- and voltage-gated K^+ (BK) channels. J Biol Chem. 2012;287(24):20509–21.
104. Zhang M, Shivakumar B, Asuncion-Chin M, Bukiya A, Dopico A. Differential contribution of individual CRAC domains to the cholesterol sensitivity of BK channels. 59th Biophysical Society Annual Meeting, Baltimore, MD, 2015. Abstract L3357.
105. Sobey CG. Potassium channel function in vascular disease. Arterioscler Thromb Vasc Biol. 2001;21(1):28–38.
106. Najibi S, et al. Enhanced role of potassium channels in relaxations to acetylcholine in hypercholesterolemic rabbit carotid artery. Am J Physiol. 1994;266(5 Pt 2):H2061–7.
107. Najibi S, Cohen RA. Enhanced role of K^+ channels in relaxations of hypercholesterolemic rabbit carotid artery to NO. Am J Physiol. 1995;269(3 Pt 2):H805–11.
108. Jeremy RW, McCarron H. Effect of hypercholesterolemia on Ca^{2+}-dependent K^+ channel-mediated vasodilatation in vivo. Am J Physiol Heart Circ Physiol. 2000;279(4):H1600–8.
109. Yoshihara S, et al. A neuroactive steroid inhibits guinea pig airway sensory nerves via Maxi-K channel activation. Int Arch Allergy Immunol. 2006;141(1):31–6.
110. Jacob MK, White RE. Diazepam, gamma-aminobutyric acid, and progesterone open K^+ channels in myocytes from coronary arteries. Eur J Pharmacol. 2000;403(3):209–19.
111. Hu XQ, et al. Pregnancy upregulates large-conductance Ca^{2+}-activated K^+ channel activity and attenuates myogenic tone in uterine arteries. Hypertension. 2011;58(6):1132–9.
112. Deenadayalu V, et al. Testosterone-induced relaxation of coronary arteries: activation of BK_{Ca} channels via the cGMP-dependent protein kinase. Am J Physiol Heart Circ Physiol. 2012;302(1):H115–23.
113. Tep-areenan P, Kendall DA, Randall MD. Testosterone-induced vasorelaxation in the rat mesenteric arterial bed is mediated predominantly via potassium channels. Br J Pharmacol. 2002;135(3):735–40.
114. Saldanha PA, et al. Long- and short-term effects of androgens in human umbilical artery smooth muscle. Clin Exp Pharmacol Physiol. 2013;40(3):181–9.
115. Han G, et al. Estrogen receptor alpha mediates acute potassium channel stimulation in human coronary artery smooth muscle cells. J Pharmacol Exp Ther. 2006;316(3):1025–30.
116. Valverde MA, et al. Acute activation of Maxi-K channels (hSlo) by estradiol binding to the beta subunit. Science. 1999;285(5435):1929–31.
117. Wellman GC, et al. Gender differences in coronary artery diameter involve estrogen, nitric oxide, and Ca^{2+}-dependent K^+ channels. Circ Res. 1996;79(5):1024–30.
118. Rosenfeld CR, et al. Calcium-activated potassium channels and nitric oxide coregulate estrogen-induced vasodilation. Am J Physiol Heart Circ Physiol. 2000;279(1):H319–28.
119. White RE, et al. Endothelium-independent effect of estrogen on Ca^{2+}-activated K^+ channels in human coronary artery smooth muscle cells. Cardiovasc Res. 2002;53(3):650–61.
120. Yu X, et al. Activation of G protein-coupled estrogen receptor induces endothelium-independent relaxation of coronary artery smooth muscle. Am J Physiol Endocrinol Metab. 2011;301(5):E882–8.
121. Tsang SY, et al. Differential regulation of K^+ and Ca^{2+} channel gene expression by chronic treatment with estrogen and tamoxifen in rat aorta. Eur J Pharmacol. 2004;483(2-3):155–62.
122. Tsang SY, et al. Contribution of K^+ channels to relaxation induced by 17beta-estradiol but not by progesterone in isolated rat mesenteric artery rings. J Cardiovasc Pharmacol. 2003;41(1):4–13.

123. Tsang SY, et al. Estrogen and tamoxifen modulate cerebrovascular tone in ovariectomized female rats. Hypertension. 2004;44(1):78–82.
124. Nagar D, Liu XT, Rosenfeld CR. Estrogen regulates {beta}1-subunit expression in Ca^{2+}-activated K^+ channels in arteries from reproductive tissues. Am J Physiol Heart Circ Physiol. 2005;289(4):H1417–27.
125. Dick GM. The pure anti-oestrogen ICI 182,780 (Faslodex) activates large conductance Ca^{2+}-activated K^+ channels in smooth muscle. Br J Pharmacol. 2002;136(7):961–4.
126. Dick GM, et al. Tamoxifen activates smooth muscle BK channels through the regulatory beta1 subunit. J Biol Chem. 2001;276(37):34594–9.
127. Dick GM, Sanders KM. (Xeno)estrogen sensitivity of smooth muscle BK channels conferred by the regulatory beta1 subunit: a study of beta1 knockout mice. J Biol Chem. 2001;276(48):44835–40.
128. Perez GJ. Dual effect of tamoxifen on arterial K_{Ca} channels does not depend on the presence of the beta1 subunit. J Biol Chem. 2005;280(23):21739–47.
129. Maher J, et al. Smooth muscle relaxation and activation of the large conductance Ca^{++}-activated K^+ (BK_{Ca}) channel by novel oestrogens. Br J Pharmacol. 2013;169(5):1153–65.
130. Holzmann S, et al. Pharmacologic differentiation between endothelium-dependent relaxations sensitive and resistant to nitro-L-arginine in coronary arteries. J Cardiovasc Pharmacol. 1994;23(5):747–56.
131. Gluais P, et al. Role of SK_{Ca} and IK_{Ca} in endothelium-dependent hyperpolarizations of the guinea-pig isolated carotid artery. Br J Pharmacol. 2005;144(4):477–85.
132. Srivastava S, et al. Phosphatidylinositol 3-phosphate indirectly activates KCa3.1 via 14 amino acids in the carboxy terminus of KCa3.1. Mol Biol Cell. 2006;17(1):146–54.
133. Srivastava S, et al. The phosphatidylinositol 3-phosphate phosphatase myotubularin-related protein 6 (MTMR6) is a negative regulator of the Ca^{2+}-activated K^+ channel KCa3.1. Mol Cell Biol. 2005;25(9):3630–8.
134. Bondarenko AI, Malli R, Graier WF. The GPR55 agonist lysophosphatidylinositol directly activates intermediate-conductance Ca^{2+}-activated K^+ channels. Pflugers Arch. 2011; 462(2):245–55.
135. Schilling T, et al. Lysophospholipids induce membrane hyperpolarization in microglia by activation of IKCa1 Ca^{2+}-dependent K^+ channels. Neuroscience. 2002;109(4):827–35.
136. Furian AF, et al. Nitric oxide and potassium channels mediate GM1 ganglioside-induced vasorelaxation. Naunyn Schmiedebergs Arch Pharmacol. 2009;380(6):487–95.
137. Weston AH, et al. Bradykinin-induced, endothelium-dependent responses in porcine coronary arteries: involvement of potassium channel activation and epoxyeicosatrienoic acids. Br J Pharmacol. 2005;145(6):775–84.
138. Large WA, Wang Q. Characteristics and physiological role of the Ca^{2+}-activated Cl^- conductance in smooth muscle. Am J Physiol. 1996;271(2 Pt 1):C435–54.
139. Pritchard HA, et al. Inhibitory role of phosphatidylinositol 4,5-bisphosphate on TMEM16A-encoded calcium-activated chloride channels in rat pulmonary artery. Br J Pharmacol. 2014;171(18):4311–21.
140. Gao SB, et al. Cerebrosides of baifuzi, a novel potential blocker of calcium-activated chloride channels in rat pulmonary artery smooth muscle cells. Cell Biol Int. 2007;31(9):908–15.
141. Sones WR, et al. Cholesterol depletion alters amplitude and pharmacology of vascular calcium-activated chloride channels. Cardiovasc Res. 2010;87(3):476–84.
142. Clapham DE. TRP channels as cellular sensors. Nature. 2003;426(6966):517–24.
143. Nilius B, Flockerzi V. Mammalian transient receptor potential (TRP) cation channels. Preface. Handb Exp Pharmacol. 2014;223:v–vi.
144. Walker RL, Hume JR, Horowitz B. Differential expression and alternative splicing of TRP channel genes in smooth muscles. Am J Physiol Cell Physiol. 2001;280(5):C1184–92.
145. Hernandez-Garcia E, Rosenbaum T. Lipid modulation of thermal transient receptor potential channels. Curr Top Membr. 2014;74:135–80.
146. Yue Z, et al. Role of TRP channels in the cardiovascular system. Am J Physiol Heart Circ Physiol. 2015;308(3):H157–82.

147. Brayden JE, et al. Transient receptor potential (TRP) channels, vascular tone and autoregulation of cerebral blood flow. Clin Exp Pharmacol Physiol. 2008;35(9):1116–20.
148. Large WA, Saleh SN, Albert AP. Role of phosphoinositol 4,5-bisphosphate and diacylglycerol in regulating native TRPC channel proteins in vascular smooth muscle. Cell Calcium. 2009;45(6):574–82.
149. Earley S. Endothelium-dependent cerebral artery dilation mediated by transient receptor potential and Ca^{2+}-activated K^+ channels. J Cardiovasc Pharmacol. 2011;57(2):148–53.
150. Earley S. TRPA1 channels in the vasculature. Br J Pharmacol. 2012;167(1):13–22.
151. Earley S, Straub SV, Brayden JE. Protein kinase C regulates vascular myogenic tone through activation of TRPM4. Am J Physiol Heart Circ Physiol. 2007;292(6):H2613–22.
152. Saleh SN, Albert AP, Large WA. Obligatory role for phosphatidylinositol 4,5-bisphosphate in activation of native TRPC1 store-operated channels in vascular myocytes. J Physiol. 2009;587(Pt 3):531–40.
153. Ju M, et al. Ins(1,4,5)P3 interacts with PIP2 to regulate activation of TRPC6/C7 channels by diacylglycerol in native vascular myocytes. J Physiol. 2010;588(Pt 9):1419–33.
154. Morales-Lazaro SL, Rosenbaum T. A painful link between the TRPV1 channel and lysophosphatidic acid. Life Sci. 2015;125C:15–24.
155. Ingueneau C, et al. Caveolin-1 sensitizes vascular smooth muscle cells to mildly oxidized LDL-induced apoptosis. Biochem Biophys Res Commun. 2008;369(3):889–93.
156. So I, et al. Lysophosphatidylcholine, a component of atherogenic lipoproteins, induces the change of calcium mobilization via TRPC ion channels in cultured human corporal smooth muscle cells. Int J Impot Res. 2005;17(6):475–83.
157. Al-Shawaf E, et al. Short-term stimulation of calcium-permeable transient receptor potential canonical 5-containing channels by oxidized phospholipids. Arterioscler Thromb Vasc Biol. 2010;30(7):1453–9.
158. Sullivan MN, et al. Localized TRPA1 channel Ca^{2+} signals stimulated by reactive oxygen species promote cerebral artery dilation. Sci Signal. 2015;8(358):ra2.
159. Slish DF, Welsh DG, Brayden JE. Diacylglycerol and protein kinase C activate cation channels involved in myogenic tone. Am J Physiol Heart Circ Physiol. 2002;283(6):H2196–201.
160. Inoue R, et al. The transient receptor potential protein homologue TRP6 is the essential component of vascular $alpha_1$-adrenoceptor-activated Ca^{2+}-permeable cation channel. Circ Res. 2001;88(3):325–32.
161. Jung S, et al. TRPC6 is a candidate channel involved in receptor-stimulated cation currents in A7r5 smooth muscle cells. Am J Physiol Cell Physiol. 2002;282(2):C347–59.
162. Earley S, et al. TRPV4 forms a novel Ca^{2+} signaling complex with ryanodine receptors and BKCa channels. Circ Res. 2005;97(12):1270–9.
163. Watanabe H, et al. Anandamide and arachidonic acid use epoxyeicosatrienoic acids to activate TRPV4 channels. Nature. 2003;424(6947):434–8.
164. Jaggar JH, et al. Calcium sparks in smooth muscle. Am J Physiol Cell Physiol. 2000;278(2):C235–56.
165. McCarron JG, et al. Origin and mechanisms of Ca^{2+} waves in smooth muscle as revealed by localized photolysis of caged inositol 1,4,5-trisphosphate. J Biol Chem. 2004;279(9):8417–27.
166. Santana LF, Navedo MF. Molecular and biophysical mechanisms of Ca^{2+} sparklets in smooth muscle. J Mol Cell Cardiol. 2009;47(4):436–44.
167. Foskett JK, et al. Inositol trisphosphate receptor Ca^{2+} release channels. Physiol Rev. 2007;87(2):593–658.
168. Seo MD, et al. Structural insights into endoplasmic reticulum stored calcium regulation by inositol 1,4,5-trisphosphate and ryanodine receptors. Biochim Biophys Acta. 2014;1853(9):1980–91.
169. Zalk R, Lehnart SE, Marks AR. Modulation of the ryanodine receptor and intracellular calcium. Annu Rev Biochem. 2007;76:367–85.
170. Fleischer S. Personal recollections on the discovery of the ryanodine receptors of muscle. Biochem Biophys Res Commun. 2008;369(1):195–207.
171. Fill M, Copello JA. Ryanodine receptor calcium release channels. Physiol Rev. 2002;82(4):893–922.
172. Vaithianathan T, et al. Subtype identification and functional characterization of ryanodine receptors in rat cerebral artery myocytes. Am J Physiol Cell Physiol. 2010;299(2):C264–78.

173. Dahan D, et al. Implication of the ryanodine receptor in TRPV4-induced calcium response in pulmonary arterial smooth muscle cells from normoxic and chronically hypoxic rats. Am J Physiol Lung Cell Mol Physiol. 2012;303(9):L824–33.
174. Silswal N, et al. Phosphatidylinositol 3,5-bisphosphate increases intracellular free Ca^{2+} in arterial smooth muscle cells and elicits vasocontraction. Am J Physiol Heart Circ Physiol. 2011;300(6):H2016–26.
175. Ellinsworth DC, et al. Interactions between thromboxane A_2, thromboxane/prostaglandin (TP) receptors, and endothelium-derived hyperpolarization. Cardiovasc Res. 2014;102(1):9–16.
176. Moss NG, et al. Thromboxane-induced renal vasoconstriction is mediated by the ADP-ribosyl cyclase CD38 and superoxide anion. Am J Physiol Renal Physiol. 2013;305(6):F830–8.
177. Chalmers S, et al. Ion channels in smooth muscle: regulation by the sarcoplasmic reticulum and mitochondria. Cell Calcium. 2007;42(4–5):447–66.
178. Narayanan D, Adebiyi A, Jaggar JH. Inositol trisphosphate receptors in smooth muscle cells. Am J Physiol Heart Circ Physiol. 2012;302(11):H2190–210.
179. Kandasamy K, et al. Lipopolysaccharide induces endoplasmic store Ca^{2+}-dependent inflammatory responses in lung microvessels. PLoS One. 2013;8(5):e63465.
180. AlSuleimani YM, Hiley CR. Mechanisms of vasorelaxation induced by oleoylethanolamide in the rat small mesenteric artery. Eur J Pharmacol. 2013;702(1–3):1–11.
181. Pantazaka E, Papadimitriou E. Chondroitin sulfate-cell membrane effectors as regulators of growth factor-mediated vascular and cancer cell migration. Biochim Biophys Acta. 2014;1840(8):2643–50.
182. Snetkov VA, et al. Mechanisms of the prostaglandin F2alpha-induced rise in $[Ca^{2+}]_i$ in rat intrapulmonary arteries. J Physiol. 2006;571(Pt 1):147–63.
183. Zheng YM, et al. Heterogeneous gene expression and functional activity of ryanodine receptors in resistance and conduit pulmonary as well as mesenteric artery smooth muscle cells. J Vasc Res. 2008;45(6):469–79.
184. Van Assche T, et al. Altered Ca^{2+} handling of smooth muscle cells in aorta of apolipoprotein E-deficient mice before development of atherosclerotic lesions. Cell Calcium. 2007;41(3):295–302.
185. Hammoud Y, Rice T, Mackrill JJ. Oxysterols modulate calcium signalling in the A7r5 aortic smooth muscle cell-line. Biochimie. 2013;95(3):568–77.
186. Adebiyi A, Narayanan D, Jaggar JH. Caveolin-1 assembles type 1 inositol 1,4,5-trisphosphate receptors and canonical transient receptor potential 3 channels into a functional signaling complex in arterial smooth muscle cells. J Biol Chem. 2011;286(6):4341–8.
187. Putney JW. Origins of the concept of store-operated calcium entry. Front Biosci (Schol Ed). 2011;3:980–4.
188. Inoue R, Mori Y. Molecular candidates for capacitative and non-capacitative Ca^{2+} entry in smooth muscle. Novartis Found Symp. 2002;246:81–90. discussion 221–7.
189. Trebak M. STIM/Orai signalling complexes in vascular smooth muscle. J Physiol. 2012;590(Pt 17):4201–8.
190. Welsh DG. The many faces of Orai. Circ Res. 2013;112(7):983–4.
191. Hopson KP, et al. S1P activates store-operated calcium entry via receptor- and non-receptor-mediated pathways in vascular smooth muscle cells. Am J Physiol Cell Physiol. 2011;300(4):C919–26.
192. Gonzalez-Cobos JC, et al. Store-independent Orai1/3 channels activated by intracrine leukotriene C4: role in neointimal hyperplasia. Circ Res. 2013;112(7):1013–25.
193. Zhang X, et al. Mechanisms of STIM1 activation of store-independent leukotriene C4-regulated Ca^{2+} channels. Mol Cell Biol. 2013;33(18):3715–23.
194. Zhang W, et al. Leukotriene-C4 synthase, a critical enzyme in the activation of store-independent Orai1/Orai3 channels, is required for neointimal hyperplasia. J Biol Chem. 2015;290(8):5015–27.

Chapter 17
Transient Receptor Potential Channels in Metabolic Syndrome-Induced Coronary Artery Disease

Stacey L. Dineen, Zachary P. Neeb, Alexander G. Obukhov, and Michael Sturek

Abstract Cardiovascular disease is the leading cause of death in the developed world. Coronary artery disease (CAD) is greatly exacerbated by underlying metabolic syndrome, which is defined as the presence of three or more of the following cardiovascular risk factors: obesity, glucose intolerance, insulin resistance, dyslipidemia, and hypertension. Alterations in the endothelial lining of the vascular wall and smooth muscle phenotype are key components in the underlying pathology. Both endothelial dysfunction and smooth muscle cell phenotype switching are mediated, at least in part, by altered intracellular Ca^{2+} handling. Transient receptor potential (TRP) channels have been implicated in CAD progression, both in endothelial dysfunction and smooth muscle phenotypic changes. Despite the widespread distribution of TRP channels in numerous cell types and the involvement in many diseases, there is a relative paucity of data on the role of TRP channels in CAD. TRP canonical (TRPC) channels are located on coronary smooth muscle (CSM) cell membranes and are involved in mediating Ca^{2+} entry. Increases in CSM TRPC expression and function are associated with CAD progression. In contrast, TRP vanilloid (TRPV) receptors are located primarily on the coronary endothelium, and function in normal physiology to increase intracellular Ca^{2+} in response to ligands or sheer stress, leading to endothelial-dependent vasodilation. Decreases in endothelial TRPV receptors function and expression are associated with progression of CAD. The loss of proper balance between CSM TRPC and endothelial TRPV contributes, at least in part, to CAD progression.

Keywords Calcium • Atherosclerosis • Sarcoplasmic reticulum • Dyslipidemia

S.L. Dineen • A.G. Obukhov • M. Sturek, Ph.D. (✉)
Department of Cellular & Integrative Physiology, Indiana University School of Medicine, 635 Barnhill Dr., Indianapolis, IN 46202, USA
e-mail: msturek@iu.edu

Z.P. Neeb
Department of Cellular & Integrative Physiology, Indiana University School of Medicine, 635 Barnhill Dr., Indianapolis, IN 46202, USA

Eli Lilly and Company, Lilly Corporate Center, Indianapolis, IN, USA

I. Levitan, A.M. Dopico (eds.), *Vascular Ion Channels in Physiology and Disease*, DOI 10.1007/978-3-319-29635-7_17

Coronary Artery Disease

The coronary circulation functions to provide oxygen and other substrates to the most tirelessly working organ in the entire human body—the heart. Myocardial oxygen consumption is high; therefore, its delivery system must be efficient and unhindered. Impediments in coronary blood flow, such as those observed in atherosclerotic coronary artery disease (CAD), can have long-lasting and drastic consequences, such as myocardial infarction and death. In the developed world, cardiovascular disease is the leading cause of death [1]. Therefore, investigations into its causes and complications are extensive and necessary. Here, we will discuss one arm of this investigation; however, it is important to recognize that many factors play a role in the development of this deadly disease.

The coronary arteries can be subcategorized as macrovascular conduit arteries and microvascular resistance arterioles [2]. During progression of CAD, macrovascular conduit arteries display classical formation of atherosclerotic plaques. Microvascular resistance arterioles, although they display robust dysfunction and remodeling [3, 4], do not exhibit the characteristics of atherosclerotic CAD, such as lipid infiltration and proliferation and migration of coronary smooth muscle (CSM). The specificity of ion channel properties in the coronary micro- and microvasculature is exemplified by the heterogeneity of L-type voltage-gated Ca^{2+} channel functional expression [5, 6] and Ca^{2+}-dependent K^{+} channel regulation by Ca^{2+} sparks [7]. This important concept of arterial specificity, therefore, prompted us to focus this chapter on the coronary macrovasculature.

Conduit coronary arteries are organized as shown in Fig. 17.1a. Two laminae, termed external (EEL) and internal (IEL) elastic laminae demarcate the borders of

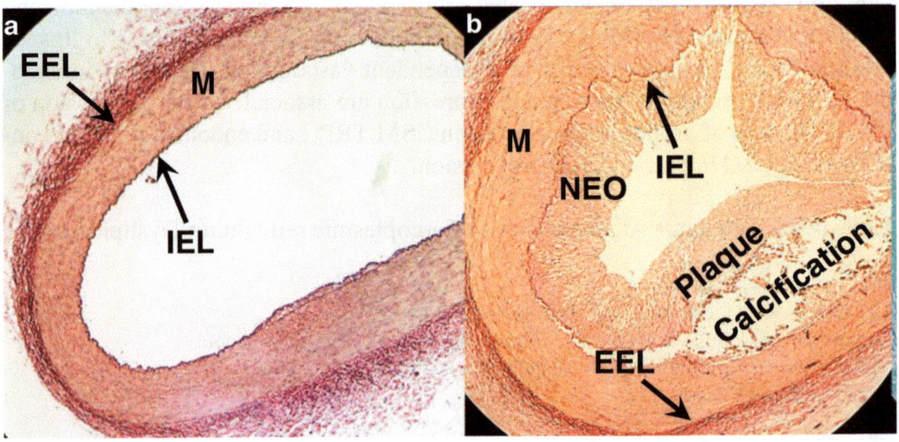

Fig. 17.1 Coronary artery ring slices stained with Verhoeff-Van Gieson stain. *EEL* External elastic lamina, *M* Media, *IEL* Internal elastic lamina. *NEO* neointima. (**a**) Healthy coronary artery ring. (**b**) Coronary artery ring with medial thickening and advanced atherosclerotic lesion (plaque) including a complex necrotic core with overt vascular calcification

the medial layer, containing CSM. In the healthy coronary artery, CSM display contractile properties, while their residence is largely restricted to the media. On the luminal side of the IEL is the intimal layer of the artery, where a single layer of endothelial cells (EC) resides. Importantly, ECs function to regulate arterial tone through a variety of intercellular signaling mechanisms to produce endothelial-derived constricting and relaxing factors that modulate CSM in the medial layer.

Atherosclerotic CAD is a progressive disease, traditionally beginning with endothelial dysfunction, lipid infiltration, and recruitment of leukocytes and macrophages into the intimal layer of the insulted region [8]. Recruited macrophages transform into lipid-laden foam cells within the forming atherosclerotic plaque. Stabilization of the forming atherosclerotic lesion may occur with the migration of CSM from the media into the neointima. This mobilization is characterized initially by phenotypic modulation of CSM from contractile to proliferative [9]. Migrating CSM display secretory properties, and, through secretion of collagen, form a cap over the atherosclerotic plaque. Plaque de-stabilization occurs with cell necrosis and the collection of lipids, cellular debris, and extracellular Ca^{2+} deposition (calcification) within the core of the forming atheroma. Figure 17.1b illustrates a coronary artery with an advanced atherosclerotic lesion. The neointima is substantially thickened into true plaque contained migratory CSM and foam cells. Plaque rupture and thrombosis often elicit hard clinical endpoints, such as myocardial infarction and sudden cardiac death.

Atherosclerotic CAD is a progressive disease, involving a variety of cell types. We will here briefly discuss endothelial dysfunction and CSM phenotypic switching in CAD.

Endothelial Dysfunction

The endothelial lining of the coronary arteries serves a variety of functions crucial to healthy vascular function. These functions include the regulation of traffic between the blood and underlying vascular tissues [10] and regulation of vascular tone [11]. Healthy endothelium mediates endothelial-dependent vasodilation through synthesis and secretion of nitric oxide and endothelium-derived hyperpolarizing factors (EDHF) to the underlying smooth muscle layer. Rises in intracellular endothelial Ca^{2+} are key processes regulating nitric oxide secretion and subsequent vasodilation.

Endothelial dysfunction is associated with a decrease in nitric oxide production [12], thereby decreasing endothelial-dependent vasodilation. Indeed, endothelial dysfunction itself is a predictor of atherosclerotic disease [13]. Lerman's group at the Mayo Clinic have extensively reported on endothelial dysfunction and coronary artery disease. They report that endothelial dysfunction is associated with vascular remodeling [14] and myocardial perfusion defects [15], and they extensively review the effects of endothelial dysfunction and cardiac pathophysiology [16]. Additionally, metabolic syndrome (MetS; defined below) contributes to endothelial dysfunction [17].

CSM Phenotypic Switching

A key component of atherosclerotic CAD is the de-differentiation and migration of CSM cells from the media into the overlying intima. The ability of CSM cells to adapt to their surrounding environment with minor phenotype changes, such as increased Ca^{2+} sensitivity, leading to changes in contractility, allows fine-tuning of their regulation of vascular tone. However, this plasticity renders them highly sensitive to environmental changes in CAD. The role of CSM phenotypic modulation in CAD progression is an area of active research, as the changes in CSM phenotype can serve both to exacerbate CAD progression and to stabilize developing atherosclerotic lesions, protecting against plaque rupture.

Intracellular Ca^{2+} handling is a crucial process in CSM, because intracellular Ca^{2+} regulates a variety of intracellular processes, including contraction, proliferation, excitation-contraction coupling, and excitation-transcription coupling. Therefore, alterations in Ca^{2+} transport are logical starting places when considering possible players in phenotypic plasticity. Elevated intracellular Ca^{2+} is necessary for induction of cell cycle protein transcription via a pathway involving nuclear factor of activated T-lymphocytes (NFAT) [18, 19]. NFAT signaling is also associated with an intact store-operated Ca^{2+} entry (SOCE) complex [19]. Therefore, alterations in intracellular Ca^{2+} levels can have significant impact on whether or not CSM remain in a quiescent state, or de-differentiate into a proliferative phenotype.

Metabolic Syndrome and Diabetes

The clustering of at least three metabolic risk factors (obesity, glucose intolerance, insulin resistance, dyslipidemia, hypertension) typically comprise the metabolic syndrome (MetS) [1]. If uncontrolled, MetS often progresses through increased fasting plasma glucose levels into overt type 2 diabetes [20]. These phenomena have recently been increasing in the developed world. In the United States alone, incidence of MetS is estimated at approximately one-third of the population [21]. Importantly, MetS and type 2 diabetes are independent risk factors for cardiovascular disease. Indeed, MetS doubles one's risk of developing cardiovascular disease [22]. Also, the chronic inflammation often observed in MetS serves to contribute to atherosclerotic CAD stimulation and/or exacerbation [23, 24]. Because MetS is a clustering of risk factors, it is difficult to determine individual roles of each risk factor in the progression of CAD. Therefore, when studying potential cellular mechanisms that underlie atherosclerotic CAD, it is interesting to consider a role for these cellular mechanisms in each MetS risk factor. For instance, the proteins reviewed in this chapter, transient receptor potential channels (TRP), are altered not only in the coronary vasculature, but also in adipose tissue and in the adrenal glands [25, 26]. As we consider potential players in the pathogenesis of atherosclerotic CAD, we will attempt to view the overall picture of MetS in the induction of atherosclerotic CAD.

Vascular Smooth Muscle Ion Channels in MetS-Induced CAD

Many have investigated molecular mechanisms underlying CSM modulation in disease, reviewed extensively by Owens et al. [9]; however, much remains undiscovered regarding ion channel regulation in atherosclerotic CAD. Intracellular Ca^{2+} regulation provides one interesting and important area of investigation, because intracellular Ca^{2+} is involved in a number of CSM functions, including contraction and phenotypic modulation.

A number of CSM Ca^{2+} transporters are known to display altered function and/or expression in MetS and diabetes-induced atherosclerotic CAD [20]. Figure 17.2 provides an overview of Ca^{2+} transport in CSM. For the purpose of this chapter, we will focus specifically on TRP channels. It should be recognized, however, that none of these transporters functions "in a vacuum;" that is to say they often work in concert relative to one another to maintain functional cytosolic Ca^{2+} levels. An example of this orchestration concerted working is SOCE.

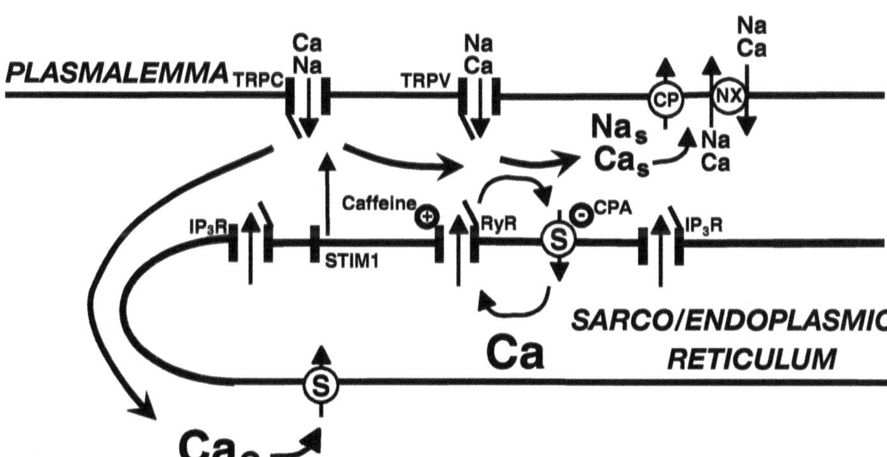

Fig. 17.2 Overview of intracellular Ca^{2+} transport. Influx mechanisms: *TRPC* Transient receptor potential canonical channel, *TRPV* Transient receptor potential villanoid channel; Extrusion mechanisms: *CP* Plasma membrane Ca^{2+} ATPase (pump), *NX* Na^+-Ca^{2+} exchanger; sarco/endoplasmic reticulum (SR/ER) Ca^{2+} sensor: *STIM1* Stromal interaction molecule 1; SR/ER Ca^{2+} release channels: IP_3R Inositol 1,4,5-trisphosphate receptor, *RyR* Ryanodine receptor; SR/ER Ca^{2+} recovery mechanism: *S* Sarco-endoplasmic reticulum Ca^{2+} ATPase. Ca_c bulk cytosolic Ca^{2+}. Ca_s localized subplasmalemmal Ca^{2+}. Na_s localized subplasmalemmal Na^+. *CPA* Cyclopiazonic acid

Store-Operated Ca^{2+} Entry

SOCE refers to a phenomenon by which depletion of the sarcoplasmic/endoplasmic reticulum (SR) Ca^{2+} store results in transmission of a signal to the plasma membrane, facilitating the influx of Ca^{2+} across the plasmalemma [27]. This phenomenon likely occurs because of a functional connection between Ca^{2+}-release channels located on the SR membrane, such as the inositol 1,4,5-trisphosphate receptor (IP_3R), a Ca^{2+}-sensing protein on the SR membrane, stromal interaction molecule (STIM) 1, and ion channels on the plasma membrane. There are currently two primary candidates for the Ca^{2+} transporter involved in SOCE. The first of these is Orai1 [28, 29], which has been demonstrated to complex with STIM1 [19] and to be involved in Ca^{2+} influx following depletion of the SR Ca^{2+} store [30]. A main argument for Orai1 being the molecular identity of the SOCE channel is the very high selectivity of Orai1 for Ca^{2+}. The second class of channels thought to mediate SOCE is the TRP channel family. TRPC channels have also been demonstrated to complex with STIM1 [31] and are activated by SR Ca^{2+} store depletion [32, 33]. TRPC channels, however, are relatively non-selective for Ca^{2+}; instead, TRP channels primarily allow Na^+ influx under physiological ionic conditions. It is this non-selectivity which called into question whether TRPC channels could be the SOCE channels in vascular smooth muscle. However, Poburko et al. [34] showed that the Na^+ entering via TRPC6 channels substantially increased Na^+ concentrations in a localized subsarcolemmal space, which in turn mediated Ca^{2+} influx via reverse mode Na^+-Ca^{2+} exchange (Fig. 17.2, Na_s, NX). The roles of TRP and Orai1 in SOCE are currently not fully understood and whether these components of the SOCE process act in concert or in competition is not fully resolved. In this chapter, we will discuss the primary role for TRP channels in SOCE in CSM; however, we recognize that Orai1 likely also plays a role in the SOCE. While the majority of data indicate that TRPC channels, rather than Orai1 channels are involved in SOCE in CSM, SOCE may be mediated by different players, depending on arterial bed.

Transient Receptor Potential Channels in MetS-Induced CAD

TRP channels were first characterized in *Drosophila* in the 1970s and 1980s [35, 36]. In 1995, the first mammalian TRP channel, TRPC1, was cloned [37, 38]. In the last 20 years, much work has been done to discover the role of the TRP family in normal and disease physiology. To date, 28 members of the mammalian TRP family have been identified (for a detailed review, see [39]). A number of the members of the TRP superfamily were proposed to be involved in the progression of atherosclerotic CAD, either directly, or through a variety of risk factors in MetS. Below, we will discuss the manner in which TRP subfamilies are implicated in and contribute to atherosclerotic CAD and MetS. We will focus our discussion on the TRPC and TRPV subfamilies in atherosclerotic CAD.

TRPC

Transient receptor potential canonical (TRPC) channels were the first mammalian TRP channels to be identified in 1995 [37, 38], and shown to be gated by depletion of the sarco-endoplasmic reticulum Ca^{2+} store, as well as being gated by other second messengers, such as diacylglycerol [40, 41]. SOCE, and therefore TRPC channels, have been implicated in smooth muscle cell growth. Within the TRPC subfamily, several candidates have been suggested as players in disease progression.

TRPC1

TRPC1 has been implicated in angiotensin II (Ang II)-induced smooth muscle cell growth in primary cultured human coronary artery smooth muscle cells [42]. Additionally, inhibition of TRPC1 results in decreased smooth muscle cell proliferation in pulmonary artery smooth muscle [43]. TRPC1 is an interesting candidate in disease progression, because it is expressed in coronary arteries in humans [42], Ossabaw miniature swine [44], and rabbits [33].

Edwards et al. [44] undertook a major study of 14 months of MetS in Ossabaw miniature swine. The pigs had robust MetS and atherosclerotic CAD was well-established. They demonstrated that coronary CSM SOCE was increased in MetS-induced CAD, as shown by fura-2 measures of intracellular free Ca^{2+} and divalent cation influx in freshly isolated CSM (Fig. 17.3). A classic protocol to assess SOCE is the depletion of the SR Ca^{2+} store with caffeine in Ca^{2+}-free extracellular solution, inhibition of the sarco-endoplasmic reticulum ATPase (SERCA) with cyclopiazonic acid, and restoration of Ca^{2+} in the bathing solution, yielding an increase in whole-cell cytosolic Ca^{2+} (Fig. 17.3a; schematically in Fig. 17.2). The peak, steady-state increase in free cytosolic Ca^{2+} is a measure of SOCE and confirmed by the inhibition with Ni^{2+} and insensitivity to the L-type voltage-gated Ca^{2+} channel inhibitor, nicardipine. Chronic exercise training attenuated SOCE (XMetS; Fig. 17.3b). The more sensitive measure of SOCE is Mn^{2+} quench of fura-2 fluorescence due to the 40-fold higher affinity of fura-2 for Mn^{2+} and the lack of Mn^{2+} transport by SERCA, the plasmalemmal Ca^{2+} ATPase (CP), and Na^{+}-Ca^{2+} exchange (NX) (Fig. 17.2). The Ca^{2+}-insensitive fura-2 excitation wavelength of 360 nm enables a quite clear index of unidirectional Mn^{2+} flux into the CSM, also inhibited by Ni^{2+} (Fig. 17.3c). Using Mn^{2+} quench, it was clear that in native atherosclerotic segments of the coronary artery Mn^{2+} influx (SOCE) was increased in CSM of MetS vs. healthy lean pigs ("Non-stent"; Fig. 17.3d). In these same pigs coronary stents were used to study restenosis, which is known to be more severe in MetS and diabetes [45]. Although the artery regions just outside the end of the stents (peri-stent regions) may show mild stenosis ("candy wrapper" effect), there was no further increase in SOCE (Fig. 17.3d). The activation of SOCE by the extreme depletion of the SR with the combination of high doses of caffeine and cyclopiazonic acid to block SERCA uncovers interesting biophysical properties, but may not have the most

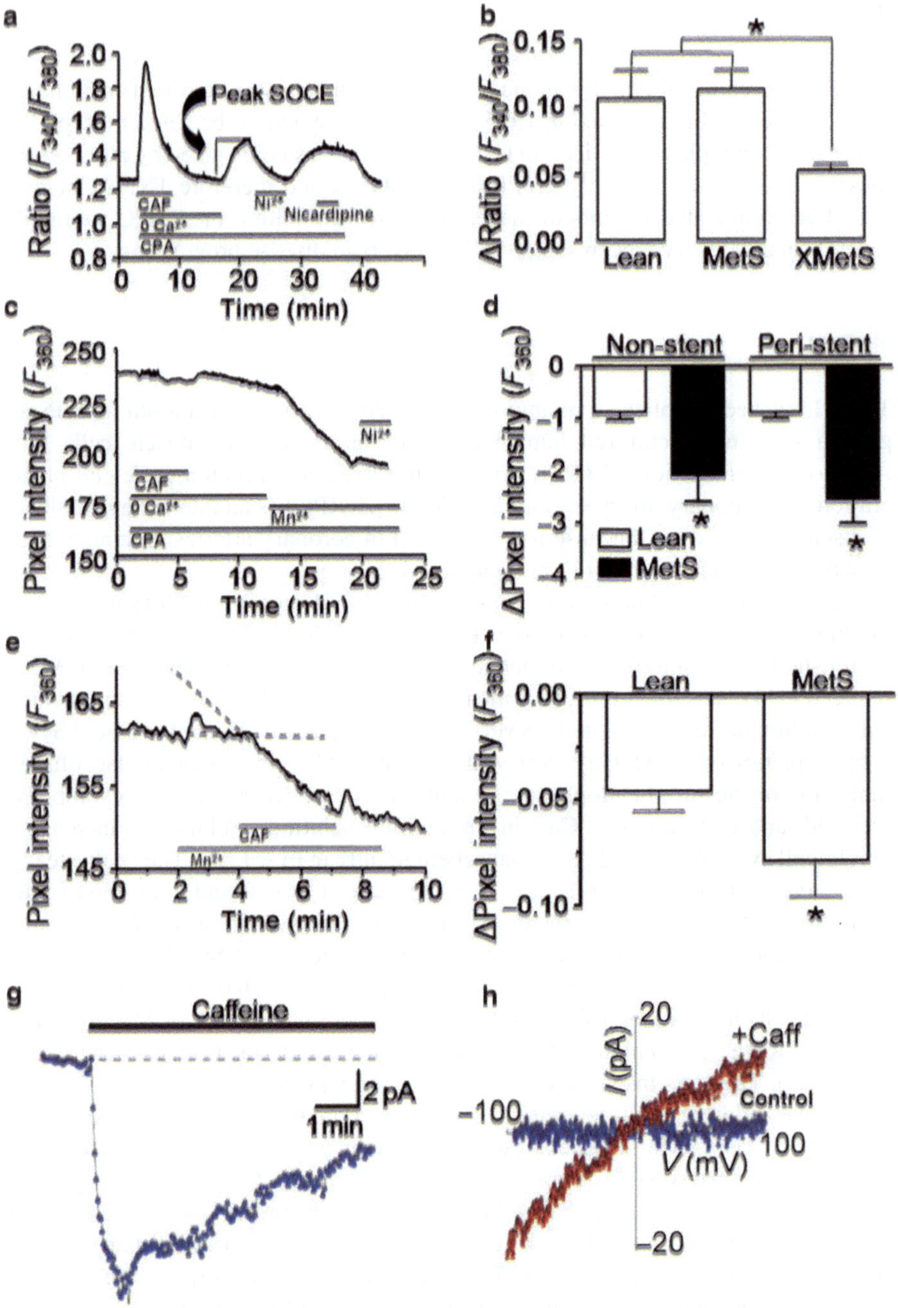

Fig. 17.3 Coronary smooth muscle store-operated Ca^{2+} entry is increased in long-term metabolic syndrome (caption modified from [44]). (**a** and **c**) Representative data from standard protocols used to assess SOCE. Peak store-depletion-mediated Ca^{2+} influx is assessed at minutes 17–20. Duration of exposure to solutions is shown by *horizontal lines*; caffeine (CAF, 5 mM); cyclopiazonic acid (CPA, 10 mM). (**b**) Exercise attenuates peak SOCE compared with Lean and MetS in

physiological relevance. However, the robust effect of MetS was further shown by the ability to resolve increased Mn^{2+} influx with caffeine-induced SR Ca^{2+} release alone (Fig. 17.3e, f).

The biophysical identity of the SOCE in MetS was shown by SR Ca^{2+} store depletion-induced ion currents measured with whole-cell patch clamp [44]. Inhibition of outward K^+ currents with Cs^+-loaded pipette revealed caffeine-induced inward current that was activated after the peak increase in myoplasmic Ca^{2+}, thus after caffeine-induced depletion of the SR Ca^{2+} store. The inward current was maintained in the presence of caffeine, which "short-circuits" SERCA to maintain depletion of the SR. This is very different from inward cation currents that are activated by the transient increase in myoplasmic Ca^{2+} and rapidly inactivate [39, 46]. The non-selective permeability was shown by the reversal potential at 0 mV (Fig. 17.3h), clearly arguing against Orai1 channels that are highly selective for Ca^{2+} [47].

Edwards et al. [44] further showed that the increase in SOCE was associated with increased TRPC1 and STIM1 mRNA and protein expression (Fig. 17.4a–g). Orai1 mRNA, but not protein, was significantly increased (Fig. 17.4d, h). Additionally, swine with MetS demonstrated greater atherosclerotic CAD as measured by

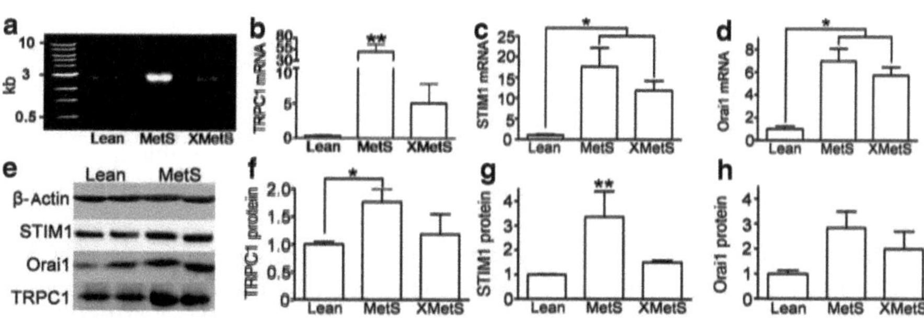

Fig. 17.4 TRPC1 and STIM1 gene and protein expression are increased in MetS and attenuated by exercise training (modified from [44]). (**a**) Representative agarose gel image quantifying TRPC1 mRNA using RT–PCR. (**b**) TRPC1 mRNA normalized to actin. STIM1 (**c**) and Orai1 (**d**) mRNA using quantitative RT–PCR normalized to actin. (**e**) Immunoblot using anti-actin, STIM1, Orai1, and TRPC1 in Lean and MetS. (**f–h**) TRPC1 (**f**), STIM1 (**g**), and Orai1 (**h**) protein expression using immunoblot analysis normalized to actin; ANOVA $P<0.10$ for (**h**). *$P<0.05$ MetS compared with Lean and XMetS; *$P<0.05$ between bracketed groups

Fig. 17.3 (continued) peri-stent sections of artery. *$P<0.05$ XMetS vs. Lean and MetS. (**c**) Mn^{2+} as a Ca^{2+} surrogate quenches fura-2 fluorescence at the isosbestic (Ca^{2+}-insensitive) 360 nm wavelength verifying divalent cation influx. (**d**) Increased SOCE in CSM from non-stent and peri-stent MetS coronary arteries compared with Lean. *$P<0.05$ vs. Lean. (**e**) Only caffeine is used to deplete the Ca^{2+} store in this SOCE protocol. (**f**) Increased SOCE in CSM from MetS swine. *$P<0.05$ vs. Lean. (**g**) Whole-cell patch-clamp data demonstrate sustained inward current by caffeine. (**h**) Current–voltage relationship of leak subtracted control (*blue*) and SOCE (*red*)

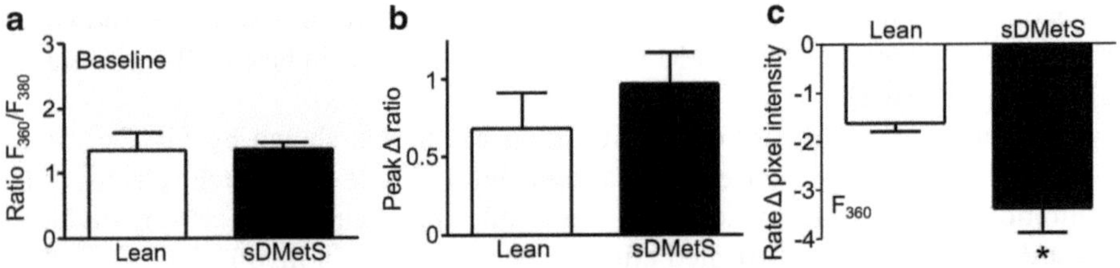

Fig. 17.5 Store-operated Ca^{2+} entry increases in short-term dyslipidemia and metabolic syndrome (sDMetS) before measurable coronary atherosclerosis. Mn^{2+} quench protocol was conducted as in Fig. 17.3c. (**a**) No difference in baseline (resting) intracellular Ca^{2+}. (**b**) Peak of the caffeine-induced Ca^{2+} transient due to Ca^{2+} release from the SR (as in Fig. 17.3c) was not different. (**c**) Mn^{2+} (divalent cation; Ca^{2+}) influx quantified as the rate of quenching of fura-2 fluorescence (as in Fig. 17.3c) was increased in sDMetS ($p<0.05$) (Neeb, Alloosh, Edwards, Long, Bratz, and Sturek, unpublished)

intravascular ultrasound and histologically by greater collagen deposition within coronary artery walls. Interestingly, CAD progression and collagen deposition, along with SOCE, TRPC1, and STIM1 expression were attenuated with exercise, highlighting the importance of exercise in the treatment of MetS and CAD. MetS has also been demonstrated to increase TRPC1 expression and activity in the adrenal medulla of Ossabaw swine [26]. Interestingly, mineralocorticoid receptor activation with aldosterone increases adrenal medullary expression of TRPC1, as well as that of TRPC5 and TRPC6, providing some insight as to a possible mechanism by which MetS contributes to CAD through increased expression and activation of TRPC channels [26].

Although the Edwards et al. study showed a clear association of increased CSM SOCE, TRPC1, and STIM1 expression with MetS and atherosclerotic CAD, consistent with the earlier studies from the Beech group [48, 49], the role of TRPC1 channels in causing CAD requires further study. Selective antagonism of TRPC1 or molecular knock-out would be highly convincing, but these are not yet feasible in large animal models. One step toward addressing cause-and-effect is to determine whether TRPC1 (SOCE) precedes CAD [50]. Figure 17.5a–c demonstrates that, In Ossabaw swine fed a hypercaloric, atherogenic diet for an abbreviated 9 weeks to elicit short-term dyslipidemic MetS (sDMetS), an increase in CSM SOCE (Mn^{2+} influx) was observed in sDMetS vs. healthy lean pigs (Fig. 17.5c). In contrast, there was no change in baseline, resting Ca^{2+} (Fig. 17.5a) or peak SR Ca^{2+} release (Fig. 17.5b), thus indicating a selective increase in SOCE. The increase in SOCE preceded overt atherosclerotic CAD development, providing rationale for a causal role of SOCE in CSM de-differentiation and subsequent proliferation/migration. Future studies will require selective agonism and antagonism of TRPC1 channels for more definitive evidence for the causal role of TRPC1 in CAD.

TRPC3

TRPC3 is expressed in human [32], rat [51], and rabbit [52] vasculature. TRPC3 participates in SOCE, and inhibition of TRPC3 is associated with a decrease in proliferation in human coronary artery cell culture [32], as well as a decrease in vascular smooth muscle proliferation in an organ culture model of stent-induced vascular injury [32]. In rats, TRPC3 expression was increased in spontaneous hypertension, and was associated with the augmented Ang II-induced contractility observed in hypertension. Interestingly, hypertensive rats treated with the Ang II type 1 receptor (AT1R) inhibitor, telmisartan, demonstrated improved systolic blood pressures, as well as decreased TRPC3 expression [51]. This is particularly interesting when one recognizes the role of the AT1R in regulation of aldosterone secretion [53], which has been implicated in elevation of other TRPC channel activity and expression in MetS [26]. Again, these data provide clues to the mechanism by which MetS risk factors, such as systemic hypertension, contribute to vascular dysfunction and disease. Additionally, elevated plasma endothelin-1 (ET-1) is observed in patients with vasospastic CAD [54, 55]. ET-1 is a potent and long-acting vasoconstrictor, and has been demonstrated to induce activation of TRPC3 through IP_3-mediated depletion of internal Ca^{2+} stores in rats [51] and rabbits [52].

TRPC5 and TRPC 6

TRPC5 and TRPC6 are also important to consider in the discussion regarding changes observed in MetS-induced CAD. In Ossabaw swine, TRP5 and TRPC6 expression is increased in the adrenal medulla in MetS, possibly mediated by elevated aldosterone [26]. In the coronary arteries of Ossabaw swine, TRPC6 mRNA is increased in MetS [44]. Additionally, TRPC6 is involved in ET-1-induced, IP_3-mediated SOCE in rabbit coronary arteries [33].

Summary

There is no doubt that expression and function of many isoforms in the TRPC subfamily are altered in vascular disease. However, it is unlikely that any one of these channels is solely responsible for the changes in smooth muscle phenotype and function in CAD. Indeed many of these isoforms co-localize in the coronary vasculature [33], and it is hypothesized that they combine as heterotetrameric channels [33].

Increased TRPC channel expression and function is implicated broadly in hypertension [51], MetS-induced CAD progression [44], vascular smooth muscle hypertrophy [56], and stent-induced vascular remodeling [32]. These changes are mediated through increased CSM proliferation [32], exacerbated coronary artery disease [44], and increased collagen deposition [44]. Importantly, many of these effects are mediated through Ang II-activated pathways, providing insight into possible mechanisms mediating MetS-induced CAD progression.

TRPV

Transient receptor potential vanilloid (TRPV) channels are ligand-gated, temperature sensitive, as well as pressure-gated. The first member of the TRPV subfamily, TRPV1, was first cloned in 1997 [57]. Involved in endothelial-dependent vasodilation, several members of this TRP subfamily are interesting players in the endothelial-dysfunction associated with initiation of atherosclerotic coronary artery disease.

TRPV1 is associated with endothelial-dependent vasodilation in porcine coronary arteries [58, 59]. Bratz et al. [59] utilized capsaicin to interrogate the role of TRPV1 on coronary artery tone in MetS in the Ossabaw miniature swine. They demonstrated that TRPV1 is associated with nitric oxide and K^+-mediated, endothelial-dependent vasorelaxation in the healthy phenotype and mediates Mn^{2+} influx in isolated endothelial cells. We wish to highlight the profound cell-selective expression of these channels in the coronary artery endothelial cells, as there was no capsaicin-induced divalent cation (Mn^{2+}) influx in CSM. In MetS capsaicin-induced vasorelaxation and endothelial Mn^{2+} influx and TRPV1 protein expression were reduced. These findings imply that the attenuation of vasorelaxation observed in MetS is mediated, at least in part, through decreased TRPV1 activity and expression. Inhibition of TRPV1 with capsazepine has been shown to decrease capsaicin-induced relaxation in isolated porcine coronary arteries, and to inhibit capsaicin-induced increases in coronary flow in isolated, perfused guinea pig hearts [58]. A central phenotype associated with MetS and type 2 diabetes is obesity. Therefore, it is interesting to note that activation of TRPV1 inhibited adipogenesis and prevented obesity in a mouse model of obesity [25], suggesting a differential TRPV1 expression and/or function in obesity.

TRPV channels are involved in endothelial-dependent vasodilation in the coronary arteries. In this capacity, they function to assist in the provision of adequate oxygen supply to the underlying myocardium. They are down-regulated in CAD, resulting in a reduced ability of the coronary circulation to adequately perfuse the heart. As such, the reduction in TRPV channel activity may contribute to myocardial infarction as a result of underlying CAD. However, TRPV1 may also be expressed in CSM, at least in dogs [60]. Thus, further studies are needed to establish the net role of TRPV1 in the regulation of coronary artery blood flow in health and metabolic syndrome.

Conclusions and Implications

It is clear that CAD progression is multifactorial and complex. It involves a variety of cells and cellular processes, all of which contribute to the pathophysiology of the disease. Risk factors, such as MetS and type 2 diabetes further complicate the mechanisms underlying CAD progression. Thus, it is important to continue rigorous investigation into the possible cellular mechanisms underlying MetS-induced CAD.

TRP channels provide an interesting and compelling potential target for the treatment of CAD. Differential roles for each subfamily and the implications of those roles are only beginning to be revealed. In CSM, TRPC channel expression is low under normal conditions, and is elevated in CAD. The result is increased SOCE, resulting in elevations in intracellular Ca^{2+}. This may contribute to the hypercontractile state observed in early stages of CAD [61, 62] and increasing CSM proliferation [19] and neointimal thickening [32]. Therefore, an increase in CSM TRPC expression and activity contributing to cytosolic free Ca^{2+} dysregulation is reasonably inferred to be pathologic in nature, potentiating CAD progression. A role of TRPC channels in leading to more complex extracellular deposition of Ca^{2+} crystals extracellularly to form clinically detectable vascular calcification (Fig. 17.1b) is completely unknown.

The differential localization of TRPC to CSM and TRPV to the endothelium, together with their opposing effects on vascular tone, suggests that a proper balance between the two TRP channel subtypes is necessary for healthy vascular function. Interruption in this balance may contribute to both endothelial dysfunction and altered CSM behavior, contributing to atherosclerotic CAD progression.

In the 20 years since the discovery of the first mammalian TRP, much progress has been made in the determination of the role these important channels play in human health and disease. Further studies should continue to elucidate potential roles for additional TRP subfamilies. Additionally, future studies should investigate the effects of individual factors in the "metabolic syndrome and diabetic milieu", such as hypertension, dyslipidemia, obesity, and insulin resistance, on coronary TRP channel function.

References

1. Go AS, et al. Heart disease and stroke statistics—2014 update: a report from the American Heart Association. Circulation. 2014;129(3):e28–292.
2. Sturek M, et al. Molecular and cellular physiology of differentiated vascular smooth muscle. In: Willerson JT et al., editors. Cardiovascular medicine. London: Springer-Verlag; 2006. p. 1511–23.
3. Knudson JD, et al. Mechanisms of coronary dysfunction in obesity and insulin resistance. Microcirculation. 2007;14:317–38.
4. Stepp DW, Pollock DM, Frisbee JC. Low-flow vascular remodeling in the metabolic syndrome X. Am J Physiol Heart Circ Physiol. 2004;286(3):H964–70.
5. Bowles DK, et al. Heterogeneity of L-type calcium current density in coronary smooth muscle. Am J Physiol. 1997;273(42):H2083–9.
6. Bowles DK, et al. Hypercholesterolemia inhibits L-type calcium current in coronary macro-, not microcirculation. J Appl Physiol. 2004;96:2240–8.
7. Mokelke EA, et al. Diabetic dyslipidemia and exercise affect coronary tone and differential regulation of conduit and microvessel K^+ current. Am J Physiol Heart Circ Physiol. 2005;288:H1233–41.
8. Libby P, Ridker PM, Hansson GK. Progress and challenges in translating the biology of atherosclerosis. Nature. 2011;473(7347):317–25.

9. Owens GK, Kumar MS, Wamhoff BR. Molecular regulation of vascular smooth muscle cell differentiation in development and disease. Physiol Rev. 2004;84:767–801.
10. van Hinsbergh VW. Endothelial permeability for macromolecules. Mechanistic aspects of pathophysiological modulation. Arterioscler Thromb Vasc Biol. 1997;17(6):1018–23.
11. Furchgott RF, Zawadzki JV. The obligatory role of endothelial cells in the relaxation of arterial smooth muscle by acetylcholine. Nature. 1980;288:373–6.
12. Oemar BS, et al. Reduced endothelial nitric oxide synthase expression and production in human atherosclerosis. Circulation. 1998;97:2494–8.
13. Gutierrez E, et al. Endothelial dysfunction over the course of coronary artery disease. Eur Heart J. 2013;34(41):3175–81.
14. Lerman A, et al. Coronary vascular remodeling in association with endothelial dysfunction. Am J Cardiol. 1998;81:1105–9.
15. Hasdai D, et al. Coronary endothelial dysfunction in humans is associated with myocardial perfusion defects. Circulation. 1997;96(10):3390–5.
16. Lerman A, Zeiher AM. Endothelial function: cardiac events. Circulation. 2005;111(3):363–8.
17. Li J, et al. Comparison of the effect of the metabolic syndrome and multiple traditional cardiovascular risk factors on vascular function. Mayo Clin Proc. 2012;87(10):968–75.
18. Hill-Eubanks DC, et al. NFAT regulation in smooth muscle. Trends Cardiovasc Med. 2003;13(2):56–62.
19. Bobe R, et al. SERCA2a controls the mode of agonist-induced intracellular Ca2+ signal, transcription factor NFAT and proliferation in human vascular smooth muscle cells. J Mol Cell Cardiol. 2011;50(4):621–33.
20. Sturek M. Ca^{2+} regulatory mechanisms of exercise protection against coronary artery disease in metabolic syndrome and diabetes. J Appl Physiol. 2011;111:573–86.
21. Ervin RB. Prevalence of metabolic syndrome among adults 20 years of age and over, by sex, age, race and ethnicity, and body mass index: United States, 2003-2006. Natl Health Stat Report. 2009;13:1–7.
22. Gami AS, et al. Metabolic syndrome and risk of incident cardiovascular events and death: a systematic review and meta-analysis of longitudinal studies. J Am Coll Cardiol. 2007;49(4): 403–14.
23. Libby P, Plutzky J. Diabetic macrovascular disease. The glucose paradox? Circulation. 2002;106:2760–3.
24. Rader DJ, Daugherty A. Translating molecular discoveries into new therapies for atherosclerosis. Nature. 2008;451(7181):904–13.
25. Zhang LL, et al. Activation of transient receptor potential vanilloid type-1 channel prevents adipogenesis and obesity. Circ Res. 2007;100(7):1063–70.
26. Hu G, et al. Canonical transient receptor potential channels expression is elevated in a porcine model of metabolic syndrome. Mol Endocrinol. 2009;23:689–99.
27. Venkatachalam K, et al. The cellular and molecular basis of store-operated calcium entry. Nat Cell Biol. 2002;4(11):E263–72.
28. Clapham DE. A STIMulus Package puts orai calcium channels to work. Cell. 2009;136(5): 814–6.
29. Hogan PG, Lewis RS, Rao A. Molecular basis of calcium signaling in lymphocytes: STIM and ORAI. Annu Rev Immunol. 2010;28:491–533.
30. Yeromin AV, et al. Molecular identification of the CRAC channel by altered ion selectivity in a mutant of Orai. Nature. 2006;443(7108):226–9.
31. Lee KP, et al. Molecular determinants mediating gating of Transient Receptor Potential Canonical (TRPC) channels by stromal interaction molecule 1 (STIM1). J Biol Chem. 2014;289(10):6372–82.
32. Koenig S, et al. A TRPC3 blocker, ethyl-1-(4-(2,3,3-trichloroacrylamide)phenyl)-5-(trifluoromethyl)-1H-pyrazole-4-c arboxylate (Pyr3), prevents stent-induced arterial remodeling. J Pharmacol Exp Ther. 2013;344(1):33–40.

33. Saleh SN, et al. Diverse properties of store-operated TRPC channels activated by protein kinase C in vascular myocytes. J Physiol. 2008;586(10):2463–76.
34. Poburko D, et al. Transient receptor potential channel 6 mediated, localized cytosolic [Na+] transients drive Na+/Ca2+ exchanger mediated Ca2+ entry in purinergically stimulated aorta smooth muscle cells. Circ Res. 2007;101(10):1030–8.
35. Minke B, Wu C, Pak WL. Induction of photoreceptor voltage noise in the dark in Drosophila mutant. Nature. 1975;258(5530):84–7.
36. Montell C, Rubin GM. Molecular characterization of the Drosophila trp locus: a putative integral membrane protein required for phototransduction. Neuron. 1989;2(4):1313–23.
37. Wes PD, et al. TRPC1, a human homolog of a Drosophila store-operated channel. Proc Natl Acad Sci U S A. 1995;92(21):9652–6.
38. Zhu X, et al. Molecular cloning of a widely expressed human homologue for the Drosophila trp gene. FEBS Lett. 1995;373(3):193–8.
39. Earley S, Brayden JE. Transient receptor potential channels in the vasculature. Physiol Rev. 2015;95(2):645–90.
40. Zitt C, et al. Cloning and functional expression of a human Ca^{2+}-permeable cation channel activated by calcium store depletion. Neuron. 1996;16:1189–96.
41. Hofmann T, et al. Direct activation of human TRPC6 and TRPC3 channels by diacylglycerol. Nature. 1999;397(6716):259–63.
42. Takahashi Y, et al. Involvement of transient receptor potential canonical 1 (TRPC1) in angiotensin II-induced vascular smooth muscle cell hypertrophy. Atherosclerosis. 2007;195(2): 287–96.
43. Sweeney M, et al. Inhibition of endogenous TRP1 decreases capacitative Ca2+ entry and attenuates pulmonary artery smooth muscle cell proliferation. Am J Physiol Lung Cell Mol Physiol. 2002;283(1):L144–55.
44. Edwards JM, et al. Exercise training decreases store-operated Ca^{2+} entry associated with metabolic syndrome and coronary atherosclerosis. Cardiovasc Res. 2010;85:631–40.
45. Gautam M, et al. Flow-activated chloride channels in vascular endothelium. Shear stress sensitivity, desensitization dynamics, and physiological implications. J Biol Chem. 2006; 281(48):36492–500.
46. Sturek M, Stehno-Bittel L, Obye P. Modulation of ion channels by calcium release in coronary artery smooth muscle. In: Sperelakis N, Kuriyama H, editors. Electrophysiology and ion channels of vascular smooth muscle cells and endothelial cells. New York: Elsevier; 1991. p. 65–79.
47. Prakriya M, et al. Orai1 is an essential pore subunit of the CRAC channel. Nature. 2006;443(7108):230–3.
48. Kumar B, et al. Upregulated TRPC1 channel in vascular injury in vivo and its role in human neointimal hyperplasia. Circ Res. 2006;98(4):557–63.
49. Li J, et al. Interactions, functions, and independence of plasma membrane STIM1 and TRPC1 in vascular smooth muscle cells. Circ Res. 2008;103(8):e97–104.
50. Hill AB. The environment and disease: association or causation? Proc R Soc Med. 1965; 58:295–300.
51. Liu D, et al. Increased transient receptor potential canonical type 3 channels in vasculature from hypertensive rats. Hypertension. 2009;53(1):70–6.
52. Peppiatt-Wildman CM, et al. Endothelin-1 activates a Ca2+-permeable cation channel with TRPC3 and TRPC7 properties in rabbit coronary artery myocytes. J Physiol. 2007;580(3): 755–64.
53. Hunyady L, Catt KJ. Pleiotropic AT1 receptor signaling pathways mediating physiological and pathogenic actions of angiotensin II. Mol Endocrinol. 2006;20(5):953–70.
54. Matsuyama K, et al. Increased plasma level of endothelin-1-like immunoreactivity during coronary spasm in patients with coronary spastic angina. Am J Cardiol. 1991;68:991–5.
55. Toyo-oka T, et al. Increased plasma level of endothelin-1 and coronary spasm induction in patients with vasospastic angina pectoris. Circulation. 1991;83:476–83.
56. Danthuluri NR, Brock TA. Endothelin receptor-coupling mechanisms in vascular smooth muscle: a role for protein kinase C. J Pharmacol Exp Ther. 1990;254:393–9.

57. Caterina MJ, et al. The capsaicin receptor: a heat-activated ion channel in the pain pathway. Nature. 1997;389(6653):816–24.
58. Gazzieri D, et al. Ethanol dilates coronary arteries and increases coronary flow via transient receptor potential vanilloid 1 and calcitonin gene-related peptide. Cardiovasc Res. 2006;70(3): 589–99.
59. Bratz IN, et al. Impaired capsaicin-induced relaxation of coronary arteries in a porcine model of the metabolic syndrome. Am J Physiol. 2008;294:H2489–96.
60. Hiett SC, et al. Mechanisms underlying capsaicin effects in canine coronary artery: implications for coronary spasm. Cardiovasc Res. 2014;103:607–18.
61. Berwick Z, et al. Contribution of electromechanical coupling between KV and CaV1.2 channels to coronary dysfunction in obesity. Basic Res Cardiol. 2013;108(5):370.
62. Owen MK, et al. Perivascular adipose tissue potentiates contraction of coronary vascular smooth muscle: influence of obesity. Circulation. 2013;128:9–18.

Chapter 18
Mitochondrial Ion Channels in Metabolic Disease

Aaron H. Truong, Saravanakumar Murugesan, Katia D. Youssef, and Ayako Makino

Abstract Mitochondria are small organelles that reside in most eukaryotic cells and are responsible for cellular energy production. They are also involved in other cellular signaling pathways including cell apoptosis. Mitochondria are composed of two phospholipid bilayers: the outer mitochondrial membrane and the inner mitochondrial membrane; and two aqueous spaces: the intermembrane space and the matrix. Like the plasma membrane, many ion channels are located in the mitochondrial membranes and play a critical role in mitochondrial functions. Increasing evidence suggests that the dysregulation of mitochondrial ion channels is implicated in the development and progression of cardiovascular diseases. In this chapter, we review the role of mitochondrial ion channels in metabolic disease.

Keywords Cardiovascular disease • Mitochondrial calcium overload • Cell apoptosis • VDAC • Coronary arterial disease

Introduction

Metabolic syndrome is a worldwide epidemic. It is a cluster of cardiovascular risk factors including increased systemic blood pressure (hypertension), high blood sugar level (diabetes), excess body fat around the waist (obesity), and abnormal cholesterol levels (hypercholesterolemia). Metabolic risk factors contribute to the development of vascular complications including coronary heart disease and stroke, which are the top two leading causes of death in the world based on the WHO report.

Mitochondria are organelles found in many eukaryotic cells which act as an important source of energy. Adenosine triphosphate (ATP) produced in the mitochondria powers most of the cellular functions including transporting the substrates across the membranes, accelerating metabolic reactions that would not occur automatically, and

---

A.H. Truong • S. Murugesan • K.D. Youssef • A. Makino (✉)
Department of Physiology, University of Arizona, Tucson, AZ 85724, USA
e-mail: aymakino@email.arizona.edu

I. Levitan, A.M. Dopico (eds.), *Vascular Ion Channels in Physiology and Disease*, DOI 10.1007/978-3-319-29635-7_18

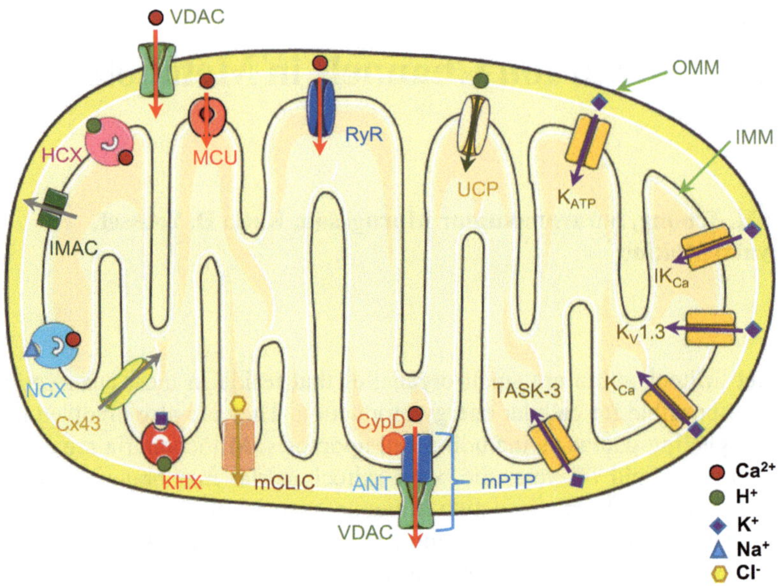

Fig. 18.1 Mitochondrial ion channels. *VDAC* voltage-dependent anion channel, *HCX* H^+/Ca^{2+} exchanger, *NCX* Na^+/Ca^{2+} exchanger, *Cx43* connexin 43, *mCLIC* mitochondrial chloride channel, *ANT* adenine nucleotide translocase, *CypD* cyclophilin D, *mPTP* mitochondrial permeability transition pore, *KHX* K^+/H^+ exchanger, *MCU* mitochondrial Ca^{2+} uniporter, UPC uncoupling protein, *RyR* Ryanodine receptor

contracting and relaxing muscle cells. Mitochondria are also key players in the determination of cell fate. Mitochondria are composed of two membranes: the inner mitochondrial membrane (IMM) and the outer mitochondrial membrane (OMM), and two aqueous spaces: the intermembrane space and the matrix. Like the plasma membrane, multiple types of ion channels are located in the mitochondrial membranes (Fig. 18.1) and they regulate not only mitochondrial function but also cellular functions.

Mitochondrial Ion Channels and Their Roles in Metabolic Syndrome

Voltage-Dependent Anion Channel (VDAC)

The VDAC is a bi-directional transporter located in the OMM. Solutes of up to 5 kDa are allowed to pass through the OMM via the VDAC pore. Three different isoforms of VDAC have been identified: VDAC1, VDAC2 and VDAC3. VDAC1 is highly expressed in most cells and its function is well-documented [110], whereas we have limited information regarding the functions of VDAC2 and VDAC3 [37, 109, 173].

VDAC1 mainly serves as a shuttle of respiratory chain substrates such as ATP [142, 150]. The ion selectivity of VDAC depends on the membrane potential; low potential is more favorable for anion transfer while high potential is favorable for cation transfer [57, 68, 150]. Ca^{2+} overload into the mitochondria is a hallmark of cell apoptosis and VDAC1 contributes to mitochondrial Ca^{2+} overload. VDAC1 overexpression increases mitochondrial calcium concentration [100] and VDAC1 siRNA transfection attenuates mitochondrial Ca^{2+} uptake and cell apoptosis induced by H_2O_2 or ceramide [39]. Recent evidence indicates that VDAC1 forms an endoplasmic reticulum (ER)-mitochondrial tethering with inositol trisphosphate (IP3) receptor and GRP75 at the mitochondria-associated membrane (MAM). This tethering facilitates Ca^{2+} transfer from the ER to the mitochondria and leads to cell apoptosis [39, 63]. Mitochondrial permeability transition pore (mPTP) is composed of VDAC1, adenine nucleotide translocator, and cyclophilin D [32, 66, 72–74]. When mPTP opens, it leads to an increase in the mitochondrial membrane permeability, and the release of proapoptotic peptides, which results in collapsing the mitochondrial membrane potential followed by uncoupling oxidative phosphorylation and ATP depletion, and eventually cell apoptosis [64, 65, 167]. VDAC2 has been reported to associate with ryanodine receptor 2, but not with IP_3R, at the sarcoplasmic reticulum (SR)-mitochondria junctions in cardiac myocytes, and to be a major contributor to the direct transfer of Ca^{2+} from the SR to the mitochondria [111].

There is no direct evidence showing that VDAC contributes to the development or progression of hypertension. Pertiz et al. demonstrated that VDAC3 expression level was decreased by high intensity exercise in spontaneous hypertensive rats (SHR) [135], but the data comparing VDAC levels between the SHR and Wistar-Kyoto rats (counterpart to SHR) without exercise, and the physiological role of VDAC3 in the regulation of blood pressure were not examined.

Earlier studies showed that VDAC expression is altered in several mouse and rat diabetic models in different cell types but the functional role of these changes were not examined. High glucose treatment increased VDAC1 protein expression and decreased VDAC2 levels in β cells compared to cells cultured in normal media [1]. VDAC1 expression levels were also upregulated in the islets isolated from type 1 diabetic mice compared to cells from control mice [171]. However, those studies did not examine whether VDAC1 expression affected the β cells' functions (e.g., insulin secretion). In cardiac myocytes, VDAC protein expression was significantly decreased in type 1 diabetic rats compared to myocytes from control rats [98, 166]. Since cardiac myocytes use mitochondrial Ca^{2+} for their contraction in addition to Ca^{2+} from the SR [159], the decreased mitochondrial Ca^{2+}, due to decreased VDAC expression, might have an adverse effect on their functions. High glucose treatment significantly increased VDAC protein levels in the myoblasts [61]. Kidneys dissected from type 1 diabetic rats exhibited an increase in VDAC1 expression compared to the controls [166].

More recently, we reported that VDAC protein expression was significantly increased in mouse coronary endothelial cells (MCECs) isolated from type 1 diabetic when compared to cells from control mice [148] (Fig. 18.2). In this paper, we demonstrated that downregulation of VDAC in diabetic MCECs decreases mitochondrial Ca^{2+} concentration and subsequently normalizes the levels of mPTP activity and mitochondrial ROS (Fig. 18.3). In addition, we found that one of the endogenous inhibitors for VDAC (HK2) is significantly decreased in diabetic MCECs (Fig. 18.4). These data suggest that increased VDAC activity contributes to mitochondria-induced cell apoptosis in diabetic MCECs (Fig. 18.5).

As described above, the expression levels of VDAC vary between different tissues/organs in diabetes; it is thus necessary to develop tissue-specific overexpression or downregulation systems to normalize the function of VDAC in diabetes.

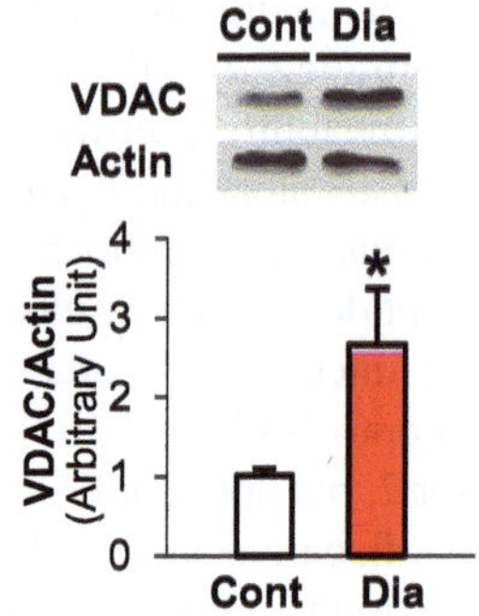

Fig. 18.2 Increased VDAC protein level in MCECs isolated from diabetic mice. Western blots showing VDAC and actin protein levels. Actin was used as a loading control. The columns show VDAC protein levels normalized by actin. Control (Cont), n=6; diabetic (Dia), n=6. Data are mean ± SE. $^*P<0.05$ vs. control

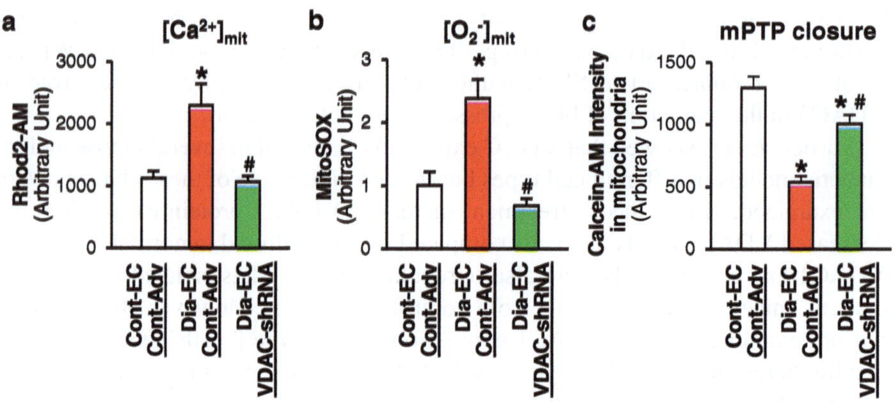

Fig. 18.3 The effect of VDAC inhibition by VDAC-shRNA adenovirus (VDAC-shRNA) on $[Ca^{2+}]_{mit}$, $[O_2^-]_{mit}$, and activity of the mitochondrial permeability transition pore (mPTP) in MCECs. (**a**) Summarized $[Ca^{2+}]_{mit}$ data. Control ECs infected with control Adv. (Cont-EC, Cont-Adv), n = 16; diabetic ECs infected with control Adv. (Dia-EC, Cont-Adv), n = 19; diabetic ECs infected with VDAC-shRNA (Dia-EC, VDAC-shRNA), n = 22. Data are mean ± SE. $^*P<0.05$ vs. Cont-EC, Cont-Adv. $^\#P<0.05$ vs. Dia-EC, Cont-Adv. (**b**) Summarized data of $[O_2^-]_{mit}$. Cont-EC, Cont-Adv., n=41; Dia-EC, Cont-Adv., n=44; Dia-EC, shRNA-VDAC, n=30. Data are mean ± SE. $^*P<0.05$ vs. Cont-EC, Cont-Adv. $^\#P<0.05$ vs. Dia-EC, Cont-Adv. (**c**) The fluorescence intensity of calcein-AM in mitochondria was measured to assess the activity of the mPTP opening. The decrease in calcein-AM intensity indicates the increase of the mPTP opening. Control ECs infected with control Adv. (Cont-EC, Cont-Adv.), n = 108; diabetic ECs infected with control Adv. (Dia-EC, Cont-Adv.), n = 133; diabetic ECs infected with VDAC-shRNA (Dia-EC, VDAC-shRNA), n = 98. Data are mean ± SE. $^*P<0.05$ vs. Cont-EC, Cont-Adv. $^\#P<0.05$ vs. Dia-EC, Cont-Adv

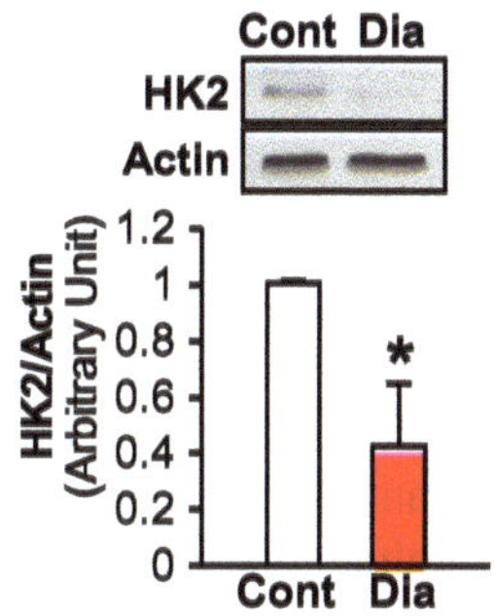

Fig. 18.4 Hexokinase II (HK2) is significantly decreased in diabetic MCECs. Western blots showing HK2 and actin protein levels. Actin was used as a loading control. The columns show HK2 protein expression levels normalized by actin. Control (Cont), n=4; diabetic (Dia), n=4. Data are mean±SE. *$P<0.05$ vs. control

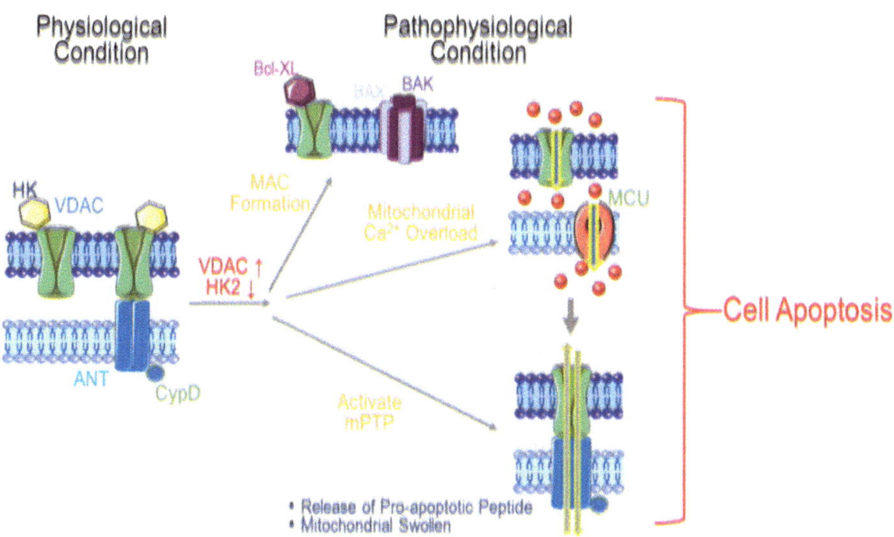

Fig. 18.5 Mitochondria-mediated cell apoptosis through VDAC

The role of MAM through VDAC-cyclophilin D interaction in maintaining insulin homeostasis is currently conversional. Tubbs et al. demonstrated that cyclophilin D is one of the components of MAM [128] and knockout mice of cyclophilin D exhibit decreased MAM fraction in the liver, abnormal glucose tolerance, and attenuated insulin signaling. In addition, VDAC1 and cyclophilin D protein expression levels in the liver were significantly decreased in the insulin-resistant obesity mice (ob/ob mice) compared to control mice [165]. On the other hand, Taddeo et al. demonstrated opposite results from the same mouse strain; cyclophilin D deletion significantly attenuated the development of insulin resistance after feeding mice a high-fat diet [163]. Therefore, it is still necessary to clarify the molecular mechanisms of cyclophilin D-VDAC interaction in the development of insulin resistance.

Mitochondrial K^+ Channels

Mitochondrial K^+ channels play an important role in maintaining the integrity of the IMM by balancing potassium concentration in the intermembrane space and matrix and maintaining the mitochondrial membrane potential. There are different potassium channels found in the IMM: ATP-sensitive K^+ (mitoK_{ATP}) channel [75], large-conductance Ca^{2+}-activated K^+ (mitoBK_{Ca}) channel [151], intermediate-conductance Ca^{2+}-activated K^+ (mitoIK_{Ca}) channel [36], small-conductance Ca^{2+}-activated K^+ (mitoSK_{Ca}) channel [42, 158], voltage-dependent potassium (mitoKv1.3) channel [162], and twin-pore TASK-3 potassium channel [143]. It has been demonstrated that the opening of mitochondrial K^+ channels leads to the depolarization of the mitochondrial membrane, reduces mitochondrial Ca^{2+} overload, and results in cardioprotection [107, 124, 125, 157].

mitoK_{ATP} was first found by Inoue et al. in the liver mitochondria [75]. Although it is still unclear whether mitoK_{ATP} is a homolog of plasma membrane K_{ATP} channels, it is proposed to be a multi-protein complex composed of sulfonylurea receptor and inward rectifier K^+ channel subunits [75, 90, 97, 112]. There are several endogenous regulatory elements for mitoK_{ATP} activity including the ATP/Mg^{2+} complex and protons (negative regulators) [14, 15] as well as reactive oxygen species (ROS, a positive regulator) [178]. Data suggest that mitoK_{ATP} channels participate in the regulation of mitochondrial volume and mitochondrial membrane potential, thereby affecting the production of ATP via inhibition of Complex I and reactive oxygen species (ROS) [95, 96]. mitoK_{ATP} channel openers are useful tools for cardioprotection during ischemic heart via (1) modulating mitochondrial Ca^{2+} concentration [70, 71, 126], (2) altering mitochondrial membrane potential [126], (3) regulating ROS production [49, 129], and (4) inhibiting mPTP opening [86].

There is evidence that chronic hyperglycemia and obesity impair the function of mitoK_{ATP} which results in decreased cardioprotection [80, 82]. Ibra et al. demonstrated that the protein expression levels of potential mitoK_{ATP} subunits, mitoKir6.1 and mitoSUR1, and mitoK_{ATP} activity were significantly decreased in the hearts of type 1 diabetic mice [46]. In contrast, mitoK_{ATP} activity was significantly increased in liver mitochondria isolated from apolipoprotein CIII transgenic mice (a mouse model of hypertriglyceridemia) compared with control mice and it was suggested that this increase might be occurring in order to reduce high energy conversion efficiency in transgenic mice [3].

mitoK_{ATP} appears to contribute to the development of vascular complications in hypertension by augmenting angiotensin II (AngII) induced ROS production. AngII leads to endothelial cell dysfunction by increasing cytosolic and mitochondrial O_2^- concentration, while the opening of mitoK_{ATP} further increases mitochondrial O_2^- concentration [120, 170, 175]. Blocking mitoK_{ATP} inhibits AngII-induced O_2^- production in cardiac myocytes and vascular smooth muscle cells [53, 84]; however, increased blood pressure induced by AngII infusion in rats was not restored by mitoK_{ATP} blocker [84]. These data suggest that the opening of mitoK_{ATP} does not influence blood pressure, but augments vascular damage by increasing mitochondrial O_2^- in hypertension.

Among mitochondrial Ca^{2+}-activated K^+ channels, mitoBK_{Ca} is the most studied channel and observed in many eukaryotic cells. The localization and functions of mitoBK_{Ca} were first characterized in mitochondria from human glioma cells in 1999 [151], followed by mitochondria isolated from cardiac myocytes in 2002 [172]. It has been confirmed that the structure and function of mitoBK_{Ca} are comparable to BK_{Ca} in the plasma membrane [9, 17, 153, 154]. It is expressed in the IMM and opening of mitoBK_{Ca} protects against damage caused by ischemia/reperfusion [125] like mitoK_{ATP}.

Nousadeghi et al. showed that the protein expression levels and open probability of mitoBK_{Ca} were significantly decreased in brain mitochondrial samples isolated from type 1 diabetic rats compared to controls [122]. Since opening mitoBK_{Ca} lowers ROS production, decreased mitoBK_{Ca} in diabetes would lead to excess ROS production and affect neuronal survival [88].

While mitoK_{ATP} and mitoBK_{Ca} were shown to have a role in metabolic syndrome, other mitochondrial K channels were not yet studied. mitoIK_{Ca} was relatively recently identified in the IMM and much less information is available regarding the expression levels of mitoIK_{Ca} in different cell types compared with mitoK_{ATP} and mitoBK_{Ca} [36, 149]. In addition, the role of mitoIK_{Ca} under physiological and pathophysiological conditions is not fully investigated. It is assumed that its function would be similar to other mitochondrial K^+ channels in terms of the regulation of mitochondrial membrane potential [151]. In regards to the cell apoptotic cascade, IK_{Ca} seems to play a different role from mitoK_{ATP} and mitoBK_{Ca} since the inhibition of IK_{Ca} does not induce cell apoptosis [149].

In 2013, two manuscripts were published from different groups demonstrating that mitoSK_{Ca} is functionally expressed in the IMM [42, 158]. Mitochondria was isolated from cardiac myocytes of guinea pigs and purified mitoSK_{Ca} was reconstructed into the lipid bilayer to test for the channel activities. This channel is apamin-sensitive and activated by an increase of Ca^{2+} concentration [158]. Another experiment was conducted using HT-22 neuronal mitoplasts and demonstrated that glutamate-induced mitochondrial depolarization was prevented by pretreatment with SK_{Ca} opener [42]. These data suggest that mitoSK_{Ca} might have a cardioprotective effect like mitoK_{ATP} and mitoBK_{Ca}, but more experimental data needs to be obtained in order to confirm its function.

mitoKv1.3 was first identified in the IMM of lymphocytes [162], followed by other cell types [16, 56, 91]. Despite the number of manuscripts describing the localization of mitoKv1.3, the functional importance is less fully investigated. The electrophysiological function of mitoKv1.3 in lymphocytes was illustrated by Szabó et al. [161, 162] showing that mitoKv1.3 regulated mitochondrial membrane potential. Further experiments are required to identify the role of mitoKv1.3 in other cell types including cardiac myocytes and vascular cells.

TASK channels [TWIK (tandem-pore domain in weak inward rectifier K^+ channels)-related acid-sensitive K^+] are members of the large subfamily of two-pore domain potassium channels [83, 92]. These channels exhibit voltage-independent leak K^+ currents, which contribute to maintaining the resting membrane potential, input resistance, and extracellular acidification [28, 44, 83, 108]. The existence of mitochondrial TASK-3 was first reported in HaCaT human keratinocytes and WM35 malignant melanoma cell lines using immunohistochemistry [143]. Later, the same group published more detailed information about the function of mito-TASK-3 [119].

Uncoupling Proteins

Uncoupling protein (UCP) is a proton carrier protein which localizes in the IMM [87]. The functions of UCPs are most studied in the brown adipose tissue where UCPs are exclusively expressed. The main role of UCP is to transport H^+ into the mitochondrial matrix in order to dissipate the proton motive force as heat instead of using H^+ for ATP production through oxidative phosphorylation. There are five UCP homologues identified in mammals. UCP1 is responsible for thermogenesis and believed to be only expressed in brown adipose tissue, although UCP1 was also detected recently in the thymus [27] and islet cells [145]. UCP2 is widely distributed in several tissues including the spleen, lung, stomach, and gonadal white adipose tissue [134], whereas UCP3 is mainly restricted to the skeletal muscle [23]. UCP4 is mainly localized in the brain [155] and UCP5/BMCP1 expresses in many tissues, with a high abundance in the brain [176].

Since the identification of UCPs, numerous studies have been done to characterize the function of UCPs in the metabolic syndrome. In the diabetic mouse model, the expression levels of most subtypes of UCP were significantly increased in the corresponding tissues: UCP1 mRNA level in brown adipose tissue [81], UCP2 protein level in the kidney [138], in islets [85], in brown and white adipose tissues [81], and UCP3 mRNA in the heart [67]. However, Kageyama et al. showed that mRNA levels of UCP2 and UCP3 were significantly decreased in the brown adipose tissue, but increased in the skeletal muscle, in type 1 diabetic rats [78]. Mice fed with high-fat diet (HF) show a significant increase in UCP1 protein expression in brown adipose tissue [25] and an increase in UCP3 expression in the heart [24] and in the skeletal muscles [30]. HF rats also exhibited a significant increase in UCP2 mRNA levels in white adipose tissue and UCP3 mRNAs in skeletal muscles [50, 105]. Obese mice exhibited a marked increase in UCP2 mRNA expression and plasma insulin level [177]. These results suggest that an increase in UCP expression levels might be the physiological response to excess fatty acids in the body in order to counteract obesity. Contrary to consistent results from animal studies, the data of UCP expression levels in obese and type 2 diabetic patients vary between samples, which makes it very difficult to reach a conclusion [10, 101, 121, 127, 152, 169]. The potential reasons behind this might be due to the different medications and different diet composition among the patients, and different sampling protocols.

In order to investigate UCP functions, knockout mice were quickly developed and the result was, in fact, contrary to many investigators' expectations: none of the knockout mice (UCP1, 2, 3) exhibited any change in their body weight compared with their wild-type [7, 24, 45, 58, 168]. On the other hand, mice that overexpressed

UCP exhibited a beneficial effect by preventing the development of metabolic syndrome. Skeletal muscle-specific UCP1 overexpression in particular emphasizes this preventative effect. This strain was first introduced in mice fed with HF diet. Body weight, plasma lipid, insulin and glucose concentration, and glucose tolerance were significantly improved in UCP-overexpressed HF mice [93]. Other researchers also used this strain to cross with transgenic mice for type 2 diabetes (KK mice) [18] and atherosclerosis (ApoE knockout mice) [55] and demonstrate resistance to the development of metabolic syndromes. UCP3 overexpression in skeletal muscle also improved glucose tolerance, reduced body weight and adipose tissue volume [31], and minimized the development of insulin resistance in HF mice [29]. In humans, polymorphisms of UCP have been implicated in the development of diabetes, obesity, and hypertension [26, 34, 35, 38, 52, 76, 147].

The basal expression levels of UCP2 in the kidney were not different between stroke-prone spontaneously hypertensive rats (SHRsp) and stroke-resistant SHR rats (SHRsr, control for SHRsp), while high-salt diet led to a significant decrease in UCP2 protein and mRNA levels in SHRsp, but not SHRsr [41]. The authors also found that miR24 and 34a were significantly decreased in high salt diet-treated SHRsp compared to SHRsp fed with normal-salt diet and implied that miR24 and 34a would be the upstream regulators of UCP2.

Mitochondrial Connexin

Connexins (Cxs) were first identified in the plasma membrane and six Cxs compose a hemichannel called connexon. The connexon forms a channel by coupling with another connexon in adjacent cells (called gap junction) and transfers small molecules and ions [59]. Cxs have 21 members in humans and 20 members in mice [115, 156]. All members of the Cx family share a common structure: four hydrophobic transmembrane domains, two extracellular loops, and three cytoplasmic domains including an intracellular loop and carboxyl and amino-terminal domains [89, 174]. Cxs are involved in many cellular processes such as homeostasis, metabolic support, electrical coupling, enhancement of tissue response, and control of cell proliferation, differentiation, survival, and apoptosis [174].

Despite the existence of many Cx subtypes in the plasma membrane, only Cx43 is found in the mitochondria in several tissues including human umbilical vein endothelial cells [94] and cardiac myocytes [22, 141]. Interestingly, stress stimulation (e.g., ischemia/reperfusion) increases the translocation of Cx43 into the mitochondria and, moreover, Cx43 reduces the damage induced by stress [94, 141]. These data suggest that mitochondrial Cx43 exerts a cytoprotective effect upon stress stimulation. It is still unclear whether the effect of Cx43 is mediated by hemichannel formation and if Cx43 directly regulates mitochondrial function and/or modulates other channels.

High-fat diet in rats significantly decreased mitochondrial Cx43 expression levels in cardiac myocytes [60]. High glucose treatment in retinal endothelial cells led to a significant decrease in mitochondrial Cx43 protein expression [114, 164], and mitochondrial Cx43 protein levels were lower in the retina of type 1 diabetic rats compared with those in control rats. It is suggested that hyperglycemia increased matrix metalloproteinase-2 in the mitochondria and led to excess degradation of Cx43 protein in the retina [114]. Based on these data, increased stress (e.g., hyperlipidemia and hyperglycemia) might lead to decreased mitochondrial Cx43 expression which results in mitochondrial dysfunction. Further experiments are required to define the functional role of Cx43 in the mitochondria by overexpressing Cx43 in disease models.

Other Channels

Mitochondrial Ca^{2+} Uniporter (MCU)

Mitochondria transport Ca^{2+} from the cytosol and accumulate it in the matrix [116, 140], and mitochondrial calcium homeostasis is important not only for energy production but also for the regulation of cytoplasmic calcium and activation of cell apoptotic pathways. The IMM expresses the MCU, which is the main transporter of Ca^{2+} from the intermembrane space into the matrix [19, 62]. The major channel-forming subunit of the MCU complex (CCDC109A) was first reported in 2011 [11, 40]. The MCU consists of two transmembrane domains and forms a complex in the IMM with many endogenous regulators for opening the pore of the MCU [79, 104]. Mitochondrial Ca^{2+} uptake 1 (MICU1), mitochondrial Ca^{2+} uptake 2 (MICU2), and MCUb are most likely negative regulators of the MCU, while mitochondrial Ca^{2+} uniporter regulator 1 (MCUR1), essential MCU regulator (EMRE), and SLC25A23 are essential for MCU activity [33, 69, 103, 136, 146]. Most of the MCU regulatory proteins were recently identified and their functions were only examined in Hela cells and HEK cells, isolated cardiac mitochondria and lipid bilayer with reconstituted proteins.

After the identification of the MCU (Table 18.1), the functional role of MCU under physiological conditions was extensively studied using siRNA/shRNA of MCU and MCU knockout mice [2, 43, 132, 137, 139]. Interestingly, but unexpectedly, MCU knockout mice did not exhibit obvious defects in any physiological functions [132]. The outcome of the authors' study implied that the systemic knockout of MCU might affect the function/expression levels of other Ca^{2+} transporters in the mitochondria to compensate for the lack of MCU.

Mitochondrial Ca^{2+} overload contributes to malfunctions in many tissues via cell apoptosis and we suspect that mitochondrial Ca^{2+} overload through excess MCU activation could be implicated in metabolic syndrome. Since the history of MCU is still very brief, the questions of whether MCU participates in the development and progression of the metabolic syndrome and whether vascular complications in the metabolic syndrome are regulated by the MCU have not yet been explored.

Table 18.1 Mitochondrial Ca^{2+} uniporter complex and its functions

Name	MW (kDa)	Other names	Functions
MCU (Mitochondrial Calcium Uniporter)	40	C10orf42 CCDC109A	Main mitochondrial Ca^{2+} transporter.
MICU1 (Mitochondrial Calcium Uptake 1)	50	CALC CBARA1 EFHA3 MPXPS	Endogenous negative regulator of MCU. Inhibition increases $[Ca^{2+}]_{mito}$.
MICU2 (Mitochondrial Calcium Uptake 2)	45	EFHA1 1110008L20Rik	Endogenous negative regulator of MCU. Inhibition lowers $[Ca^{2+}]_{mito}$.
MCUb (Mitochondrial Calcium Uniporter b)	40	CCDC109B	Paralog of MCU with lower expression. No channel activity. Overexpression reduces $[Ca^{2+}]_{mito}$.
MCUR1 (Mitochondrial Calcium Uniporter Regulator 1)	40	CCDC90A C6orf79	Endogenous positive regulator of MCU. Inhibition diminishes mitochondrial Ca^{2+} uptake.
EMRE (Essential MCU Regulator)	10	C22orf32 SMDT1 DDDD DJ186O1.1	Endogenous positive regulator of MCU. Inhibition decreases mitochondrial Ca^{2+} uptake.
SLC25A23 solute carrier family 25A23	50	APC2 MCSC2 SCaMC-3	Binding site with MCU. Inhibition decreases mitochondrial Ca^{2+} uptake.

Mitochondrial Ryanodine Receptor (mRyR)

The RyR is a Ca^{2+}-releasing channel in the sarcoplasmic reticulum in muscle cells. It was also identified in the IMM and plays an important role in the mitochondrial Ca^{2+} uptake [4, 20]. The activation of mRyR is tightly controlled by ryanodine concentration and cytosolic Ca^{2+} concentration [48]. The sensitivity of mRyR to Ca^{2+} is higher than that of MCU [144]. Electrophysiological experiments and immunoblotting demonstrate that mRyR could be similar or the same as RyR1 [21]. To the best of our knowledge, there is no report examining the role of mRyR in the metabolic syndrome.

Mitochondrial Ca^{2+}/H^+ Antiporter

The negative membrane potential in the IMM drives Ca^{2+} entry and it is mostly mediated through the MCU as described above. The MCU uptakes Ca^{2+} into the IMM at mM range, while mitochondrial Ca^{2+}/H^+ antiporter is able to import Ca^{2+} at nM concentrations. A recent manuscript presented the potential candidate for mitochondrial Ca^{2+}/H^+ antiporter, leucine zipper-EF-hand containing transmembrane protein 1 (LETM1) [102]. LETM1 has two EF hand domains in the intermembrane space and catalyzes the one-to-one electronic exchange of Ca^{2+} for H^+. The increase of mitochondrial Ca^{2+} through LETM1 is pH sensitive and is inhibited by ruthenium red. Interestingly, LETM1 not only imports Ca^{2+} into the matrix through the IMM, but can also extrude Ca^{2+} from the IMM when mitochondrial Ca^{2+} concentration is high [102].

Park et al. demonstrated that LETM1 protein expression is significantly decreased in white adipose tissue from HF mice and ob/ob mice compared with that from their control mice [133]. In the manuscript, however, the authors were not focusing on the role of LETM1 as a mitochondrial Ca^{2+}/H^+ antiporter, rather they aimed to investigate its inhibitory effect on PKB activity. Therefore, there was no functional data related to mitochondrial Ca^{2+} transport in their manuscript.

Mitochondrial Na^+/Ca^{2+} Exchanger (NCLX)

The NCLX is another major transporter of mitochondrial Ca^{2+} which releases Ca^{2+} from the mitochondrial matrix [6, 106, 130]. NCLX is located in the IMM and is primarily active in excitable cells. It transports not only Na^+ and Ca^{2+}, but also Li, which makes this channel's character distinguishable from the plasma membrane Na^+/Ca^{2+} exchanger (NCLX is the abbreviation of Na/Ca/Li exchanger) [131]. Mitochondrial Ca^{2+} concentration is mainly determined by the balance between influx through the MCU and efflux via NCLX [118]. In ischemia, NCLX acts as a key regulator of mitochondrial Ca^{2+} accumulation [117].

Babsky et al. reported that NCLX is more susceptive to the change in the outside (cytosolic) Na^+ concentration in diabetic cardiac myocytes compared with controls [8], but it is still unknown whether NCLX expression level is altered in any metabolic syndromes.

Mitochondrial K^+/H^+ Exchanger

The IMM exhibits high negative membrane potential (−150 to −180 mV) and this electrical gradient generates the driving force of cation uptake into the matrix. K^+ accumulates in the mitochondrial matrix through a uniport system and the concentration of K^+ is tightly regulated by exchangers, including a K^+/H^+ exchanger, to prevent an increase of mitochondrial osmotic pressure and swelling [19, 113]. The mitochondrial function of exchanging K^+ with H^+ in the IMM was reported as early as 1966 [113], however, the molecular details were not revealed until recently.

Schweyen et al. identified yeast Mdm38/human LETM1 as an essential component of the K^+/H^+ exchanger [51, 123]. It has to be noted that LETM1 also serves as a Ca^{2+}/H^+ antiporter located in the IMM [77] (refer to section "Mitochondrial Ca^{2+}/H^+ Antiporter").

Inner Membrane Anion Channel (IMAC) and Mitochondrial Chloride Channels (mClC, mCLIC)

IMAC was first reported in 1986 [54]. The initial finding was based on the experiment of anion flux from the IMM during mitochondrial swelling. To date, IMAC was partially characterized, unlike other mitochondrial ion channels. IMAC's function is inhibited by protons and Mg^{2+} [12, 13, 54], and its activity is dependent on voltage changes. Blockage of IMAC decreases mitochondrial depolarization induced by ROS [5]. As potential candidates of IMAC, chloride channels are the most studied channels. There are two classes of chloride channels identified in the IMM; the voltage-dependent chloride channel (ClC) and the chloride intracellular channel (CLIC). Among ClC family, only ClC-Nt was found in the IMM and proposed as a candidate for IMAC [99]. Fernandez et al. first demonstrated the existence of CLIC4 in the IMM and found that its activity is involved in mitochondria-induced apoptosis [47]. Since CLIC4 is also expressed in the cytoplasm and nucleus, its translocation under stress conditions has been proposed [160].

Conclusion

During these past few decades, many ion channels were identified in the mitochondrial membrane; however our understanding of the functional role of each mitochondrial channel in pathophysiological condition (e.g., metabolic syndrome) lags far behind. UCPs are the most investigated mitochondrial ion channels in the field of metabolic syndrome because UCPs help burn fatty acid during heat generation and might protect from obesity. mitoK channels are also tested for cardioprotection during ischemic/reperfusion injury and exhibit beneficial effects, but still there is space to explore the functions of new mitoK (e.g., TASK-3).

Mitochondrial Ca^{2+} homeostasis is critical for energy production and cell contractility in muscle cells and mitochondrial Ca^{2+} overload is the hallmark of apoptosis in all cell types. Maintaining the appropriate concentration of mitochondrial Ca^{2+} is a "must-do" in any cell type. The recent identification of MCU and Ca^{2+}/H^+ antiporter (LETM1) has dramatically increased our understanding of mitochondrial channels and may enable the future discovery of potential therapeutic approaches in metabolic disease.

Acknowledgment This work was supported by the grant of HL115578 (A. Makino) from the National Institutes of Health. We would like to thank Annisa Westcott for her critical review of this chapter.

References

1. Ahmed M, Muhammed SJ, Kessler B, Salehi A. Mitochondrial proteome analysis reveals altered expression of voltage dependent anion channels in pancreatic beta-cells exposed to high glucose. Islets. 2010;2:283–92.
2. Alam MR, Groschner LN, Parichatikanond W, Kuo L, Bondarenko AI, Rost R, Waldeck-Weiermair M, Malli R, Graier WF. Mitochondrial Ca^{2+} uptake 1 (MICU1) and mitochondrial Ca^{2+} uniporter (MCU) contribute to metabolism-secretion coupling in clonal pancreatic beta-cells. J Biol Chem. 2012;287:34445–54.
3. Alberici LC, Oliveira HC, Patricio PR, Kowaltowski AJ, Vercesi AE. Hyperlipidemic mice present enhanced catabolism and higher mitochondrial ATP-sensitive K^+ channel activity. Gastroenterology. 2006;131:1228–34.
4. Altschafl BA, Beutner G, Sharma VK, Sheu SS, Valdivia HH. The mitochondrial ryanodine receptor in rat heart: a pharmaco-kinetic profile. Biochim Biophys Acta. 2007;1768:1784–95.
5. Aon MA, Cortassa S, Marban E, O'Rourke B. Synchronized whole cell oscillations in mitochondrial metabolism triggered by a local release of reactive oxygen species in cardiac myocytes. J Biol Chem. 2003;278:44735–44.
6. Arco AD, Satrustegui J. New mitochondrial carriers: an overview. Cell Mol Life Sci. 2005;62:2204–27.
7. Arsenijevic D, Onuma H, Pecqueur C, Raimbault S, Manning BS, Miroux B, Couplan E, Alves-Guerra MC, Goubern M, Surwit R, Bouillaud F, Richard D, Collins S, Ricquier D. Disruption of the uncoupling protein-2 gene in mice reveals a role in immunity and reactive oxygen species production. Nat Genet. 2000;26:435–9.
8. Babsky A, Doliba N, Doliba N, Savchenko A, Wehrli S, Osbakken M. Na^+ effects on mitochondrial respiration and oxidative phosphorylation in diabetic hearts. Exp Biol Med. 2001;226:543–51.
9. Balderas E, Zhang J, Stefani E, Toro L. Mitochondrial BKCa channel. Front Physiol. 2015;6:104.
10. Bao S, Kennedy A, Wojciechowski B, Wallace P, Ganaway E, Garvey WT. Expression of mRNAs encoding uncoupling proteins in human skeletal muscle: effects of obesity and diabetes. Diabetes. 1998;47:1935–40.
11. Baughman JM, Perocchi F, Girgis HS, Plovanich M, Belcher-Timme CA, Sancak Y, Bao XR, Strittmatter L, Goldberger O, Bogorad RL, Koteliansky V, Mootha VK. Integrative genomics identifies MCU as an essential component of the mitochondrial calcium uniporter. Nature. 2011;476:341–5.
12. Beavis AD. Properties of the inner membrane anion channel in intact mitochondria. J Bioenerg Biomembr. 1992;24:77–90.
13. Beavis AD, Powers M. Temperature dependence of the mitochondrial inner membrane anion channel: the relationship between temperature and inhibition by magnesium. J Biol Chem. 2004;279:4045–50.
14. Bednarczyk P, Dolowy K, Szewczyk A. Matrix Mg^{2+} regulates mitochondrial ATP-dependent potassium channel from heart. FEBS Lett. 2005;579:1625–32.
15. Bednarczyk P, Dolowy K, Szewczyk A. New properties of mitochondrial ATP-regulated potassium channels. J Bioenerg Biomembr. 2008;40:325–35.
16. Bednarczyk P, Kowalczyk JE, Beresewicz M, Dolowy K, Szewczyk A, Zablocka B. Identification of a voltage-gated potassium channel in gerbil hippocampal mitochondria. Biochem Biophys Res Commun. 2010;397:614–20.
17. Bednarczyk P, Koziel A, Jarmuszkiewicz W, Szewczyk A. Large-conductance Ca^{2+}-activated potassium channel in mitochondria of endothelial EA.hy926 cells. Am J Physiol Heart Circ Physiol. 2013;304:H1415–27.
18. Bernal-Mizrachi C, Weng S, Li B, Nolte LA, Feng C, Coleman T, Holloszy JO, Semenkovich CF. Respiratory uncoupling lowers blood pressure through a leptin-dependent mechanism in genetically obese mice. Arterioscler Thromb Vasc Biol. 2002;22:961–8.

19. Bernardi P. Mitochondrial transport of cations: channels, exchangers, and permeability transition. Physiol Rev. 1999;79:1127–55.
20. Beutner G, Sharma VK, Giovannucci DR, Yule DI, Sheu SS. Identification of a ryanodine receptor in rat heart mitochondria. J Biol Chem. 2001;276:21482–8.
21. Beutner G, Sharma VK, Lin L, Ryu SY, Dirksen RT, Sheu SS. Type 1 ryanodine receptor in cardiac mitochondria: transducer of excitation-metabolism coupling. Biochim Biophys Acta. 2005;1717:1–10.
22. Boengler K, Dodoni G, Rodriguez-Sinovas A, Cabestrero A, Ruiz-Meana M, Gres P, Konietzka I, Lopez-Iglesias C, Garcia-Dorado D, Di Lisa F, Heusch G, Schulz R. Connexin 43 in cardiomyocyte mitochondria and its increase by ischemic preconditioning. Cardiovasc Res. 2005;67:234–44.
23. Boss O, Samec S, Paoloni-Giacobino A, Rossier C, Dulloo A, Seydoux J, Muzzin P, Giacobino JP. Uncoupling protein-3: a new member of the mitochondrial carrier family with tissue-specific expression. FEBS Lett. 1997;408:39–42.
24. Boudina S, Han YH, Pei S, Tidwell TJ, Henrie B, Tuinei J, Olsen C, Sena S, Abel ED. UCP3 regulates cardiac efficiency and mitochondrial coupling in high fat-fed mice but not in leptin-deficient mice. Diabetes. 2012;61:3260–9.
25. Boudina S, Sena S, Sloan C, Tebbi A, Han YH, O'Neill BT, Cooksey RC, Jones D, Holland WL, McClain DA, Abel ED. Early mitochondrial adaptations in skeletal muscle to diet-induced obesity are strain dependent and determine oxidative stress and energy expenditure but not insulin sensitivity. Endocrinology. 2012;153:2677–88.
26. Bulotta A, Ludovico O, Coco A, Di Paola R, Quattrone A, Carella M, Pellegrini F, Prudente S, Trischitta V. The common −866G/A polymorphism in the promoter region of the UCP-2 gene is associated with reduced risk of type 2 diabetes in Caucasians from Italy. J Clin Endocrinol Metab. 2005;90:1176–80.
27. Carroll AM, Haines LR, Pearson TW, Fallon PG, Walsh CM, Brennan CM, Breen EP, Porter RK. Identification of a functioning mitochondrial uncoupling protein 1 in thymus. J Biol Chem. 2005;280:15534–43.
28. Chapman CG, Meadows HJ, Godden RJ, Campbell DA, Duckworth M, Kelsell RE, Murdock PR, Randall AD, Rennie GI, Gloger IS. Cloning, localisation and functional expression of a novel human, cerebellum specific, two pore domain potassium channel. Brain Res Mol Brain Res. 2000;82:74–83.
29. Choi CS, Fillmore JJ, Kim JK, Liu ZX, Kim S, Collier EF, Kulkarni A, Distefano A, Hwang YJ, Kahn M, Chen Y, Yu C, Moore IK, Reznick RM, Higashimori T, Shulman GI. Overexpression of uncoupling protein 3 in skeletal muscle protects against fat-induced insulin resistance. J Clin Invest. 2007;117:1995–2003.
30. Chou CJ, Cha MC, Jung DW, Boozer CN, Hashim SA, Pi-Sunyer FX. High-fat diet feeding elevates skeletal muscle uncoupling protein 3 levels but not its activity in rats. Obes Res. 2001;9:313–9.
31. Clapham JC, Arch JR, Chapman H, Haynes A, Lister C, Moore GB, Piercy V, Carter SA, Lehner I, Smith SA, Beeley LJ, Godden RJ, Herrity N, Skehel M, Changani KK, Hockings PD, Reid DG, Squires SM, Hatcher J, Trail B, Latcham J, Rastan S, Harper AJ, Cadenas S, Buckingham JA, Brand MD, Abuin A. Mice overexpressing human uncoupling protein-3 in skeletal muscle are hyperphagic and lean. Nature. 2000;406:415–8.
32. Crompton M, Virji S, Doyle V, Johnson N, Ward JM. The mitochondrial permeability transition pore. Biochem Soc Symp. 1999;66:167–79.
33. Csordas G, Golenar T, Seifert EL, Kamer KJ, Sancak Y, Perocchi F, Moffat C, Weaver D, de la Fuente Perez S, Bogorad R, Koteliansky V, Adijanto J, Mootha VK, Hajnoczky G. MICU1 controls both the threshold and cooperative activation of the mitochondrial Ca^{2+} uniporter. Cell Metab. 2013;17:976–87.
34. D'Adamo M, Perego L, Cardellini M, Marini MA, Frontoni S, Andreozzi F, Sciacqua A, Lauro D, Sbraccia P, Federici M, Paganelli M, Pontiroli AE, Lauro R, Perticone F, Folli F, Sesti G. The −866A/A genotype in the promoter of the human uncoupling protein 2 gene is associated with insulin resistance and increased risk of type 2 diabetes. Diabetes. 2004;53:1905–10.

35. de Luis Roman DA, Aller R, Izaola Jauregui O, Gonzalez Sagrado M, Conde Vicente R, de la Fuente Salvador B, Romero Bobillo E. Relation of −55CT polymorphism of uncoupling protein 3 gene with fat mass and insulin resistance in morbidly obese patients. Metab Clin Exp. 2010;59:608–12.
36. De Marchi U, Sassi N, Fioretti B, Catacuzzeno L, Cereghetti GM, Szabo I, Zoratti M. Intermediate conductance Ca^{2+}-activated potassium channel (KCa3.1) in the inner mitochondrial membrane of human colon cancer cells. Cell Calcium. 2009;45:509–16.
37. De Pinto V, Guarino F, Guarnera A, Messina A, Reina S, Tomasello FM, Palermo V, Mazzoni C. Characterization of human VDAC isoforms: a peculiar function for VDAC3? Biochim Biophys Acta. 2010;1797:1268–75.
38. de Souza BM, Brondani LA, Boucas AP, Sortica DA, Kramer CK, Canani LH, Leitao CB, Crispim D. Associations between UCP1–3826A/G, UCP2–866G/A, Ala55Val and Ins/Del, and UCP3–55C/T polymorphisms and susceptibility to type 2 diabetes mellitus: case-control study and meta-analysis. PLoS One. 2013;8:e54259.
39. De Stefani D, Bononi A, Romagnoli A, Messina A, De Pinto V, Pinton P, Rizzuto R. VDAC1 selectively transfers apoptotic Ca^{2+} signals to mitochondria. Cell Death Differ. 2012;19:267–73.
40. De Stefani D, Raffaello A, Teardo E, Szabo I, Rizzuto R. A forty-kilodalton protein of the inner membrane is the mitochondrial calcium uniporter. Nature. 2011;476:336–40.
41. Di Castro S, Scarpino S, Marchitti S, Bianchi F, Stanzione R, Cotugno M, Sironi L, Gelosa P, Duranti E, Ruco L, Volpe M, Rubattu S. Differential modulation of uncoupling protein 2 in kidneys of stroke-prone spontaneously hypertensive rats under high-salt/low-potassium diet. Hypertension. 2013;61:534–41.
42. Dolga AM, Netter MF, Perocchi F, Doti N, Meissner L, Tobaben S, Grohm J, Zischka H, Plesnila N, Decher N, Culmsee C. Mitochondrial small conductance SK2 channels prevent glutamate-induced oxytosis and mitochondrial dysfunction. J Biol Chem. 2013;288:10792–804.
43. Drago I, De Stefani D, Rizzuto R, Pozzan T. Mitochondrial Ca^{2+} uptake contributes to buffering cytoplasmic Ca^{2+} peaks in cardiomyocytes. Proc Natl Acad Sci U S A. 2012;109:12986–91.
44. Duprat F, Lesage F, Fink M, Reyes R, Heurteaux C, Lazdunski M. TASK, a human background K^+ channel to sense external pH variations near physiological pH. EMBO J. 1997;16:5464–71.
45. Enerback S, Jacobsson A, Simpson EM, Guerra C, Yamashita H, Harper ME, Kozak LP. Mice lacking mitochondrial uncoupling protein are cold-sensitive but not obese. Nature. 1997;387:90–4.
46. Fancher IS, Dick GM, Hollander JM. Diabetes mellitus reduces the function and expression of ATP-dependent K^+ channels in cardiac mitochondria. Life Sci. 2013;92:664–8.
47. Fernandez-Salas E, Sagar M, Cheng C, Yuspa SH, Weinberg WC. p53 and tumor necrosis factor alpha regulate the expression of a mitochondrial chloride channel protein. J Biol Chem. 1999;274:36488–97.
48. Fill M, Copello JA. Ryanodine receptor calcium release channels. Physiol Rev. 2002;82:893–922.
49. Forbes RA, Steenbergen C, Murphy E. Diazoxide-induced cardioprotection requires signaling through a redox-sensitive mechanism. Circ Res. 2001;88:802–9.
50. Frier BC, Jacobs RL, Wright DC. Interactions between the consumption of a high-fat diet and fasting in the regulation of fatty acid oxidation enzyme gene expression: an evaluation of potential mechanisms. Am J Physiol Regul Integr Comp Physiol. 2011;300:R212–21.
51. Froschauer E, Nowikovsky K, Schweyen RJ. Electroneutral K^+/H^+ exchange in mitochondrial membrane vesicles involves Yol027/Letm1 proteins. Biochim Biophys Acta. 2005;1711:41–8.
52. Fukuyama K, Ohara T, Hirota Y, Maeda K, Kuno S, Zenibayashi M, Teranishi T, Kouyama K, Maeda E, Sakamoto N, Kasuga M. Association of the −112A>C polymorphism of the uncoupling protein 1 gene with insulin resistance in Japanese individuals with type 2 diabetes. Biochem Biophys Res Commun. 2006;339:1212–6.
53. Garciarena CD, Caldiz CI, Correa MV, Schinella GR, Mosca SM, Chiappe de Cingolani GE, Cingolani HE, Ennis IL. Na^+/H^+ exchanger-1 inhibitors decrease myocardial superoxide production via direct mitochondrial action. J Appl Physiol. 2008;105:1706–13.

54. Garlid KD, Beavis AD. Evidence for the existence of an inner membrane anion channel in mitochondria. Biochim Biophys Acta. 1986;853:187–204.
55. Gates AC, Bernal-Mizrachi C, Chinault SL, Feng C, Schneider JG, Coleman T, Malone JP, Townsend RR, Chakravarthy MV, Semenkovich CF. Respiratory uncoupling in skeletal muscle delays death and diminishes age-related disease. Cell Metab. 2007;6:497–505.
56. Gazula VR, Strumbos JG, Mei X, Chen H, Rahner C, Kaczmarek LK. Localization of Kv1.3 channels in presynaptic terminals of brainstem auditory neurons. J Comp Neurol. 2010;518:3205–20.
57. Gincel D, Zaid H, Shoshan-Barmatz V. Calcium binding and translocation by the voltage-dependent anion channel: a possible regulatory mechanism in mitochondrial function. Biochem J. 2001;358:147–55.
58. Gong DW, Monemdjou S, Gavrilova O, Leon LR, Marcus-Samuels B, Chou CJ, Everett C, Kozak LP, Li C, Deng C, Harper ME, Reitman ML. Lack of obesity and normal response to fasting and thyroid hormone in mice lacking uncoupling protein-3. J Biol Chem. 2000;275:16251–7.
59. Goodenough DA, Goliger JA, Paul DL. Connexins, connexons, and intercellular communication. Annu Rev Biochem. 1996;65:475–502.
60. Gorbe A, Varga ZV, Kupai K, Bencsik P, Kocsis GF, Csont T, Boengler K, Schulz R, Ferdinandy P. Cholesterol diet leads to attenuation of ischemic preconditioning-induced cardiac protection: the role of connexin 43. Am J Physiol Heart Circ Physiol. 2011;300:H1907–13.
61. Gu Y, Ande SR, Mishra S. Altered O-GlcNAc modification and phosphorylation of mitochondrial proteins in myoblast cells exposed to high glucose. Arch Biochem Biophys. 2011;505:98–104.
62. Gunter KK, Gunter TE. Transport of calcium by mitochondria. J Bioenerg Biomembr. 1994;26:471–85.
63. Hajnoczky G, Csordas G, Yi M. Old players in a new role: mitochondria-associated membranes, VDAC, and ryanodine receptors as contributors to calcium signal propagation from endoplasmic reticulum to the mitochondria. Cell Calcium. 2002;32:363–77.
64. Halestrap AP, Richardson AP. The mitochondrial permeability transition: a current perspective on its identity and role in ischaemia/reperfusion injury. J Mol Cell Cardiol. 2015;78C:129–41.
65. Hausenloy DJ, Yellon DM. The mitochondrial permeability transition pore: its fundamental role in mediating cell death during ischaemia and reperfusion. J Mol Cell Cardiol. 2003;35:339–41.
66. Haworth RA, Hunter DR. The Ca^{2+}-induced membrane transition in mitochondria. II. Nature of the Ca^{2+} trigger site. Arch Biochem Biophys. 1979;195:460–7.
67. Hidaka S, Kakuma T, Yoshimatsu H, Sakino H, Fukuchi S, Sakata T. Streptozotocin treatment upregulates uncoupling protein 3 expression in the rat heart. Diabetes. 1999;48:430–5.
68. Hodge T, Colombini M. Regulation of metabolite flux through voltage-gating of VDAC channels. J Membr Biol. 1997;157:271–9.
69. Hoffman NE, Chandramoorthy HC, Shanmughapriya S, Zhang XQ, Vallem S, Doonan PJ, Malliankaraman K, Guo S, Rajan S, Elrod JW, Koch WJ, Cheung JY, Madesh M. SLC25A23 augments mitochondrial Ca^{2+} uptake, interacts with MCU, and induces oxidative stress-mediated cell death. Mol Biol Cell. 2014;25:936–47.
70. Holmuhamedov EL, Jovanovic S, Dzeja PP, Jovanovic A, Terzic A. Mitochondrial ATP-sensitive K^+ channels modulate cardiac mitochondrial function. Am J Physiol. 1998;275:H1567–76.
71. Holmuhamedov EL, Wang L, Terzic A. ATP-sensitive K+ channel openers prevent Ca^{2+} overload in rat cardiac mitochondria. J Physiol. 1999;519(Pt 2):347–60.
72. Hunter DR, Haworth RA. The Ca^{2+}-induced membrane transition in mitochondria. I The protective mechanisms. Arch Biochem Biophys. 1979;195:453–9.
73. Hunter DR, Haworth RA. The Ca^{2+}-induced membrane transition in mitochondria. III. Transitional Ca^{2+} release. Arch Biochem Biophys. 1979;195:468–77.

74. Hunter DR, Haworth RA, Southard JH. Relationship between configuration, function, and permeability in calcium-treated mitochondria. J Biol Chem. 1976;251:5069–77.
75. Inoue I, Nagase H, Kishi K, Higuti T. ATP-sensitive K^+ channel in the mitochondrial inner membrane. Nature. 1991;352:244–7.
76. Ji Q, Ikegami H, Fujisawa T, Kawabata Y, Ono M, Nishino M, Ohishi M, Katsuya T, Rakugi H, Ogihara T. A common polymorphism of uncoupling protein 2 gene is associated with hypertension. J Hypertens. 2004;22:97–102.
77. Jiang D, Zhao L, Clapham DE. Genome-wide RNAi screen identifies Letm1 as a mitochondrial Ca^{2+}/H^+ antiporter. Science. 2009;326:144–7.
78. Kageyama H, Suga A, Kashiba M, Oka J, Osaka T, Kashiwa T, Hirano T, Nemoto K, Namba Y, Ricquier D, Giacobino JP, Inoue S. Increased uncoupling protein-2 and -3 gene expressions in skeletal muscle of STZ-induced diabetic rats. FEBS Lett. 1998;440:450–3.
79. Kamer KJ, Sancak Y, Mootha VK. The uniporter: from newly identified parts to function. Biochem Biophys Res Commun. 2014;449:370–2.
80. Katakam PV, Jordan JE, Snipes JA, Tulbert CD, Miller AW, Busija DW. Myocardial preconditioning against ischemia-reperfusion injury is abolished in Zucker obese rats with insulin resistance. Am J Physiol Regul Integr Comp Physiol. 2007;292:R920–6.
81. Kato H, Ohue M, Kato K, Nomura A, Toyosawa K, Furutani Y, Kimura S, Kadowaki T. Mechanism of amelioration of insulin resistance by beta3-adrenoceptor agonist AJ-9677 in the KK-Ay/Ta diabetic obese mouse model. Diabetes. 2001;50:113–22.
82. Kersten JR, Montgomery MW, Ghassemi T, Gross ER, Toller WG, Pagel PS, Warltier DC. Diabetes and hyperglycemia impair activation of mitochondrial K(ATP) channels. Am J Physiol Heart Circ Physiol. 2001;280:H1744–50.
83. Kim Y, Bang H, Kim D. TASK-3, a new member of the tandem pore K^+ channel family. J Biol Chem. 2000;275:9340–7.
84. Kimura S, Zhang GX, Nishiyama A, Shokoji T, Yao L, Fan YY, Rahman M, Abe Y. Mitochondria-derived reactive oxygen species and vascular MAP kinases: comparison of angiotensin II and diazoxide. Hypertension. 2005;45:438–44.
85. Kjorholt C, Akerfeldt MC, Biden TJ, Laybutt DR. Chronic hyperglycemia, independent of plasma lipid levels, is sufficient for the loss of beta-cell differentiation and secretory function in the db/db mouse model of diabetes. Diabetes. 2005;54:2755–63.
86. Korge P, Honda HM, Weiss JN. Regulation of the mitochondrial permeability transition by matrix Ca^{2+} and voltage during anoxia/reoxygenation. Am J Physiol Cell Physiol. 2001;280:C517–26.
87. Krauss S, Zhang CY, Lowell BB. The mitochondrial uncoupling-protein homologues. Nat Rev Mol Cell Biol. 2005;6:248–61.
88. Kulawiak B, Kudin AP, Szewczyk A, Kunz WS. BK channel openers inhibit ROS production of isolated rat brain mitochondria. Exp Neurol. 2008;212:543–7.
89. Kumar NM, Gilula NB. The gap junction communication channel. Cell. 1996;84:381–8.
90. Lacza Z, Snipes JA, Miller AW, Szabo C, Grover G, Busija DW. Heart mitochondria contain functional ATP-dependent K^+ channels. J Mol Cell Cardiol. 2003;35:1339–47.
91. Leanza L, Henry B, Sassi N, Zoratti M, Chandy KG, Gulbins E, Szabo I. Inhibitors of mitochondrial Kv1.3 channels induce Bax/Bak-independent death of cancer cells. EMBO Mol Med. 2012;4:577–93.
92. Lesage F, Barhanin J. Molecular physiology of pH-sensitive background K(2P) channels. Physiology (Bethesda). 2011;26:424–37.
93. Li B, Nolte LA, Ju JS, Han DH, Coleman T, Holloszy JO, Semenkovich CF. Skeletal muscle respiratory uncoupling prevents diet-induced obesity and insulin resistance in mice. Nat Med. 2000;6:1115–20.
94. Li H, Brodsky S, Kumari S, Valiunas V, Brink P, Kaide J, Nasjletti A, Goligorsky MS. Paradoxical overexpression and translocation of connexin43 in homocysteine-treated endothelial cells. Am J Physiol Heart Circ Physiol. 2002;282:H2124–33.

95. Light PE, Kanji HD, Fox JE, French RJ. Distinct myoprotective roles of cardiac sarcolemmal and mitochondrial KATP channels during metabolic inhibition and recovery. FASEB J. 2001;15:2586–94.
96. Liu D, Slevin JR, Lu C, Chan SL, Hansson M, Elmer E, Mattson MP. Involvement of mitochondrial K^+ release and cellular efflux in ischemic and apoptotic neuronal death. J Neurochem. 2003;86:966–79.
97. Liu Y, Sato T, Seharaseyon J, Szewczyk A, O'Rourke B, Marban E. Mitochondrial ATP-dependent potassium channels. Viable candidate effectors of ischemic preconditioning. Ann N Y Acad Sci. 1999;874:27–37.
98. Lumini-Oliveira J, Magalhaes J, Pereira CV, Moreira AC, Oliveira PJ, Ascensao A. Endurance training reverts heart mitochondrial dysfunction, permeability transition and apoptotic signaling in long-term severe hyperglycemia. Mitochondrion. 2011;11:54–63.
99. Lurin C, Guclu J, Cheniclet C, Carde JP, Barbier-Brygoo H, Maurel C. CLC-Nt1, a putative chloride channel protein of tobacco, co-localizes with mitochondrial membrane markers. Biochem J. 2000;348(Pt 2):291–5.
100. Madesh M, Hajnoczky G. VDAC-dependent permeabilization of the outer mitochondrial membrane by superoxide induces rapid and massive cytochrome c release. J Cell Biol. 2001;155:1003–15.
101. Mahadik SR, Lele RD, Saranath D, Seth A, Parikh V. Uncoupling protein-2 (UCP2) gene expression in subcutaneous and omental adipose tissue of Asian Indians: relationship to adiponectin and parameters of metabolic syndrome. Adipocyte. 2012;1:101–7.
102. Mailloux RJ, Harper ME. Uncoupling proteins and the control of mitochondrial reactive oxygen species production. Free Radic Biol Med. 2011;51:1106–15.
103. Mallilankaraman K, Doonan P, Cardenas C, Chandramoorthy HC, Muller M, Miller R, Hoffman NE, Gandhirajan RK, Molgo J, Birnbaum MJ, Rothberg BS, Mak DO, Foskett JK, Madesh M. MICU1 is an essential gatekeeper for MCU-mediated mitochondrial Ca^{2+} uptake that regulates cell survival. Cell. 2012;151:630–44.
104. Marchi S, Pinton P. The mitochondrial calcium uniporter complex: molecular components, structure and physiopathological implications. J Physiol. 2014;592:829–39.
105. Matsuda J, Hosoda K, Itoh H, Son C, Doi K, Tanaka T, Fukunaga Y, Inoue G, Nishimura H, Yoshimasa Y, Yamori Y, Nakao K. Cloning of rat uncoupling protein-3 and uncoupling protein-2 cDNAs: their gene expression in rats fed high-fat diet. FEBS Lett. 1997;418:200–4.
106. McCormack JG, Denton RM. Mitochondrial Ca^{2+} transport and the role of intramitochondrial Ca^{2+} in the regulation of energy metabolism. Dev Neurosci. 1993;15:165–73.
107. McCully JD, Levitsky S. The mitochondrial K(ATP) channel and cardioprotection. Ann Thorac Surg. 2003;75:S667–73.
108. Meadows HJ, Randall AD. Functional characterisation of human TASK-3, an acid-sensitive two-pore domain potassium channel. Neuropharmacology. 2001;40:551–9.
109. Menzel VA, Cassara MC, Benz R, de Pinto V, Messina A, Cunsolo V, Saletti R, Hinsch KD, Hinsch E. Molecular and functional characterization of VDAC2 purified from mammal spermatozoa. Biosci Rep. 2009;29:351–62.
110. Messina A, Reina S, Guarino F, De Pinto V. VDAC isoforms in mammals. Biochim Biophys Acta. 2012;1818:1466–76.
111. Min CK, Yeom DR, Lee KE, Kwon HK, Kang M, Kim YS, Park ZY, Jeon H, Kim do H. Coupling of ryanodine receptor 2 and voltage-dependent anion channel 2 is essential for Ca^{2+} transfer from the sarcoplasmic reticulum to the mitochondria in the heart. Biochem J. 2012;447:371–9.
112. Mironova GD, Grigoriev SM, Skarga Y, Negoda AE, Kolomytkin OV. ATP-dependent potassium channel from rat liver mitochondria: inhibitory analysis, channel clusterization. Membr Cell Biol. 1997;10:583–91.
113. Mitchell P. Chemiosmotic coupling in oxidative and photosynthetic phosphorylation. Biol Rev Camb Philos Soc. 1966;41:445–502.

114. Mohammad G, Kowluru RA. Novel role of mitochondrial matrix metalloproteinase-2 in the development of diabetic retinopathy. Invest Ophthalmol Vis Sci. 2011;52:3832–41.
115. Molica F, Meens MJ, Morel S, Kwak BR. Mutations in cardiovascular connexin genes. Biol Cell. 2014;106:269–93.
116. Montero M, Alonso MT, Carnicero E, Cuchillo-Ibanez I, Albillos A, Garcia AG, Garcia-Sancho J, Alvarez J. Chromaffin-cell stimulation triggers fast millimolar mitochondrial Ca^{2+} transients that modulate secretion. Nat Cell Biol. 2000;2:57–61.
117. Murphy E, Cross H, Steenbergen C. Sodium regulation during ischemia versus reperfusion and its role in injury. Circ Res. 1999;84:1469–70.
118. Murphy E, Eisner DA. Regulation of intracellular and mitochondrial sodium in health and disease. Circ Res. 2009;104:292–303.
119. Nagy D, Gonczi M, Dienes B, Szoor A, Fodor J, Nagy Z, Toth A, Fodor T, Bai P, Szucs G, Rusznak Z, Csernoch L. Silencing the KCNK9 potassium channel (TASK-3) gene disturbs mitochondrial function, causes mitochondrial depolarization, and induces apoptosis of human melanoma cells. Arch Dermatol Res. 2014;306:885–902.
120. Nazarewicz RR, Dikalova AE, Bikineyeva A, Dikalov SI. Nox2 as a potential target of mitochondrial superoxide and its role in endothelial oxidative stress. Am J Physiol Heart Circ Physiol. 2013;305:H1131–40.
121. Nordfors L, Hoffstedt J, Nyberg B, Thorne A, Arner P, Schalling M, Lonnqvist F. Reduced gene expression of UCP2 but not UCP3 in skeletal muscle of human obese subjects. Diabetologia. 1998;41:935–9.
122. Noursadeghi E, Jafari A, Saghiri R, Sauve R, Eliassi A. Impairment of brain mitochondrial charybdotoxin- and ATP-insensitive BK channel activities in diabetes. Neuromolecular Med. 2014;16:862–71.
123. Nowikovsky K, Reipert S, Devenish RJ, Schweyen RJ. Mdm38 protein depletion causes loss of mitochondrial K^+/H^+ exchange activity, osmotic swelling and mitophagy. Cell Death Differ. 2007;14:1647–56.
124. O'Rourke B. Evidence for mitochondrial K^+ channels and their role in cardioprotection. Circ Res. 2004;94:420–32.
125. O'Rourke B. Mitochondrial ion channels. Annu Rev Physiol. 2007;69:19–49.
126. O'Rourke B. Myocardial K(ATP) channels in preconditioning. Circ Res. 2000;87:845–55.
127. Oberkofler H, Dallinger G, Liu YM, Hell E, Krempler F, Patsch W. Uncoupling protein gene: quantification of expression levels in adipose tissues of obese and non-obese humans. J Lipid Res. 1997;38:2125–33.
128. Paillard M, Tubbs E, Thiebaut PA, Gomez L, Fauconnier J, Da Silva CC, Teixeira G, Mewton N, Belaidi E, Durand A, Abrial M, Lacampagne A, Rieusset J, Ovize M. Depressing mitochondria-reticulum interactions protects cardiomyocytes from lethal hypoxia-reoxygenation injury. Circulation. 2013;128:1555–65.
129. Pain T, Yang XM, Critz SD, Yue Y, Nakano A, Liu GS, Heusch G, Cohen MV, Downey JM. Opening of mitochondrial K(ATP) channels triggers the preconditioned state by generating free radicals. Circ Res. 2000;87:460–6.
130. Palty R, Hershfinkel M, Sekler I. Molecular identity and functional properties of the mitochondrial Na^+/Ca^{2+} exchanger. J Biol Chem. 2012;287:31650–7.
131. Palty R, Ohana E, Hershfinkel M, Volokita M, Elgazar V, Beharier O, Silverman WF, Argaman M, Sekler I. Lithium-calcium exchange is mediated by a distinct potassium-independent sodium-calcium exchanger. J Biol Chem. 2004;279:25234–40.
132. Pan X, Liu J, Nguyen T, Liu C, Sun J, Teng Y, Fergusson MM, Rovira II, Allen M, Springer DA, Aponte AM, Gucek M, Balaban RS, Murphy E, Finkel T. The physiological role of mitochondrial calcium revealed by mice lacking the mitochondrial calcium uniporter. Nat Cell Biol. 2013;15:1464–72.
133. Park J, Li Y, Kim SH, Yang KJ, Kong G, Shrestha R, Tran Q, Park KA, Jeon J, Hur GM, Lee CH, Kim DH, Park J. New players in high fat diet-induced obesity: LETM1 and CTMP. Metab Clin Exp. 2014;63:318–27.

134. Pecqueur C, Alves-Guerra MC, Gelly C, Levi-Meyrueis C, Couplan E, Collins S, Ricquier D, Bouillaud F, Miroux B. Uncoupling protein 2, in vivo distribution, induction upon oxidative stress, and evidence for translational regulation. J Biol Chem. 2001;276:8705–12.
135. Petriz BA, Almeida JA, Gomes CP, Pereira RW, Murad AM, Franco OL. NanoUPLC/MS(E) proteomic analysis reveals modulation on left ventricle proteome from hypertensive rats after exercise training. J Proteomics. 2015;113:351–65.
136. Plovanich M, Bogorad RL, Sancak Y, Kamer KJ, Strittmatter L, Li AA, Girgis HS, Kuchimanchi S, De Groot J, Speciner L, Taneja N, Oshea J, Koteliansky V, Mootha VK. MICU2, a paralog of MICU1, resides within the mitochondrial uniporter complex to regulate calcium handling. PLoS One. 2013;8:e55785.
137. Qiu J, Tan YW, Hagenston AM, Martel MA, Kneisel N, Skehel PA, Wyllie DJ, Bading H, Hardingham GE. Mitochondrial calcium uniporter Mcu controls excitotoxicity and is transcriptionally repressed by neuroprotective nuclear calcium signals. Nat Commun. 2013;4:2034.
138. Qiu W, Zhou Y, Jiang L, Fang L, Chen L, Su W, Tan R, Zhang CY, Han X, Yang J. Genipin inhibits mitochondrial uncoupling protein 2 expression and ameliorates podocyte injury in diabetic mice. PLoS One. 2012;7:e41391.
139. Quan X, Nguyen TT, Choi S, Xu S, Das R, Cha S, Kim N, Han J, Wiederkehr A, Wollheim CB, Park K. Essential role of mitochondrial Ca^{2+} uniporter in the generation of mitochondrial pH gradient and metabolism-secretion coupling in insulin-releasing cells. J Biol Chem. 2014;290(7):4086–96.
140. Rizzuto R, Bernardi P, Pozzan T. Mitochondria as all-round players of the calcium game. J Physiol. 2000;529(Pt 1):37–47.
141. Rodriguez-Sinovas A, Boengler K, Cabestrero A, Gres P, Morente M, Ruiz-Meana M, Konietzka I, Miro E, Totzeck A, Heusch G, Schulz R, Garcia-Dorado D. Translocation of connexin 43 to the inner mitochondrial membrane of cardiomyocytes through the heat shock protein 90-dependent TOM pathway and its importance for cardioprotection. Circ Res. 2006;99:93–101.
142. Rostovtseva T, Colombini M. VDAC channels mediate and gate the flow of ATP: implications for the regulation of mitochondrial function. Biophys J. 1997;72:1954–62.
143. Rusznak Z, Bakondi G, Kosztka L, Pocsai K, Dienes B, Fodor J, Telek A, Gonczi M, Szucs G, Csernoch L. Mitochondrial expression of the two-pore domain TASK-3 channels in malignantly transformed and non-malignant human cells. Virchows Arch. 2008;452:415–26.
144. Ryu SY, Beutner G, Dirksen RT, Kinnally KW, Sheu SS. Mitochondrial ryanodine receptors and other mitochondrial Ca^{2+} permeable channels. FEBS Lett. 2010;584:1948–55.
145. Sale MM, Hsu FC, Palmer ND, Gordon CJ, Keene KL, Borgerink HM, Sharma AJ, Bergman RN, Taylor KD, Saad MF, Norris JM. The uncoupling protein 1 gene, UCP1, is expressed in mammalian islet cells and associated with acute insulin response to glucose in African American families from the IRAS Family Study. BMC Endocr Disord. 2007;7:1.
146. Sancak Y, Markhard AL, Kitami T, Kovacs-Bogdan E, Kamer KJ, Udeshi ND, Carr SA, Chaudhuri D, Clapham DE, Li AA, Calvo SE, Goldberger O, Mootha VK. EMRE is an essential component of the mitochondrial calcium uniporter complex. Science. 2013;342:1379–82.
147. Sasahara M, Nishi M, Kawashima H, Ueda K, Sakagashira S, Furuta H, Matsumoto E, Hanabusa T, Sasaki H, Nanjo K. Uncoupling protein 2 promoter polymorphism −866G/A affects its expression in beta-cells and modulates clinical profiles of Japanese type 2 diabetic patients. Diabetes. 2004;53:482–5.
148. Sasaki K, Donthamsetty R, Heldak M, Cho YE, Scott BT, Makino A. VDAC: old protein with new roles in diabetes. Am J Physiol Cell Physiol. 2012;303:C1055–60.
149. Sassi N, De Marchi U, Fioretti B, Biasutto L, Gulbins E, Franciolini F, Szabo I, Zoratti M. An investigation of the occurrence and properties of the mitochondrial intermediate-conductance Ca^{2+}-activated K^+ channel mtKCa3.1. Biochim Biophys Acta. 2010;1797:1260–7.
150. Shoshan-Barmatz V, Ben-Hail D. VDAC, a multi-functional mitochondrial protein as a pharmacological target. Mitochondrion. 2012;12:24–34.

151. Siemen D, Loupatatzis C, Borecky J, Gulbins E, Lang F. Ca^{2+}-activated K channel of the BK-type in the inner mitochondrial membrane of a human glioma cell line. Biochem Biophys Res Commun. 1999;257:549–54.
152. Simoneau JA, Kelley DE, Neverova M, Warden CH. Overexpression of muscle uncoupling protein 2 content in human obesity associates with reduced skeletal muscle lipid utilization. FASEB J. 1998;12:1739–45.
153. Singh H, Lu R, Bopassa JC, Meredith AL, Stefani E, Toro L. MitoBK(Ca) is encoded by the Kcnma1 gene, and a splicing sequence defines its mitochondrial location. Proc Natl Acad Sci U S A. 2013;110:10836–41.
154. Singh H, Stefani E, Toro L. Intracellular BK(Ca) (iBK(Ca)) channels. J Physiol. 2012;590:5937–47.
155. Smorodchenko A, Rupprecht A, Sarilova I, Ninnemann O, Brauer AU, Franke K, Schumacher S, Techritz S, Nitsch R, Schuelke M, Pohl EE. Comparative analysis of uncoupling protein 4 distribution in various tissues under physiological conditions and during development. Biochim Biophys Acta. 2009;1788:2309–19.
156. Sohl G, Willecke K. An update on connexin genes and their nomenclature in mouse and man. Cell Commun Adhes. 2003;10:173–80.
157. Soltysinska E, Bentzen BH, Barthmes M, Hattel H, Thrush AB, Harper ME, Qvortrup K, Larsen FJ, Schiffer TA, Losa-Reyna J, Straubinger J, Kniess A, Thomsen MB, Bruggemann A, Fenske S, Biel M, Ruth P, Wahl-Schott C, Boushel RC, Olesen SP, Lukowski R. KCNMA1 encoded cardiac BK channels afford protection against ischemia-reperfusion injury. PLoS One. 2014;9:e103402.
158. Stowe DF, Gadicherla AK, Zhou Y, Aldakkak M, Cheng Q, Kwok WM, Jiang MT, Heisner JS, Yang M, Camara AK. Protection against cardiac injury by small Ca^{2+}-sensitive K^{+} channels identified in guinea pig cardiac inner mitochondrial membrane. Biochim Biophys Acta. 2013;1828:427–42.
159. Suarez J, Hu Y, Makino A, Fricovsky E, Wang H, Dillmann WH. Alterations in mitochondrial function and cytosolic calcium induced by hyperglycemia are restored by mitochondrial transcription factor A in cardiomyocytes. Am J Physiol Cell Physiol. 2008;295:C1561–8.
160. Suh KS, Mutoh M, Nagashima K, Fernandez-Salas E, Edwards LE, Hayes DD, Crutchley JM, Marin KG, Dumont RA, Levy JM, Cheng C, Garfield S, Yuspa SH. The organellular chloride channel protein CLIC4/mtCLIC translocates to the nucleus in response to cellular stress and accelerates apoptosis. J Biol Chem. 2004;279:4632–41.
161. Szabo I, Bock J, Grassme H, Soddemann M, Wilker B, Lang F, Zoratti M, Gulbins E. Mitochondrial potassium channel Kv1.3 mediates Bax-induced apoptosis in lymphocytes. Proc Natl Acad Sci U S A. 2008;105:14861–6.
162. Szabo I, Bock J, Jekle A, Soddemann M, Adams C, Lang F, Zoratti M, Gulbins E. A novel potassium channel in lymphocyte mitochondria. J Biol Chem. 2005;280:12790–8.
163. Taddeo EP, Laker RC, Breen DS, Akhtar YN, Kenwood BM, Liao JA, Zhang M, Fazakerley DJ, Tomsig JL, Harris TE, Keller SR, Chow JD, Lynch KR, Chokki M, Molkentin JD, Turner N, James DE, Yan Z, Hoehn KL. Opening of the mitochondrial permeability transition pore links mitochondrial dysfunction to insulin resistance in skeletal muscle. Mol Metab. 2014;3:124–34.
164. Trudeau K, Muto T, Roy S. Downregulation of mitochondrial connexin 43 by high glucose triggers mitochondrial shape change and cytochrome C release in retinal endothelial cells. Invest Ophthalmol Vis Sci. 2012;53:6675–81.
165. Tubbs E, Theurey P, Vial G, Bendridi N, Bravard A, Chauvin MA, Ji-Cao J, Zoulim F, Bartosch B, Ovize M, Vidal H, Rieusset J. Mitochondria-associated endoplasmic reticulum membrane (MAM) integrity is required for insulin signaling and is implicated in hepatic insulin resistance. Diabetes. 2014;63:3279–94.
166. Turko IV, Murad F. Quantitative protein profiling in heart mitochondria from diabetic rats. J Biol Chem. 2003;278:35844–9.

167. Vander Heiden MG, Chandel NS, Li XX, Schumacker PT, Colombini M, Thompson CB. Outer mitochondrial membrane permeability can regulate coupled respiration and cell survival. Proc Natl Acad Sci U S A. 2000;97:4666–71.
168. Vidal-Puig AJ, Grujic D, Zhang CY, Hagen T, Boss O, Ido Y, Szczepanik A, Wade J, Mootha V, Cortright R, Muoio DM, Lowell BB. Energy metabolism in uncoupling protein 3 gene knockout mice. J Biol Chem. 2000;275:16258–66.
169. Vidal H, Langin D, Andreelli F, Millet L, Larrouy D, Laville M. Lack of skeletal muscle uncoupling protein 2 and 3 mRNA induction during fasting in type-2 diabetic subjects. Am J Physiol. 1999;277:E830–7.
170. Wojtera E, Konior A, Fedoryszak-Kuska N, Beresewicz A. Obligatory role of intraluminal O_2– in acute endothelin-1 and angiotensin II signaling to mediate endothelial dysfunction and MAPK activation in guinea-pig hearts. Int J Mol Sci. 2014;15:19417–43.
171. Xie X, Li S, Liu S, Lu Y, Shen P, Ji J. Proteomic analysis of mouse islets after multiple low-dose streptozotocin injection. Biochim Biophys Acta. 2008;1784:276–84.
172. Xu W, Liu Y, Wang S, McDonald T, Van Eyk JE, Sidor A, O'Rourke B. Cytoprotective role of Ca^{2+}-activated K^+ channels in the cardiac inner mitochondrial membrane. Science. 2002;298:1029–33.
173. Xu X, Decker W, Sampson MJ, Craigen WJ, Colombini M. Mouse VDAC isoforms expressed in yeast: channel properties and their roles in mitochondrial outer membrane permeability. J Membr Biol. 1999;170:89–102.
174. Yeager M, Harris AL. Gap junction channel structure in the early 21st century: facts and fantasies. Curr Opin Cell Biol. 2007;19:521–8.
175. Yu KY, Wang YP, Wang LH, Jian Y, Zhao XD, Chen JW, Murao K, Zhu W, Dong L, Wang GQ, Zhang GX. Mitochondrial KATP channel involvement in angiotensin II-induced autophagy in vascular smooth muscle cells. Basic Res Cardiol. 2014;109:416.
176. Yu XX, Mao W, Zhong A, Schow P, Brush J, Sherwood SW, Adams SH, Pan G. Characterization of novel UCP5/BMCP1 isoforms and differential regulation of UCP4 and UCP5 expression through dietary or temperature manipulation. FASEB J. 2000;14:1611–8.
177. Zhang CY, Baffy G, Perret P, Krauss S, Peroni O, Grujic D, Hagen T, Vidal-Puig AJ, Boss O, Kim YB, Zheng XX, Wheeler MB, Shulman GI, Chan CB, Lowell BB. Uncoupling protein-2 negatively regulates insulin secretion and is a major link between obesity, beta cell dysfunction, and type 2 diabetes. Cell. 2001;105:745–55.
178. Zhang DX, Chen YF, Campbell WB, Zou AP, Gross GJ, Li PL. Characteristics and superoxide-induced activation of reconstituted myocardial mitochondrial ATP-sensitive potassium channels. Circ Res. 2001;89:1177–83.

Erratum to: Calcium Mobilization via Intracellular Ion Channels, Store Organization and Mitochondria in Smooth Muscle

John G. McCarron, Susan Chalmers, Calum Wilson, and Mairi E. Sandison

Erratum to:
I. Levitan, A.M. Dopico (eds.), *Vascular Ion Channels in Physiology and Disease*, https://doi.org/10.1007/978-3-319-29635-7_11

The chapter 'Calcium Mobilization via Intracellular Ion Channels, Store Organization and Mitochondria in Smooth Muscle' has now been made available open access under a CC BY 4.0 license.

The updated original online version for this chapter can be found at https://doi.org/10.1007/978-3-319-29635-7_11

I. Levitan, A.M. Dopico (eds.), *Vascular Ion Channels in Physiology and Disease*, https://doi.org/10.1007/978-3-319-29635-7_19

Index

A
Adenosine triphosphate (ATP), 195, 397
ADP-ribose, 43, 57
A-kinase anchoring proteins (AKAPs), 146
Alkaline phosphatase, 145
2-Aminoethoxydiphenyl borate (2-APB), 371
4-Aminopyridine (4-AP), 197
Aneuploidy, 257
Aneurysmal subarachnoid hemorrhage (aSAH), 192
Angiogenic switch, 257
ANK. *See* Ankyrin repeats (ANK)
Ankyrin repeat domain (ARD), 262
Ankyrin repeats (ANK), 262
Arachidonic acid (AA), 5, 54
ARD. *See* Ankyrin repeat domain (ARD)
Arginine vasopressin (AVP), 200
Arterial myocytes, lysosomal TRPML channels, 216–218
  TRPML1 channel, 216–217
  TRPML1-mediated Ca^{2+} release, 217–218
Arterial myogenic responsiveness
  intraluminal pressure, 132
  K^+ channel expression, 133
  myogenic contraction, 133
  TMEM16A/ANO1 chloride channels, 132
Autophagic flux, 225
Autophagolysosomes (APLs), 225
Autophagosome-lysosome fusion, 217
Autophagosomes (APs), 224

B
BAECs. *See* Bovine aortic endothelial cells (BAECs)
Barrier function, 37
Basic fibroblast growth factor (bFGF), 52, 54, 57
Basilar artery myocytes, 200–202
Bevacizumab, 258
BK channels, KCNMA1/slo1
  cholesterol, 362
  fatty acids, 358, 360
  homotetramers, 357
  leukotrienes, 362
  phosphoinositides, 358
BK_{Ca} channel function, 145, 146
BK_{Ca} expression, 134
Bovine aortic endothelial cells (BAECs), 331, 332
Bradykinin receptors, 247
BTEC, 54

C
Ca^{2+}, 216, 235
  CICR, 218
  concentrations, 243
  role of mitochondria in modulating signals, 248
  signaling-related cellular activities, 217
  waves, 245
Ca^{2+} channel, 39
  and functional role in ECs, 56, 57
  angiogenesis, 52–54
  CNG, 43, 44
  immune function, 54–56

I. Levitan, A.M. Dopico (eds.), *Vascular Ion Channels in Physiology and Disease*, DOI 10.1007/978-3-319-29635-7

Ca^{2+} channel (*cont.*)
inflammation, 54–56
mechanisms, 38
migration, 52–54
oxidative stress, 43
permeability, 44–50
plasma membrane (PM), 38
SOCE (*see* Store-operated calcium entry (SOCE))
vascular remodeling, 52–54
Ca^{2+} channels, PASMC
expression and function, 312, 313
intracellular concentration, 303
ROCE and SOCE, 304–306
VDCC, 304
Ca^{2+} responses, 247
Ca^{2+} stores, 240, 242
arrangement, 235
methods used to investigate, 238–241
structure of, 235–238
subcompartments, 238, 243
in tracheal myocytes, 237
Ca^{2+}-activated K^+ channels (BK_{Ca}), 134–135, 161–162, 196. *See also* Transient receptor potential (TRP) channels
arterial myocytes, 160
BK β1 and Rab11A, 159, 160
ER retention sequence, 158
KCNMA1 gene, 158
myocyte relaxation and eventual vasodilation, 158
S-palmitoylation, 158
STREX exon, 158
vascular myocytes, 159
ZERO variant channels, 158
γ subunits, 158
Ca^{2+}-induced Ca^{2+} release (CICR), 7, 218, 219, 234
Ca^{2+}-release-activated Ca^{2+} channels (CRAC channels), 275
Ca^{2+}-sensitive K^+ (BK) channels, 175–178
CAD. *See* Coronary artery disease (CAD)
Caenorhabditis elegans, 262
Caffeine, 239, 242, 243
Calcium sparklets, 16, 18
Calcium sparks, 69
Calcium waves, 7, 11, 12
Calsequestrin (CSQ), 247
Cardiovascular Alterations, 109–110
$Ca_V1.2$. *See* L-type voltage-gated Ca2+ (CaV1.2) channels
Cell apoptosis
$mitoK_{ATP}$ and $mitoBK_{Ca}$, 403
oxidative phosphorylation and ATP depletion, 399
VDAC, 401
VDAC1, 399
Cell function
permeability, 44–50
Cell migration, 283, 290
Cell proliferation, 277, 288
Cell-adhesion molecules, 55
Cell-attached mode, 241
Ceramides, 351, 355, 357
Cerebral vasospasm, 192–194, 203–207
aSAH pathology, 206
celecoxib, 206
electrocorticographic activity, 206
fibromyalgia, 205
retigabine, 204
CFTR. *See* Cystic fibrosis transmembrane conductance regulator (CFTR)
Cholesterol (CLR), 352
Chronic inflammatory diseases, 261
CICR. *See* Ca^{2+}-induced Ca^{2+} release (CICR)
Cl^- channels, PASMC
Bestrophin-3, 308–309
CFTR, 309–310
expression and function, 312–315
intracellular and extracellular concentrations, 306
transmembrane protein 16 (TMEM16)/ anoctamin (ANO), 306–308
volume-sensitive Cl- channels (ClC-3), 309
Cl–Ca activation
cholesterol, 366
phosphoinositides, 366
sphingolipids, 366
CNG, 43, 52
Conducted vasomotor responses, 19
Confocal microscopy, 225
Connexins (Cxs), 405, 406
Coronary arterial myocytes (CAMs), 216
Coronary arterial smooth muscle cells (CASMCs), 221
Coronary Artery Bypass Graft, 283–284
Coronary artery disease (CAD), 280
atherosclerosis, 382, 383
coronary circulation functions, 382
CSM phenotypic switching, 384
endothelial dysfunction, 383
laminae, 382
macrovascular conduit arteries, 382 (*see also* Metabolic syndrome-induced coronary artery disease (MetS-induced CAD))
microvascular resistance arterioles, 382
risk factors, 392
Coronary heart disease, 225
Coronary smooth muscle (CSM), 385

atherosclerosis, 383
contractile properties, 383
MetS-induced CAD (*see* Metabolic syndrome-induced coronary artery disease (MetS-induced CAD))
phenotypic switching, 384
proliferation and migration, 382
Cortical spreading depression (CSD), 206
CRAC activation domain (CAD), 41
CRAC channels, 278
CSM. *See* Coronary smooth muscle (CSM)
Cyclic ADP-ribose (cADPR), 217
Cyclic nucleotide-gated (CNG) channel, 43, 44
Cyclopiazonic acid (CPA), 237
Cystic fibrosis transmembrane conductance regulator (CFTR), 309, 310
Cytosolic tail domain (CTD), 357

D
Diacylglycerol (DAG), 4, 353
1,2-Dioctanoyl-glycerol (DiC8), 353
Dyslipidemia, 384, 390, 393

E
Eicosanoids, 356
Electrical conduction, 82–88
Electrophysiology techniques, 42
Endoplasmic reticulum (ER) Ca^{2+} store, 39
Endothelial barrier function, 44
Endothelial cells (ECs), 37, 80, 328
Ca^{2+} channel (*see* Ca^{2+} channel)
Ca^{2+} signaling, 38
characterization, 50
Kir channels (*see* Inwardly-rectifying K^+ channels (Kir))
homeostasis, 38
physiological and pathological processes, 38
physiological and pathophysiology function, 37, 38
Endothelial dysfunction, 107–108, 115
Endothelial ion channels, 6, 13–25
agonist-induced Ca^{2+} influx, 19
cultures of porcine coronary artery, 14
electrophysiologically, 14
human umbilical vein endothelial cells, 16
hyperpolarization, 19
IP_3R, 14
murine pulmonary microvascular endothelial cells, 16
pharmacologically, 14
plateau phase, 6, 14, 15
TRP family, 15
TRPA1 channels, 17, 18
TRPC family, 15, 16
TRPC1, 15, 16
TRPC3, 17
TRPC4, 16
TRPC6, 18
TRPV3 channels, 18
TRPV4 channels, 16, 17
VGCC (*see* Voltage-gated Ca^+ channels (VGCC))
agonist-induced Ca^{2+} signals, 7–14
IP_3 receptors (*see* Inositol-1,4,5-trisphosphate (IP_3) receptors)
RyR, 13–14
and Ca^{2+} signaling, 4, 5
and cell-cell communication, 12
cell volume regulation, 4
cell-cell communication, 4
hyperpolarization
$K_{Ca}1.1$ channels, 21–22
$K_{Ca}2.3$ and $K_{Ca}3.1$ channels, 20–21
$K_{IR}2.X$ channels, 22–25
intracellular Ca^{2+}, 4
methacholine-induced global Ca^{2+} transients in arteriolar endothelial cells, 4, 6
plasma membranes, 4
plasmalemmal ion channel activity, 4
vasodilators, 4
vasomotor tone, 4
Endothelial Kir channels, cholesterol, 334–342
BAECs and HAECs, 331, 332
electrophysiological recordings, 332
hypercholesterolemia, 331
in vitro and *in vivo*, 332
Kir2 channels
cholesterol-binding motifs, 338, 340–342
cholesterol-sensitivity belt, 338, 339
$PI(4,5)P_2$, 342
purification, 337
sterol analysis, 336
transmembrane and cytoplasmic domains, 337
two-way molecular switch, 338
progenitor cells and cardiomyocytes, 333
silent channel hypothesis
cavelin-1 (Cav-1), 335
cholesterol-rich domains, 336
evidence, 334
Endothelial progenitor cells (EPCs), 54, 287
endothelial tube preparation, 89

Endothelium, 69–70, 80–82
microvascular, 55
oxidative damage, 43
and smooth muscle layers of human arteries, 44
vascular tone control, 50–52
Endothelium-dependent hyperpolarization (EDH)-, 102, 104
Endothelium-dependent vasodilatation, 4, 11, 14–17
Endothelium-derived hyperpolarization factor (EDHF)-, 102
Epoxides of arachidonic acid (EETs), 4
Epoxyeicosatrienoic acids (EET), 361
11,12 Epoxyeicosatrienoic acid (11,12 EET), 369
Erratic blood flow, 256

F
Fas ligand (FasL), 225
Fatty acids, 351, 354
FGF. *See* Fibroblast growth factor (FGF)
Fibroblast growth factor (FGF), 258
Fluo-4/genetic Ca^{2+} sensors, 11

G
Gating Modulators, 114–115
G-protein coupled receptors (GPCR), 39
GPCR-dependent and GPCR-independent mechanisms, 49
and induce Ca^{2+} signaling, 56
PAR-1, 44
in plasma membrane, 39
SOCE, 49
Green fluorescent protein (GFP), 240

H
HAECs. *See* Human aortic endothelial cells (HAECs)
HEK. *See* Human embryonic kidney (HEK)
Heterogeneity BKCa subunits
amino terminus, 135
cerebral blood vessels, 139
molecular composition, 139
mRNA expression levels, 138
non-vascular tissues, 138
pathophysiological states, 139
pre-mRNA, 137
STREX, 137
voltage and Ca^{2+}, 140–141
voltage-dependent gating, 137
ZERO, 137
α-subunit amino acid sequence, 135
β accessory subunits, 138
β1 subunit, 137
Hexokinase II (HK2), 401
HMVECs, 53, 54
HPAEC. *See* Human pulmonary arterial endothelial cells (HPAEC)
Human aortic endothelial cells (HAECs), 331, 332
Human dermal microvascular ECs (HDMECs), 41
Human embryonic kidney (HEK), 263
Human microvascular ECs (HMVECs), 53
Human pulmonary arterial endothelial cells (HPAEC), 41, 260
Human umbilical vein endothelial cells (HUVECs), 41–43, 45, 48, 49, 53, 54, 56, 261, 287, 288
Ca^{2+}-selective SOC activity, 49
knocked-down Orai1, 48
Orai1 mRNA and protein, 42
SOCE, 56
TRPC1, TRPC3, TRPC4 and TRPC6, 41
TRPM7, 43
VEGF-induced HUVEC proliferation, 53
15-Hydroxyeicosatetraenoic acid (15-HETE), 356, 370–371
20-Hydroxyeicosatetraenoic acid (20-HETE), 354
Hypertension, 170–172
Hypoxia, 53

I
Immuno-detection of BK_{Ca} channels, 142–143
Inner membrane anion channel (IMAC), 409
Inner mitochondrial membrane (IMM), 398
Inositol 1,4,5 trisphosphate receptors (IP_3)
Ca^{2+} waves, 11
calcium pulsars, 11
characteristics, 7
cytoplasmic N-terminus, 7
and DAG, 5, 7
endothelium-dependent vasodilators, 11
expression of transcripts, 11, 13
Fluo-4/genetic Ca^{2+} sensors, 11
isoforms, 7, 11, 12
MEJs, 11, 12
MEPs, 11, 12
microvascular endothelial ion channels, 7–10
mouse mesenteric resistance vessels, 11
pharmacological studies, 7

pharmacology, 7–10
second order hamster cremaster arteriole, 6, 7
tetrameric Ca^{2+} release channels, 7
Inositol trisphosphate receptor (IP_3R), 234, 235, 244–245, 399
activity in smooth muscle, 248
arrangement on store, 238
Ca^{2+} concentration, 241
Ca^{2+} pool, 243
Ca^{2+} stores structure, 235, 238, 239
mitochondrial control, 248
Inositol-1,4,5-trisphosphate (IP_3), 4
Intermediate (IK) channels
EET, 366
endothelial and VSM cells, 365
lysophospholipids, 366
phosphoinositides, 366
sphingolipids, 366
Intermediate-conductance K_{Ca} (IK_{Ca}) channel, 20
Inwardly-rectifying K^+ Channels (Kir), 330, 331
Ca^{2+}-sensitive K^+ channels, 328
dyslipidemia, 328 (*see also* Endothelial Kir channels, cholesterol)
flow-sensitivity
cell proliferation and migration, 331
lung ischemia, 331
vasodilation, 330
Xenopus oocytes, 330
hypercholesterolemia, 328
Kir2 channels, 328, 329
Kir6/ATP-dependent Kir channels, 329
shear stress, 328
Ion channel trafficking, 156–157
Ca^{2+}-activated potassium BK channels, 157–160
$Ca_V1.2$ channels, 154–156
Kv channels (*see* Voltage-gated (Kv) channels)
myocytes, 153
Ion channels, 4
advanced age, electrical conduction, 91
arterioles and resistance arteries, 80
blood flow control, 80
ECs, 80–82
electrical conduction, 86, 87
endothelial cell (*see* Endothelial ion channels)
endothelial dysfunction, 92, 93
gap junctions, 88
intracellular channels, 134
mouse skeletal muscle, 134
myogenic reactivity, 134
pressure-induced regulation, 132
Simple Cable Theory, 82
SKCa/IKCa, 90
SMCs, 80
Ion channels, smooth muscle cells. *See* Pulmonary arterial hypertension (PAH)
IP_3R. *See* Inositol trisphosphate (IP_3R)
Ischemia, 72–74

K

K^+ channels (Kir), 351–352
K^+ channels, PASMC
ATP-sensitive K^+ channels (K_{ATP}), 302–303
Ca^{2+}-activated K^+ channels (K_{Ca}), 300–302
expression and function, 310–312
families of, 300
inwardly rectifying K^+ channels (K_{ir}), 302
membrane potential, 299
two-pore domain K^+ channels (K_{2P}), 303
voltage-gated K^+ channels (Kv), 300, 301
K_{ATP} channels, 68, 195
K_{Ca} and Endothelium-Dependent Vasodilation, 106–107
K_{Ca} channels, 103
$K_{Ca}1.1$ channel complex, 105
KCa1.1 Channels, 21, 22
$K_{Ca}2$, 103, 104
KCa2.3 and KCa3.1 Channels
agonist-mediated events, 21
blockade, 21
calmodulin, 20
cellular distribution, 21
distinct genes, 20
distribution, 20
$G\alpha_q$-protein coupled receptors, 20
IK_{Ca}, 20
MEJs, 20
MEPs, 20
mouse carotid arteries and cremaster muscle microcirculation, 20
rat mesenteric arteries, 21
rat middle cerebral arteries, 20
shear-stress-induced dilation, 21
SK_{Ca}, 20
transmission, 20
voltage insensitive, 20
$K_{Ca}3.1$ expression, 111
$K_{Ca}3.1/K_{Ca}2$ inhibitors, 118
KCNN3, 103, 105, 106
KIR2.X Channels
in arteriolar endothelial cells, 22, 23
Ba^{2+}-sensitive K_{IR} currents, 23

KIR2.X Channels (*cont.*)
blood vessels, 25
conductance, 22
current-voltage relationship, 22
endothelial cells of resistance arteries, 25
inward rectifier, 23
inward-rectification, 22
isoforms, 22
physiological function, 22
rat mesenteric arteries, 25
resistance arteries and arterioles, 22
skeletal muscle arterioles, 23–25
tetramers, 22
Kv channels. *See* Voltage-gated (Kv) channels
Kv7 channel activators, 199

L
Late endosomes or lysosomes (LEL), 217
Leucine-rich repeat containing (LRRC), 141
Lipid signaling, 367
Lipophilic Ca^{2+} indicators, 243
Lipopolysaccharide (LPS) modulation, 370
LPS, 55, 57
L-type voltage-gated Ca2+ (CaV1.2) channels
CCT fragment overexpression, 156
C-terminus-dependent conformational rearrangement, 155
distal N-terminus, 156
E1c, smooth muscle, 155
endothelin-1, vasoconstrictor pathway, 156
in arterial myocytes, 154
in vitro, proteolytic enzymes, 156
IP_3R-mediated sarcoplasmic reticulum (SR) Ca^{2+} release, 156
membrane depolarization, 154–155
pregabalin (Lyrica™), 155
transcriptional upregulation, 155
$\alpha_2\delta$-1-4 subunits, 155
Luminal regulation, 243
Lysophosphatidic acid (LPA), 353
Lysosomal acid lipase (LAL) activity and, 224
Lysosomal Ca^{2+} store, 222
Lysosomal trafficking, 220, 221
Lysosomal TRPML1
in Apoptosis, 225
and Autophagic Flux, 224–225
implication of dysfunction in Atherosclerosis, 222–225
in regulation of vascular tone and vasomotor response, 218–222
Lysosome (Lyso)-SR Junction, 219–221
Lysosome exocytosis, 217

M
MCOLN1, 216
MEJs, 11, 12, 21
Membrane lipids
K_V channels, 356
TRPA1 channels, 368
MEPs, 11, 12, 16–18, 20
Mesenteric artery, 239
Messenger RNA (mRNA), 196
Metabolic syndrome (MetS), 385
CAD (*see* Metabolic syndrome-induced coronary artery disease (MetS-induced CAD))
and diabetes, 384–385
Metabolic syndrome-induced coronary artery disease (MetS-induced CAD), 386
intracellular Ca^{2+} transport, 385
store-operated Ca^{2+} entry (SOCE), 386
TRP channels (*see* Transient receptor potential canonical (TRPC) channels)
Methacholine-induced global Ca^{2+} transients in arteriolar endothelial cells, 4, 6
Methyl-β-cyclodextrin (MβCD), 371
MetS. *See* Metabolic syndrome (MetS)
MetS-induced CAD. *See* Metabolic syndrome-induced coronary artery disease (MetS-induced CAD)
MicroRNAs (miRNAs), 182
Mitochondria
Ca^{2+} signals, 248
Ca^{2+} uptake, 248
control of IP_3R, 248
and IP_3R, 248
Mitochondrial BK channel (mitoBK), 177
Mitochondrial Ca^{2+} uniporter (MCU), 406
Mitochondrial calcium activated potassium, 68
Mitochondrial calcium overload
K^+/H^+ exchanger, 408–409
uniporter complex, 407
Mitochondrial depolarization
and cerebral ischemia, 72, 73
endothelium, 69, 70
insulin resistance, 72
mitoKATP Channels and Cerebral Vascular Tone, 68
morphological features, 64
perivascular and parenchymal nerves, 71
ROS production, 67
signaling events, 67–68
VSM, 64, 69
Mitochondrial K+ channels
AngII, 402
HT-22 neuronal mitoplasts, 403

hyperglycemia and obesity, 402
IMM, 402
mitoKv1.3, 403
structure and function, 403
TASK channels, 403
type 1 diabetic rats, 403
Mitochondrial Ryanodine Receptor (mRyR), 407
Ca^{2+}/H^+ antiporter, 408
Mouse cremaster arterioles, 11
Mucolipidosis Type IV (MLIV), 216
Muscarinic receptors, 247
Myoendothelial gap junctions (MEJs), 11
Myoendothelial junctions (MEJs), 11, 12
Myoendothelial projections (MEPs), 11, 12
Myometrial cells, 218

N

Na^+/Ca^{2+} exchanger (NCLX), 408
NAADP. *See* Nicotinic acid adenine dinucleotide phosphate (NAADP)
NEC. *See* Normal endothelial cells (NEC)
Negative-Gating Modulation, 118, 119
Neointimal hyperplasia, 276, 283, 288
Neurons, 64
Neurosearch compound, 114
Nicotinic acid adenine dinucleotide phosphate (NAADP), 216–218, 221
Niemann-Pick type C1(NPC1) diseases, 223
Non-small cell lung cancer (NSCLC), 258
Nonsteroidal anti-inflammatory drug (NSAID), 198
Normal endothelial cells (NEC), 257
NSCLC. *See* Non-small cell lung cancer (NSCLC)

O

Oleoylethanolamide (OEA)-induced relaxation, 370
Orai, 371–372
in endothelial cells, 288
in vascular smooth muscle cells, 284–285
Orai channels, 42, 278–279
in vascular remodelling, 279–280
ORAI families, 15
Orai proteins, 277
Orai1, 39, 42, 45, 49, 53, 55, 56
Outer mitochondrial membrane (OMM), 398
Oxidative stress
Ca2+ channel, 43
Oxidized low-density lipoprotein (oxLDL), 368

P

PAH. *See* Pulmonary arterial hypertension (PAH)
Palmitoyl-L-carnitine, 353
Pancreatic acinar cells, 240
Pan-Kv7 channel, 206
Pathogen-associated molecular patterns (PAMPs), 55
Pericytes, 256
Permeability
barrier function in pulmonary arteries, 44
EC, 49
endothelial barrier function, 44
endothelium, 44
GPCR-dependent and GPCR-independent mechanisms, 49
inflammatory mediators, 44
monolayer, 45
Orai channels, 49
Orai1, 49
Orai1 and STIM1, 45, 48
proteinase-activated receptor-1 (PAR-1) GPCR, 44
ROS, 49
SOCE, 49
thrombin, 45, 48
thrombin on $[Ca^{2+}]_i$ in LECs, 45, 46
thrombin/PAR-1 agonist peptide, 45
TRPC1 channels, 44
TRPC4 activation of store-operated Ca^{2+} entry (SOCs) regulates microvessel permeability, 45, 47
TRPC4-deficient mice, 45
TRPC6, 45
TRPM2, 49, 51
Phenotypes, 110–111
4β-Phorbol 12-myristate 13-acetate (PMA), 202
Phosphatase and tensin homolog (PTEN), 45
Phosphatidylinositol (3,4,5)-trisphosphate receptors
CLR, 371
LPS action, 370
Phosphatidylinositol 4,5 bisphosphate (PIP_2), 202, 351
Phosphoinositide 3,5-bisphosphate ($PI\text{-}3,5P_2$), 370
Phosphoinositides, 351, 353
Phospholipase C (PLC) activity, 39
Phospholipids, 351
Phosphorylation, 141, 143
PIGF. *See* Placental growth factor (PIGF)
PIP_2-induced BK activation, 358
PKD. *See* Polycystic kidney disease (PKD)
Placental growth factor (PIGF), 258

Platelet-derived growth factor (PDGF), 52, 276, 285
Polycystic kidney disease (PKD), 261
Polyunsaturated FAs (PUFAs), 355
Porcine brain capillaries, 23
Positive-Gating Modulation, 116–118
Potassium channels, 343
  artery myocytes, 196
  celecoxib, 198
  cerebral arteries, 196
  cerebral vasospasm, 195, 197
  COX inhibition, 198
  glibenclamide, 196
  hypoxic conditions, 195
  KCNQ gene products, 197
  Kir channels (*see* Inwardly-rectifying K^+ channels (Kir))
  linopirdine, 197
  physiological voltages, 195
  retigabine and flupirtine, 198
  vasculature, 200
  VDCCs, 194
PRD. *See* Proline-rich domain (PRD)
Pro-angiogenic factors, 257
Proline-rich domain (PRD), 262
Prostacyclin, 352
Prostanoids, 361
Protein kinase C- (PKC-), 200
Proteinase-activated receptor (PAR), 44
Pulmonary arterial hypertension (PAH), 299
  adventitia, media and intima, 297
  blood flow, 297
  categorization, 296
  cell proliferation and apoptosis, 297
  definition, 296
  development and progression, 314–315
  invasive and noninvasive methods, 296
  IPAH patents and PH models, 310, 311
  PASMC (*see* Pulmonary artery smooth muscle cell (PASMC))
  vascular remodeling, 297–298
  vasoconstriction, 298
Pulmonary artery smooth muscle cell (PASMC), 53, 299, 303–310
  Ca^{2+} Channels (*see* Ca^{2+} Channels, PASMC)
  Cl^- channels (*see* Cl^- channels, PASMC)
  ion channels, 299
  K^+ channels (*see* K^+ channels, PASMC)
  membrane potential, 299
Pulmonary hypertension, 173
Purkinje neurons, 243

Q
Quantal calcium release, 246

R
Rat uterine myoctes, 246
Reactive oxidative species (ROS), 216
Reactive oxygen species (ROS), 43, 263, 402
Receptor tyrosine kinase inhibitors (RTKIs), 258
Receptor-operated Ca^{2+} entry (ROCE), 52, 304
  mechanism, 39
Renal outer medullary potassium channel (ROMK), 67
ROCE. *See* Receptor-operated Ca^{2+} entry (ROCE)
ROS. *See* Reactive oxygen species (ROS)
RTKI. *See* Receptor tyrosine kinase inhibitors (RTKIs)
Ryanodine, 239, 243
Ryanodine receptors (RyRs), 134, 217, 218, 234, 244–245
  arrangement on store, 238
  Ca^{2+} concentration, 241
  Ca^{2+} pool, 243
  Ca^{2+} stores structure, 235, 238, 239
  Ca^{2+} transients in arteriolar endothelial cells, 13, 14
  Ca^{2+}-sensitive-Ca^{2+}-release channels, 13
  caffeine-induced Ca^{2+} transients, 13
  endoplasmic and sarcoplasmic reticulum, 369
  expression of transcripts, 13
  immunofluorescence, 13
  isoforms, 13
  in macrovascular endothelial cells, 13
  mouse mesenteric resistance arteries, 13
  transcripts, 13
RyR. *See* Ryanodine receptors (RyR)

S
Sarco/endoplasmic reticulum Ca^{2+}-ATPases (SERCA), 233–234
Sarcoplasmic reticulum (SR), 65, 217, 385–387, 389, 390
Sea urchin eggs, 243
Senicapoc, 112
Severe combined immune deficiency (SCID), 42
Shear stress, 263
Single-nucleotide polymorphisms (SNP), 105
SKA-31, 114, 117

Small (SK) channels
 endothelium-dependent relaxation, 365
 epoxyeicosatrienoic acids, 366
 lysophospholipid, 366
 phosphoinositides, 366
Small-conductance, K_{Ca} (SK_{Ca}) channel, 20
Smooth muscle, 247
 functions, 233
Smooth muscle cells (SMCs), 80, 218
SOC channels, 276
SOCE, 275–277, 289
Spasmogen-induced constriction, 202–203
Sphingolipids, 351, 357
Sphingosine (SPH), 355
spontaneous transient outward currents (STOC), 109
Spontaneously hypertensive rats (SHR), 171
STIM proteins
 STIM1, 41, 42
 STIM2, 42
STIM1, 41, 42, 45, 49, 53, 55, 56
Stim1 Proteins, 371–372
STIM1-Orai Activation region (SOAR), 41
Store-operated Ca^{2+} (SOC) influx, 371
Store-operated Ca^{2+} channels, 276
Store-operated Ca^{2+} entry (SOCE), 304–306, 312
Store-operated calcium entry (SOCE)
 in angiogenetic processes, 54
 Ca^{2+} entry pathway, 39
 description, 39
 endoplasmic reticulum (ER) Ca^{2+} store, 39
 in endothelial cells, 288
 in HUVEC, 56
 and I_{CRAC}, 45
 mechanism, 39, 45, 52, 53
 and NFAT, 55
 and NO synthesis, 52
 orai channels, 41, 42
 physiological conditions, 39
 and ROCE, 39
 SOCE inhibitor BTP-2, 54 (*see* Store-operated Ca^{2+} entry (SOCE))
 STIM, 41, 42
 in T lymphocytes, 45
 TRPC channels, 40, 41
 in vascular remodelling, 279–280
 in vascular smooth muscle cells, 280–282, 284–285
Strokes, 64, 72–74
Stromal interaction molecule (STIM), 15
Stromal interaction molecule 1 (STIM1), 279
Sulfonylurea receptor (SUR) subunits, 195

T
TASK-3 potassium channel, 402
TEC. *See* Tumor Endothelial Cells (TEC)
TEM. *See* Tumor endothelial markers (TEM)
Testosterone (19 C), 365
Thrombin, 44, 46
Toll-like receptors (TLRs), 55
Tracheal myocytes, 237
TRAM-34, 112
Trans-Golgi-network (TGN), 217
Transient receptor potential (TRP) channels, 216, 259
 arterial contractility, 162
 classification, 161
 diacylglycerol, 368
 HEK293 cells, 161
 homologous region, 216
 intracellular Ca^{2+} store depletion, 161
 lysophospholipid, 367
 non-selective cation channels, 161
 oxidized lipids, 368
 phosphoinositides, 367
 pressure-induced PKC activation, 162
 proteins, 216
 subfamilies, 367
 TRPA, 261
 TRPC, 259–260
 TRPC3 C-terminal calmodulin and IP3R binding (CIRB) domain, 161
 TRPM, 260
 TRPP, 261
 TRPV1, 261
 TRPV4, 262, 265, 266
Transient receptor potential (TRP) family, 15, 256
Transient receptor potential channels (TRPC), 387–390
 cell-type and tissue-specific manner, 40
 description, 387
 expression, 41
 homomeric/heteromeric structures, 40
 isoforms (TRPC1 to 7), 40
 PLC, 40
 stoichiometry and biophysical properties, 40
 superfamily of ion channels, 40
 tetramers, 40
 transmembrane domain, 40
 TRPC1
  angiotensin II (Ang II), 387
  short-term dyslipidemia and metabolic syndrome (sDMetS), 390
  STIM1 gene and protein expression, 389
  store-operated Ca^{2+} entry (SOCE), 387–389

Transient receptor potential channels (TRPC) (*cont.*)
TRPC3, 391
TRPC5 and TRPC 6, 391
in vascular remodelling, 279–280
in vascular smooth muscle cells, 280–282
Transient receptor potential mucolipin (TRPML) channel, 215
defects, 224
genetic defects of TRPML1, 215
Transient receptor potential vanilloid (TRPV) channels, 392
Transluminal angioplasty, 194
TRP ankyrin (TRPA) family, 261
TRP canonical channels (TRPC), 259, 260
TRP melastatin channels (TRPM), 260
TRP polycystin (TRPP), 261
TRP vanilloid (TRPV) family, 261
TRPC1, 39–42, 44, 45, 48, 51, 52, 54
TRPM2, 43, 49, 51
TRPM7, 43, 54
TRPV channels. *See* Transient receptor potential vanilloid (TRPV) channels
TRPV4, 262–267
Tumor, 258–259
angiogenesis, 256, 257, 266
approaches to target, 258–259
extracellular matrix, 256
promoting hormones, 256
vasculature, 256
Tumor endothelial cells (TEC), 257–258, 264, 267
Tumor endothelial markers (TEM), 257
Tumor necrosis factor (TNF), 225
Tumors
oxygen and nutrient demands, 255
turbid blood flow, 256
Turbid blood flow, 256
Tyrosine kinase-linked receptor activators, 234

U
Uncoupling protein (UCP), 404, 405
Uterine smooth muscle cells, 218

V
Vascular Ca^{2+} channels, 178–183
Vascular endothelial growth factor (VEGF), 39, 42, 44, 52–54, 57, 256, 258, 287–289
Vascular normalization principle, 258
Vascular remodelling, 276
Orai channels, 279–280
SOCE, 279–280
TRPC channels, 279–280
Vascular smooth muscle (VSM), 64
CLR interaction, 364
indirect and direct mechanisms, 360
ion channels, 350
lipid and ion channel protein, 350
LTB4 docking, 363
plasmalemma, 350
tissue proliferation, 350
Vascular smooth muscle cells (VSMCs), 132
Bestrophin-3, 308, 309
Ca^{2+}-activated K+ channels (KCa), 302
Ca^{2+}-permeable ion channels, 170
cellular mechanisms, 170
cystic fibrosis transmembrane conductance regulator (CFTR), 309
growth factor activation of Orai1, 285–286
ionic conductance, 310
K^+ channels, 170
Orai, 284
other activators of TRPC, 282–283
RyR isoforms, 13
small-molecule inhibitor of SOCE, 287–288
SOCE, 280, 281, 284
TRPC channels, 280, 281
transmembrane protein 16 (TMEM16)/ anoctamin (ANO), 308
voltage-gated K+ channels (Kv), 300
volume-sensitive Cl^- channels (ClC-3), 309
Vascular tone control, 50–52
Vasopermeability, 37
VDCC. *See* Voltage-dependent Ca^{2+} channels (VDCC)
VEGF. *See* Vascular endothelial growth factor (VEGF)
Voltage-activated (delayed rectifier), 196
Voltage-activated potassium channel
cerebral aneurysms, 192
impaired cerebral autoregulation, 192
symptoms, 192
Voltage-dependent anion channel (VDAC), 398–401
Voltage-dependent Ca^{2+} channels (VDCCs), 193, 300, 302, 304, 305, 310
Voltage-gated (Kv) channels
abnormal depolarization, 172
aortic-banded hypertensive rats, 173
arterial K_v2 channels, 173
Caveolin-1 interaction, 157
homo-/hetero-tetramers, 157

hypoxia-induced pulmonary hypertension, 173
membrane depolarization, 156, 172
oxyhemoglobin, 157
pore-forming structure, 172
pore-forming structures, 174
postsynaptic density-95 (PSD95), 157
Src-dependent tyrosine phosphorylation, 157
Voltage-gated Ca^{+} channels (VGCC), 19
ceramides, 355
cytochrome P450-mediated metabolism, 354
EET-induced channel inhibition, 355
fatty acids, 354
lipophilic vitamins, 356
L-PC, 353
lysophospholipids, 353
phenylalamines, 353
phosphoinositides, 353
PUFAs, 355
subfamilies, 353
Voltage-gated Na+ channels (VGSC)
ceramides, 356
fatty acids, 356
plasma membrane depolarization, 356
Voltage-gated potassium channels (KV)
ceramides, 357
eicosanoids, 356
membrane depolarization, 356
sphingolipids, 357
Voltage-operated Ca^{2+} channels (VOCC), 134
VSMC. *See* Vascular smooth muscle cells (VSMC)

W
Wave progression and store arrangement, 237
Wistar-Kyoto (WKY) rats, 171

X
Xenoestrogens, 365

MIX
Papier aus verantwortungsvollen Quellen
Paper from responsible sources
FSC® C105338

If you have any concerns about our products,
you can contact us on
ProductSafety@springernature.com

In case Publisher is established outside the EU,
the EU authorized representative is:
Springer Nature Customer Service Center GmbH
Europaplatz 3, 69115 Heidelberg, Germany

Printed by Libri Plureos GmbH
in Hamburg, Germany